Lawrence K. Mag[nature]

BIOLOGY AND SOCIETY

BIOLOGY AND SOCIETY

Paul R. Ehrlich
Richard W. Holm
Irene L. Brown
Department of Biological Sciences
Stanford University

McGraw-Hill Book Company

New York • St. Louis • San Francisco • Auckland
Düsseldorf • Johannesburg • Kuala Lumpur • London
Mexico • Montreal • New Delhi • Panama • Paris
São Paulo • Singapore • Sydney • Tokyo • Toronto

BIOLOGY AND SOCIETY

Copyright © 1976 by McGraw-Hill, Inc. All rights reserved. Printed in the United States of America. No part of this publication may be reproduced, stored in a retrieval system, or transmitted, in any form or by any means, electronic, mechanical, photocopying, recording, or otherwise, without the prior written permission of the publisher.

Library of Congress Cataloging in Publication Data

Ehrlich, Paul R
    Biology and society.

    Includes index.
    1. Biology. 2. Biology—Social aspects. I. Holm, Richard W., joint author. II. Brown, Irene L., joint author. III. Title.
QH308.2.E4   574   75-30919
ISBN 0-07-019147-6

1 2 3 4 5 6 7 8 9 0   MURM   7 9 8 7 6

This book was set in Melior by York Graphic Services, Inc.
The editors were William J. Willey, Janet Wagner,
and Richard S. Laufer;
the designer was Jo Jones;
the production supervisor was Thomas J. LoPinto.
The drawings were done by Thomas Brennan and Rick Rodrigues.
Cover illustration by John Sovjani.
The printer was The Murray Printing Company; the binder, Rand McNally & Company.

Preface	vii
How to use the learning guides	ix

CONTENTS

1 LIFE IN THE CITY	1

The city as an ecosystem • Urban ecology • The impact of urbanization

2 LIFE IN THE COUNTRY	34

Some principles of ecosystems in general • The farm as an ecosystem • Rural ecology • The impact of the farm

3 LIFE IN NATURAL ECOSYSTEMS	56

The self-sufficient life • Components of natural ecosystems • Some natural ecosystems • The stability of ecosystems

4 THE EARTH, THE MOON, THE SUN, AND THE STARS	83

The universe • The solar system • The earth

5 THE DISTRIBUTION OF LIFE	110

High and low on the earth • North and south on earth • Freshwater habitats of the biomes • East and west on the earth • Dispersal • People and biogeography

6 THE KINDS OF LIVING THINGS	149

Cells, the basic units of life • Classification of organisms • Some major kinds of animals • Some major kinds of plants • Some Protista • Some Monera • Viruses

7 BEING YOUNG AND GROWING OLD	206

Life cycles of organisms • Development • Aging and death

8 THE MARCH OF GENERATIONS	228

Reproductive strategies • Reproduction in human beings • Reproduction in other animals and plants • Reproduction and the formation of populations

9 THE STUDY OF HEREDITY	257

Heredity and pedigrees • Mendelian genetics • Some complexities of heredity • How heredity works

10 THE ORIGIN OF DIVERSITY	287

The origin of diversity • Variability in populations • Natural selection • Differentiation of populations

11 THE HISTORY OF LIFE 313

The first signs of life • Life in the early seas • Plants move onto land • Animals move onto the land • The triumph of the mammals, birds, arthropods, and flowering plants • Human evolution

12 OBTAINING ENERGY 345

Energy and cell structure • Photosynthesis • Respiration • Obtaining energy from other organisms • Human nutrition

13 USING ENERGY 379

Synthesis • Maintenance • Doing things

14 THE STEADY STATE 394

Homeostasis • The problem of coordination • The problem of protection

15 BEHAVIOR AND SURVIVAL 425

Behavior and coevolution • Behavior in reproduction • Social behavior • Conceptual behavior and language

16 CULTURE AND SURVIVAL 455

Hunting and food gathering • The agricultural revolution • The city and the origins of technological society

17 SOCIETY IN THE FUTURE 471

Human prospects • Ecological prospects • Cultural prospects

18 YOUR ROLE IN THE FUTURE 511

You as an individual • You as a citizen • You and your career

Appendix 529

Glossary 530

Index 543

PREFACE

Biology and Society is designed to do two things. One is to provide a broad, basic background in biology: to explain how living things function, evolve, and interact with each other and their nonliving surroundings. The second is to put human societies and their activities into their biological context: to show how human beings function, evolve, and interact with each other and the other living things with which they share the nonliving environment. People are very much a part of nature and not only influence, but are influenced by, the living and nonliving world.

This book was written for the student who wishes to understand the principles of biology in the context of problems confronted by a person in the world today. Only by understanding the operation of biological systems can a person make socially responsible decisions, decisions which affect everyone's way of life and the world of the future. For example: Do we need stricter air-pollution controls? Is our city's sewage treatment plant doing a satisfactory job? Should petroleum be used at ever-increasing rates for fuel? What aspects of the growth of our city affect the photosynthetic productivity of its surroundings? Should our hospitals provide genetic counseling? All of these and many many more vital questions can be answered responsibly only if basic biological principles are understood.

All of the fundamental topics usually included in beginning biology courses are presented in this book; however, it differs somewhat in organization and emphasis from other, more conventional texts. Simple concepts of physics and chemistry are introduced early, in the context of the city. In general, we have tried to minimize the use of technical terms and to use illustrations extensively for summaries of complex material and for supplemental material. A glossary of technical terms is provided at the end of the book, as is a table of metric-English conversions. Measurements are generally given in metric units and, wherever helpful, the English equivalents appear in parenthesis. The metric system, universally used by scientists, will soon be used in the United States, and it seems logical for people to begin to think now in metric units.

In addition to the customary material from biology, we have included sections on our species' cultural evolution, nomadic life, agriculture, urbanization, and the growth of technology. These topics have often suffered from not having been considered in a biological context. Biology is the study of living things and their interdependencies; hence, biology is the most fundamental study people can engage in. Furthermore, biology and other sciences are carried out by human beings, one species of living things out of several million species, on one planet of one star out of trillions of stars.

We are unique among living beings on this planet in our ability to examine our own nature. Yet we all too often fail to realize that all our studies and cultural advances are colored by our sensory and mental capacities—themselves part of the whole of nature. It is possible to change one's point of view and to look at the world in a new perspective. In this book we have tried to enable the reader to examine the world with the enlarged perspective of biological understanding.

Many friends and colleagues have been generous in providing data and illustrative material. We are indebted to all those persons who read part or all of the book and made helpful suggestions. Especially useful were critical comments by Burton S. Guttman and several student reviewers. Invaluable assistance in the preparation of the manuscript was given by Jane Bavelas, Dee Stead, and Darryl Wheye. John Hendry prepared the learning objectives at the beginning of each chapter, and the review questions.

Paul R. Ehrlich

Richard W. Holm

Irene L. Brown

HOW TO USE THE LEARNING GUIDES

LEARNING OBJECTIVES

Each chapter of this text begins with a set of learning objectives. As the heading at the top of each set states, these are *some* learning objectives—not an exhaustive list of them. They are intended to help you study and review the chapter material; they do *not* specify everything you can or "should" learn about the topics covered in the chapter or what you will "be responsible for" on tests. But, if you can do all the things specified in the objectives, you will be well on your way toward a solid grasp of the chapter material.

You may find it helpful to turn to the learning objectives often. We suggest that you look them over before you read the chapter. You might also look at them now and again, as you study the chapter, to see how well you are grasping some of the main points. And, finally, you should be able to fulfill all the objectives for a chapter before you consider your study and review of that chapter complete.

QUESTIONS FOR REVIEW

Each chapter ends with some review questions. Their purpose, like that of the learning objectives, is to help you master the chapter material. Although you should consider the learning objectives more important than the review questions and make it your business to master the objectives first, you should also give your attention to the review questions. Some of them cover important items that do not fit neatly into the learning objectives. Some of them cover, in slightly different ways, topics that are included in the learning objectives because it is sometimes helpful to look at a concept from different angles. A few of the review questions ask you to apply facts and concepts from the text in ways that you might find enjoyable as well as instructive.

BIOLOGY AND SOCIETY

LIFE IN THE CITY

SOME LEARNING OBJECTIVES

The objectives at the beginning of each chapter are intended to help you study and review the chapter material. They are not an exhaustive list of everything you "should know" about this material. See page ix at the front of this book for a fuller discussion of the nature and purpose of the learning objectives and for a word about the review questions at the end of each chapter. After you have studied this chapter, you should be able to:

1. Demonstrate some things you have learned about the city as an ecosystem by
   a. Stating the determining difference between complete (natural) and incomplete ecosystems, saying which kind the city is, and giving an example of the other kind.
   b. Naming the two most general kinds of city inputs, stating the text's definition of each, and giving some specific examples of each.
   c. Listing several useful and several useless and/or harmful city outputs.
2. Demonstrate your knowledge of certain important properties of matter, energy, and living material by
   a. Stating the text's definition or description of wastes, matter, energy, atom, molecule, ion, chemical compounds, and chemical energy.

   b. Listing at least seven elements that are important in living material, including the five that together make up almost 98 percent of the weight of the human body and at least one trace element.
   c. Distinguishing between the two meanings of the term *organic compound* and stating which meaning is used in this book.
   d. Explaining why carbon is such an important element.
   e. Stating whether work is done with only evenly distributed heat or with only unevenly distributed heat, and supporting your answer by describing what happens on either side of a steam- or gasoline-engine piston that causes the piston to move.
   f. Naming the form of energy some of which is *always* produced as a *byproduct* whenever energy of any kind is put to work.
3. Demonstrate your knowledge of urban ecology by
   a. Explaining why cities cannot be self-sufficient in food production.
   b. Naming the original (ultimate) source of most of the city's energy inputs.
   c. Describing the role of plants in the energy inputs of the city and its inhabitants.
   d. Stating three reasons why the city is a difficult environment for plants.
4. Demonstrate your understanding of the impact of urbanization by
   a. Explaining, with examples, why some problems of city inputs and outputs become more difficult as cities grow bigger.
   b. Giving one reason why urban renewal is sometimes a failure.
   c. Stating three urban problems caused by automotive transportation (particularly the automobile).
   d. Listing some negative impacts of cities on (1) their own environment, including their inhabitants; (2) the surrounding environment.
   e. Listing some beneficial impacts of cities on (1) their own inhabitants; (2) the surrounding environment and its inhabitants.

CHAPTER ONE

One of the most obvious characteristics of human beings is their tendency to gather together in cities. Cities seem to have appeared as soon as agriculture became efficient enough so that one farmer could support more than one family, and thus some people were freed from the task of growing food. Ruins of the houses, temples, and shops of ancient cities date back to over 5,000 years ago; apparently these cities once flourished as centers of commerce, religion, government, and the arts. In ancient times cities were limited in their size, since all food had to be brought into the city on the backs of people or animals—or in carts pulled by animals. Cities could not grow beyond the capacity of these modes of transportation to feed them; the largest ancient cities such as Rome probably never had more than 1,000,000 inhabitants. Following the industrial revolution and the accompanying development of fuel-powered machines for transporting food, cities have become larger and larger. For example, in 1973, New York City had over 8,000,000 inhabitants.

Cities are and have been the source of civilization; many volumes have been written about the great cities and their roles in various societies. Yet, cities are places of great contradictions; penthouse apartments a few blocks from slums; creative jobs and dull assembly lines; loneliness in the midst of throngs of people. One of the most pressing problems of the world today is how cities can accommodate the vast numbers of people migrating to them. This phenomenon, **urbanization,** has drawn attention to both the positive and negative impacts cities have on their people and their environments. In the developed countries such as the United States, cities have become so large that they have run together into vast urban-suburban complexes. In the underdeveloped countries, many cities are growing at such a rate that they will double their populations in 15 years. Already their people are ill-fed and poorly housed. Perhaps a study of the biology of cities will help us understand some of their problems.

THE CITY AS AN ECOSYSTEM

An **ecosystem** consists of all the organisms of an area and their nonliving environment. **Ecology** is the branch of biology which deals with the relationships of organisms to each other and to their nonliving environment. Since we view the city as an ecosystem in this chapter, we can refer to our studies as **urban ecology.** Cities are "artificial" ecosystems in the sense that they are almost entirely constructed and dominated by people. Forests and deserts are examples of "natural" ecosystems in that they can maintain themselves without the help of people; later in this book we shall study some natural ecosystems in detail. It is important to remember that dividing ecosystems into natural and artificial for convenience of study does not alter the fact that the human species is part of nature. What people do in constructing cities is natural for them as a species. The extent of the impact of the activities of people on the ecological systems of which they are a part makes it important to apply the concepts of ecology to human affairs. The science of ecology puts people in perspective as part of nature and not above it.

1-1 City dwellers are often estranged from their resource base

In the city, people seem to be independent of nature. Everywhere people in the cities look they see evidence of human strength and cleverness. Tall buildings soar into the sky (Fig. 1-1). The very earth itself is covered with solid sheets of paving. Motor vehicles, ranging in size from tiny scooters to huge trucks, move in unending streams down the streets and avenues. At night, the city is bathed in electric light and even the air shows the effects of human inventiveness. It reeks of the power of people's machines; take a deep breath of city air and you will inhale waste products of thousands of heating, manufacturing, and transportation devices. Step into a modern office building; there, people have conquered the seasons; the temperature is kept hovering around 21°C (70°F) winter and summer. People in the city have evidence all about them that makes it seem they are independent of nature and even isolated from it.

The city is designed by people for people. In fact, in the central part of big cities most of the living things are either people or the pets or pests of people. Unfortunately, the nonhuman animals living in a highly artificial ecosystem tend to be health hazards for people. Roaches and rats may carry disease, and rats have been known to

FIG. 1-1 Aerial view of a part of Manhattan island, New York City. The many tall buildings depend on inputs of energy for construction, heating, cooling, ventilating, lighting, cleaning, etc. Notice also the piers used by ships transporting people and materials into and out of the city. (*Fairchild Aerial Surveys, Inc., Editorial Photocolor Archives.*)

bite babies. Even people's pets are a problem in the city; dog manure on sidewalks is unpleasant and may also carry diseases. In addition, abandoned dogs have gone wild in many cities. These *feral* (wild) dogs encourage rats by overturning garbage cans. They occasionally attack children, and some have rabies. So city officials pass dog laws and try, usually unsuccessfully, to get rid of the rats and roaches. People who live in the center of the city may have to go to the suburbs, the park, or the zoo to see a variety of animal life. As for plants, perhaps a few trees are growing along the streets, and some areas of the city may have lawns. Generally there are parks, but some cities are replacing the grass and shrubs with Astroturf and other plastic plants. In restaurants and hotels you may have to pinch the plants to find out if they are plastic or alive.

People in the city don't have the chance to see many kinds of organisms relating to each other and to the nonliving environment; they see people relating to people, to buildings, and to transportation devices. Yet cities need plants, which, as you will shortly discover, can help reduce summer temperatures, purify the air, and muffle street noises. City dwellers must learn that their people-designed environment is not really independent of the rest of nature. Even with an abundance of trees and grass, *cities are incomplete ecosystems because they are not self-sustaining.* For food, industrial raw materials, and energy cities must draw on the ecosystems that surround them, and they must also rely on these ecosystems for the absorption of wastes.

1-2 The city requires inputs of energy and materials

The flows of things the city needs, called **inputs,** are often not apparent to city dwellers. Some inputs are hidden under the city streets; water, for example, flows in through giant pipes, often driven by powerful pumps. Natural gas, used in heating and as a power source for industry, also flows in through pipes. Electricity flows through wires which are now underground in some cities. Many of the inputs of the city occur at night when they will not conflict with the traffic of people coming and going to and from the city. Food, especially, arrives late at night or early in the morning at market distribution centers in the city. Gasoline and diesel oil to run automobiles, trucks, construction equipment, and other machinery arrive in tanker trucks and railroad tank cars. Coal arrives in railroad cars for use in heating and for use in generating electric power. Raw materials for industry arrive in trucks and trains. And, of course, trucks and trains bring in the flow of food without which the people of the city would soon starve.

Matter The inputs of the city are of two general kinds: materials and energy. Materials are made up of **matter,** which is anything that has weight and takes up space. All matter is made up of one or more of the nearly 100 naturally occurring chemical **elements,** some of which are shown in Table 1-1. Matter may enter the city as an ele-

TABLE 1-1 ELEMENTS IMPORTANT IN LIVING MATERIAL

ELEMENT	SYMBOL	APPROXIMATE PERCENTAGE OF EARTH'S CRUST	APPROXIMATE PERCENTAGE OF HUMAN BODY
Hydrogen	H	0.1	9.5
Boron	B	Trace	Trace
Carbon	C	0.03	18.5
Nitrogen	N	Trace	3.3
Oxygen	O	46.6	65.0
Fluorine	F	0.03	Trace
Sodium	Na	2.9	0.2
Magnesium	Mg	2.1	0.1
Phosphorus	P	0.1	1.0
Sulfur	S	0.05	0.3
Chlorine	Cl	0.05	0.2
Potassium	K	2.6	0.4
Calcium	Ca	3.6	1.5
Manganese	Mn	0.1	Trace
Iron	Fe	5.0	Trace
Cobalt	Co	Trace	Trace
Copper	Cu	0.01	Trace
Zinc	Zn	Trace	Trace
Selenium	Se	Trace	Trace
Molybdenum	Mo	Trace	Trace
Iodine	I	Trace	Trace

Adapted from Keeton, William T., *Biological Science,* 2d ed., Norton, New York, 1972.

ment, such as iron, or it may arrive in the form of complex combinations of different elements such as wood. The smallest subdivision of an element that still has the properties of that element is an **atom**. Atoms are extraordinarily small particles—so small that they cannot be seen through the best microscopes that modern technology has developed. For instance, a cube of pure iron that is a centimeter (about ½ in.) on each side has about 60,200,000,000,000,000,000,000 iron atoms in it. Since no one has actually seen an atom, descriptions of atomic structure are really hypotheses based on the way atoms seem to behave. One hypothesis suggests that an atom consists of a central area that has a haze of electrical charges around it. The central part of the atom, which is called the nucleus, is composed of particles called **protons** and **neutrons**. Each proton has one positive charge, but the neutrons are, as their name implies, neutral, that is, uncharged. The electrical charges outside the nucleus are called **electrons**. Each electron carries one negative charge, and since the number of electrons in an atom equals the number of protons, the atom as a whole is electrically neutral. Different kinds of atoms have different numbers of protons. For instance, the smallest atom, hydrogen, has one proton, whereas oxygen has 8, sodium 11, and chlorine 17.

When two or more atoms are chemically joined together, the combination is called a **molecule.** A molecule may be constructed from atoms of one element. For instance, under normal circumstances the atoms in pure oxygen (O) are found in pairs, and each pair is called an oxygen molecule (O_2). Occasionally three oxygen atoms unite to form a molecule called ozone (O_3). Ozone is a gas that is found primarily in the outer layer of the atmosphere, where it absorbs much of the deadly ultraviolet radiation from the sun and thus makes life possible on earth.

Chemical **compounds** are composed of two or more different kinds of elements joined by chemical bonds. Thus, water (H_2O) is a compound composed of one hydrogen (H) and two oxygen atoms (Fig. 1-2a); common table salt (Fig. 1-2b) is made up of the elements sodium (Na) and chlorine (Cl) and has the formula NaCl. When sodium chloride is dissolved in water, the sodium and chlorine separate, with sodium losing an electron and becoming positively charged and chlorine gaining an electron and becoming negatively charged because neither changed its number of protons. Such charged particles are called **ions.** The sodium ion is written as Na^+ and the chlorine ion—called chloride—is written as Cl^-. Some of the roles of ions in living things will be described later.

When certain kinds of chemical bonds are formed, energy is released. Chemical reactions involve the rearrangement of atoms in molecules. If a molecule with atoms joined by relatively weak bonds enters into a reaction in which the atoms are rearranged to form stronger bonds, *there is a net release of energy.* Thus it is a convenient shorthand to speak of energy as "stored" in chemical bonds. It should always be remembered that energy is not contained in the bonds but may be released as old bonds are broken and new ones are formed.

Chemists generally place all molecules into one of two broad classifications: organic and inorganic. **Organic** substances contain the element carbon (C), whereas **inorganic** substances do not. Examples of organic materials are coal, which is made up mostly of pure carbon, and proteins, fats, and carbohydrates, which are carbon-containing compounds. Examples of inorganic matter are iron, water, and sodium chloride. However, in more common usage (and in the usage of this book) carbon compounds synthesized by living things in the normal course of their activities are called organic compounds and carbon compounds synthesized in laboratories or factories are called synthetic compounds.

Carbon is an extremely important element because it can combine with many other atoms to form long chains or rings. Thus it can form a great variety of compounds, the simplest of which is methane (Fig. 1-2c). An example of carbon's chain or ring-forming ability is illustrated in Fig. 1-2d. The letters in the diagrams are the symbols for the elements, and the lines that join the letters represent chemical bonds. The organic compound shown in the figure is glucose, a type of sugar that is found in human blood. Glucose is actually relatively

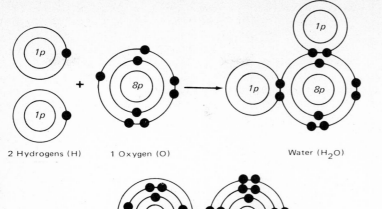

FIG. 1-2 Ways of indicating chemical structure: (*a*) formation of water from hydrogen and oxygen; (*b*) sodium (a very reactive metal) and chlorine (a poisonous gas) in the compound sodium chloride (table salt) which dissolves in water forming sodium ions and chloride ions; (*c*) the simple carbon compound methane, formed of four hydrogen atoms and one carbon atom; (*d*) two ways of representing the six-carbon sugar glucose.

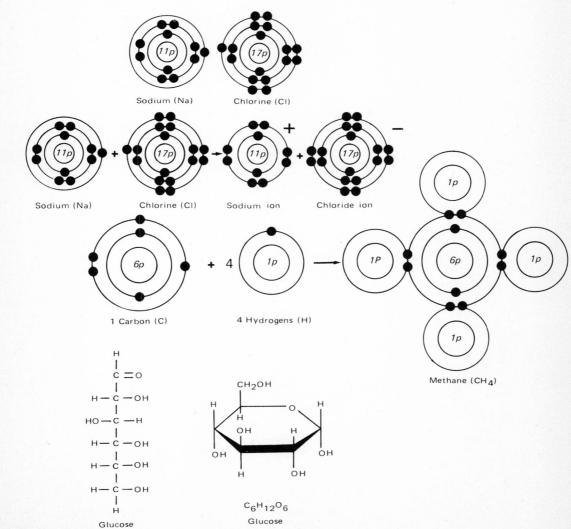

small for an organic molecule; other carbohydrates contain many more carbons. Some organic molecules, such as proteins, may even contain thousands of carbons, linked together in a long chain; such large molecules are called **macromolecules.**

Energy The second general input required by cities is energy. **Energy** is the stored ability to do work, meaning work in the very broadest sense; for example, to run an automobile, to hammer a nail, to chew a sandwich. In fact, a lot of energy is used to obtain more energy; energy is used to drill oil wells and mine coal and to transport these fuels to the city. Energy comes in a number of different forms—electrical, wind, light, heat, and chemical, to name a few. Energy is required for the formation of chemical bonds. Release of energy from chemical bonds is what we mean by chemical energy. Chemical energy is vital to people; for example, all animals, including humans, obtain their energy from the molecules of their food. In addition, many of the inanimate objects of the city are driven by the chemical energy released from burning coal, oil, and gas.

No matter what form energy takes, it always behaves in ways described by the **laws of thermodynamics.** Thermodynamics is the science of heat (*thermo*) and its relationships to other forms of energy. The first law of thermodynamics states that *energy can never be created or destroyed but only transformed* from one type to another. The second law is difficult to define simply, but, among other things, it implies that, in all transformations of energy, there is a loss of useful energy, that is, a loss of energy available to do work. This unavailable energy is in the form of *evenly distributed* heat. (Work can be done with heat only if it is *unevenly distributed,* as in the case of a boiler of hot steam leading to a cooler turbine.) Let us look at some examples of these laws in action.

Some cities get their electrical energy from hydroelectric plants that are located long distances away from urban centers. Water that falls as rain on high mountains contains potential energy which can be released into the energy of movement as it runs downhill. This energy can be used to drive a turbine (Fig. 1-3), which converts the energy of water movement into electrical energy. However, not all of the energy of water movement captured by the turbines is converted to electricity. The transformation of energy requires work, and whenever energy is put to work some of it is *always* converted to unavailable energy in the form of heat. Thus when the potential energy in water is transformed, only part of it is transformed to electricity; the rest is released as heat. More familiar examples of this phenomenon can be found in the home. For instance, the primary function of an electric light bulb is to produce light, but it produces heat as a by-product. Some devices seem at first glance to be an exception to this rule. Refrigerators seem to use energy to produce cold. The refrigerator is indeed made cold by the use of energy, but the proportion of heat energy in the *entire system* (the refrigerator plus the room containing it) is increased. In the process of cooling the inside of the refrigerator, the room is warmed—in fact, more heat

FIG. 1-3 Turbines at a hydroelectric plant in Wisconsin. These huge turbines which dwarf the workers who watch over their operation provide electricity for a large city. (*Daniel S. Brody, Editorial Photocolor Archives.*)

is released into the room than is removed from the refrigerator! Air conditioning is another process that generates large amounts of heat. You have undoubtedly walked in front of the street side of a room air conditioner and noticed the blast of hot air (Fig. 1-4).

If a hot gas from an appliance, boiler, or some other energy transformer is captured within a container it can also be used to do work because the heat is unevenly distributed. For instance, exploding gasoline produces hot gases inside the cylinders of an automobile engine, and the resulting pressure pushes pistons that drive the automobile. Or, in the case of a steam engine, the hot, high-pressure steam moves a piston in a cylinder (Fig. 1-5). However, if there were steam on the other side of the piston at the same temperature as that in the cylinder (heat evenly distributed), the piston would not move, the energy in the steam would remain unavailable, and no work would be done. The steam engine was one of the inventions that made massive industrialization possible. Manufacturers were no longer dependent on the energy of water turning a mill wheel, and they could also haul their goods long distances with the steam locomotives. Today in the United States most steam engines are used to generate electricity (Fig. 1-6).

Unfortunately, heat given off by electrical appliances and machines is not generally made use of; it could be used for heating a room or building. Instead, it usually disperses throughout the atmosphere, from which it cannot be reharnessed to do work. This is an example of energy waste. Energy can never be destroyed, but energy use does make large amounts of energy unavailable to do work.

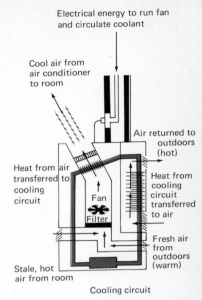

FIG. 1-4 A room air conditioner set in a window requires an input of electricity. While cooling the air of the room, it transfers heat from the room to the outdoors. The heat from the motor which drives its fan also goes into the air outside the room. The cooling circuit is indicated in color.

FIG. 1-5 The invention of efficient steam engines of many kinds was the major turning point which led to the industrial revolution. The steam engine was used not only for transporting materials and people, as was this locomotive, but also for pumping water from mines and for powering many kinds of machines in factories. Steam engines require inputs of air, water, and a fuel such as coal, wood, or oil. They produce outputs of smoke and heat as well as useful work. Since the burning of the fuel takes place outside the cylinder, they are often referred to as external combustion engines, as opposed to internal combustion engines such as those used in automobiles.

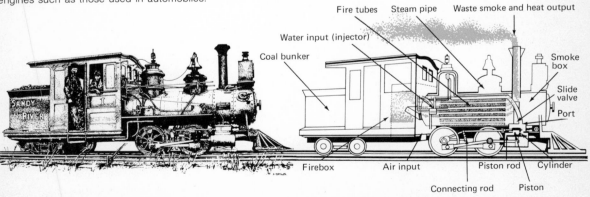

FIG. 1-6 Coal is often used to produce steam, which can be used to drive steam turbines, and these in turn drive electrical generators. The tall smoke stacks distribute the by-products of combustion of the coal high above the ground so that they will not obviously contaminate nearby houses or farms. *(Courtesy of J. Paul Kirouac.)*

1-3 Outputs of the city

Generally the material and energy inputs of the city are changed into some other form. Some inputs are transformed into parts of the city; for example, iron is melted and converted to steel which is made into the girders that become part of the city's buildings (Fig. 1-7), and food is transformed into body tissue and body energy.

But many of the inputs are transformed into beneficial or detrimental **outputs** of the city. Beneficial outputs include typewriters, tractors, sewing machines, etc., which flow out of the city to other cities and to towns and farms. People who buy these devices pay money for them which becomes an input to the city. These *economic transformations* make it possible for people in the city to buy other inputs which they need. The seemingly useless, and often harmful, outputs of the city are *wastes*. Wastes are by-products of almost all the transformations of matter and energy. For example, when gasoline is burned in an automobile engine, part of its chemical energy does work and the other part is converted to unusable heat. Furthermore, as the gasoline is burned, water vapor and other gases are produced, and these are wastes. If leaded gas is used, then lead is a waste by-product that settles on streets, buildings, and plants. Many

FIG. 1-7 Steel girders are being assembled as part of this building under construction. Here material input to a city is becoming part of the city. (*Bruce Anspach, Editorial Photocolor Archives.*)

manufacturing processes also expel harmful gases and large amounts of heat into the air and waterways.

1-4 The city is dependent on inputs and outputs

Natural ecosystems can usually maintain themselves on inputs of sunlight and water only. By contrast, cities collapse without continual inputs and outputs of energy and of a wide variety of kinds of matter (see Fig. 1-8). Thus an urban area is dependent on the surrounding countryside, which provides the city with most of its inputs and absorbs most of its outputs.

Failure of energy input The 1965 electric power failure in New York City is an example of how city life is disrupted by a loss of energy input. A failure of the system for supplying electrical energy took place over several states, and New York City, losing a major energy input, was paralyzed. Traffic lights would not work and monumental traffic jams occurred. Subway trains and elevators stopped moving and people were trapped in them. People searched for flashlights and matches. All electrically powered machinery from computers to vacuum cleaners ground to a halt and food began to spoil as refrigerators failed. Emergency generators were switched on at hospitals and air-

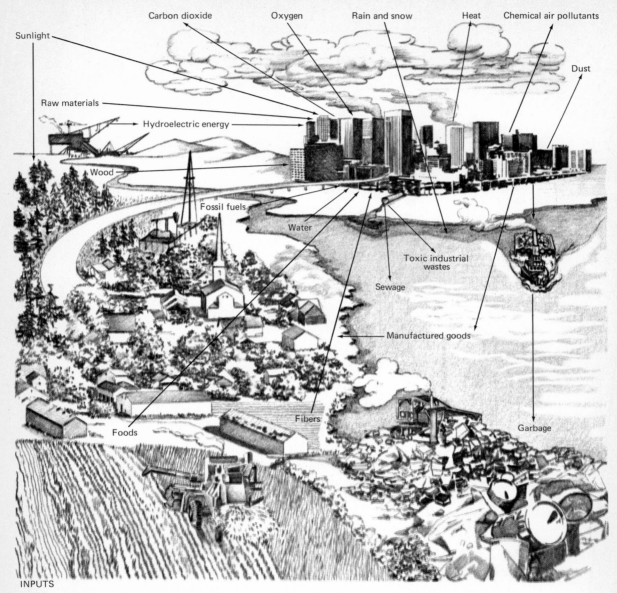

FIG. 1-8 Some of the inputs and outputs of a city.

ports fortunate enough to have them. People who were in New York City during this great power failure are not apt to forget the inconveniences they suffered and the sight of a great city paralyzed by lack of an energy input.

Failure of material input Throughout history, cities have been besieged by armies with the intent of starving them into submission. During the Second World War, the Russian city of Leningrad was almost completely isolated from the rest of the Soviet Union for three years. A trickle of food and other supplies got into the city in spite of the efforts of the besieging German armies, but these were not enough to prevent starvation. In the biting cold of the Russian winters, the people burned their furniture to keep warm and ate anything they could get that might have some nutritional value. Leningrad did not surrender even though about 1 million to 3 million inhabitants died of starvation, or from cold and disease because they were so weakened by hunger. The tragic suffering of the people of Leningrad is a lesson in the dependence of cities on inputs of matter and energy.

1-5 How we measure inputs and outputs

In this book you will come across a number of units of measure with which you may not be familiar. These are the terms used in the metric, or decimal, system to describe mass-weight and length. For example, the metric equivalent for an ounce (weight) is the **gram** (symbolized as g), which equals 0.035 ounces; and the measure for length is the **meter** (m), which is 39.4 in. The prefix **kilo** means 1,000. Thus 1 kilogram (1 kg) equals 1,000 g or 35 ounces (2 pounds, 3 ounces), and 1 kilometer (km) equals 1,000 m, or 3,280 feet (a little over six-tenths of a mile).

The familiar measure of temperature in the United States is the Fahrenheit scale on which water freezes at 32°F and boils at 212°F. In this book, temperature measurements are also given in degrees Celsius (formerly called degrees centigrade). The Celsius, or centigrade, scale is divided into 100 degrees between the freezing point of pure water at 0°C and the boiling point at 100°C.

American scientists have always used the metric system, and in a few years all American citizens will too. In fact, a number of canned goods are currently labeled in both ounces and grams. A table showing these new units of measure and their old American equivalents appears in the Appendix.

URBAN ECOLOGY

Now that we have studied some of the basic requirements of the city as an ecosystem, let us study the ecology of cities in more depth. In order to keep all the networks of inputs and outputs of cities working together properly, a great number of skilled persons must be doing their jobs both in and out of the cities. Many living things, both in the city environment and elsewhere, must carry out their life processes successfully if the city is to survive and be a pleasant place to live. For example, as we shall see, green plants are directly and indirectly involved in supplying the city with many of its materials and often with some of its energy.

1-6 Plants provide materials required by other organisms

Green plants and algae contain green pigments called chlorophylls which are capable of capturing the energy in sunlight. They use the sun's energy to convert carbon dioxide from the air and water from the soil into organic compounds which make up the substance of plants and algae and which are used as food by other organisms. This process is called **photosynthesis.** Scientists long ago proved that plants take very little out of the soil except water (Fig. 1-9) and do indeed make themselves out of air and water and sunlight. The oxygen we breathe is a by-product of the process of photosynthesis and in Chap. 12 we shall discuss this process in more detail.

Because growth of plants and algae is dependent on the sun's energy, a great deal of the earth's surface must be used for food production. Sunlight arrives spread out over the earth rather than concentrated in a few spots. Furthermore, not all the sunlight that arrives at a given spot on the earth can be used in photosynthesis; some is reflected back into space by the atmosphere, some heats the air and the ground, and some evaporates water from oceans and lakes. Finally, plants themselves reflect some of the sunlight and can use only about 2 percent of the energy of the sun that falls on them. Food production therefore occupies much space on the earth's surface because the sun's energy is spread out over the earth and only a fraction of what arrives at any one spot can actually be used by plants.

Food chains Even more space is needed for food production because people often do not eat only plants. They also eat meat from animals such as cattle, which in turn have consumed plants. More space is needed to produce meat than plants. For example, 100 kilograms (220 pounds) of grass eaten by a cow may result in only 10 kilograms (22 pounds) of meat for people to eat. Sometimes people consume animals that are even further removed from the green plants or algae. When we eat fish such as tuna, we are eating a big fish that fed on small fish that fed on still smaller fish that fed on tiny animals that fed on algae, photosynthesizers of the sea. At each step of such a food chain, energy is lost so that the person who eats the tuna obtains only a fraction of the food energy captured by photosynthesis in the algae. Figure 1-10 shows how energy is utilized in a terrestrial food chain.

2.6 Kilograms of tree

Planted in

90.9 Kilograms of soil

Watered for 5 years

Produces

76.8 Kilograms of tree

90.898 Kilograms of soil

FIG. 1-9 This experiment, performed by van Helmont in the early seventeenth century, showed that plants make most of their substance out of the air and not out of the soil, as most people had believed previously.

FIG. 1-10 The energy in foods that organisms eat comes from the sun. Energy is lost via reflection and absorption in the atmosphere before the sun's light even gets to the green plants at the base of the food chain. (See Fig. 4-12.) Then energy is lost at each step in a food chain, so that the weasel at the end of this food chain can receive only a small portion of the energy captured by photosynthesis by the plants. For simplicity, in this figure, the energy actually utilized by the green plants equal 100 percent. (*Based on data of Golley, 1969.*)

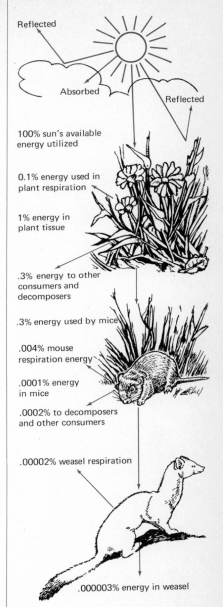

Dependency of cities Because so much land (and water) surface is needed for the production of food, cities cannot be self-sufficient in food production. Terrible famine occurred in Leningrad when it was cut off from its food supply. American cities get their food from all over the world, but most of it is produced in the United States. It takes about 1 **hectare** (approximately 2 acres) of land, including pasture for cattle, to feed one person at the level of affluence of the United States. In 1970, New York City had about 10,000 people per square kilometer (about 26,000 people per square mile). If the people in New York City got all their food from the United States, the people in 1 square kilometer of the city would require about 100 square kilometers (38 square miles) to feed them. Of course, people in New York City also eat foods from the ocean and from many other countries so that even more of the surface of the earth is devoted to feeding 1 square kilometer of New Yorkers.

Food-gathering webs New York City, then, like all other large cities, lies at the center of a food-gathering web (Fig. 1-11). Fishes caught, often in distant seas, move to the city in refrigerated ships, trains, and trucks. Grains, vegetables, meat, and milk move to the city from farms, ranches, and dairies. Sometimes the food is processed (cooked, canned, bottled, mixed, frozen, wrapped) before it reaches the city; sometimes the food is processed in the city. It may be handled by many people—shippers, wholesalers, distributors—before it arrives on the shelf of a grocery store in the city. All these processes require energy in addition to the energy of sunlight used in photosynthesis in producing the food. Transportation, preservation, and processing of food all require energy.

1-7 Almost all the energy that powers the city comes indirectly from the sun

The sun is the original source of most of the energy that drives the city's industry, powers the trucks and trains, and heats its homes. And plants and algae have played a vital role in transforming the sun's energy into forms that can be used by people's machines. This was accomplished not by the plants of today but by those that lived millions of years ago.

Transformation of energy by plants and algae A lump of coal or a barrel of oil is the transformed matter of ancient living things. In the distant past, some 350 million years ago, green plants and other organisms growing in great swamps or shallow oceans died and did

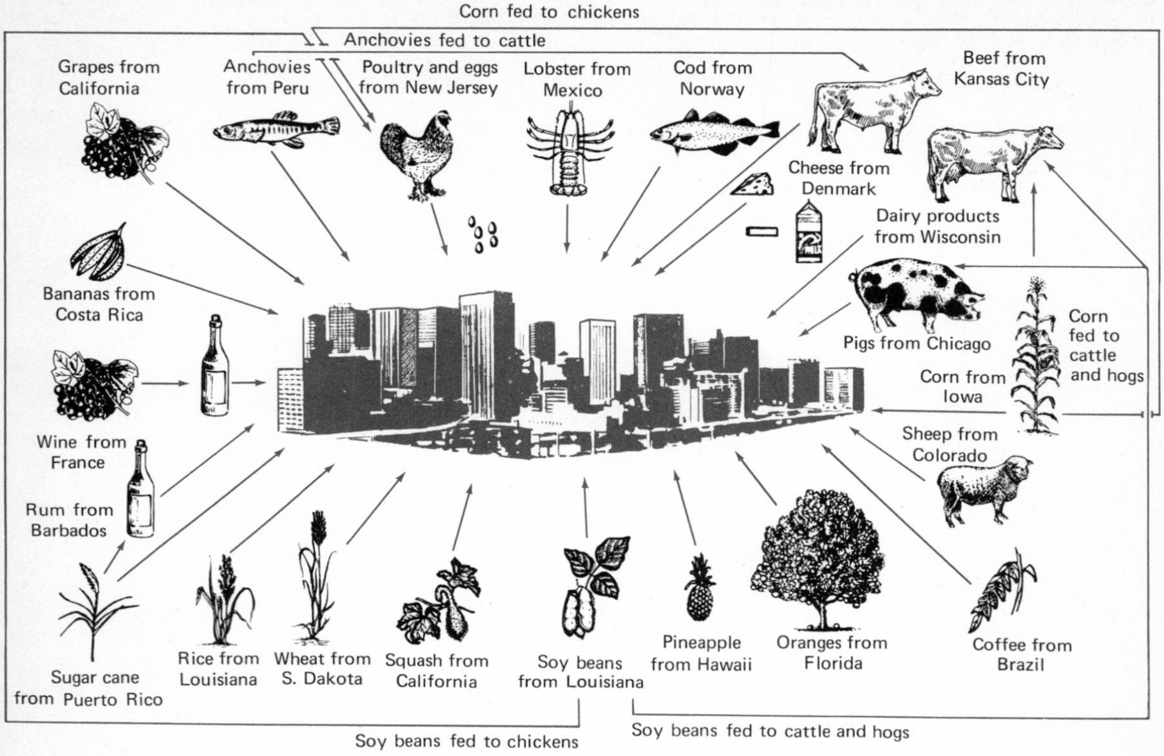

FIG. 1-11 A city may be thought of as the center of a vast food-gathering web. Some of the foods that enter New York City are illustrated, and some of their sources are given.

not decay completely. In ways that are not fully understood, these masses of vegetation and other dead organisms were slowly converted into coal, oil, and natural gas. Such remains of life of the distant past are known as **fossils;** that is why we refer to these materials as **fossil fuels.** The sun's energy millions of years ago was stored in chemical bonds by ancient living things, and it is released when their remains are used for fuel today. Thus, the sun powers automobiles, trucks, and trains and is the source of the electricity which flows from fossil-fueled steam power plants.

Hydroelectric power Even the electricity which flows from hydroelectric plants has its source in the sun's energy (Fig. 1-12). The water behind a dam which contains the hydroelectric generators fell on the highlands as rain. The rain came from clouds which formed as the sun's heat evaporated the water from land and sea. These clouds were transported to the highlands by the winds and the circulation of the atmosphere, which are also powered by the sun (see Chap. 4). In fact, the sun is the major driving force of the earth's weather and winds. Even when electric power is generated by a windmill, the ultimate source of energy is the sun. There are ways of making electricity directly from sunlight using solar cells, but at present these are too expensive to power cities.

Nonsolar energy The only energy the city receives that does not have its source in the sun is nuclear energy, which comes from radioactive elements such as uranium and plutonium. A tiny fraction of the electric energy used in cities today comes from nuclear power plants, which use the heat released from controlled nuclear reactions to heat fluids which turn turbines and produce electricity, much as steam-operated plants do. Because supplies of fossil fuels will eventually run out and because energy demand keeps increasing, some people want to build more nuclear-fired electric generating plants quickly. Many other people have doubts about the safety of such plants and worry about the disposal of their radioactive waste products, which must be isolated from living things for hundreds of thousands of years.

1-8 Plants provide side benefits for the city

Urban planners who want to replace living plants with plastic ones seem to think that the city does not need to have living plants in it. Actually, plants do many useful things in a city even if they are not producing food for people. Smog contains some gases that, in low concentrations, can be used as nutrients by plants. Thus plants can absorb some air pollutants. They also improve the quality of the air by giving off oxygen and woodsy-smelling compounds, such as those given off by pine trees. Evaporation of water from plants cools the air, and the leaves of plants catch falling dust particles. Trees and shrubs also muffle the noise of heavy street traffic. The roots of plants—even weeds on vacant lots—help to hold soil in place and reduce the amount of soil particles blown into the air and washed into storm sewers. It seems people have a psychological need for plants; city dwellers who can afford it have weekend cabins in the woods, apartment dwellers grow plants in pots indoors, and almost all cities have parks with lawns and trees. A study of people who grew up in urban areas showed that the feature of their childhood environment they mentioned most often in interviews was the presence or absence of lawns in their neighborhoods. Many families who can afford it move away from the city centers to the suburbs where there is more plant life. In the suburbs there are often enough plants of different kinds to provide shelter and food for animal life. Children, especially, seem to enjoy encounters with wild creatures such as song birds, rabbits, lizards, toads, and even insects.

Unfortunately, the city is a difficult environment for green plants. Since the soil of the city is covered mostly with buildings and pavement, there is little space for green plants to grow. Those that do exist in the city often do not get as much energy as they need for vigorous growth because the haze and smog that cover many big

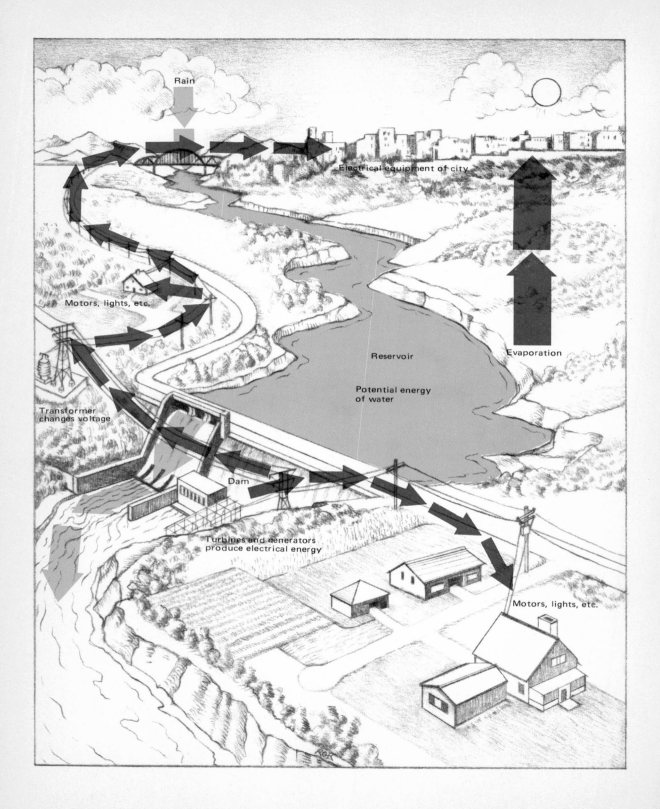

FIG. 1-12 Diagram of hydroelectric power. The sun evaporates water from lakes, seas, cities, forest, and fields. This water condenses high in the atmosphere and is returned to the surface as precipitation. Rain that falls on highlands or water from melted snow runs back to the sea, releasing energy which can be used to drive turbines. Generators driven by the turbines convert the energy of moving water into electrical energy and heat. To ensure a steady flow of water and to increase the potential energy available, dams are built across rivers. The lake which forms behind a large dam often covers thousands of hectares of land. The useful life of hydroelectric dams is expected to be not much more than 200 years because the silt trapped in lakes eventually fills them up.

cities screen out some of the sunlight. Air pollution can also inhibit plant growth. Even though the plants use some of the gases in smog, high concentrations of other gases damage plants. Some plants are more sensitive to pollution than others; snapdragons, alfalfa, and spinach, for example, do poorly in polluted air. Well-meaning people do not always help the situation either. Uninformed about the city's need for greenery, they spray plants with herbicides, pull them out, plow them under, and cover them with asphalt. Generally, highly urbanized areas have few different kinds of plants; that is, they are low in **plant diversity.** Often the kinds **(species)** of plants found in the city are growing there because they were chosen by city planners who know they will survive under the difficult growing conditions of city streets.

THE IMPACT OF URBANIZATION

Although most of the people in the world still live in the country or in small towns, more and more people are moving to cities each day. This trend toward urbanization creates numerous environmental and social problems even though it provides jobs and presumably better lives for many people. The Congress of the United States recognized that the activities of human beings are having major detrimental impacts on the environment and in 1969 passed the National Environmental Policy Act (NEPA). Under this act federally funded projects cannot be implemented until an environmental impact statement (EIS) is filed, which details the effects on the environment of such projects. Many states have followed suit with similar laws relating to state and privately funded projects. Most of the projects which have come under NEPA have been either in or near cities or for their benefit.

1-9 Cities today are different from cities of the past

Origins While some ancient cities seem to have been planned from the start, most grew from small towns that happened to be located in especially favorable places. Towns that lay at the crossings of overland trade routes or on the banks of navigable rivers or had harbors or religious shrines attracted merchants, scholars, artisans, and government officials. In addition, most ancient (and modern) cities were founded on fertile plains because many foods could not be transported far from their source of production without refrigeration, canning, or other modern methods of preservation. As a city grows, however, it spreads out across the agricultural land that feeds it. Food then must be brought from greater and greater distances, and the character of agriculture that remains around the city changes. Instead of growing crops such as grains that take a great deal of space, the farmers grow more vegetables. They also raise more cattle and poultry, feeding them with grain grown far from the city. Frequently, farmers near the city sell some of their land to city dwellers for second homes or for suburban developments. Some of these people grow a little food, but often they allow the land to return to forest or they plant only ornamental plants. Thus the food inputs of the city come from ever greater distances as the city grows in size. In the modern city, food inputs often depend on several transportation links. For example, chickens consumed in New York City may be raised in New Jersey on fish meal from Peru and grain from the Midwestern United States.

Waste disposal problems As cities get larger, the problems of keeping them clean and disposing of their wastes become more difficult. Some ancient cities solved these problems; for example, inhabitants of ancient Rome had water-borne sewage systems and public baths. During the Middle Ages these health-supporting public services disappeared from the cities of England and Europe. Sewage and garbage were dumped in the streets, and bathing was considered to be bad for one's health. European cities of the Middle Ages were places of stench and sickness; it is not surprising that a great plague, the Black Death, swept through Europe during this period, killing about one-fourth of its inhabitants. Eventually, these people began to improve their sanitary facilities, and in many modern urban-industrial cities, complex social agencies take care of removing sewage, disposing of garbage, and sweeping the streets.

Slums One would expect modern American cities today to be clean and attractive places to live. Some parts of some cities are. Other parts of many cities are rundown and filthy; garbage collects on vacant lots; people live in apartments without heat or running water. Halls stink of urine; toilets are stopped up; rats bite babies. Alleys and doorways are lined with passed-out winos and junkies. Crime rates are high, and people are afraid to walk the streets for fear of being robbed, raped, or mugged. In 1972, in cities of over 250,000 inhabitants, there were 19.7 murders per 100,000 in the city itself, but only 4.6 per 100,000 in the suburbs. Furthermore, there were twice as many burglaries per 100,000 in the urban areas as in the suburban areas. Slum areas often occur in the older parts of the central city where landlords find they can make a profit by not repairing buildings because their tenants are poor and cannot afford to live anywhere else. Various government agencies such as the Federal Housing Administration (FHA) and the Department of Housing and Urban

Development (HUD) have failed to help people in the inner cities.

Failure of "urban renewal" One method of trying to alleviate the problems of the central city is to tear down old buildings and relocate the tenants in modern high-rise apartments. But many of these "model" communities were such disastrous failures that they now stand empty or have been torn down. Pruet-Igoe in St. Louis was a complex of 43 tall buildings built in 1954 for $36 million. By 1971 it had become an "urban jungle" and only 600 of its 2,800 housing units were occupied. In 1972 it cost $39 million to demolish part of Pruet-Igoe and shorten and redesign the remaining buildings. In Brazil a group of people living in tar paper shacks were moved into a new modern apartment building and their tar paper shacks were demolished. The next day they moved out of the apartments and began building new tar paper shacks. A few such projects have been successful; however, it seems clear that urban-renewal projects which disrupt families and the neighborhood structure are likely to fail.

New problems There are a great many problems in the modern city that never existed in the city of the past. For example, automobiles and other vehicles make possible speedy transport of food and other goods into the city, and they allow people greater mobility; yet they have created many inconvenient and harmful side effects. In 1932, many American cities had excellent rail transport systems. In that year, it has been reported, a major automobile company began buying up rail systems, converting them to buses and selling them back to the operators. The buses of those days were noisy and foul-smelling and turned patrons away. The public transit systems failed, and people bought many more automobiles. Although some cities in the United States have public transit, automobile traffic seems to be a problem in all. As fast as city officials build wider streets and better highways, traffic gets worse. Parking lots take up space that could be used for other purposes, and it becomes difficult for pedestrians to stroll in the city since automobiles have priority. In many cities, most air pollution comes from automobiles, and in areas of heavy traffic, pollutants often reach dangerous levels. Carbon monoxide (CO) has been measured at several hundred **parts per million (ppm)** of air in Los Angeles traffic jams. It has been shown that 100 ppm causes dizziness and 1,000 ppm is fatal. The automobile has also made commuting long distances possible. Parents often leave for work so early and return home so late that they see their children only on weekends; they also miss the opportunity to enjoy the space and greenery that was their reason for moving to the suburbs.

Workforce Modern cities are extremely vulnerable to workers who do not do their jobs. Sometimes workers in the city simply *cannot* do their jobs; for example, shortages of fuel or breakdowns of trains may not allow those who live outside the city to commute in to work. Or people in the city may decide for one reason or another not to report for work; they may be protesting some injustice or striking for higher wages. In 1966, a strike shut down all subway and bus service in New York City for twelve days. Some New Yorkers could not get to work while others could not get home. The estimated loss to the city's business was 1 billion dollars. In 1971, striking bridge tenders in New York City left twenty-seven drawbridges locked in open position. As you can imagine, severe traffic jams resulted. A few years ago a strike of garbage collectors created a potential health hazard in New York City. In San Francisco a strike of city employees nearly stopped all city business: Nurses, clerks, bus and streetcar drivers all refused to work. In November 1974, firefighters in Montreal went on strike. Fires were allowed to burn and over fifty families lost their homes. Even strikes of workers outside the city can affect it seriously. In the winter of 1973 to 1974, coal miners in England staged a work slowdown at the same time that oil imports were cut off. The lack of fuel and light left the shops and factories of London functioning only three days a week. Thus modern cities are vulnerable to the breakdown of work forces due to complex social and economic factors.

1-10 Modern industrialized cities use massive amounts of energy

Every electrical appliance has a measurable power consumption. **Power** is the rate of doing work; that is, power is the amount of work done in a given unit of time. A common measure of power is the **watt.** If a 1,000-watt iron or toaster is used for an hour, it consumes 1,000 watt-hours of energy. That amount of energy is called 1 **kilowatt-hour.** Table 1-2 gives the estimated energy consumption of some common appliances in terms of kilowatt-hours. An average electric-power generating plant produces 500 million watts. This may seem like a lot in comparison to the small amount used by appliances; however, if we add up the average wattage of all the appliances listed in Table 1-2, we get almost 60,000 watts. Only ten thousand households with all these appliances turned on at once could use up the entire output of an average power plant. Fortunately, not every household has every appliance it contains turned on at once! Nevertheless, the energy needs of a large city are immense—consider the number of households that are crowded into the city, add in the lights and air conditioning required by office buildings, and include the huge amount of energy required to run urban industries.

Americans use more energy per person than any other people in the world, and the total amount used in the United States has so far continued to increase every year (Fig. 1-13). Each resident uses each year, on the average, an amount of energy about the same as that contained in 11 metric tons (24,200 pounds) of coal. This is roughly twice that used on the average by Europeans, and almost 200 times the per person (usually referred to as **per capita**) energy use in Nigeria. In the past, the United States *was* richly supplied with easily available energy resources, but we have used up a good deal of them. Today we import about 11 percent of the energy we use. In the winter of 1973 to 1974, a "mini-energy crisis" occurred.

TABLE 1-2 ENERGY CONSUMPTION OF ELECTRIC APPLIANCES

APPLIANCE	AVERAGE WATTAGE	ESTIMATED USE	ESTIMATED KILOWATT-HOURS/YR CONSUMED
Broiler	1,436	80 min/week	100
Carving knife	92	104 min/week	8
Coffee maker	894	20 min/day	106
Deep fat fryer	1,448	66 min/week	83
Dishwasher	1,201	50 min/day	363
Food blender	386	6 min/day	15
Food mixer	127	17 min/day	13
Food waste disposer	445	11 min/day	30
Frying pan	1,196	26 min/day	186
Hot plate	1,257	12 min/day	90
Range	12,207	16 min/day	1,175
Roaster	1,333	25 min/day	205
Toaster	1,146	5.6 min/day	39
Food freezer (15 cu ft)	341	40% of the time	1,195
Food freezer (frostless 15 cu ft)	440	46% of the time	1,761
Refrigerator-freezer (14 cu ft)	326	40% of the time	1,137
Refrigerator-freezer (frostless 14 cu ft)	615	34% of the time	1,829
Clothes dryer	4,856	34 min/day	993
Iron (hand)	1,008	23 min/day	144
Washing machine (automatic)	512	33 min/day	103
Washing machine (nonautomatic)	286	44 min/day	76
Air conditioner (room)	1,566	10% of the time	1,389
Bed covering	177	2.3 hr/day	147
Dehumidifier	257	40% of the time	377
Fan (attic)	370	2.2 hr/day	291
Fan (roll-about)	171	2.2 hr/day	138
Hair dryer	381	42 min/week	14
Heat pump	11,848	3.7 hr/day	16,003
Heater (radiant)	1,322	22 min/day	176
Humidifier	117	16% of the time	163
Shaver	14	6 min/day	0.5
Sun lamp	279	66 min/week	16
Toothbrush	7	117 min/day	5
Water heater (standard)	2,475	4.7 hr/day	4,219
Water heater (quick recovery)	4,474	2.9 hr/day	4,811
Radio	71	3.3 hr/day	86
Television (b and w)	237	4.2 hr/day	362
Television (color)	332	4.1 hr/day	502
Clock	2	100% of the time	17
Sewing machine	75	2.8 hr/week	11
Vacuum cleaner	630	84 min/week	46

SOURCE: Edison Electric Institute, national averages.

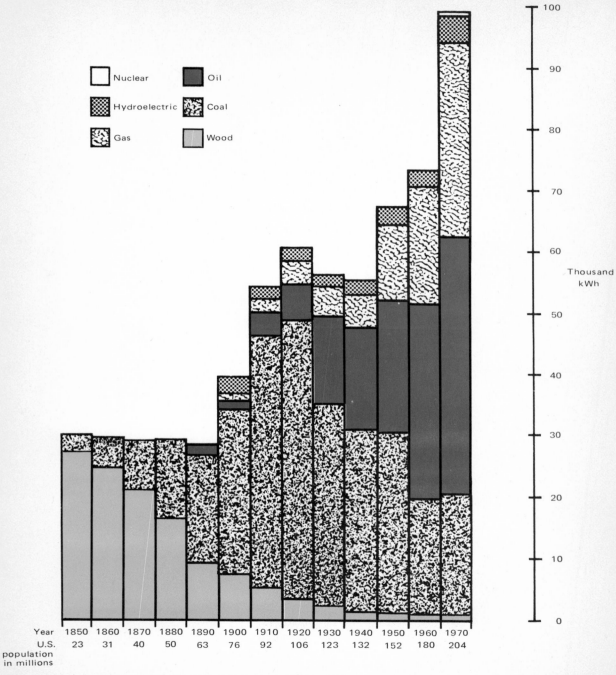

FIG. 1-13 Per capita energy use in the United States from 1850 to 1970.

This was a relatively minor affair in the United States, caused by oil company greed, poor planning, governmental mismanagement, and (to a small degree) the 1973 Arab-Israel War. Government officials began attempts to achieve "energy independence" for the United States; new offshore oil wells were approved, the Alaska pipeline was started, and pollution controls were relaxed so that fuels such as high-sulfur coal could be burned. The nuclear-power industry urged speedy construction of more electricity-producing reactors.

At the same time, some people favored reducing our energy consumption. In the winter of 1973 to 1974, the United States government asked citizens to lower their thermostats and turn off unnecessary lights. In many areas, gasoline stations were closed by law on Sundays, and drivers were expected to keep their speeds down to 88 kilometers per hour (55 miles per hour). This sudden shortage of a fuel which people had considered a basic necessity led to a highly irrational response on the part of many. Gasoline thefts became common, as did cases of people poisoned by inhaling gasoline while trying to siphon it from one car to another. Several people were shot in arguments at gas station lines. The social impact was greatest in cities, such as Los Angeles, which have little public transportation so that people must drive their own cars if they are to get to work.

1-11 Cities have negative impacts on the environment

The city could be viewed as a sort of vacuum cleaner, sucking in materials, energy, and people from a wide area. Energy is often moved long distances to cities (Fig. 1-14); oil, gas, and coal are rarely found near the places where they will be burned. Los Angeles receives hydroelectric power from Hoover Dam on the Colorado River, 400 kilometers (250 miles) away. Other cities in the Southwest receive electrical energy from the Four Corners plant on the Hopi reservation in New Mexico. Thus the impact of the pollution of this coal-burning plant is far from the people who use the electricity. Oil spills from breakups and collisions of tankers, or leaking undersea oil wells, often occur many miles from the places where the oil will be used. Energy production for cities, and also for their industries, may have widespread negative environmental impacts. The beaches, as well as the birds, of Santa Barbara, California, and Cornwall, England, have been fouled with oil. Sweden is complaining about acid rain resulting from the drift of polluted air from the industrial midlands of England. The effects of cities on environments a good distance from them are often negative. Tourists do not come to polluted beaches, and acid rain impairs crop production.

Thermal pollution Cities produce heat from many sources. Electrical generating plants and factories use water for cooling and dump it back into rivers and lakes several degrees hotter than when they took it in. Aerial photos taken with special heat-sensitive film show bands of thermal (heat) pollution extending many miles downstream from cities. Many fish and shellfish do not thrive in heated waters, and

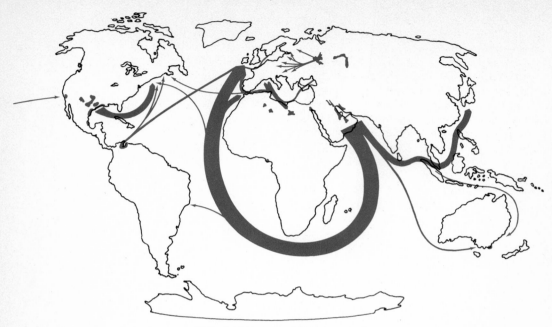

FIG. 1-14 Energy of all kinds moves in complex ways over the earth's surface. This map shows the major movements of crude oil in 1967. The dark spots indicate major producing areas at that time. The thinnest arrows represent flows of approximately 180 billion barrels. The broader arrows indicate proportionately greater flows. Most of this crude oil is carried in huge ships called supertankers. (*After* Oxford Economic Atlas of The World, *4th ed., D. B. Jones (ed.),* Oxford University Press, London, 1972.)

some cannot survive at all. On the other hand, some species are benefitted by increased heat and are killed when power plants are turned off resulting in suddenly cooler water.

Our machines produce heat; our bodies produce heat; and our trash as it rusts or decays produces heat. Heat production due to human activities now exceeds five percent of the incoming solar radiation at the surface in some local areas which are tens of thousands of square kilometers in size. Such local heat production has already affected the local weather of cities, which tend to be 2 to 3°C (4 or 5°F) hotter than the surrounding countryside. If the trend to more heat discharge continues, climatic patterns over whole areas of the earth may be changed.

Planting more greenery can reduce the heating effect. In cities rain falls mostly on pavement and buildings and runs off into the sewers. The cooling effect produced by evaporating water is therefore lost. Where there are more trees, lawns, and vacant lots with weeds, the water soaks into the soil and is gradually evaporated by plants; in fact evaporation from a single, properly tended tree can produce a cooling effect approximately equal to five room-size air conditioners running 24 hours a day, and trees do not use up electricity.

Air pollution Cities produce a great deal of dust and smoke (see Fig. 1-6). Air circulation in the city sometimes traps this pollution in a huge dust dome that is visible for hundreds of kilometers away. Particles of dust and soot in the air often cause moist clouds to produce more precipitation (rain, snow, or hail) downwind of cities. Since 1925, there has been a 31 percent increase in precipitation in the LaPorte, Indiana, region, downwind of Chicago. There has also been a 38 percent increase in thunderstorms, and a 245 percent increase in hail. All these increases are the result of the great increase in industrial pollution from the Chicago-Gary steel-production complex.

The chemicals in polluted air may also injure crops that are planted far away from urban areas; Sweden's complaint about crop loss due to acid rains is an example.

Urban waste water pollutes waterways Cities are also major sources of waste water. Wastes pour into sewer systems from toilets, sinks, showers, and industrial plants. Some industrial plants bypass the sewer systems and pour their wastes directly into waterways. On the edges of cities, where cattle are fattened for market in feed lots, manure often enters waterways in huge quantities. Where septic tanks take the place of sewers, human wastes may pollute groundwater. During storms dirt from the streets and soil from vacant land also wash into waterways.

Some of the components of waste water are inorganic—for example, some kinds of soil particles. These settle to the bottom of waterways, eventually blocking shipping channels which require dredging. Other components of polluted water are organic materials—for example, human and animal wastes, garbage, and remains from slaughter houses. The organic materials are **biodegradable;** that is, living organisms such as bacteria can break them down into simpler chemical substances. Most of these decay organisms are microscopic and are often called microorganisms of decay. These microorganisms derive energy and substance from organic materials during the process of decay. While they are performing this very essential process of breaking down organic wastes (and even whole dead plants and animals), they often produce gases such as hydrogen sulfide, which smells like rotten eggs. This explains the foul odors arising from polluted lakes, rivers, and streams.

In natural ecosystems, there are **detritus-**based food chains, that is, chains at the base of which is decaying organic material. Scavenger organisms such as vultures, certain beetles, and some other insects specialize in eating decaying flesh; the microorganisms in such a food chain feed on both the dead plants and animals and on the wastes of the scavengers. Usually these food chains are able to decompose all the wastes that occur in natural ecosystems, breaking down complex organic compounds to simple compounds and elements that are useful for the growth of plants or other organisms. There are even bacteria which can use hydrogen sulfide for energy. In artificial ecosystems such as cities, detritus-based ecosystems often become overloaded. Even though the presence of organic materials in water is quite natural, when there is too much for decay organisms to process, these materials become pollutants. The organisms of decay are a waste-disposal system; up to a point this system can purify a body of water of organic materials—that is, it can convert them into substances that are not harmful to people or to other organisms. The water pollution that surrounds many of our cities today results from overloading a natural system.

Sewage treatment plants are designed to create conditions favorable for microorganisms of decomposition (Fig. 1-15). In fact, sewage plants can now be made so efficient that the water they produce as an output is drinkable. Unfortunately, such sophisticated sewage treatment plants are very expensive to build and operate, and therefore very few of them exist. Sewage plants are usually constructed so that they require energy to operate, although a few use the methane gas produced by decomposition as their energy source. Public health officials attempt to design waste disposal systems which can accept some overloads and partial breakdowns. However, storms and disasters which break down the waste disposal systems of cities can lead to epidemic disease, especially when the drinking water supply becomes contaminated by sewage.

But even under ideal conditions microorganisms cannot solve all of our waste-disposal problems. For one thing, many organic compounds produced by industries biodegrade very slowly. For another, some of the inorganic wastes which enter our water supplies are poisonous in relatively small amounts. For instance, the element mercury (Hg) is a by-product of various industrial processes, and very small quantities of this element can lead to brain damage and even death. Until recently it was thought that the mercury which escaped

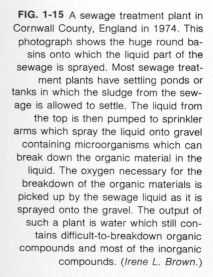

FIG. 1-15 A sewage treatment plant in Cornwall County, England in 1974. This photograph shows the huge round basins onto which the liquid part of the sewage is sprayed. Most sewage treatment plants have settling ponds or tanks in which the sludge from the sewage is allowed to settle. The liquid from the top is then pumped to sprinkler arms which spray the liquid onto gravel containing microorganisms which can break down the organic material in the liquid. The oxygen necessary for the breakdown of the organic materials is picked up by the sewage liquid as it is sprayed onto the gravel. The output of such a plant is water which still contains difficult-to-breakdown organic compounds and most of the inorganic compounds. (*Irene L. Brown*.)

into the waterways simply fell to the bottom and remained there harmlessly. Recently, however, people have learned that this comforting idea is wrong. Microorganisms cannot, of course, decompose mercury, or any other element for that matter. But, especially in polluted water, they can combine mercury with other elements, such as carbon and hydrogen, to form the organic compound methyl mercury (CH_3Hg). Methyl mercury is much more dangerous than pure mercury because it can become bound to the proteins and fats that make up the tissues of the body. In fact, there is a disorder called Minamata disease which results from eating fish containing methyl mercury. The disease is named after a town in Japan, where between 1956 and 1961 forty-three people died and sixty-eight were permanently disabled from methyl mercury poisoning. This tragedy resulted from the mistaken belief of industrialists and pollution-control officers that inorganic mercury fell to the bottom of waste water and remained in the inorganic form after it was discharged into rivers or the sea.

Some people estimate that as many as one-half million chemical compounds enter the waters of the earth as pollutants. Many of these are not biodegradable and most of them originate in cities. Industry uses over 17 percent of the total amount of water consumed in the United States. Much of the water used by industry is returned to lakes and rivers mixed with various poisons, which are, like mercury, by-products of industrial processes. Cleaning up water polluted by industrial processes is a difficult job. Pollutants should be removed from waste water before it is returned to the river, lake, or sea. Waste water can be held in large tanks or collecting ponds and treated with chemicals that form compounds with the pollutants. In some cases these compounds are nontoxic or even useful substances. In other cases, the new compounds are insoluble and sink to the bottom, leaving the water relatively clean.

1-12 Cities have negative impacts on their inhabitants

Effects of pollution Many people who have grown up in polluted cities do not seem to realize that their environment could be any other way. They have always had days of choking air, rivers unfit for swimming, noise everywhere they go, and garbage in the streets. Is not all this just part of life in the city? As long as city dwellers think so, little is apt to be done to improve their environment. Perhaps a realization of the hidden costs of pollution will inspire people to demand a cleaner environment. Pollution costs human lives and health. People who live in polluted areas of cities have more respiratory diseases and a higher death rate. People who work in certain kinds of factories may be exposed to pollutants which cause diseases after long-term, low-level exposure. For example, by October 1974, 26 workers in plastic factories who were exposed to vinyl chloride had died of a rare form of liver cancer. Pollution costs us in damage to crops and in weather changes caused by pollution; pollution costs when it kills food fishes. Pollution leads to more frequent cleaning bills, more frequent painting of structures. In some areas, acids in the air are slowly dissolving the stone of priceless works of art. These costs, plus the obvious degradation in the quality of our lives when we must breathe choking, eye-watering air and forgo water sports because our lakes, rivers, and streams are polluted (Fig. 1-16), should inspire people to be willing to pay the cost of controlling pollution at its source. Perhaps someday we will be able to treat all waste as well as nature does so that it is no longer waste but reusable. For example, properly treated sewage makes excellent fertilizer and is so used in many countries.

Concentration of human problems Not all people who move to the city or who live there benefit from doing so. Just as cities concentrate materials and energy so do they concentrate human problems. Many people who deviate from the norms considered acceptable in small towns move to the city for its greater tolerance of differences. Some of these people may have problems in the city too. Many workers who move to the city cannot find jobs or can find only low-paying ones. Many immigrants to America found that they could afford to live only in the slums of the central city. Many blacks and other "racial" or ethnic minorities also find themselves in ghettos or slums. Some of the people who cannot get legitimate jobs in the city find they can make money by prostitution, gambling, selling illegal drugs, or stealing. Today, all large cities have some kind of police force. In spite of this, many people in cities commit crimes and many become victims of crimes.

Effects of crowding Increased population density seems to be associated with increased mental illness, according to studies by psychologist Jonathan Freedman and his associates. Since people have a limited amount of time, dialogues between them become shorter and less profound as they increase in number. In midtown Manhattan a person can pass 220,000 people within a 10-minute walk from his or her office. Even for a person who likes rich and varied sensory experiences, the city can be overwhelming. In fact, in the city most interactions between people are highly formalized—one talks to strangers only when necessary to ask the time or a direction.

In short-term (3-hour) experiments, performance at various tasks was not affected by crowding. In experiments which measured aggression, however, crowding did have an effect. These experiments indicated that men tended to be more aggressive when they were crowded, but that women got less aggressive. Furthermore, mixed groups of men and women do not show an increase in aggressiveness when crowded. Here is a reason why jury rooms should be spacious and juries should be composed of both men and women. Certainly, international negotiating teams deciding on matters of war and peace should meet in spacious rooms and include both men and women.

Some people have claimed that high crime rates, riots, and other urban problems are an inevitable result of people living together closely. They point to experiments with rats and other animals where

FIG. 1-16 In this polluted stream there is an overgrowth of algae which uses up so much oxygen that few fish can survive. Disease bacteria and viruses may be present in such polluted streams, and this makes them unsafe for fishing and other recreational activities. People often dump rubbish into streams, and no one bothers to clean them up since the stream is not a pleasant place to be. Such streams not only look bad and smell bad, but they are health hazards to people and wildlife. (*Courtesy of J. Paul Kirouac.*)

crowding alone has led to abnormal behavior and a general breakdown of social systems. Even though, as we have seen, crime rates are higher in the city than in the suburbs, this does not mean that crowding itself is the cause. There are many other differences between the city and the suburbs. High crime rates *do* seem to be associated with the rates at which people are moving into cities. Those cities with the highest rates of in-migration tend to have the highest crime rates. One of the greatest problems facing the city is the threat of epidemic diseases, which spread more rapidly under conditions of crowding. And, of course, almost all the problems of inputs and outputs are made more difficult as population density grows.

1-13 Cities have beneficial impacts on their surroundings

Sources of manufactured goods and new ideas Cities also have beneficial impacts. Cities are more than just huge vacuum cleaners sucking the countryside dry of resources and spewing out pollutants. They are sources of creativity and progress. Manufactured goods and creative ideas are beneficial outputs of cities. The complicated machines of modern civilization cannot be made by one person working alone. Automobiles, locomotives, television sets, computers, radio transmitters, etc., require the knowledge and skills of many people working together for their production. Cities bring together in one place people of many different talents and skills who can set up factories and produce these sophisticated machines. Also, in a city, there are usually several factories making goods such as shoes or clothing. Therefore, there is a great variety of different kinds of shoes and clothing produced. Not only do the buyers not have to make their own, they can choose among expensive or cheap and colorful or plain styles. Even artists and writers who work alone often seek out the city for inspiration or companionship or for the more open moral standards it offers. Most great cities have "artists' quarters."
Centers of cultural change When people all lived as hunters and food gatherers, social roles were stereotyped and cultural change occurred slowly, if at all. People of the same age and sex had similar jobs in society. Men hunted and were warriors; women gathered food, cooked, and cared for children. Children learned their future roles at an early age, and the games they played were those which prepared them for their future roles. Each generation learned the taboos and traditions of the tribe, and in turn, taught them to the next generation. Studies of present-day hunting and gathering tribes show that while they actually spend little time in the collection of food, they have few material possessions. They have to move continually to follow the game and the fruits in season. Therefore they can own only what they can carry conveniently.
Centers of cultural interaction About 10,000 years ago, when agriculture was invented, the social structure began to change. People could settle down in one place, build permanent dwellings, and accumulate

tools and furniture. As the techniques of agriculture grew more sophisticated, one farmer could provide for more than one family. This allowed some people to do other things than farm. A person especially skilled at making tools might do just that and trade the tools for food. A person skilled at making clothing might give up farming and trade clothing for food. Agriculture thus led to a large-scale division of labor in society and eventually, 5,000 or 6,000 years ago, to the establishment of the first cities. In cities, large numbers of people, often from very different cultures, come together and interact. Traditional ideas conflict with other traditional ideas. Bright young people synthesize new ideas from the conflicting ideas of their elders. The great numbers of people in a city make possible the accumulation of many people with similar talents. Scholars gather together in universities; musicians form symphony orchestras. If only one person in 1,000 has the talent to become a good musician, then a community of 100,000 or so people will be necessary to produce a symphony orchestra.

Cities have, naturally, become the centers of change in society, and the centers of civilization itself. The cities today have the great libraries, the great symphony orchestras, the operas, the great art museums, the studios of TV networks, the great publishing houses, and the headquarters of most enterprises, commercial and governmental. Although rapid transport systems and nationwide instant communication networks are starting to soften the dividing line between city and country, we must still think of ideas and change as major outputs of cities.

1-14 Cities have beneficial impacts on their inhabitants

Just as cities accumulate energy and materials from the surrounding countryside, so do they accumulate people, from their surroundings and often from the entire world. Today most large American cities have people from all over the world living in them. These people come to the city because they have business there or because they believe they will find advantages in the city not to be found in their own cities, towns, or countries. People who like opera or who need the books offered by a large library or who just like crowds and excitement often prefer to live in cities. As agriculture in the United States became more mechanized, there was a steady decrease in the number of farm jobs available. People migrated to the city to find work. In 1920, 30 percent of Americans lived on farms. In 1970, only 4.8 percent remained. People immigrating to America from other countries often remained in cities because that is where the jobs were.

Many people are enthusiastic about living in the city. Those people who find good jobs and pleasant places to live do benefit from the city environment. All over the world the trend is for more and more people to move to cities. It seems that people believe they will find better opportunities in cities. If this vast process of urbanization is to continue to lead to benefits for people and to attractive cities, which do not have too much negative impact on their surroundings or their inhabitants, city planners must cope with the many problems created by ever-growing cities.

QUESTIONS FOR REVIEW

1 What do ecosystem, ecology, and urban ecology mean?
2 Why is thermal pollution unavoidable in cities, and why is it usually so much more concentrated there than elsewhere?
3 What are some of the things that can happen to cities when inputs are blocked?
4 What are three basic units of measure in the metric system, and what is the prefix which means *thousands* in that system?
5 What do these terms mean: power, watt, kilowatt hour, metric ton, per capita?
6 What is one big advantage of the metric system over the English system? (Hint: Do these calculations in your head—How many meters are there in 7 kilometers? How many feet are there in 7 miles?)
7 What are three reasons why the waste-disposal problems of the city cannot be completely solved by microorganisms?
8 "Everybody complains about the weather but nobody does anything about it." In what sense is this old saying untrue—at least in regard to people who live or work in cities? In what sense is it especially true in cities?
9 Can you list at least four "hidden costs" of pollution?
10 "The sun powers trucks, TV sets, hydroelectric generators, sailboats, cabbages, and people." What does this mean? Give an explanation for each case.
11 Division (specialization) of labor was a key element in the establishment of the first cities, and it is still a great strength of the city. Why is it also a potential *weakness* of the city—that is, how may it threaten the city's functioning, and inconvenience or even endanger its inhabitants?
12 Is the worldwide trend toward migration *into* or *out of* cities? What are some reasons for this trend?

READINGS

Davis, K.: "The Urbanization of the Human Population," *Scientific American,* **213**(3):41–53, offprint 659, 1965.
George, C. J., and D. McKinley: *Urban Ecology—In Search of an Asphalt Rose,* McGraw-Hill, New York, 1974.
Hall, P.: *The World Cities,* World University Library, McGraw-Hill, New York, 1966.
Hamblin, D. J.: *The Emergence of Man: The First Cities,* Time-Life Books, New York, 1973.
Jacobs, J.: *The Death and Life of Great American Cities—The Failure of Town Planning,* Random House, New York, 1961.
Reid, T. S., and W. B. Watt: "Basic Chemistry for Biology," *Biocore,* unit I, McGraw-Hill, New York, 1974.
Schmid, J. A.: "The Environmental Impact of Plants and Animals," *Ekistics,* **218**:53–61, 1974.
Watt, K. E. F.: "Ecology," *Biocore,* unit XXI, McGraw-Hill, New York, 1974.

2
LIFE IN THE COUNTRY

SOME LEARNING OBJECTIVES

After you have studied this chapter, you should be able to:
1. Demonstrate your knowledge of the basic principles of ecosystems by doing the following:
   a. Define the roles of producers, consumers, and decomposers in a food chain.
   b. State approximately how much energy is lost in each step of a food chain.
   c. Explain why little loss of matter occurs in a natural ecosystem.
   d. Give examples of how people cause gains and losses of matter in an ecosystem.
2. Show your understanding of farm ecosystems by
   a. Listing the land and climatic requirements for a farm ecosystem.
   b. Defining subsistence, swidden, slash and burn, small, and corporate farming, and by listing some merits and drawbacks of each.

   c. Listing some reasons why industrialized farms require greater inputs of energy than other farm ecosystems.
   d. Comparing loss and gain of matter in an industrialized farm ecosystem with loss and gain in a well-managed family farm.
3. Demonstrate your knowledge of the relationship between plants and the ecosystem by doing the following:
   a. Explain why water-borne sewage systems cause a loss of matter from the ecosystem.
   b. List some of the kinds of organisms (including plants) that contribute to soil buildup, and state the role of each.
   c. State what farming practices can prevent or retard soil deterioration.
   d. Describe two ways in which pesticides disrupt the relationship between plants and insects.
4. Demonstrate your knowledge of three farm outputs (soil, fertilizers, and pesticides) by
   a. Listing two types of erosion and stating the negative impacts of each.
   b. Defining eutrophication and biological oxygen demand.
   c. Defining biological amplification.
5. List some of the positive and negative social impacts of farms.

In Chapter 1, you learned how cities are almost totally dependent on inputs of matter and energy, especially food from farms. This chapter will look at various kinds of farms as ecosystems and consider the inputs, outputs, and interrelationships involved in rural ecology. Rural ecology is closely bound up with urban ecology; while cities have been growing in population and spreading out over the farmland that feeds them, farms have been declining in population in the industrialized countries. Because of technological developments, fewer and fewer people have been needed as farm workers. Population growth and increasing affluence have led to increasing demands for farm products such as food and fiber, and the area of the earth's surface devoted to agriculture has increased as has the output of food per hectare of land. In spite of this, food output has failed to keep up with demand; localized famines occurred during the early 1970s, and food prices reached new highs. An understanding of the inputs, outputs, and impacts of various kinds of farms is essential to an understanding of society's most basic problem: how to feed itself without destroying the capacity of the earth to sustain future generations.

SOME PRINCIPLES OF ECOSYSTEMS IN GENERAL

In order to appreciate how rural ecosystems function, it is helpful to understand some of the processes that go on naturally in all ecosystems.

2-1 Food chains are part of ecosystems

Animals eat plants, and then they are eaten by other animals or they die and are eaten by bacteria. A feeding sequence such as one in which a plant which is eaten by an insect which is eaten by a bird which is eaten by a hawk is called a **food chain.** Usually in natural ecosystems there are several kinds of birds eating several kinds of insects which eat several kinds of plants. This kind of feeding relationship is called a **food web** (Fig. 2-1). Organisms can be classified by the roles they play in these systems. The photosynthesizers, green plants and algae, are **producers;** the eaters of photosynthesizers or the eaters of the eaters of photosynthesizers are **consumers;** the microorganisms that break down waste products and bodies of dead producers and consumers are **decomposers.** Within these groups, organisms play special roles. For example, green plants store energy and matter in their bodies. A green plant may die and pass its energy and matter on to decomposer bacteria and fungi or it may be eaten by a type of consumer called a **herbivore** (plant eater), such as a caterpillar. The insect may be eaten by another kind of consumer called a **carnivore** (meat eater), such as a **predatory** bird. The bird may be eaten by another carnivore, a hawk. The hawk ends up with some of the energy and matter stored by the producer several steps down the food chain. Many consumers, such as human beings, feed

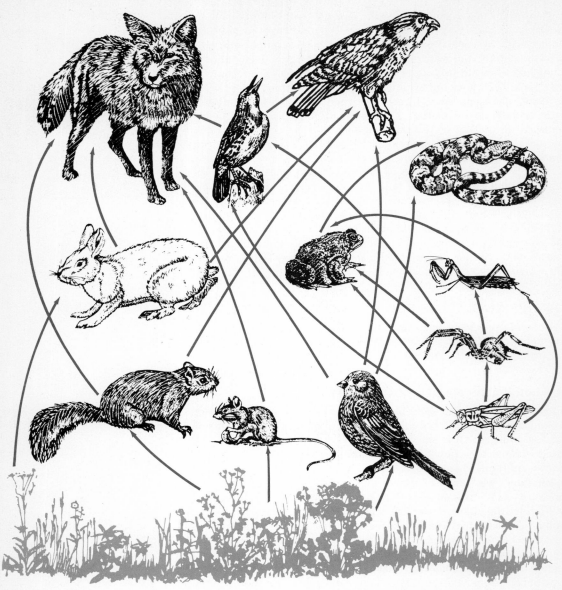

FIG. 2-1 In most places, food chains are woven into food webs. The arrows point from the food source to the animal that utilizes it.

on both plant and animal materials and are called **omnivores** (all-eating). Organisms that live in or on other organisms and consume some of their nutrients or tissues are called **parasites** and their victims are **hosts.** (In order to survive, a parasite must be able to take what it needs from its host without killing it; for the death of the host spells the end of the parasite's food supply.) People can become the hosts of many parasites: mosquitos, lice, fleas, intestinal worms, to name a few. Plants can also host parasites—for example, the southern corn blight is a fungus parasite of corn. Finally, most ecosystems have **scavenger** organisms, such as vultures, which specialize in feeding off the decomposing bodies of dead herbivores or carnivores.

Energy losses In both food chains and food webs, energy is lost at every step because, as you will remember, every time energy is used to do work, some of it is converted to heat. It takes a lot of work by the consumer organisms in a food chain to find, eat, digest, and absorb other organisms, and all organisms must work to maintain their own bodies and grow and reproduce. At each step in this food chain, useful energy stored as chemical bonds is converted to heat energy, which is helpful in speeding up some life processes but which itself cannot sustain life. Eventually, the heat is dissipated into the environment. Thus energy in ecosystems is not recycled, and ecosystems must have a continual input of energy from the sun or from other ecosystems in order to survive.

Although it varies from system to system, usually about one-tenth of the useful energy in each level of a food chain is passed on to the next level. In a typical situation, it would take 100 kilograms (220 pounds) of green plants to produce 10 kilograms (22 pounds) of insects, which would produce 1 kilogram (2.2 pounds) of insect-eating birds. As a general rule, food chains operate with about 10 percent efficiency, and in most of them there is only enough energy at the base to support four or five links (see Fig. 1-10).

Consequences of energy losses A term that is frequently used in discussions of food chains is **biomass,** which means the combined weights of all the individuals in a group of organisms. Normally, the biomass of green plants in an area will be much greater than the biomass of all the herbivores in the area, and the biomass of all the herbivores will be greater than that of all the carnivores. This relationship can be represented graphically as a pyramid (Fig. 2-2). This is the reason that meat eaters need more space for food production than plant eaters (Sec. 1-6). Corn is often fed to cattle and then the cattle are eaten by people. In a world in which 10 million to 20 million people starve to death each year, this seems a poor way of arranging things, especially since much of the world's land is *not* suitable for growing corn but *is* suitable for grazing cattle on plants that people cannot eat. Fish and chickens have about the same nutritional value, yet 90 percent of the world's fish meal catch is exported from undeveloped countries to developed countries, where it is used to feed cattle and poultry. Much of the meat, milk, eggs, and broiler chickens

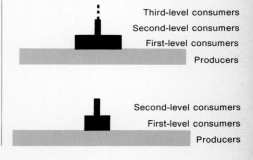

FIG. 2-2 The biomass of producers and consumers in an ecosystem takes the form of a pyramid. A very large base of green plants (producers) is needed to support successively less biomass of consumers. *a* illustrates data from an eelgrass community in the North Sea; *b* depicts data from a Wisconsin lake. (*Adapted from Eugene P. Odum,* Fundamentals of Ecology, *2d ed., Saunders, Philadelphia, 1953.*)

in America and Western Europe are produced from this annual import of about 2 million metric tons of protein *from* the countries that need it most *to* the countries that need it least.

2-2 Matter cycles through ecosystems

Recycling of essential elements In natural ecosystems, each of the chemical elements necessary for life is part of a cycle. When green plants photosynthesize, matter from the air and the soil enters them and becomes part of their substance. As their energy and substance are passed along food chains, the elements are passed along, usually in the form of organic compounds. Eventually, through the action of decomposers, the atoms of these elements find their way back into the nonliving earth, air, or water and are ready to be used again in living things. Unlike energy which flows through ecosystems and is lost as heat, matter cycles—that is, elements are never used up: their atoms are merely rearranged in different compounds as they go through chemical transformations from one part of the cycle to the next (Fig. 2-3). When they are not part of living systems, nitrogen and oxygen exist mainly as gases. When carbon, phosphorus, and sulfur are not in living things, they are often solids. These latter elements are found in the environment as compounds: carbon in the atmosphere combines with oxygen as carbon dioxide, and sulfur

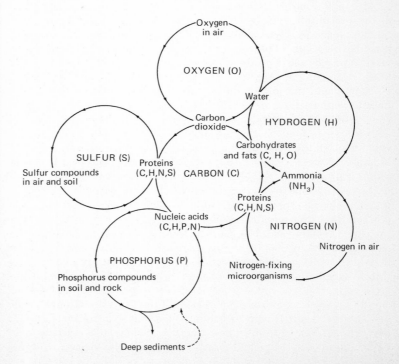

FIG. 2-3 The major cycles of chemical elements. Elements cycle between the living and nonliving, usually as compounds with other elements. In this diagram, the carbon cycle is placed at the center, and the major carbon compounds found in living things are indicated at the intersection of the carbon cycle and the cycles of other elements appearing in these compounds (nucleic acids are necessary for protein synthesis).

combines with oxygen as sulfur dioxide or with hydrogen as hydrogen sulfide. Nitrogen cannot be used by green plants unless it is combined with other elements.

Interference with cycling of elements The human population is currently overloading some cycles with more of certain compounds than they can use. Overloading the nitrogen cycle with nitrogen fertilizers has caused many problems such as runoff of nitrogen compounds into wells; this process is one of the important harmful impacts of farms (see Sec. 2-10). Burning high-sulfur coal in England has led to damaging sulfuric acid rains in Scandinavia. Burning of fossil fuels has increased the carbon dioxide content of the earth's atmosphere by about 10 percent since 1850. Furthermore, the human species can produce compounds that cannot be used or broken down by any organism. Certain plastics and pesticides cannot be broken down at all; others can be broken down only very slowly. The elements that make up these products are effectively taken out of their cycles. These products then become wastes, and often pollutants if they are harmful to living systems. The human species seems to be the only species that produces "true wastes," that is, substances which other organisms cannot use at all.

THE FARM AS AN ECOSYSTEM

Farms, of course, are artificial ecosystems, constructed and dominated by people. Different kinds of farms vary, however, in how artificial they are. No farms can maintain themselves without the help of people; but some farms do less damage to local natural ecosystems than others, and some need fewer inputs and outputs than others.

2-3 All farm ecosystems depend on land

Need for large areas of land Because photosynthesis is a process requiring sunlight, and sunlight is spread out over the surface of the globe, farms also must be spread out over the surface of the globe. On farms, people harness the power of photosynthesis to produce food and fiber for themselves and food for the animals they eat as well as to export food and fiber to other places. In farming country, fields often stretch from horizon to horizon (Fig. 2-4), giving an impression of limitless land for human use. This impression is misleading, however, since more than half the land surface of the earth is not suitable for any kind of agriculture. In 1965, people were already using about one-fourth of the 13.5 billion hectares (33 billion acres) of land surface of the earth for crops or for grazing. Most of the land not in use in 1965 lacks qualities which make it really desirable for farmland.

Need for special kinds of land If animals are to be grazed on land, it must have suitable weather conditions and suitable grasses must grow there. If plants are to be cultivated, land must be **arable;** that is, it must not be too wet, too dry, too steep, too hot, or too cold.

FIG. 2-4 These wheat fields in Manitoba, Canada, illustrate the vast amount of land devoted to the production of grain in North America. (*Freelance Photo Guild.*)

There must be sufficient sunlight and a long enough growing season, and if the food is to be sold, there must be a way to transport it to markets. The land also must be **fertile,** or if it does not contain the elements plants need to grow, inputs of fertilizer must be obtainable. **Natural fertilizers** can often be obtained by recycling the unused portions of crops and such things as manure or bone meal (ground up bones, which provide nitrogen and phosphorus). Sometimes nearby rocks can be mined for limestone, used to make soils less acid. **Synthetic fertilizers** are compounds made especially for use as fertilizers. They usually require inputs of large amounts of energy to produce and often must be transported long distances.

2-4 Subsistence farmers are not estranged from their resource base

Subsistence farming was the earliest method of farming, and is still practiced in parts of the world today. Subsistence farmers grow just enough food for their own and their family's needs. This kind of farming is hard work, and subsistence farmers have to be aware of the natural conditions that make successful farming possible. They have to know when to plant what crop, how much to store for food, and how much to save for seed. Subsistence farmers are well aware that their survival depends on sun, rain, and their own skills. They know that bad weather or carelessness can result in their starvation.

Swidden agriculture Swidden agriculture got its name from the old Norse word for "singe," because it often involves burning of trees cut down to form a clearing. Once the trees are burned, a garden is planted and harvested, and gradually the forest is allowed to return to the clearing. When they weed their gardens, swidden farmers are careful not to pull up the tree seedlings that sprout. One tribe of people refers to the tree seedlings as "mother of gardens," and rightly so. The roots of the tree seedlings help hold the soil in place, while the leaves protect the garden plants from downpours and store soil elements which might otherwise be lost. The tree seedlings soon grow so large that the swidden farmers find it difficult to harvest and weed their gardens and must abandon them before they deplete the soil. They then move on and create another clearing. Swidden farming has been practiced in places and times as far apart as medieval England and present-day New Guinea. It requires inputs of labor, sunlight, and water, and as long as the same clearings are not cut too often, it may be able to support its practitioners indefinitely. Unfortunately, it does *not* produce outputs of surplus food abundant enough to feed growing cities.

Slash-and-burn agriculture In the tropics, farmers often "slash and burn" an area of forest and then farm it until it no longer produces a satisfactory crop. These farmers do not allow the trees to grow back until they are done with the land; then it is often impossible for trees to grow there, and the area becomes a wasteland as the soil washes away or turns to **laterite,** a rocklike substance impervious to cultivation. The farmers move on to new areas of forest, destroying them in turn. Slash-and-burn agriculture requires the usual inputs of labor, sun, and water. But it also requires continuing inputs of new lands and produces outputs of barren, useless land.

Agriculture on small farms Almost all farms in the United States once were family farms owned, or at least farmed, by one family with perhaps a few hired workers. The number of family farms has gone down greatly in the last century, however. Although family farms were partly subsistence farms, they needed some inputs from the city and they also produced outputs for the city. People who lived on the farm grew diverse crops for their own consumption. They produced their own vegetables, chickens, eggs, etc.; however, they often bought cloth, salt, furniture, and so forth from a nearby town which had imported them from the city. These family farms usually produced outputs in the form of extra vegetables, eggs, etc., and many specialized in one kind of output to sell, such as apples, corn, or dairy products. Before the industrial revolution people and animals did all the work on farms. In addition, farm animals produced manure, which was used as fertilizer. Carefully managed family farms actually improved the fertility of the soil in many areas by adding organic matter to the soil.

The transition to industrialized farms In 1910, there were only about 1,000 tractors on farms in the United States; 40 years later there were over 4,500,000. About this time, corporations started taking over family farms as part of the post-World War II technology boom. Much of the farm machinery which replaced human and animal labor became so big and expensive that only corporations could afford to buy it. Corporations also influenced agricultural colleges, and these colleges engaged in research which would benefit them. This research resulted in expensive new pesticides, fertilizers, seed strains, and even bigger machines. Because family farmers could not afford to compete with corporations, they had to sell out or make contracts with them. In 1974, 2,000 family farms closed down every week in the United States, and in 1974 there were just about half as many farms as in 1949. In 1974, almost half the farmland in California was controlled by corporations.

2-5 Corporate farmers are estranged from their resource base

A business corporation is a group of people bonded together with the common objective of making profits. In agriculture, what might be called **agribusiness** corporations have developed. These corporations own not only farms but also factories that make farm machines, fertilizers, and pesticides. They often own processing, packaging, and distribution facilities. This system of ownership is known as *vertical integration* and is very efficient for the corporation because it can make its profits at any or all links in this long chain of business. Therefore, the people in agribusiness often have little knowledge or interest in sustaining the natural processes that are needed for the growth of crops in the field or animals on the range. They are satisfied that all is well as long as the corporation as a whole shows a profit. **The high yields of corporate farms** Corporate farms produce high yields of produce per hectare with little human labor. In the middle of the eighteenth century, it took 75 to 95 percent of the human population working on farms to provide food and fiber for the whole population. In the middle of the twentieth century in the United States only 15 percent of the people fed the entire country and there was food left over for export. In 1970, only 4.8 percent of the United States population worked on farms. Between 1950 and 1970 farmers in the United States more than doubled their production per hectare of most crops, but there seems to be good evidence that such high yields are not sustainable. They require enormous inputs of materials and energy, which are becoming increasingly in short supply, and such use of the land may damage it because of loss of organic material. For example, extensive studies of fields in Ohio which have been used to grow corn for 41 years show that the soil has lost over one-half of its organic material. On this land, reasonable applications of synthetic fertilizers improved yields, but they alone could not sustain them as the organic material in the soil continued to waste away.

2-6 Most farms in the United States are now industrialized

Inputs to industrialized farms Industrialized farms use the technology and products that have resulted from the industrial revolution. Energy arrives in the form of electricity, gasoline, and diesel fuel to run the farm machinery as well as in the form of sunlight to power

FIG. 2-5 Large amounts of fossil fuel power these huge combines harvesting hybrid grain sorghum in Texas. (*U.S. Department of Agriculture, Department of Information.*)

photosynthesis. In fact, in 1960, agriculture used the refined products of 18 billion gallons of crude petroleum and 22 billion kilowatthours of electricity. In 1960, as much finished steel was used to make farm machines as was used to make passenger cars, and steel making also requires energy. In 1974, the energy value of the food produced by farms in the United States was much *less* than the energy value of the fossil fuels used to produce it (Fig. 2-5).

Inputs are required from many sources to maintain industrialized farms. For example, rainwater from streams and ponds is used in irrigation systems. This requires inputs of energy for pumps and construction. Inputs of information are necessary, as are inputs of materials. Farm advisors and veterinarians who learn their trades in the city explain to the farmer the latest ways to fertilize, kill insects, get rid of weeds, and inoculate farm animals against disease. They supervise the animals' supplements of antibiotics, hormones, pesticides, and tranquilizers. Farmers often purchase their own food as well as high-protein supplements for their farm animals. They must purchase special high-yield, disease-resistant seeds too. In addition to these inputs, they receive cultural inputs in the form of newspapers, books, and radio and television programs.

Outputs of industrialized farms The high yield per hectare of industrialized farms is a highly desirable output; however, this also results in some highly undesirable outputs. Dust is sent billowing into the air by huge farm machines. Pesticides and fertilizers pollute water and air, heat and carbon dioxide from burning fossil fuels enter the environment, and soil washes into waterways. People displaced from their farms make their way to the cities looking for employment.

RURAL ECOLOGY

Now that we have studied some of the types of farms and their inputs and outputs, let us look at some of the interrelationships that occur on farms. In order for people to harness photosynthesis for their use, they must disrupt natural ecosystems to some extent; however, many natural ecological relationships and processes must be allowed to proceed if the earth is to produce and to continue to be fit to produce for generations to come. Energy must continue to flow through food chains, and the cycles of the elements must be maintained.

2-7 Green plants need an appropriate inorganic environment in order to photosynthesize and grow

Water Green plants require a great deal more water than that used as a raw material in photosynthesis. In a process called **transpiration,** water constantly passes upward from their roots through their stems and out through tiny pores. Water helps give plants rigidity; if you have ever forgotten to water a potted plant, you may have found it limp and wilted. Its cells lacked the internal water pressure, called **turgor,** which kept them firm. Provided you had not neglected your plant too long, it probably "picked itself up" and appeared as healthy as ever when you watered it again. But, wilted plants cannot grow; therefore it is important that farmers and others who raise plants see that their plants have the water they need to transpire.

It takes about 750 liters (200 gallons) of water to produce a single pound of dry rice, and this water must be there when the plants need it for growth. When rain comes too early or too late in the growing season, crops fail. In many areas, if rains do not come at all, the crops are doomed; in other areas farmers may be able to provide their plants with inputs of **irrigation** water; in some areas, crop plants are totally dependent on irrigation water.

Nutrients from the soil Carbon dioxide and water provide plants with carbon, hydrogen, and oxygen. Plants, however, need many other nutrients; most of these pass into the roots of the plant along with water from the soil. Table 2-1 lists some of the elements that are essential to the life of plants. These elements are usually not available to plants until they are combined with other elements in appropriate compounds. For example, air is 78 percent nitrogen, but plants cannot use pure nitrogen. They can use only nitrogen combined with other elements. A group of soil microorganisms is engaged in combining nitrogen with oxygen; they are part of the nitrogen cycle (see Fig. 2-3). Synthetic fertilizers usually supply only three of the major nutrients, nitrogen (N), phosphorus (P), and potassium (K); however, some soils produce poor crops because they are deficient in one of the elements needed in only tiny amounts, the **trace elements.**

Depletion of nutrients Even if the soil were originally rich in nutrients, years of farming can remove them. Today, plants harvested on farms are taken away and eaten in cities or they are eaten by cattle which are then taken away and eaten in cities. Thus the nutrients from the farmland often end up in sewage, washing out to sea from a city thousands of miles from the farm. For this reason, farming is sometimes said to "mine the soil." A soil depleted of nutrients is

TABLE 2-1 ESSENTIAL MINERALS FOR HIGHER PLANTS

ELEMENT	NUMBER OF GRAMS (APPROX.) NEEDED TO GROW 2,500 KILOGRAMS OF CORN	FUNCTION
	Macronutrients	
Nitrogen (N)	72,600	Component of proteins and other compounds
Phosphorus (P)	18,100	Component of many compounds, such as nucleic acids
Potassium (K)	57,000	Essential but its exact role is not well understood
Sulfur (S)	34,000	Component of some proteins, vitamins, etc.
Magnesium (Mg)	23,000	Component of chlorophyll
Calcium (Ca)	23,000	Influences permeability of membranes
	Micronutrients	
Iron (Fe)	900	Component of energy-transport compounds of cells
Manganese (Mn)	136	Helper to many proteins involved in photosynthesis
Boron (B)	27	Function unknown
Chlorine (Cl)	27	Function unknown
Sodium (Na)	27	Function unknown
Zinc (Zn)	Trace	Necessary for synthesis of growth substance
Copper (Cu)	Trace	Component of proteins involved in energy use
Molybdenum (Mo)	Trace	Essential for combination of N and O by nitrogen-fixing bacteria

said to lack fertility; unless its nutrients are replaced, it will grow only poor crops or none.

Recycling of nutrients When the waste parts of a crop and the wastes from the people and animals that eat the crop are returned to the soil, the fertility of the soil can be kept high. In areas with water-borne sewage systems, nutrients are usually not returned to the soil; however, there are some areas in Europe where fields are being sprayed with liquid sewage, and some forest areas of the United States have been treated with sewage liquids in experiments. Most attempts to return nutrients which would be lost from city sewage have used the sludge, the solids that settle out during treatment. Dissolved nutrients are lost this way, but some fields in England have been successfully treated with sewage sludge for over a hundred years. There is always the danger that sewage used as fertilizer will transmit diseases and parasites, and since many industrial firms dump their wastes into sewage systems, there is the added danger of harmful industrial wastes which might damage the soil, the plants, or the people and animals who eat the plants. In mainland China, a system apparently has been devised to treat sewage so that it may be used as fertilizer without spreading parasites and disease. The rest of the world needs to concentrate on designing safe systems for recycling the vast quantities of nutrients now washed away as sewage.

2-8 Green plants have many interrelationships with other living things

Soil organisms Soil organisms are necessary in soil which is to support plant life; fertile soil is teeming with life. Soil is very complex; it consists of tiny rock fragments, remains of dead plants, animals, microorganisms, and billions of other living things (Fig. 2-6). In some pastures, more than 10 million small animals may exist in each square meter of soil; these are mostly worms, insects, and relatives of insects. A teaspoon of fertile soil may contain 5 billion bacteria, 1 million protozoa, and 200,000 algae and fungi. The earthworms and other animals stir up the soil and let in air, making it a better place for roots to grow. The microorganisms play complex roles which scientists are only just beginning to understand. Certain fungi often form special associations called **mycorrhizae,** with roots, and help transfer nutrients from the soil to the plant. Other microorganisms transform nitrogen into compounds plants can use; still others transform other essential elements from forms that plants cannot use into forms that they can use. These microorganisms are thus important parts of the cycles of the elements. Decomposer organisms produce **humus,** which is a complex organic product of decay composed especially of plant material. Abundant humus is needed to preserve the texture of the soil and hold moisture, and it also plays a part in the release of nutrients to plant roots. Peat moss is a good example of humus, and it is often added to garden soils to improve their ability to support plant life.

Soil formation Fertile soils form very slowly in nature, often over a few decades or even many centuries. Rock is broken down slowly by weather and organisms; organic material builds up slowly in the form of humus as generation after generation of organisms grows and dies. Unfortunately, although soil is formed very slowly, it can be lost very rapidly. Almost all farming produces erosion of soil as the soil is washed or blown away from plowed fields. By 1952 over 60 percent of the world's tilled agricultural soil had lost half or more than half of its original humus.

Reducing erosion Contour plowing (Fig. 2-7) can help keep soil from washing away; windbreaks of trees or hedges can prevent wind

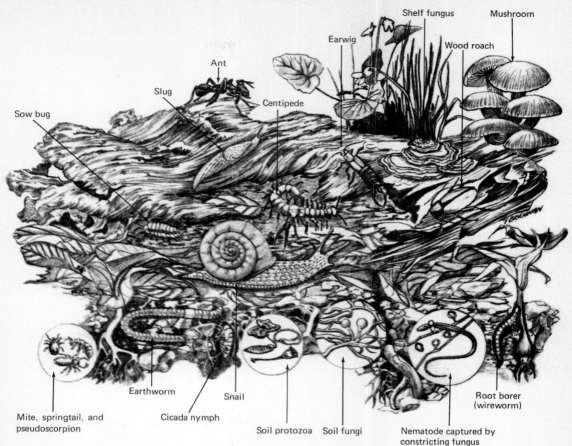

FIG. 2-6 Soil is a complex mixture of living and nonliving things, and its structure and composition gradually change with time due to the action of wind, water, heat and cold, and the living things themselves.

erosion. Cover crops planted when the land is not being used for a regular crop can help hold the soil in place as well as add organic material when they are plowed into the soil. Fertilizers made up of organic material, such as manure, can help maintain humus which will hold water and prevent runoff. While some synthetic fertilizers may be beneficial in certain areas at times, we have already noted how much organic material can be lost from fields fertilized with *only* synthetic fertilizers (Sec. 2-5), over a long period of years. Many farmers believe that the use of more natural fertilizers can help prevent such losses. Many farms in Europe, England, and China, which have maintained their soil (and its fertility) over hundreds or perhaps thousands of years, have been treated exclusively with natural fertilizers.

FIG. 2-7 Contour plowing—plowing at right angles to the slope of the land—reduces erosion by water. (*U.S. Department of Agriculture.*)

Insects and disease Crop plants are particularly vulnerable to attacks by insects and disease because they are usually grown in **monocultures** (all one kind of plant), and they are often weakened in their resistances to insects and disease by special breeding which makes them taste better to us (and to insects). Even in a monoculture there are usually predatory or parasitic insects which eat crop-destroying insects; however, most insecticides kill beneficial insects more readily than they do pests and leave the crop more vulnerable to the plant eaters. The industrialized farmer then has to apply more insecticide.

Insects as pollinators Many plants will not set fruit and seeds unless they are pollinated (Chap. 6) by insects which fly from flower to flower. Honeybees are our most valuable pollinators, yet they are vulnerable to many insecticides. Beekeepers are now able to rent their bees out as pollinators for considerable sums of money. A great variety of other insects are important as pollinators also.

THE IMPACT OF THE FARM

The modern industrialized farm is the center of a web of inputs and outputs (Fig. 2-8). Sunlight, carbon dioxide, water, fossil fuels, fertilizers, pesticides, and so forth are essential inputs to the industrialized farm. Food, moreover, is not the only output of such farms; on farms energy and materials are also transformed into a diverse array of waste products, soil from erosion, fertilizer runoff, pesticide derivatives, food that has to be destroyed because it contains too much pesticide or hormones. And farms alter their original environments by removing the habitats of many species of plants and animals native to a region.

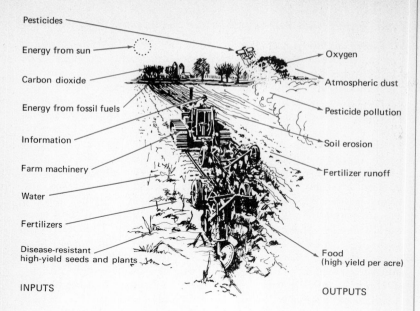

FIG. 2-8 The inputs and outputs of a modern industrialized farm.

2-9 Soil is a major output of farms

The negative impact of soil erosion by water What impact does the soil, which is constantly being eroded from farms by water, have after it leaves the farm? It is often washed into lakes, rivers, and estuaries where it may make the water too muddy to support fish and other aquatic life; it may be washed out to sea; or it may be stopped in its course by a dam. If it is stopped by a dam, it will gradually fill in the lake behind the dam. This process, **silting,** occurs behind all dams, and for this reason, they are temporary structures because eventually silting makes them useless. There is one positive impact of soil erosion by water, however. The soil may be deposited in river deltas, such as the Mississippi delta, where it may again be used in agriculture. (A delta is a usually triangular deposit of soil at the mouth of a river.) Soil erosion occurs even without farms, but farms greatly speed up the process.

The impact of wind erosion Soil that is swept by the wind may well be the most damaging form of erosion from farms. Dust from farms has a worldwide impact as an important component of air pollution. When agriculture is practiced in dry areas on land which is not really suitable, great amounts of wind erosion result; large areas of the United States became a "dust bowl" in the 1930s owing to improper farming techniques applied to marginal land in the Midwest and Southwest. Recently, much marginal land has been brought back into production, and this may lead to more wind erosion. Agricultural "hazes" are common in Africa and southern Asia. This dust in our atmosphere cools the planet by reflecting some of the sun's light back into space and thus changes the weather in a way that may be having adverse effects on agriculture.

2-10 Fertilizers and pesticides are major outputs of farms

As previously noted, humus is lost from modern industrialized farms. As the humus content of the soil drops, the soil loses some of its ability to hold water, and water moves through it rapidly, washing out valuable nutrients. This washing-out process is called **leaching.** As the humus content of the soil drops, the soil loses its ability to break nutrients down slowly and supply them gradually to the plants. Thus, as humus content drops, farmers depend more and more on synthetic fertilizers, and they must employ such fertilizers in ever greater quantities because they are rapidly washed away.

Effects of excess fertilizer runoff The nitrogen, phosphorus, and potassium washed away from farms end up in streams, lakes, and rivers, *fertilizing them.* Suddenly the many algae living in the water are presented with a superabundance of nutrients, and they grow rapidly into great patches called *algal blooms.* Algal blooms cover such large areas that they often clog waterways and drains. They use up so much of the oxygen in the water that they begin to die and then wash up on beaches, making the beaches ugly and ill-smelling. As they are being decomposed, the bacteria that break them down use up even more oxygen so that finally there is so little oxygen left in the water that fish and other aquatic organisms die. This process, called **eutrophication,** is overfertilization and is often enhanced by sewage and industrial runoff. The biological oxygen demand (BOD) of water is a measure of how far eutrophication has proceeded; it measures how much oxygen the bacteria in the water need to utilize the organic materials present.

Effects of pesticides There are many ways that agricultural pesticides escape the sites of their application. They may drift away before they reach the crop, they may evaporate from the field, they may be blown away with dust, they may wash away with water, or they may be carried away as residues on the produce from the farm. Studies have shown that 75 percent of the insecticide sprayed by crop-dusting airplanes (Fig. 2-9) never even reaches the crop—it just drifts away. Since 60 percent of all insecticides used in agriculture are applied by airplane, this is undoubtedly a major way insecticides get into the atmosphere. When pesticides which escape from farms are short-lived ones, such as some of the organophosphate insecticides, only people and animals in nearby areas are apt to be poisoned. (Organophosphates are carbon compounds containing phosphorus.) Many organophosphates break down quickly into nonharmful substances. But, if the pesticide is a long-lived one, it may have harmful effects far away from its point of application and long after it is applied.

Chlorinated hydrocarbons (compounds containing carbon, hydrogen, and chlorine) are among the most common long-lived pesticides. And the most infamous chlorinated hydrocarbon is DDT; it is so long-lasting and mobile that it is found all over the world—in seals and penguins in the Antarctic, and in human beings everywhere. DDT is not very soluble in water but is very soluble in fats, such as those found in organisms. Microorganisms living in water that contains

FIG. 2-9 Pesticides released from a plane during crop dusting may spread great distances beyond the target area. (*Grant Heilman, Freelance Photo Guild.*)

DDT soon accumulate it in their bodies, and when they are eaten by other organisms, they pass on their DDT to their consumers. Since most organisms cannot readily break down the carbon-chlorine bonds in DDT, nearly the same amount of DDT gets passed on up the food chain, but at each step it is concentrated in a smaller biomass (Fig. 2-10). Notice that the concentrations of DDT in Fig. 2-10 are expressed in parts per million (ppm). If you have trouble picturing 1 part per million, you can visualize 1 milligram in 1 kilogram of material or a lump about two-thirds the size of a bouillon cube dissolved in a cube of liquid 1 meter square (Fig. 2-11). This process by which compounds become concentrated as they pass along food chains, known as **biological amplification,** can occur with any element or compound which is stored in living tissue rather than being degraded or gotten rid of (excreted). Pollutants do *not* become evenly diluted throughout the environment, and surprisingly small quantities of pollutants can have harmful effects. Birds such as brown pelicans and peregrine falcons that feed high on food chains are harmed by DDT because it interferes with their ability to lay eggs with shells of normal thickness. The parents break their eggs when they sit on them to keep them warm.

Limitations on use of pesticides DDT has been partially banned for agricultural use in some countries such as the United States; however, nowhere has it been banned for public health use, such as mosquito population control, and it is still widely used throughout the world. No one is certain what the long-term effects will be of the DDT which is concentrated in human beings throughout the world. Aldrin and dieldrin, two other chlorinated hydrocarbons, also concentrate in food chains and are found in the tissues of almost every human being tested in the United States. In August 1974, the Environmental Protection Agency suspended manufacture of aldrin and dieldrin because these had been shown to cause cancer in laboratory mice at a concentration close to that present in human beings (0.3 ppm). As

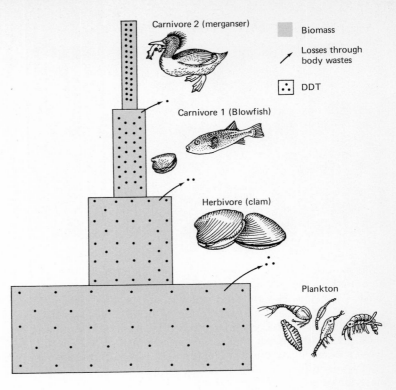

FIG. 2-10 DDT becomes concentrated in food chains, as almost the same amount of DDT is present in increasingly smaller biomass. (*Data from G. Woodwell.*)

more research is done on pesticides, we can expect a great restriction on their usage; they are proving largely ineffective for their tasks in the way they have been used, and their impact on nontarget organisms usually seems to be detrimental.

2-11 Farms have social impact

While cities are centers of action and change, farms and rural areas have traditionally been centers of conservative values. People in rural areas had less contact with new ideas and different cultures than people in urban areas, and therefore they were apt to cling to the old well-tested values of the past and resist poorly planned and rapid change. Now that farmers have cultural inputs of modern communications and many can easily travel and visit cities, the traditional conservatism of farming communities is changing. Since this is a time of accelerating change all over the earth, more openness to new ideas may help industrialized farmers adjust to changes such as fertilizer and energy shortages, while traditional farmers may accept more efficient farming methods with greater readiness.

Negative impacts People on farms often live far from their neighbors and may have difficulty getting the benefits of resources that require the cooperative effort of many people, for example, universities,

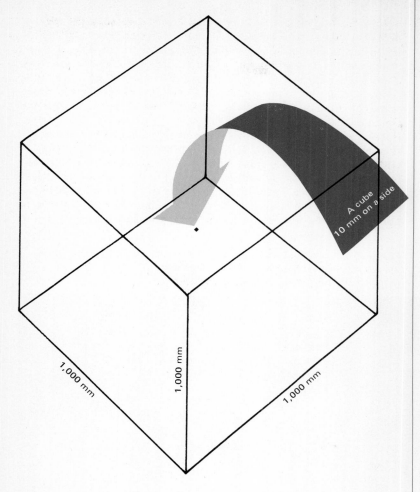

FIG. 2-11 This diagram helps a person visualize what one part per million represents. Although a 10-mm cube in a 1,000-mm cube may seem small, it may contain an enormous number of molecules of a pollutant. For example, air with 1 ppm SO_2 has 1 molecule of SO_2 per million molecules of air. At standard temperature and pressure, the large cube in this illustration would contain about 2.7×10^{25} molecules of gas. If one in a million were SO_2, it would contain about 270,000,000 molecules of SO_2. Whether such a level is damaging depends on the nature of the pollutant. Some are damaging in the parts per billion range (ppb), and some are believed to cause cancer when only one molecule gets into a cell.

libraries, art galleries, symphonies, and so forth. Protective services, such as police and fire departments offer, are spread thin. Medical help is usually farther away than it is in the city, and specialists and special treatment facilities may be hundreds of miles away. Rural people often lack the stimulation of being with different kinds of people: people of different races, people with different interests. The transition from family farms to industrialized farms has changed farming from a way of life to a way of making a living, and many small farmers must hold down jobs as well as work their farms.

Poverty One of the shocking facts about life in America is that many of our citizens are hungry and that many of these hungry people *live on farms*. Many of our farms are too small or their soil is too worn out or eroded away to support a family. For a number of reasons, people on these farms are not able to do much to improve their lot.

Fertilizers, farm machinery, and livestock cost money; and sometimes depleted soils cannot be restored. Rural poverty (Fig. 2-12) can be just as degrading as the crowding, filth, and high crime rates of the urban ghettos. Often people from these depleted farms become migratory laborers, harvesting other farmers' crops, or they move to the central city hoping to better their lot only to find themselves not much better off than before—still hungry and unable to cope with the complexities of city life.

Positive impacts Rural families often enjoy a much richer family life than most city dwellers. Often rural individuals have more contact with people of all age groups. There are opportunities in small towns for persons with special skills, so that not everybody in rural areas has to be a farmer. Farm towns need doctors, mechanics, bankers, plumbers, and businessmen. Farm people are less susceptible to pollution-related illness and to epidemic diseases since they live in

FIG. 2-12 In the region of southeastern United States known as Appalachia there is wide-spread poverty. This family lives in the hills of Tennessee. (*Kenneth Murray, Nancy Palmer Photo Agency.*)

a cleaner environment and have fewer contacts with other people who may be carrying disease.

Rural areas allow their inhabitants to have close and rich contacts with other living things. Rural people have the opportunity to see that the fate of the human species is rooted in the land. They understand the importance of the sun and the rain and the soil in providing our food. In spite of tractors, pesticides, and synthetic fertilizers, rural people still have contact with a way of life that stretches thousands of years into the past before cities even existed. Rural people on family farms may be self-sustaining in food production and thus not so dependent on others as are city people. It is true that many farmers spend their spare time reading farm journals which tell them how to "dominate" nature with a new pesticide or a new machine. Nevertheless, they at least have the opportunity to see that humanity cannot dominate but must live in harmony with the rest of nature.

QUESTIONS FOR REVIEW

1 What is rural ecology?
2 Can you define these terms: food chain, food web, producer, consumer, decomposer, herbivore, carnivore, omnivore, parasite, host, scavenger?
3 Why is energy lost from an ecosystem—that is, why is energy not recycled through it?
4 Why are there usually fewer carnivores in a given area than there are herbivores?
5 Why may it be said that "farming mines the soil"? What can be done to minimize the losses from such "mining"?
6 What are two beneficial roles that insects play in agricultural areas?
7 Does soil erosion have any positive impacts?
8 "All dams are temporary structures." Why?

READINGS

Clark, W.: "U.S. Agribusiness Is Growing Trouble as Well as Crops," *Smithsonian,* **5**(10):59–65, 1975.
Ehrlich, P. R., J. P. Holdren, and R. W. Holm (eds.): *Man and the Ecosphere—Readings from Scientific American,* Freeman, San Francisco, 1971.
Farb, P.: *Living Earth,* Harper, New York, 1959.
French, M.: *Worm in the Wheat,* John Baker, London, 1969.
Lord, R.: *The Care of the Earth: A History of Husbandry,* Nelson, New York, 1962.
Pimentel, D.: "Food Production and the Energy Crisis," *Science,* **182**:443–449, 1973.
Rappaport, R. A.: "The Flow of Energy in an Agricultural Society," *Scientific American,* **224**(3):116–132, offprint 666, 1971.
Stadtfeld, C. K.: *From the Land and Back: What Life Was Like on a Family Farm and How Technology Changed It,* Scribner, New York, 1972.
Steinhart, J. S., and C. E. Steinhart: "Energy Use in the U.S. Food System," *Science,* **184**:307–316, 1974.
Watt, K. E. F.: "Ecology," *Biocore* unit XXI, McGraw-Hill, New York, 1974.
Wong, E.: "Agribusiness Plows under the Family Farm," *Environmental Action,* **6**(9):3–6, 1974.

LIFE IN NATURAL ECOSYSTEMS

SOME LEARNING OBJECTIVES

After you have studied this chapter, you should be able to

1. Choose one of the hunting and gathering societies or the Pacific-island society discussed in this chapter and explain why, in terms of its ecosystem inputs, outputs, and transformations, it may be called "self-sufficient."

2. Explain, with examples, why a modern agricultural-technological society is *not* "self-sufficient."

3. Define the terms ecosystem, biological community, and population, and explain how they differ.

4. State why time and place must be specified in any discussion of a biological community, name one biological community mentioned in the text, and name (define) two other biological communities not mentioned.

5. Explain—in terms of some inputs, outputs, and transformations—why a city ecosystem and a farm ecosystem, taken together, may be viewed as a larger ecosystem. In your explanation, include a brief discussion of the carbon cycle.

6. Explain why the whole earth can be considered a single ecosystem.

7. Explain how an "open" ecosystem differs from a "closed" one; rearrange these four systems—a

farm, the planet earth, a city, a forest—in order from "most closed" to "most open"; and support your rearrangement by listing some inputs and outputs of each.

8. Name three main types of natural ecosystems, briefly describe some major features of each, and list some impacts of modern agricultural-technological society on each.

9. Name an important characteristic of stable ecosystems and list some reasons why this characteristic helps produce stability.

10. Explain, in terms of food chains and food webs, why a single-crop farm is a less stable ecosystem than a temperate-zone forest.

11. Describe some of the things that can happen to unstable ecosystems.

12. Compare and contrast a temperate-zone forest with a grassland in terms of their (*a*) average annual rainfalls; (*b*) vertical complexity; (*c*) plant diversity and plant types; (*d*) animal diversity and animal types, including birds; (*e*) amounts of energy and nutrients stored in their "standing crop" (living plants) and in their soils.

13. Demonstrate your knowledge of the ocean as an ecosystem by
   a. Stating whether it is more "open" or more "closed" and more stable or less stable than a farm, a city, a balsam-fir forest, the planet earth.
   b. Naming the kind of organism that accounts for most of the photosynthesis that takes place in oceans.
   c. Stating two factors which prevent the kind of organism just referred to above from thriving equally well throughout the ocean's breadth and depth; and stating where in the ocean it *can* thrive.
   d. Stating whether oceanic food chains tend to be longer or shorter than terrestrial ones, giving the main reason why this is so, and explaining why the length of oceanic food chains is of great practical importance to fishers and their customers.

In the preceding chapters we studied two major kinds of artificial ecosystems, cities and farms. Now let us put these systems into perspective in the rest of nature. First, we shall look at some of the ways people lived before cities and farms were invented. Then we shall look at some peoples of the present or recent past who still practice these ways or who have only recently lost them. We shall consider these peoples in the context of their ecosystems: their impact on their ecosystems and the ecosystem's impact on their way of life. Then we shall look at the major natural ecosystems of earth: forests, grasslands, and seas. Cities and farms have, of course, destroyed much of the original forest and grassland ecosystems, and outputs of cities and farms such as industrial wastes, pesticides, sewage, and soil from erosion have become major inputs to the seas. We shall describe these natural systems and the impact modern technological society is having on them.

THE SELF-SUFFICIENT LIFE

There have been many societies of human beings which lived in relative harmony with their environment and were able to survive with only the natural sun and rain from outside their local ecosystem. Today most of these peoples are extinct or they have had contact with modern civilization and have lost or are losing the knowledge and skills that enabled them to survive as self-sufficient societies.

3-1 Our early ancestors did not practice agriculture

The practice of agriculture provides food surpluses which allow populations to grow. Before human beings took up farming some 10,000 years ago there were an estimated 5 million people alive at any one time; today there are some 4 billion, about 800 times as many people present on the face of the earth. We know a great deal about how those preagricultural peoples lived and how they interacted with their environment because, although they left no written descriptions of their lives, they did leave records in the form of tools and artwork on cave walls (Fig. 3-1). From this record we can determine that these peoples led a hunting and gathering life—hunting for animals and gathering fruits, seeds, berries, roots, and other plant materials.

Hunting and gathering methods Different ecosystems and different levels of technological advancement led to many different styles of hunting and gathering. The earliest human beings living in Africa several million years ago probably roamed as groups over the countryside, with each individual gathering food only for himself or herself. Slightly modified sticks and stones would have been used as tools to help dig up roots, to kill small, relatively helpless game, and to serve as weapons of defense against dangerous large animals. Such hunting and gathering methods were a far cry from those of more recent human beings who lived 100,000 years or so ago. In these groups, specialized hunting parties of males efficiently attacked large

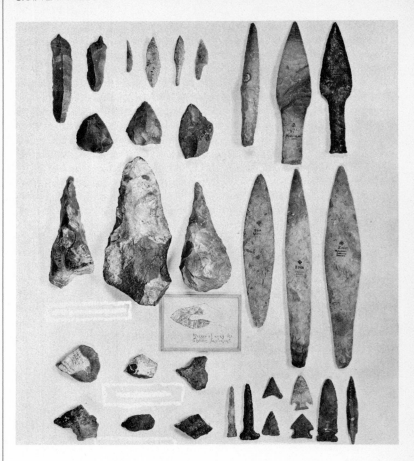

FIG. 3-1 These stone tools were used by a people who had not yet discovered agriculture. (*American Museum of Natural History.*)

game animals, using finely made weapons with wooden handles and stone blades. A variety of strategies were used, such as driving herds into lakes where animals could be killed relatively easily, or driving them over cliffs to their deaths.

The impact of prehistoric hunters and gatherers The peoples of several million years ago probably had little impact on their ecosystems with their small numbers and simple technology. However, vast piles of bones left by more recent hunters have led some scientists to believe that human populations had a large effect on animal life even before agriculture. In the last few hundred thousand years, many species of large animals became extinct all over the earth. Although some people believe that changes in climate were primarily responsible, others are convinced that human hunting activities played an important part in these extinctions. Whether or not pre-agricultural peoples were directly responsible for the extinction of many animal species, they undoubtedly had a profound influence on both the plant and animal life of the planet through their use of fire.

Controlled burning was unquestionably one of the methods used by early hunters to drive game onto killing grounds.

3-2 Some peoples live by hunting and food gathering

A few groups of people still exist in much the same way as did our ancestors hundreds of thousands of years ago. Some bushmen in southern Africa still lead a wandering life, living by eating roots, berries, insects, and other small game, and storing water in empty ostrich egg shells. The Mouti pygmies of Zaire live in the forest and gather plant foods. They also kill small animals and gather honey. A few Australian aborigines are still wanderers, having no permanent shelters. They live in temporary camps behind simple brush or rock windbreaks and are able to find food and water in country where a nonaborigine would quickly die of thirst or hunger (Fig. 3-2). Recently tribes of hunters and gatherers were discovered in the middle of New Guinea and in the Philippines; these people had had no previous contact with modern civilization.

Life in hunting and gathering societies One may be tempted to feel sorry for these human beings who live, by our standards, a life of great hardship. But, ask yourself, are our standards the ones by which to judge? Consider for example the aborigines; true, wandering aborigines lack many of the things we think of as comforts. They have no houses, little in the way of clothing, no toys for their children, no

FIG. 3-2 Australian aborigines are among the few people who are still wandering hunters and gatherers. (*American Museum of Natural History.*)

means of transportation except their own two feet. They live, on the average, shorter lives than Americans. *But,* they live in a society with a rich culture and religious beliefs and a great artistic tradition (Fig. 3-3). Aborigines know their culture thoroughly, and each individual knows his or her place in it. Aborigines are aware of their precise biological relationship to very large numbers of people, while Americans are usually only aware of their relationships to close relatives (brothers, sisters, parents, grandparents, uncles, aunts, and first cousins); that is, the aborigines have a much more extensive system of **kinship** than Americans or Europeans. Aborigines also relate to parts of the natural world in a protective fashion based on their religion. Presently, 1 percent of the entire world's known reserve of uranium ore is at a place called Gabo Djang in aborigine territory. They have steadfastly refused to sell since in their religion Gabo Djang is the "dreaming place of the great green ants" who will turn into man-eating monsters and ravage the world if their sacred place is desecrated. The aborigines live in harmony with themselves, each other, and their environment. We have great difficulty doing any of these things; are we in a position to judge them or say that our ways are better than theirs?

3-3 Until recently some North American peoples lived a hunting and gathering life similar to that of our ancestors

The Plains Indians Until the middle of the nineteenth century much of North America was occupied by advanced hunters and food gatherers, for example, the Plains Indians. The Plains Indians' economy depended on the vast herds of buffalo which roamed the prairies (grasslands) of the central part of the continent—an estimated 40 million of these animals in 1830. The Indians depended on buffalo meat for food, on their hides for shelter and clothing, and on their droppings for fuel. Their hunting did not constitute a threat to the ecosystem any more than did depredation by wolves or other meat eaters. Only the "surplus" animals were harvested.

The appearance of European people changed the entire prairie ecosystem. Buffalo were killed in immense numbers, mostly for their tongues and hides (Fig. 3-4). It is estimated that 70 percent of the meat was left to rot. So common were buffalo bones that it became profitable to gather them and ship them east for grinding into fertilizer. In less than 50 years, a population of 40 million buffalo was reduced to a herd of 26 animals! With the buffalo went the way of life of the Plains Indian, as much a victim of the hunter as of the cavalryman. Surviving Indians were forced to take up herding or farming, mostly on reservations.

The Netsilik Eskimos Some groups of Eskimos persisted in hunting and food gathering until the 1960s. At the end of the nineteenth century and through the first half of the twentieth century, however, more and more Eskimo groups moved into permanent settlements

FIG. 3-3 An Australian aboriginal bark painting of animals in a billabong (a dead-end channel of a river) sacred to the Dhalawangu tribe of Arnhem land. The animals are totems (emblems) of the tribe, and the hatching represents the mud and weeds of the billabong. Painted by the well-known artist Yarjgarrin in 1964. (*P. R. Ehrlich.*)

FIG. 3-4 In the days when buffalo were still abundant, special trains were run for the benefit of hunters. Here train passengers slaughter buffalo during an excursion on the Kansas-Pacific Line. (*Culver Pictures, Inc.*)

and became trappers. They also became economically dependent on the *kabloonak,* their word for "big eyebrows," the white men. The Netsilik Eskimos, whose name means "people of the seal," were the last to give up a nomadic (wandering) life.

The last nomads on our continent depended for their existence upon seals (Fig. 3-5). Seals provided them and their dogs with most of their food; seal skins were the raw materials for their clothes, seal oil fueled their fires, and seal bones were fashioned into tools. The Netsilik Eskimos traveled over the ice-covered sea in a prolonged seal hunt, moving their belongings on wooden sleds pulled by dogs harnessed with seal or walrus hide. Camps were made when the snow was suitable for building igloos.

The process of seal hunting required great skill and patience. Each seal has a series of holes through the ice which it uses for breathing. The Eskimo men, with the help of their dogs, found as many of these holes as they could, and then each hunter stationed himself by a hole. He explored the shape of the hole so that he would know in which direction to thrust his spear. Then he refilled most of the hole with snow so that the seal would not know the hole had been disturbed. A small Y-shaped bone with a bit of down (soft feather) attached to it was placed in the entrance to the hole as an indicator. This warned the hunter when the seal came to the hole to breathe (Fig. 3-6).

With the hole prepared, the hunter settled back for what was usually a long, cold period of remaining motionless, for the slightest disturbance would warn the seal. Often a snow windbreak was prepared to protect the hunter poised over the hole. With luck the seal would return to the hole while the hunter was there and could make his thrust. If his aim was true, the detachable head of his harpoon with a line attached would be buried deep in the seal. Then the hole could be cleared and the dead or dying seal heaved up onto the ice.

FIG. 3-5 Seals are used by the Netsilik Eskimos for a great variety of things.

When a seal was killed, the hunting party immediately had a victory ceremony in which the fresh liver of the seal was shared among all the hunters, who considered it a great delicacy. Then when the party returned to camp, the seal was divided up by the wife of the successful hunter according to predetermined rules. Little boys received a first taste of the meat, and little girls were treated to some fresh blood. Then the remainder of the seal was divided among families according to the rules of the party.

The Netsilik, like other Eskimos, have a close family life. Eskimo mothers are especially close to their children, who are usually nursed until they are about three years old. Babies are physically close to their mothers, who carry them on their backs for much of their early lives. Fathers, grandparents, and other adults spend a great deal of time playing with the children and passing on the history and culture of their people, often in the form of stories.

The closeness of the families was in part a natural result of the cramped living quarters of the igloo, where three generations often slept on a single snow shelf, snug in warm animal skins. The children also learned when young to be accustomed to cold. Babies often played outside in weather which would bring all outdoor life in an American city to a halt.

The life of these nomadic Eskimos was spiced with other diversions than the hunt. Large igloos were constructed, with ice window panes, for community affairs. Dancing to the music of special drums, as well as various types of contests, was enjoyed by the entire group. In many ways life was perilous. But, as with the Australian aborigines, for those living it, it was rich and meaningful. Unfortunately, by 1970 this last group of nomadic Eskimos had moved into a permanent settlement and its way of life is no more.

3-4 Life on a tropical island may also be a self-sufficient life

Life on Ifaluk Seemingly the very opposite of the harsh life of the Netsilik Eskimo is that led by the legendary South Sea Islander. The people of Ifaluk Atoll (Fig. 3-7) in the Caroline Islands of the western Pacific (Fig. 3-8) seem to live an idyllic life. When they were studied in 1953, some 260 people lived on Ifaluk, which had a land area of about 1.3 square kilometers (0.5 square mile). They gathered their food over a much larger area, including the extensive reefs of the atoll. Atolls are more or less circular coral reefs that may or may not protrude above the ocean surface in places to form islands (Fig. 3-9). Each person on the atoll got his or her protein from fishes—somewhat more than 45 kilograms (100 pounds) per year were caught for each inhabitant. Other staple foods include coconut milk, coconut meat, coconut sap, and the starchy vegetables breadfruit and taro (Fig. 3-10).

Unlike the Netsilik, the Ifaluk islanders were not strictly hunters and gatherers. On their atoll, the growth of desirable plants such as

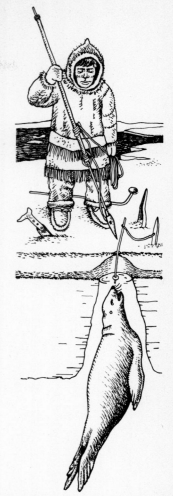

FIG. 3-6 Eskimo harpooning a seal at the seal's breathing hole in the ice.

FIG. 3-7 An aerial view of Ifaluk Atoll. (*U.S. Naval Air Station, Guam.*)

FIG. 3-8 Ifaluk is one of many small islands in the western Pacific Ocean.

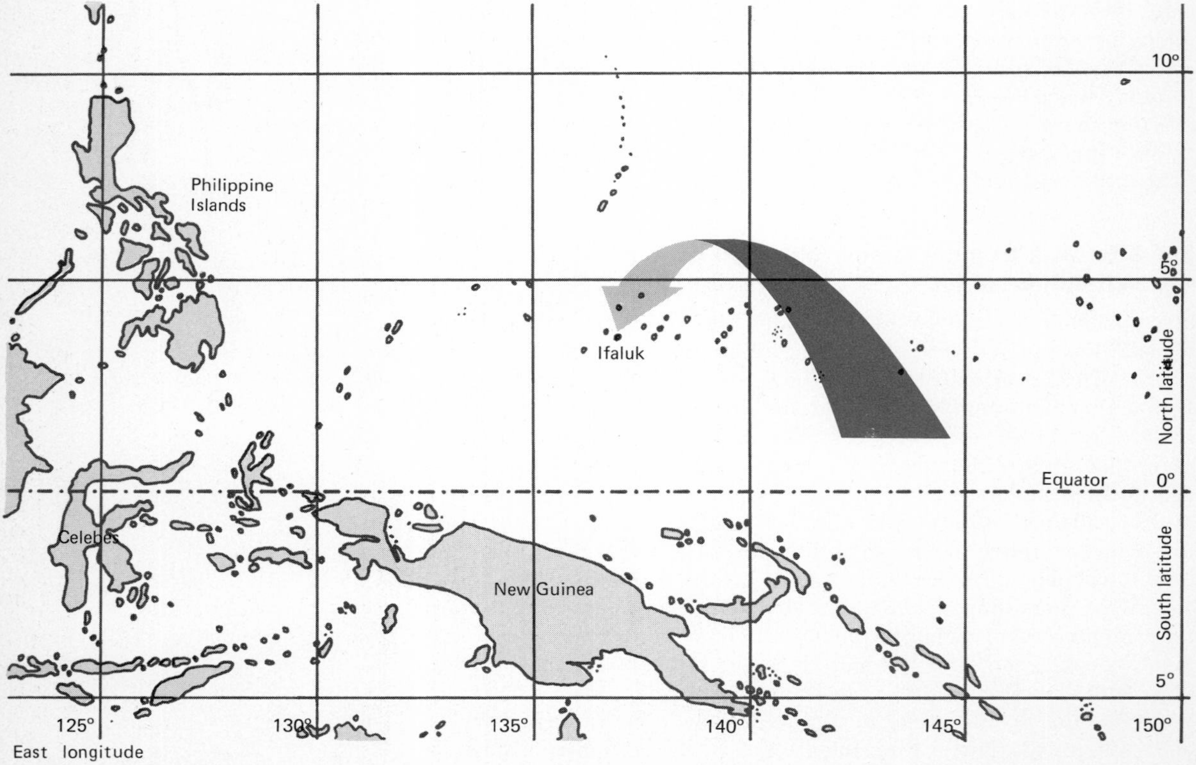

coconut palms, breadfruit trees, and taro plants was encouraged by cutting down weeds. Regular taro gardens were maintained (Fig. 3-11), and sap was gathered from the cut flower stalks of coconut palms. The young men who made the rounds of sap gathering had to climb each tree, empty the coconut shell which had been placed under the cut stalks to gather the dripping sap, and take another slice off the stalk to encourage the flow of sap.

Although the people of Ifaluk spent a great deal of their time obtaining food and doing other chores, such work seemed to be thoroughly enjoyed by all participants—including the chiefs of the islands (Fig. 3-12). This certainly distinguishes Ifaluk culture from ours!

Ifaluk society On Ifaluk much time was also spent in recreation, especially in large dance celebrations honoring gods. In contrast to the men of our society, the men of Ifaluk were more interested in personal ornaments than were the women. They often wore flowers, feathers, beads, and perfume. Tattooing was widely practiced, with men, women, and children wearing tattoos. This urge to make one's skin more decorative is widespread in people; the Eskimos practice it also, and so do many Americans.

As in Eskimo society, children were pampered on Ifaluk. They were given a great deal of attention and were encouraged to take part in adult activities as early as possible. Children were treated almost as if they belonged to the entire community. Adoption was very common, with the foster mother staying with the natural mother during childbirth and thus being with her adopted infant from the very start of life. Whenever an Ifaluk child was in distress, it would be comforted by the nearest adult.

3-5 Self-sufficient cultures are adapted to their ecosystems

We tend, of course, to think of Pacific Islanders as being carefree and happy-go-lucky, and in some circumstances, they are. The impression given by South Seas music today, extolling tropical living, love, flowers, and the sea, does not give the complete picture, however. Delightful as the life on Ifaluk may seem from the foregoing account, the islanders, like most of the others in the Pacific, have faced continual problems of overpopulation. Great as the fish supply of the reefs is, it is not unlimited. Furthermore, fierce storms periodically ravage the island and greatly reduce the supplies of edible plants. A typhoon in 1907 created such havoc on Ifaluk that some of the islanders had to move elsewhere temporarily.

With limited space and limited food, South Seas people often practiced infanticide (killing of newborn infants). Eskimo groups also killed female babies and permitted old people to starve when times were difficult. South Sea Islanders also developed rigid systems of taboos, the violation of which was punishable by death. Taboos are restrictions on certain acts or behaviors, such as the prohibition

FIG. 3-9 Formation of an atoll. Atolls are formed by the underwater deposition of calcium by algae around the skeletons of tiny animals which make up coral. As the coral mass grows, it eventually reaches the surface. Soil can then form, and drifting and imported plants can become established. One theory of atoll formation suggests that the coral formation occurs around a volcano and continues to grow as the volcano sinks from the surface because of geological changes.

FIG. 3-10 The staple foods of Ifaluk are taro, breadfruit, and coconuts.

among the Ifaluks of sexual intercourse during the periods of fishing. Interisland warfare was common in many places, including the Hawaiian Islands.

We must always be very careful in interpreting what people of other societies do. Men "loafing" around talking and twisting fibers (as Ifaluk men commonly did) may be making the twine that is critical

FIG. 3-11 The root crop taro must be tended. (*From Marston Bates and Donald Abbott,* Coral Island, *Scribner, New York, 1958.*)

FIG. 3-12 Much of the food of the Ifaluk islanders comes from the sea. (*From Marston Bates and Donald Abbott,* Coral Island, *Scribner, New York, 1958.*)

to the island economy as the raw material for fish nets. The making of beads, feather capes, and ornaments may have very special religious significance, and may be confined to certain groups within the society. The complexities of "primitive" culture often are not obvious to untrained observers, and they are often closely tied to the regulation of population size.

Modern people depend more and more on technology to shape the environment. In the process, modern people sometimes seem to lose sight of what technology was once thought to do—make human life more secure and rewarding. There is no question about what an Netsilik Eskimo harpoon or drum is designed to do; both clearly play an important role in the life of the entire community. Can we say the same thing about a Cadillac or electric scissors? Today, people are often asked to change their way of life so that machines can be more efficient; for example, machines such as computers do jobs like billing, accounting, and many phases of banking. People are put out of work and other people with complaints have to try to communicate with computers. So-called primitive peoples can help us to understand ourselves and put our "advanced" culture into perspective.

COMPONENTS OF NATURAL ECOSYSTEMS

Ecosystems have both living and nonliving components. We have already studied (Chap. 2) the many roles played by living things in ecosystems as producers, consumers, decomposers, and more specifically as herbivores, predators, etc. We also looked at some of the nonliving parts of the environment: energy from the sun, which flows through ecosystems, and elements from the soil and air, which cycle through ecosystems. We pointed out that cities and farms can be considered artificial ecosystems because they need large inputs and outputs of materials from other systems and that natural ecosystems are for the most part self-sustaining.

3-6 Biological communities are the living parts of ecosystems

A **biological community** is all the organisms found at a particular place at a particular time: plants, animals, microorganisms, and any other living things present. In discussing a community it is necessary to specify the time and place and extent of the area under consideration. If, for instance, you studied the biological community of the New York metropolitan area, which includes suburban areas in New York State, New Jersey, and Connecticut, you would find many more kinds of organisms than if you restricted your study to Manhattan Island. If you studied the New York metropolitan area in the winter, you would find fewer organisms than if you studied it in summer. Populations in biological communities Within a community, the individuals of each kind of organism make up a population. When discussing populations it is important to make clear the limits of the unit under discussion—otherwise you give the impression that you are speaking of all the individuals of that kind. Thus, if we write "the population of blue whales is shrinking" or "the human population now numbers 4.0 billion," we are writing about all the blue whales and all the people that live on earth. If we were to write, however, "the population of New York in 1970 was 7,539,000 people," it is not clear whether we mean New York City or New York State. It is common for two different communities to have the same species of organism but for the population size of that species to be very different in the two communities. For example, New York City and rural New York State both have human populations but in 1970 New York City had about 8 million people and all the rural areas of New York State only about 2 million.

3-7 Ecosystems involve interactions

Interactions between city and farm ecosystems If we think of biological communities as simply collections of populations of different species of organisms, they may seem little different from collections of stamps or toy soldiers. But this view changes when we view these organisms in the setting of their physical environment and attempt to understand the relationships among the organisms and between each organism and its environment. If we consider the city and the farm to be part of a larger community (Fig. 3-13), we discover that inputs and outputs of each go to or come from the other. We can consider the city-farm combination as an ecosystem and study its inputs, outputs, and transformations.

People in the city community are links in food chains which originate on a farm, while most of the manufactured items used on a farm are produced in the city. Photosynthesis on a farm produces a surplus of oxygen, and the city uses a great deal more oxygen than it produces. Automobiles and other machines of the city that burn fuel chemically combine the fuel with oxygen. The combination of

LIFE IN NATURAL ECOSYSTEMS   69

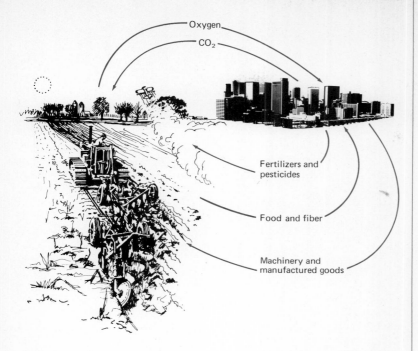

FIG. 3-13 Interrelationships between cities and farms. Cities could not survive without food and fiber from farms; modern, industrialized farms could not survive without cities as markets for their produce and as sources of machinery and fuels.

a substance with oxygen is one form of the process of **oxidation.** Oxidation may occur slowly, as when iron combines with oxygen and forms rust, or rapidly, as when gunpowder combines with oxygen in an explosive burning. Animals and plants use a relatively controlled form of oxidation when they "burn" their fuel. We call this fuel "food" when speaking of animals.

In a city, machines and people both use oxygen, but they produce no oxygen, and there are few oxygen-producing green plants in the city. Therefore the city is an oxygen consumer. The process of burning fuels in both people and machines produces an important by-product, carbon dioxide. A city produces a great deal of carbon dioxide, and carbon dioxide is one of the raw materials of photosynthesis. Since little photosynthesis goes on in a city, a city produces a surplus of carbon dioxide. This works out very well if there is enough circulation of air between cities and agricultural areas, because farms are carbon dioxide consumers. In fact, this is a good example of one of the cycles of elements, the carbon cycle (Fig. 3-14). Carbon moves from a city to a farm in CO_2, is incorporated into food on a farm in the process of photosynthesis, and returns to a city in food.

The global ecosystem We have already noted that it is important to specify the extent of area being considered when discussing a community or a population. The same is true for ecosystems; if we just say "ecosystem" we are referring to the ecological system of the entire planet. *All* living organisms of the earth are part of a gigantic

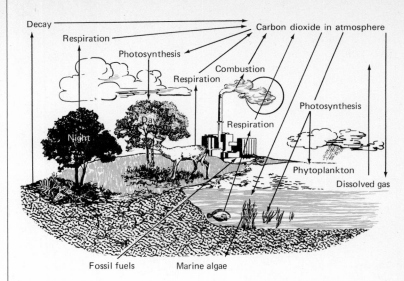

FIG. 3-14 The carbon cycle. Burning fossil fuels, the decay of organisms, and the use of foods for energy by organisms (called respiration) produce CO_2, which goes into the air and water. Green plants and algae convert CO_2 into organic molecules which become part of their bodies. When they die or are eaten the carbon returns to the air and water as CO_2, thus completing the cycle. Photosynthetic organisms respire also, when it is dark.

system. The only significant input to the global ecosystem is the energy from the sun. As we have seen elements important in living things cycle through the system. Some cycle very slowly so that part of the supply may be unavailable to living things for many millions of years. Phosphorus, for example, forms sediments of rock in the form of phosphates on the bottom of the sea. Except in certain coastal areas this becomes available again only when mountain building brings sediments to the surface. Elements in other cycles such as the carbon cycle make the rounds much faster. It is important to note that except for energy from the sun, the earth is a closed ecosystem; cities and farms, on the other hand, are open ecosystems in that they must receive inputs and give off outputs of both matter and energy in various forms.

SOME NATURAL ECOSYSTEMS

Forests, grasslands, and the seas are natural local ecosystems which are intermediate in "openness" between the global ecosystem and cities and farms.

3-8 Temperate zone forests

In those parts of the earth called the temperate zones—the parts in between the heat of the tropics and the cold of the polar regions—large areas were once covered by forest. Some of these forests still exist, but these forests today are mostly ones that have grown up after the original virgin forests had been cut down. Such forests are called **second growth.** Very little virgin temperate zone forest remains, but for the purposes of our discussion, virgin forest and undisturbed second growth show many of the same features.

A typical temperate zone forest consists primarily of trees that drop their leaves every fall, remain leafless throughout the winter, and grow new leaves in the spring. Trees that lose all their leaves for part of the year are called **deciduous.** The trees of a temperate zone deciduous forest lose their leaves and have a nongrowing season during the winter. In some tropical areas, there are deciduous trees that lose their leaves during the dry season.

Diversity of plants and animals Temperate zone deciduous forests usually have many different kinds of shrubs and herbs growing under the trees. Usually these plants do a great deal of their growing, and flowering, early in the spring before the trees grow new leaves. Great diversity in species is characteristic of the trees in a temperate deciduous forest.

The diversity of plants in the forest is matched by the diversity of animals. A wide variety of herbivores is present: deer, squirrels, mice, seed-eating birds, and above all, thousands of species of insects. These herbivores support a rich carnivore community: wolves and cougars (until they were exterminated over much of their range), a variety of weasellike animals, frogs, snakes, hawks, insect-eating birds, insect-eating insects, and so on (Fig. 3-15). Also found in the temperate forest are omnivores such as bears and racoons, which will feed on fish or berries with equal gusto.

Period of inactivity Temperate zone forests are ecosystems that are found in conditions of high rainfall, 100 centimeters (40 inches) or more a year, a moderate summer climate, and a prolonged, four- to six-month period of **dormancy** during which the system ceases to function. In the dormant period, virtually all photosynthesis ceases, so that the energy input from the green plants to the system is turned off. The vast majority of organisms, including the wealth of decomposers in the soil, turn down their life processes to a very low level, depending on stored energy for whatever "work" they continue to do. A few animals remain active, such as squirrels, which have carefully hoarded energy in the form of stored nuts. The simultaneous shedding of leaves by all of the forest plants, followed immediately by a period when the decomposers are inactive, makes conditions ideal for the development of a rich humus. This humus helps prevent nutrients from being washed from the soil by rains.

An important characteristic of temperate zone forests is that large amounts of energy and nutrients are stored in the trees themselves, and large amounts of nutrients are stored in the soil. The amount of energy in the trees would be obvious to you if you could be near a forest that was undergoing rapid oxidation—that is, if you could see a forest fire.

Figure 3-16 shows a small part of the aftermath of the most destructive forest fire in American history. The fire occurred in Idaho and Montana in 1910. It was started by lightning following a year of drought when no rain fell from April until August in the affected

FIG. 3-15 Some forest dwellers. Deer and many insects are herbivores. Some insects eat other insects (praying mantis). Many birds eat insects (flycatchers), but some eat fruits and seeds or other birds and mammals (jays). Snakes and frogs are carnivorous. Many small animals are herbivorous (squirrels) or omnivorous (bears, raccoons), but some, such as weasels, are carnivorous.

area. The lightning started thousands of separate fires; many were put out and firefighting crews were relaxing when a hot dry wind from the southwest, a chinook, started blowing with hurricane force on August 20, 1910. The thousands of small fires that remained were fanned to a fury, and the flames swept across the forest as fast as 70 miles per hour. This created a fire storm, great masses of burning gases turned into whirlwinds that ripped up huge trees like toothpicks in advance of the arrival of the flames. Cities as far away as Denver were covered with ashes, and 4,800 kilometers (3,000 miles) away in Boston people were perplexed by mysterious dark clouds that blotted out the sun. Several towns burned and over 89 people were killed. Three million acres of timber burned in about 48 hours, enough to keep a sawmill operating for a hundred years. After the fire inferior tree species invaded the area, an epidemic of bark beetles broke out, and much soil erosion occurred. This fire led to improved regulations on logging and better fire protection for forests. It is now recognized that some forests need periodic fires to clear the undergrowth and that some tree seeds will not germinate without fire; therefore, today some forest fires are allowed to burn. They are watched carefully, however, and it is unlikely that a disaster like the 1910 Idaho fire could occur again.

Civilization's relationship with temperate zone forests Western civilization had its roots around the Mediterranean Sea, which has its own special climate and vegetation, but many of its activities have cen-

FIG. 3-16 A small part of the aftermath of the most destructive forest fire in American history. (*R. H. MacKay, U.S. Department of Agriculture, Forest Service.*)

tered in areas of northern and central Europe and eastern North America where deciduous forests were the primary natural community. Much of Chinese civilization also developed in areas of deciduous forest. The consequences for the forest were the same in both cases—forests were very largely cut down. The complex deciduous forest ecosystems were replaced by much simpler agricultural ecosystems.

Human intervention in the deciduous forest had many consequences. Agricultural ecosystems generally lack the humus-building properties of the forest, so that the closed nutrient cycle—soil to plants and back to soil—is opened up. The problem of nutrient loss is made worse in the agricultural system by the absence of trees to break the force of wind and rain.

The natural stability of the temperate zone forest Perhaps even more important is the reduction in diversity that occurs when an agricultural system replaces a temperate forest system. A single species of plant now occurs where previously there were dozens. Most herbivores of the forest cannot feed on the crop and so the diversity of herbivores drops precipitously, as does the diversity of carnivores. What difference does it make, anyway, if the diversity is reduced? Unfortunately, it makes a great deal of difference because, at least in part, *it is the diversity of natural ecosystems that gives them their* **stability.**

A temperate zone forest will, if undisturbed by fire or people, persist year in and year out. If it is burned over, or cleared and then left alone, it will gradually reestablish itself in a process known as **ecological succession** (Fig. 3-17). Therefore the forest ecosystem is **stable**—if it is disturbed, it will tend to return to its undisturbed state. What happens, on the other hand, if an agricultural ecosystem, such as a cornfield, is left undisturbed? The answer is that it will be destroyed by pests and weeds. It will not reestablish itself without human intervention. When it is disturbed, it does not return to its original state; it is **unstable.**

Why does diversity produce stability? Suppose a disease infects one species of forest tree and kills the entire population of that tree. The forest will be modified, but increasing numbers of other tree species will fill the gaps. Or perhaps an entirely new tree species will invade the forest and replace the lost one. Any herbivore living only on the now-extinct tree will be wiped out, but it will be just one of many herbivores. And any predator dependent on that herbivore alone will also disappear—but it will be just one of many. In fact, very few herbivores feed on one kind of plant, and few predators feed on only a single species of prey. Finally, because of the mixture of trees in the forest, individuals of the same tree species are less crowded together than are, say, the tomato plants in a garden. The greater diversity leads to greater spacing; this makes it less likely that a serious epidemic will occur and much less likely that an entire species will be wiped out of the community.

Instability of agricultural ecosystems In a cornfield or tomato patch

Time zero: bare field

2 years later: grasses

20 years later: grasses and shrubs

100 years later: pine forest

150 years later: oak and hickory forest replacing pines

FIG. 3-17 Ecological succession. When a temperate forest is destroyed by fire or is cleared by people, it may gradually return to its original state. Over a period of years, grasses and herbs grow and hold the soil in place. Shrubs may then be able to grow and trees may eventually establish themselves. Most of the forests of the United States are second-growth forests that reestablished themselves after being cut down for lumber.

the situation is reversed. The plants crowded together are all members of the same species; that is, they form a monoculture. If a disease kills most or all of the plants in a monoculture, that effectively destroys the ecosystem. All the herbivores in such an ecosystem are normally dependent on the same plant, and all the carnivores are dependent on those herbivores. Of course, if the monoculture is limited in size, both herbivores and carnivores may be able to find food in neighboring natural ecosystems.

Even in the absence of catastrophic disease, agricultural ecosystems are much more subject to destruction than are temperate zone forests. For instance, the destruction of the monoculture by a population explosion of a herbivore is much more probable than the destruction of a temperate zone forest by such an outbreak. Compare the simplified diagrams of food chains in a monoculture and a forest (Fig. 3-18). Notice that the most important insect herbivore in the monoculture is eaten by a single species of bird. What would happen if a disease or bad weather greatly reduced the numbers of that bird? The insect would multiply unchecked until it had destroyed its food supply, the crop plant making up the monoculture. The herbivores would eventually starve, but it would be too late to save the monoculture.

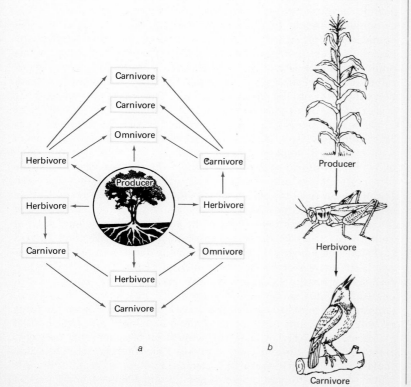

FIG. 3-18 Food chains in a temperate forest (a) are more complex than those of a monoculture created by farming (b).

Contrast these events with those in the temperate zone forest. Although the diagram shows only part of the complexity, notice how the food chains are interlinked with one another. If, for instance, the population of a carnivore is greatly reduced by disease, this does not necessarily lead to an outbreak in a herbivore, because other carnivores also feed in part on that herbivore. As one carnivore population declines, other populations will increase and compensate for the loss. *In general, the more food chains there are in an ecosystem, and the more they are interlinked into food webs, the more stable that system will be.*

3-9 Grasslands

In general appearance, grasslands seem more like agricultural ecosystems than temperate zone forests. From the point of view of diversity, however, they are closer to forests. Large areas of central North America, the central part of the Eurasian land mass, and western Africa are occupied by relatively treeless grasslands. Grasslands, of course, lack the vertical structure of forests, but the diversity of plant species may be as high as or higher than in the forest. The plant diversity is not reduced, as a casual examination might indicate, because there usually is a very rich assortment of grass species and a wide variety of other **herbaceous** (nonwoody) plants.

Conditions leading to grasslands Because of the lack of vertical complexity (that is, trees and shrubs) in a grassland ecosystem, the bird diversity is more limited than it is in most forests. The mammals tend to be burrowers, such as the prairie dog (a relative of the squirrel) of the American West, or large grazing animals like the American bison (buffalo) and the African antelopes. The grassland also supports a great diversity of insects.

Grasslands occur in regions where annual rainfall averages 25 to 75 centimeters (10 to 30 inches), but where that rainfall is uneven, with drought in some years and abundance in others. Periodic lack of water is thought to be one of the main reasons why temperate forests do not invade areas such as the prairies of the American Middle West. Another reason may be the occurrence of periodic fires, the presence of large herds of grazing animals such as bison, and the intervention of human beings. The action of all these may prevent young trees from becoming established. Human beings are known to have created very large grasslands by clearing forests and repeatedly burning areas. Some scientists feel that human beings are responsible for the existence of all major grasslands, as will be discussed later in Chap. 5.

Grassland soils It is common for grasslands to have large amounts of both energy and nutrients in their humus-rich soils. In contrast to forests, relatively little in the way of either energy or nutrients is tied up in the standing crop of plants. The soil of grasslands makes them extremely rich areas for farming, as long as crops with low water requirements are grown. It should come as no surprise to you that the major crops grown in what once were grasslands are grasses—wheat, corn, and barley, for instance. These cultivated grasses require relatively little water in comparison with nongrass crops such as peas, beans, cabbages, and tomatoes.

Where grasslands are too dry for successful growing of crops, people often utilize them by replacing the large native grazing animals with domesticated animals. Unhappily, if the tight network of grassroots that holds grassland soil in place is broken up by plowing, or if overgrazing kills the grasses and the hooves of animals break up the soil, heavy rains wash away the unprotected topsoil or wind blows it away; thus erosion may be rapid and ruinous. In addition, as the vegetation becomes sparser, less water is absorbed by plant roots and returned to the atmosphere via transpiration. Instead, large amounts of water run down through the layers of soil, dissolving many of the nutrients, and carrying the nutrients away with it. This latter process is called soil leaching. Careless farming in a grassland area may quickly lead to disaster. The perils of substituting an agricultural ecosystem for a grassland ecosystem are quite similar to those of replacing a forest.

3-10 The oceans

The basic structure of oceanic ecosystems is similar to that of forest or grassland. The major superficial difference is in the photosynthesizers. Rather than having the sun's energy captured by familiar green plants, such as oak trees or grasses, most photosynthesis is carried on in the ocean by algae called **phytoplankton** (Fig. 3-19). **Plankton** is the name for very small organisms that live out their lives floating in water. **Zooplankton** are small floating animals, some of which make their living as herbivores, that is, by eating phytoplankton. Exceedingly small plankton are called nanoplankton.

It has been estimated that 75 percent of the earth's photosynthesis is carried out in the oceans. Phytoplankton have basically the same requirements as green plants on land. They must have sunlight, water, and CO_2 in order to carry on photosynthesis, and a supply of the other nutrients, such as N, K, and P, which are necessary for their life processes. Sunlight can penetrate only a hundred meters or so into the oceans, and so it is in the upper layers that the phytoplankton function. They have water in abundance, and CO_2 from the atmosphere is dissolved in that water.

Poor supply of nutrients The availability of nutrients is another matter, as there is no surface for soil formation in the ocean. The remains of dead organisms drop to the bottom, often miles below the area in which the phytoplankton carry on photosynthesis. If they have not been eaten before they reach the depths, they are broken down by decomposers. The nutrient products of decomposition, however, can only become available to phytoplankton if they are returned to the surface by upwelling currents. Such currents do exist in many places, especially along the western edges of continents. It

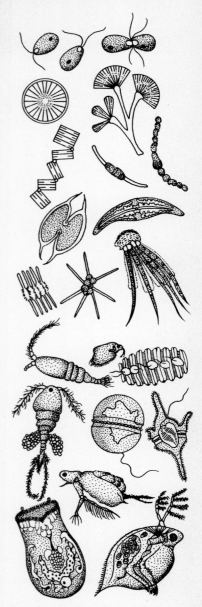

FIG. 3-19 Some types of plankton. Some plankton, called phytoplankton, are photosynthetic. They provide the base of food webs in lakes and oceans. Other plankton are animals and are known as zooplankton.

is only in those places, and in areas receiving nutrients washed from the land by great rivers, that the full productivity of the oceans is seen. Few people realize that most of the open oceans support relatively little life because the water is very poorly supplied with nutrients.

Oceanic food chains (Fig. 3-20), like **terrestrial** (land) food chains, have both herbivores and carnivores. The herbivores range in size from the tiny zooplankton to blue whales (Fig. 3-21), the largest animal which has ever lived. The blue whales have sievelike plates of whalebone in their mouths with which they strain plant and animal plankton from the water; thus it is more technically correct to call blue whales omnivores rather than herbivores. Oceanic carnivores include some zooplankton, many sizes of bony fishes, squid, sharks, seabirds, and killer whales.

Oceanic food chains A general characteristic of oceanic food chains is that they tend to be longer than terrestrial food chains. The main reason for this is the small size of the individual phytoplankton which permits the existence of a community of zooplankton herbivores. These tiny herbivores introduce a step in the food chains between the plants and the small fishes. The small fishes are roughly the size of the insect herbivores in terrestrial ecosystems. The length of oceanic food chains has great practical importance. The larger phytoplankton may support sizable fish as herbivores. Thus, relatively large populations of these fishes may occur which people can harvest. If, however, the phytoplankton are small, another carnivore link may exist in the chain between the floating plants and the fishes we eat. In Chapter 1 we noted how much energy is lost by eating fishes high on a food chain.

Impact of the human population People are not yet attempting to farm oceanic ecosystems, but their activities are having a serious impact on these systems. Part of this impact comes from people's

FIG. 3-20 Food chains in the ocean may be longer than those of land communities. Plankton are consumed by a fish called *Chromis,* which may be eaten by another kind of fish called a jack. Jacks may be eaten by sharks, which in turn may be eaten by killer whales.

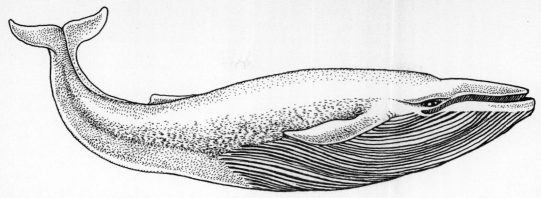

FIG. 3-21 A blue whale. Blue whales are the largest animals ever to have lived, reaching a length of 30 meters (100 feet). Not even dinosaurs and mammoths reached the size of this animal, which is rapidly disappearing as a result of whaling.

fishing activities, especially when fishes are caught more rapidly than the populations can replenish themselves. Such overfishing has already resulted in the disappearance of once-valuable fishing industries. Table 3-1 gives the date of decline of seven species which had not recovered as of 1968. The other aspect of human impact comes from pollution of the oceans. Pollution may lead to the extinction of entire groups of marine organisms. At the moment, DDT alone, moving from water into the fat of phytoplankton and animals and concentrated by its passage up oceanic food chains, threatens the existence of seabirds as a group. If a wide variety of seabird species become extinct, that would simplify oceanic ecosystems, and might reduce their stability.

DDT, mercury, and probably other pollutants seem to reduce the rate at which phytoplankton carry on photosynthesis. This poisoning

TABLE 3-1 OVERFISHING

SPECIES	APPROXIMATE DATE OF DECLINE
Antarctic blue whales	1935
Eastern Asian sardines	1945
California sardine	1946
Northwest Pacific salmon	1950
Atlanto-Scandian herring	1961
Barents sea cod	1962
Antarctic fin whales	1962

Data from Ricker, W. E., "Food from the Sea," in *Resources and Man: A Study and Recommendations by the Committee on Resources and Man.* Freeman, San Francisco.

does not affect all phytoplankton species equally, so that, as pollution increases, some species will be affected more than others and the composition of phytoplankton communities could change. There are many ways in which this might affect people. For instance, if in certain coastal areas small species of phytoplankton began to replace large species, food chains could be lengthened and there would be fewer larger fishes for people to harvest (remember, every step added to a food chain reduces the amount of energy available at the top).

The oceans have been called the "ultimate dump" on the planet earth. Sooner or later most of the output of other ecosystems ends up in the oceans. People have greatly increased the rate of output of many of these systems and have added to the output poisons of their manufacture. Because people depend very heavily on the highly nutritious food from the sea, it seems most unwise to treat the oceans as a dump.

THE STABILITY OF ECOSYSTEMS

The human population depends very heavily on natural ecosystems continuing to function properly. If, for instance, oceanic ecosystems lost much of their stability, it might become impossible to plan rational harvests of the seas. Population explosions of fishes might occur with little predictable pattern, with the populations dying off before people could mobilize to harvest surpluses. Similarly, relatively unstable populations such as those involved in vast plagues of locusts (Fig. 3-22) could become very common as terrestrial ecosystems become less and less stable.

FIG. 3-22 In a plague of locusts, the insects darken the sky and eat all plant materials in their path. Often there is no way for the people to survive except by eating the locusts. (*P. W. Hay, National Audubon Society.*)

3-11 Simple ecosystems are less stable than diverse ones

People are already creating unstable ecosystems on a large scale with their farming activities. About 5 percent of all photosynthetic productivity on land now occurs in agricultural ecosystems! As people create simple ecosystems, they also create ideal conditions for pest outbreaks and thus force themselves into the role of stabilizers for their ecosystems. Instead of there being a built-in ecological stabilizer, people use tools of various kinds. In many cases these tools, such as insecticides like DDT, which people employ in an attempt to stabilize artificial ecosystems, produce serious problems—often including instability in other ecosystems!

Growth of monocultures People have now gone a long way toward removing the natural land systems of the planet earth and replacing them with monocultures. These monocultures grow larger every year. We have only to look at history to see the great danger in this. In the late eighteenth and early nineteenth centuries, the agricultural base of Ireland was converted into a huge monoculture of potatoes. The population thrived until, around 1840, a fungal blight invaded the monoculture and caused two successive crop failures. About one-eighth of the population, more than 1 million Irish people, starved to death, and another 2 million had to migrate. The resulting migration had dramatic social effects, especially in the United States. It is important that every human being know this basic fact of biology: *simple ecosystems are unstable!*

The ecosystems of people are imbedded in a matrix of natural ecosystems; these natural ecosystems provide us with valuable "public services." They dispose of our waste products and maintain the flow of energy and cycles of matter which sustain us. It is clear that we cannot go on indefinitely destroying the diversity of the systems that allow us to live.

Reducing our impact on natural systems We can learn a great deal about the self-sufficient life from hunting and gathering peoples; however, it is not possible for the present human population of 4 billion to survive without agriculture. If hunting and gathering are the only methods of obtaining food used, the most favorable ecosystems can support only four persons per square kilometer (ten persons per square mile), and the Australian desert can support only about 0.4 person per square kilometer (one per square mile). The earth in 1974 had 29 people per square kilometer of land surface and 61 people per square kilometer of arable and potentially arable land. Although we cannot go back to hunting and gathering, we can learn from the hunters and gatherers a nonexploitive attitude toward the environment.

QUESTIONS FOR REVIEW

1 What happened about 10,000 years ago which caused a rapid growth in the human population?

2 Why should we think twice before we feel sorry for hunting and gathering peoples of the past who led such a hard life by our standards?

3 Why was the Plains Indians' heavy dependence on the buffalo and the Netsilik Eskimos' heavy dependence on the seal so much less ecologically disruptive than is modern agriculture's often heavy dependence on a single-species crop?
4 What do these terms mean: oxidation, photosynthesis, virgin forest, second growth, deciduous, monoculture, complex ecosystem, simple ecosystem, ecological stability?
5 What is the only significant input to the global ecosystem? Does this ecosystem have any significant outputs? If so, what?
6 What has been called "the ultimate dump" of planet earth, and why is treating it as a dump such a foolish and dangerous business?
7 What very important fact of biology was demonstrated by the great Irish potato famine of the 1840s?
8 Natural ecosystems provide valuable "public services." What does this statement mean?
9 What important lesson can we learn from nonagricultural peoples about our relationship to (impact on) natural ecosystems?

READINGS

Balikci, A.: *The Netsilik Eskimo,* Natural History Press, Garden City, N.Y., 1970.
The Biosphere, a *Scientific American* book, Freeman, San Francisco, 1970.
Cole, L. C.: "The Ecosphere," *Scientific American,* **198**:(4), offprint 144, 1958.
Farb, P. (ed.): *The Forest,* Life Nature Library, New York.
Idyll, C. P.: "The Anchovy Crisis," *Scientific American,* **228**(6), offprint 1273, 1973.
Iverson, J.: "Forest Clearance in the Stone Age," *Scientific American,* **194**(4), offprint 1151, 1956.
Kemp, W. B.: "The Flow of Energy in a Hunting Society," *Scientific American,* **225**(3), offprint 665, 1971.
Lenski, G.: *Human Societies: A Macrolevel Introduction to Sociology,* McGraw-Hill, New York, 1970 (see especially chap. 7, "Hunting and Gathering Societies").
Nicholson, M.: *The Environmental Revolution—A Guide for the New Masters of the Earth,* McGraw-Hill, New York, 1970.
The Ocean, a *Scientific American* book, Freeman, San Francisco, 1969.
Watt, K. E. F.: "Ecology," *Biocore* unit XXI, McGraw-Hill, New York, 1974.
Whittaker, R. H.: *Communities and Ecosystems,* 2d ed., Macmillan, New York, 1975.
Woodwell, G. M.: "Toxic Substances and Ecological Cycles," *Scientific American,* **216**(3), offprint 1066, 1967.

4

THE EARTH, THE MOON, THE SUN, AND THE STARS

SOME LEARNING OBJECTIVES

After you have studied this chapter, you should be able to

1. Explain why some astronomers think it is likely that life exists elsewhere in the universe; why, if it does, some extraterrestrial species may be more advanced technologically than we are; and why it is *not* likely that intelligent life exists elsewhere in our own solar system.
2. List the nine known planets of our star in order of their increasing distance from it, and name two other types of other natural bodies orbiting our star.
3. Describe the seasonal, lunar, and daily cycles.
4. Describe the "greenhouse effect" in terms of terrestrial and atmospheric absorption, radiation, and reflection; state the crucial importance of this effect to living things.
5. Demonstrate your understanding of how certain astronomical phenomena affect climate and weather by
   a Naming two factors which are mainly responsible for the differential heating of the earth's atmosphere.
   b Explaining why the earth is warmer at the equator than at the poles.
   c Describing the *global* pattern of air circulation caused by equatorial warming and polar cooling.
   d Naming at least three things which affect *local* weather patterns.
6. Demonstrate some of the things you have learned about the periodic behavior of animals by describing some effects of
   a The seasonal cycle on plants and animals.
   b The daily cycle on plants and animals.
   c The lunar cycle on two different animals.
7. Further demonstrate your knowledge of periodic, or cyclical, behavior by
   a Describing what is meant by a "biological clock."
   b Naming two astronomical "compasses" and a kind of animal that uses each.
   c Describe how two kinds of animals use time-compensated sun-compass orientation.
8. State what a circadian cycle is and two ways in which it differs from daily cycles, explain what "jet lag" is and how its effects are overcome, and say what type of cycle is involved in "jet lag."

In the last chapter we placed cities and farms in the larger context of natural ecosystems. We considered how some societies other than ours have related to their ecosystems, and we studied the impact of our modern technological society on the major ecosystems of earth. In this chapter we will place the systems we have studied so far in the context of the universe, the solar system, and the earth as a whole.

People have long had a deep, even mystical, interest in the moon, the sun, and the stars; almost all societies have had beliefs and traditions related to these heavenly bodies. When agriculture was developed, people had a very practical reason to study the skies; they needed to be able to predict the seasons, to know when to plant and when to harvest. The changing positions of the sun, moon, and stars in the skies were used to formulate calendars. As early as 3000 B.C. the Mesopotamians had a calendar, and ancient Egyptians used the first appearance of the star Sirius in the morning sky to indicate the time of the flooding of the Nile and thus the fertile period of the land.

Astronomy, the scientific study of the heavenly bodies, was early associated with astrology, a method of prediction based on the belief that the stars influence human affairs. In spite of this impetus, it took many generations of earthbound astronomers to work out our present picture of the solar system and the universe. In October 1957, the Soviet Union launched *Sputnik I* into orbit around the earth; for the first time an object built by human beings was in space and sending back information to earth. Since then people have orbited the earth in space capsules and walked on the moon; automated space probes have visited and radioed back information from our four nearest planetary neighbors. People who have seen pictures of earth taken from space and followed the information reported from other planets have gained a new perspective on our home, earth, the only planet known certainly to sustain life.

THE UNIVERSE

The universe contains millions of galaxies each made up of billions of stars (Fig. 4-1). Most stars are mixtures of gases in which energy-producing reactions continually go on; our sun is a medium-sized star in a fairly typical galaxy.

4-1 Galaxies are made up of billions of stars

Galaxies seem to take three main shapes, spiral (whirlpool-shaped), elliptical (egg-shaped), and irregular (glob-shaped). They are composed of varying numbers of different kinds of stars as well as interstellar dust and gas. The galaxies other than ours are all so far away that with small telescopes they appear as if they were just stars in our galaxy. It was not until 1925 and the invention of the 100-inch telescope situated on Mount Wilson in California that the existence of galaxies beyond our own was confirmed. Our sun is a star located

FIG. 4-1 A spiral galaxy. A galaxy consists of billions of stars being born, stars dying in nuclear explosions, and stars existing in a vigorous state. The universe is composed of millions of galaxies. Until the twentieth century and improved telescopes, it was thought that galaxies were clouds of dust and gas at the edges of the Milky Way. (*U.S. Naval Observatory.*)

near the edge of the Milky Way, a flattened spiral galaxy (Fig. 4-2). If it is clear enough when we look into the night sky, we can see a dense band of stars and interstellar dust which is the central portion of the Milky Way viewed edge-on. Our sun is one of about 100 billion stars orbiting the hub of the galaxy and completes an orbit once every 200 million years.

Some stars have other objects orbiting them. Our star, the sun, has nine planets circling it, and most astronomers see no reason why most stars should not have planets. There should also be regions around most stars where planets could have water, gaseous atmospheres, and temperature and radiation similar to those of earth. The other stars are so far away that we cannot confirm the existence of planets around them; however, there are double stars and stars whose positions appear to be affected by the gravity of unseen bodies orbiting them.

4-2 Life elsewhere in the universe

Many scientists think that because the universe is so large, life *must* have originated many times, on many different planets of many

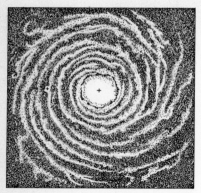

FIG. 4-2 The galaxy in which our solar system is located is the Milky Way. Its name comes from its appearance to us as we view it from near one edge. On a very clear night, when city lights do not interfere, the galaxy can be seen as a bright band extending across the sky. A large cross shows the center of the galaxy and a small cross to the left the position of our sun, 30,000 light-years from the center.

different stars. Harlow Shapley of Harvard University calculated that there well may be 100 million planets on which life has arisen. The other astronomers, I. S. Shklovskii and Carl Sagan, believe that Shapley's estimate is too low; they believe there may be 100 trillion planets with life in the universe.

Science fiction writers have written about many different forms that life might take on other planets. These forms range from bug-eyed monsters (Fig. 4-3) to forms quite similar to our own. Many biologists believe that living systems based upon the same chemical elements as ours would develop into life forms like those on earth. If this is so, some planets may be in the early states of evolution of earthlike life forms and others may be in later stages than our earth is. This raises the possibility that on some of the millions of planets which may have life, there are species with much more advanced technology than ours. Shapley estimated that there may be at least 100,000 planets with technologies more advanced than ours. The nearest planet with a technological civilization could be anywhere between ten and several thousand light-years away from us. A light-year is the distance which light, moving at about 299,793 kilometers per second (186,300 miles per second), travels in a year (9,460 billion kilometers or 5,878 billion miles).

We do not now have the ability to visit our technologically advanced planetary neighbors if they exist. If they have superior technology, however, *they* may be able to visit us. Their cultural values, nevertheless, might be such that they would not want to visit us! There *is* much speculation that space visitors may have visited earth in the distant past. The presence of space visitors is one possible explanation for some of the mysterious artifacts of past civilizations. It is more likely, however, that such extremely distant civilizations might try to communicate by radio waves. Some of our radio telescopes have devoted time to attempts to pick up meaningful signals from space; as of 1975 the results were negative.

Even though we are not able to visit planets of other solar systems, we can attempt to communicate by sending rockets carrying information. *Pioneer 10,* one of our space probes, sent back information from a fly-by of Jupiter in 1973 and then left our solar system, the first object constructed by human beings to do so. It carried an engraved plate (Fig. 4-4) designed to tell a technologically advanced species where it came from and something about us.

THE SOLAR SYSTEM

4-3 The energy given off by the sun is produced by the conversion of mass to energy

Our sun radiates energy into space in many forms: as light we can see and as invisible radiation such as infrared (heat) waves, ultraviolet waves, radio waves, and x-rays (Fig. 4-5). The energy given off by the sun is produced by the fusion of nuclei (cores) of the atoms

FIG 4-3 Artists and writers have speculated what creatures from another planet or solar system might look like.

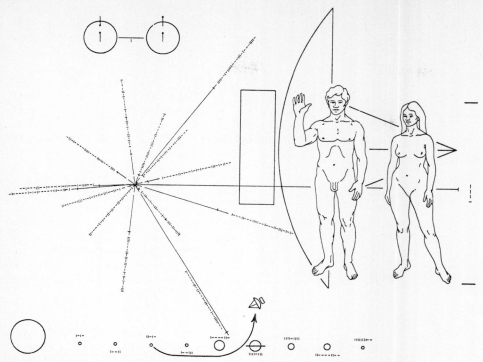

FIG. 4-4 The space probe *Pioneer 10* contains a metal plate engraved with this figure. *Upper left*, the hydrogen atom, a time measure. *Center left*, time interval notations for the location of our solar system in space. The 14 lines represent time intervals corresponding to pulsars, stars which produce radiation in pulses. *Bottom*, our solar system and the flight of *Pioneer 10*. *Right center*, human beings superimposed over the outline of the probe showing relative size. (*After Linda Salzman Sagan.*)

of some elements to produce other, lighter elements. Three protons (which are the nuclei of three hydrogen atoms) fuse to produce an unstable form of helium ($_3$He). The two $_3$He nuclei fuse to produce stable helium ($_4$He) and two protons which can start the cycle over. Two $_3$He nuclei are heavier than one $_4$He plus two protons, so that mass is lost and energy is given off. Carbon nuclei are also involved in energy-producing fusion reactions in the sun, and in total 4.5 million tons of matter is converted to energy every second in the sun. Einstein's famous formula $E = mc^2$ is used to calculate the amount of energy this much matter can be converted into (E = energy, m = mass, and c^2 = the speed of light squared); in watts, it turns out to be 380 million billion billion watts each second. In the sun these energy-producing reactions take place under enormous pressure and at a temperature of about 14 million°C (25 million°F); nevertheless, people hope to control this kind of fusion reaction on earth some

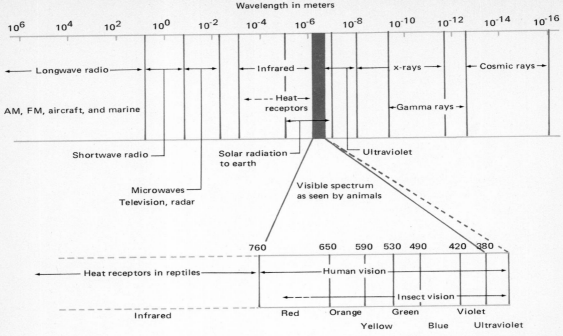

FIG. 4-5 The electromagnetic spectrum. The energy that comes from the sun and other stars is in the form of electromagnetic rays which vary greatly in wavelength. Gamma rays have very short wavelengths while radio waves have very long wavelengths. In the middle of the range is the radiation we call light, which appears to be white because it is composed of short wavelengths (violet, blue), intermediate wavelengths (green, yellow), and long wavelengths (orange, red). Human beings are not able to see shorter wavelengths called ultraviolet, although other animals may; nor do we see wavelengths longer than red, called infrared, although we feel them as heat when they touch us. (Visible wavelengths are in nanometers.)

day for the purpose of producing electricity. Of course, people have already produced *uncontrolled* fusion reactions on earth—hydrogen bombs; producing a controlled reaction is much more difficult.

The sun like other stars is evolving. Over billions of years stars are born and die; during their lifetimes they change, that is, evolve. On the sun, the $_4$He produced cannot be used again; it is a kind of ash which in about 5 billion years will cause the sun to swell and get enormously hotter. When this happens, if any life still exists on earth it will be exterminated, since the temperature at the surface of the earth will approach 537°C (1000°F).

Effects of the sun's energy on the earth Energy from the sun must travel about 150 million kilometers (around 93 million miles) to get to the earth; there some of the energy is absorbed and the rest reflected, creating the earthshine which astronauts see from space. We can feel the infrared radiation which warms our bodies as well as the earth. We see the visible light rays, and fair-skinned people, especially, may get a painful sunburn from the ultraviolet rays if they stay too long in the sun. Most people do not notice the other kinds of radiation from the sun; however, vast solar storms called sunspots have profound effects on earth. They produce changes in the atmosphere which affect our electrical and radio transmissions, and they may be involved in some complex way with rainfall patterns on earth.

4-4 The planets and other bodies of our solar system.

It was obvious to many of our ancestors that the sun circled around the earth and not the other way around; after all they saw the sun move across the sky every day. There were other observations, however, that were to change this belief. Ages ago, people had found patterns in the stars which are today called constellations (Fig. 4-6). But they found that some starlike bodies seemed to move through the constellations in regular patterns instead of remaining in the same positions relative to each other as did the other stars. These starlike bodies were called *planets* from the Greek word for "wanderer," and it took $2\frac{1}{2}$ centuries and five major astronomers (Copernicus, Brahe, Kepler, Galileo, and Newton) to finally produce a satisfactory model of the motions of these "wanderers" and thus "discover" the solar system. There was much controversy before people accepted that the planets, including the earth, circled the sun. Figure 4-7 shows the spacing of the planets of the solar system and their relative sizes. The planets move around the sun in approximately the same plane but at different speeds; Mercury, the closest to the sun, circles the sun in 88 earth days, Pluto, the farthest away, requires 248 earth years!

Moons, asteroids, and comets Many of the planets have moons, and there are a large number of smaller bodies called asteroids in the solar system. Most asteroids remain in an orbit between Mars and Jupiter; however, some like Icarus have orbits which bring them within a million or so miles of earth, close enough to have caused some people a fright. The great meteor crater in Arizona may have been caused by an asteroid hitting the earth about 5,000 years ago; however, statistics suggest that substantial asteroids collide with the earth only around once every 100,000 years. Asteroids do collide with each other, and the fragments of these collisions which fall through the earth's atmosphere are called meteorites if they do not burn up completely on their way through the atmosphere. Only about 7 of the 500 of these "shooting stars" which fall to earth each year are found and studied.

Comets are also part of the solar system; their structure is poorly understood and their orbits are often uncertain. They have frightened people in many times and places with their bright tails. One comet, Halley's, is due to appear again in 1986; in the past it often appeared

FIG. 4-6 A map of the sky as seen in the north temperate region showing the star patterns (constellations).

in years of wars and invasions, leading people to think of it as a sign of bad luck.

Mercury Mercury is the smallest planet and nearest the sun. The United States space probe, *Mariner 10*, passed by Mercury twice in 1974; pictures it radioed to earth revealed Mercury to be a rugged maze of craters on one side and smooth on the other. It has only a thin atmosphere of helium and argon, and its temperatures range

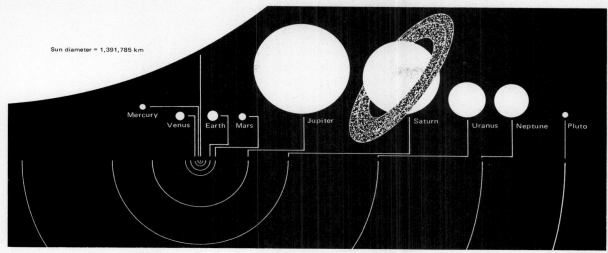

FIG. 4-7 Diagram of our solar system showing the relative sizes (*above*) and spacing (*below*) of the planets. The sizes and orbits are drawn to scale.

from 504°C (940°F) during the day to −212°C (−350°F) during the night; these conditions are entirely unsuitable for life.

Venus Venus has been considered as a place where life might exist; it is somewhat smaller than earth, and earth-based telescopes showed it to be covered with dense clouds. In 1969 space probes from the Soviet Union passed through the heavy atmosphere of Venus and landed on the surface. The information they transmitted back showed that conditions on Venus are extreme by earth standards. The atmospheric pressure at the surface is about 90 times that at the earth's surface, and the atmosphere is largely carbon dioxide. The temperature was even more severe than the pressure—about 500°C (932°F). It was probably these high temperatures that caused the probes to stop transmitting after a short time. If life exists on Venus, it must be most unlike life on earth. Earth creatures do live in ocean depths at pressures of over 1,000 atmospheres; however, even the hardiest earth organism is destroyed by heating it to 500°C.

Earth Although some parts of earth have extreme pressures and temperatures, a good part of its surface is just right for life. From space it appears mostly blue; in photos it is hard to distinguish the continents from the oceans because of the bright reflections of sunlight from the clouds that usually cover about half of the earth. The earth is the only place in the solar system where people can walk free and unencumbered by space suits or space capsules which carry miniature earth environments with them. We shall discuss our earth in detail later in this chapter.

The single natural satellite of our earth, the moon, is about one-quarter the size of the earth. Between 1969 and 1972 four manned United States space flights landed on the moon. The men who walked on the moon described it as a lifeless, awesome landscape much like some of the high deserts of earth; they took extensive photos and brought back to earth many samples of its crust for study. The oldest moon rocks so far analyzed are estimated to be around 4.6 billion years old, older than any rocks found on earth so far; none of the samples showed any evidence that life had ever existed on the moon.

When the first group of astronauts returned from the moon, they were quarantined from the rest of earth's people for a period of time just in case they may have been contaminated by some unknown disease from the moon. Tests showed that the astronauts were not contaminated by the moon; however, they had contaminated the moon. They left behind earth bacteria because their space suits were not sealed sufficiently to prevent earth bacteria which live in and on all people from escaping. It is not likely that any earth bacteria will survive the temperature extremes and lack of atmosphere on the moon.

Mars Mars is somewhat more than half the size of earth and has two moons. The Martian day is about the same length as an earth day, but the Martian year is twice as long as that of earth. "Men from Mars" have been characters in many science fiction novels, and there was a time when some astronomers believed that Mars had a life form with technology. They "saw" lines on the Martian surface they thought were canals built by Martians. There is still great dispute as to whether Mars has life on it; it is *certain,* however, that the so-called canals of Mars are not canals. They are illusions caused by difficulties of telescope viewing through the earth's atmosphere. The Mariner space probes sent back photos showing a very different picture of Mars. *Mariner 9* sent back over 7,000 photos of Mars taken during 698 orbits of the planet in 1971 and 1972. About one-half of the surface is covered with craters; there are also volcanic regions, deep rifts and chasms, and some sinuous channels with branching tributaries. These channels resemble branching rivers enough to suggest that they once contained running water. The polar regions suggest the presence of glaciers, and there seems to be evidence of wind erosion. The polar ice caps (Fig. 4-8) may be made up of frozen water, frozen carbon dioxide, or both. There is evidence of frozen subsurface water. This, combined with the possibility of past free-flowing water, strengthens the possibility that there are at least simple organisms living on Mars, perhaps just beneath the crust where they would be protected from solar radiation, drying out, and extremes of temperature.

Jupiter Jupiter is 11 times the diameter of earth and has 13 moons. The four largest moons of Jupiter are believed to have atmospheres. Until recently Jupiter was believed to be too cold for life to exist there; however, two space probes, *Pioneer 10* (1973) and *Pioneer 11* (1974) have recently revealed new information about the temperature and composition of Jupiter. Jupiter *is* cold, $-146°C$ ($-230°F$), at its surface; but one layer of its cloudy atmosphere is about room temperature, and air-borne microorganisms might survive there. The atmos-

FIG. 4-8 A composite photograph of Mars by *Mariner 9* in 1971 and 1972. The receding north polar ice cap, as well as other surface features of the planet, can be seen clearly. (*NASA.*)

phere of Jupiter has methane, ammonia, hydrogen, and water and is similar to the atmosphere on earth at the time life was beginning here; in fact, in 1974 scientists at Ames Research Center found earth bacteria that could survive conditions as alkaline as those on Jupiter. Some form of life might exist on one of Jupiter's moons; Callisto has a polar ice cap, and Io plays a role in mysterious bursts of radio waves which periodically emanate from Jupiter. *Pioneer 11* revealed that the huge red spot visible in photographs of Jupiter is nothing more than an immense hurricane; nevertheless, many mysteries about Jupiter remain. For example, at its core Jupiter seems to be six times as hot as the surface of the sun, and in spite of its cold exterior it radiates twice as much heat as it receives from the sun.

Saturn, Uranus, Neptune, and Pluto *Pioneer 11,* now called *Pioneer-Saturn,* is scheduled to fly by Saturn in 1979; perhaps it will reveal conditions hospitable to life. Presently it seems unlikely that life of the kind found on earth could exist on Saturn or for that matter on Uranus, Neptune, or Pluto. As far as we can tell now, either the chemical constitution of these planets is unsuitable or they are too cold for life. It is just remotely possible that there are other kinds of life than that on earth, perhaps based on different chemical compounds; however, it is difficult for earthbound biologists to hypothesize a chemical system dramatically different from ours that would provide the properties we define as life.

For the present it is probably best to assume that we are the only intelligent beings in our solar system and that any life which may exist on other planets of our system is microscopic. The other planets, their moons, and our moon are not places where we could live without bringing along our own life-support systems.

THE EARTH

Earth is one of the smallest planets of our solar system, and our sun is only a medium-sized star. There are stars 100 times smaller and 1,000 times larger; some of these last may be a million times brighter than our sun. In order to maintain some perspective about ourselves, we should appreciate where we, our planet, our solar system, and our galaxy fall in the scheme of things. The relative sizes of a series of things are shown in Fig. 4-9. As *things,* we are neither very large nor very small, although we are toward the small end of the scale.

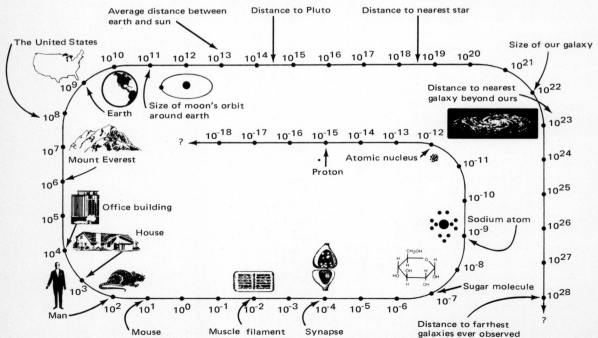

FIG. 4-9 Relative sizes in terms of orders of magnitude. Orders of magnitude are a series of powers of 10. Thus, $10^0 = 1$, $10^1 = 10$, $10^2 = 100$, $10^3 = 1,000$, etc. Each dimension of a given order of magnitude is multiplied by 10 in the next. The universe is immense compared with a mouse, yet a mouse is immense compared with an atom. Negative powers of 10 indicate smallness. Thus, $10^{-1} = 0.1$, $10^{-2} = 0.01$, $10^{-3} = 0.001$, etc.

Earth seems to be a smallish planet revolving around an average sun in an ordinary way; the only thing really special about it is that we live on it.

4-5 The composition of the earth is well known

The composition of the earth is related to its origin. Data from rocks brought back from the moon by our astronauts suggest that the moon was formed from rocky chunks of matter which were in orbit around the earth. This does not seem to fit with most theories of the origin of the planets and their moons. One most commonly accepted view is that the entire solar system originated from a vast cloud of cosmic dust and gases which was formed of a variety of chemical elements. These elements began to come together and condense, forming clusters; the process of condensation produced heat and pressure. In the main cluster, heat and pressure made possible the beginning of the nuclear reactions of the sun.

Smaller clusters condensed into the planets of the solar system and were trapped by the gravitation of the sun. In the process of condensation of the earth, the heavier elements, such as iron and nickel, sank to the middle of the planet, forming its core. Lighter elements such as hydrogen, oxygen, carbon, and nitrogen remained in the surface layers. Other elements crucial for life were also concentrated in the outer crust of the earth (Fig. 4-10).

The composition of the atmosphere Figure 4-11 shows a diagram of the atmosphere of the earth. It is most dense near the surface of the planet and gradually thins to nothingness. Above the highest layer of the troposphere, called the **tropopause,** relatively little stirring of the atmosphere occurs. Below this point, the troposphere is characterized by winds constantly moving the atmosphere about the earth in patterns that determine the climate.

At sea level the weight of the column of atmosphere above a single square inch of surface is about 15 pounds. Or, to put it another way, atmospheric pressure is 15 pounds per square inch. In the metric system, atmospheric pressure is usually given in millibars (1,000 dynes per square centimeter). One atmosphere is equal to 1,013 millibars. Often weather maps have lines of equal pressure in millibars. As one goes to higher elevations, atmospheric pressure becomes less and less.

The atmosphere is made up primarily of the gases nitrogen (78 percent) and oxygen (21 percent). The remainder is mainly the gas argon, but there are also some rarer gases, for example, carbon dioxide, neon, and helium. Water vapor is also present in the air. A form of oxygen (O_3), called ozone, forms a layer about 50 kilometers up in the atmosphere. Ozone absorbs much of the ultraviolet light from the sun. If the ozone layer did not exist, most animals and plants would suffer fatal sunburns.

The composition of organisms Animals, plants, and microorganisms are all made up of basically the same chemical elements. They repre-

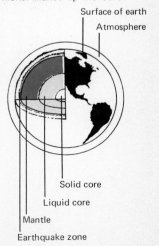

FIG. 4-10 A cross section of the earth. The way that earthquake shock waves travel to various parts of the earth shows that this planet is made up of layers of varying density. Density increases with depth, and it is thought that an extremely dense alloy of iron and nickel makes up the core.

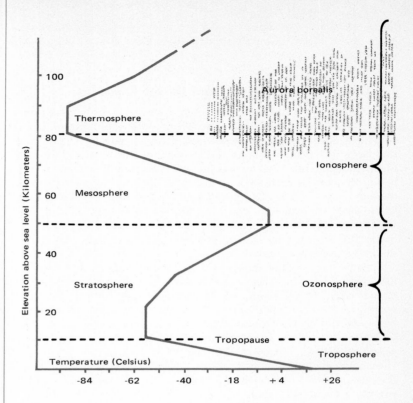

FIG. 4-11 The atmosphere of the earth can be divided into several layers. Weather occurs in the troposphere. The stratosphere contains a layer of ozone (O_3) which absorbs much of the sun's ultraviolet radiation. The ionosphere is the layer in which the northern and southern lights (aurora borealis and aurora australis) can be seen. These are patches of white or colored light in varying patterns in the night sky.

sent a special arrangement of the same kinds of atoms found in the earth's crust. And these atoms eventually are returned to the crust. Organisms are made up principally of water (about 85 percent). Other components of organisms are inorganic molecules and organic molecules in great abundance. The inorganic molecules are usually dissolved in water and play critical roles in the lives of organisms. Many of the essential nutrient elements (see Chap. 2) are used in organisms as inorganic ions or compounds. The organic molecules (by our definition, carbon compounds made by living cells) also serve a variety of functions in organisms which are discussed later in Chap. 12.

4-6 The earth is warmed by the sun

Most of the energy from the sun reaches the earth in the form of visible and ultraviolet light. These are forms of radiation described by physicists as having short wavelengths. Except for those absorbed by ozone, these wavelengths pass freely through the atmosphere.

Very little energy from this short-wavelength radiation is absorbed before it reaches the earth's surface. When it does reach the surface, however, some of it is absorbed and the earth is warmed. All molecules are always in motion, those of gases more so than those of solids. Absorption of light by a substance causes its molecules to move faster. What we call **heat** is the speeding up of molecules. As it warms up, the earth radiates energy, just as energy is radiated from a hot pan removed from a stove or fire. But this energy is in a different form from that of the incoming energy of the sun. It is long-wavelength radiation known as **infrared radiation.** The atmosphere is only slightly transparent to infrared radiation; both water in the atmosphere and carbon dioxide (CO_2) absorb outbound infrared radiation and are warmed by it. The atmospheric water and CO_2 in turn reradiate infrared, about half toward outer space and half back toward the surface. This leads to a sort of "trapping" of heat near the surface of the earth (Fig. 4-12).

The greenhouse effect The way heat is trapped near the surface of the earth is very similar to one way in which heat is trapped in a greenhouse. The glass in a greenhouse is transparent to light, which enters the greenhouse and warms the plants and soil inside. The plants and soil then radiate energy in the form of infrared waves, which are absorbed by the glass. The warmed glass then reradiates the infrared, sending some outward and some back into the greenhouse. Thus the greenhouse traps heat. On a sunny day it is always warmer inside a greenhouse than outside. This way of heat-trapping by a greenhouse and the heat-trapping of the earth are similar; therefore trapping of long-wavelength radiation is sometimes called the **greenhouse effect.**

The greenhouse effect is the reason why cloudy nights are warmer than clear nights (other things being equal). At night the earth's surface radiates heat accumulated during the day. Clouds are made up of water vapor and droplets, and water is a very good absorber of radiation. Because there is more water in a cloudy sky than in

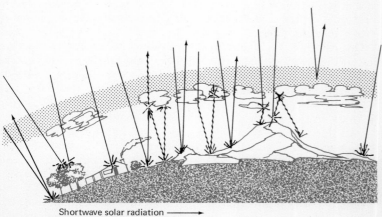

FIG. 4-12 Incoming shortwave radiation from the sun is absorbed by objects on the earth's surface. The objects are thus heated by the radiation. The longer heat waves (infrared) are absorbed by water, carbon dioxide, and particles in the atmosphere (shaded), causing the atmosphere to warm. Eventually all the solar radiation absorbed is radiated back into space, and therefore the temperature of our atmosphere does not continually rise.

a clear sky, the greenhouse effect is stronger. More outgoing infrared radiation is absorbed by the atmosphere and reradiated toward the earth. Indeed, the greenhouse effect is one of the things that makes life possible on our planet. If the trapping of heat did not occur, the surface of the earth would have an average temperature around −23°C (−10°F) instead of about +16°C (+60°F).

Reflected radiation Another important factor in the **heat balance** of the earth is the amount of light that is reflected by our planet. Some of the short-wavelength radiation coming to us is reflected by the atmosphere—especially by the upper surfaces of the clouds which, as we have seen, cover about one-half of the surface of the earth at any given time. They reflect back into space some 60 percent of the radiation that reaches them from the sun. Clouds are formed of water vapor and droplets that reflect some wavelengths and transmit others.

In outline, then, the factors that affect the average temperature of the earth's surface are simple. Short-wavelength energy from the sun is partly reflected by the earth and partly absorbed at its surface. The fraction that is absorbed warms the surface, which then radiates long-wavelength energy back toward space. Part of the energy is trapped by the greenhouse effect.

4-7 Weather is caused by the uneven heating of the earth by the sun

The *average* temperature of the earth is only *part* of the story of climate. Many other factors determine how the energy received from the sun does the work of driving the global weather system. These factors are not entirely understood, but it is known that **differential heating** is very important, especially the degree of contrast between the equator and the poles. This differential, hot at the equator and cold at the poles, is not produced by a difference in distances from the sun; it is produced because some parts of the earth absorb more energy than do others.

One cause of this is that the reflectivity of the surface varies from place to place. The land surface reflects between 10 and 30 percent of the light that reaches it, and the seas reflect 5 to 15 percent. Forests and farmlands reflect less light than do deserts.

The tilted axis As you know, the earth turns once on its axis every 24 hours and circles the sun one time for about every 365 turns on its axis (Fig. 4-13). A turn on the axis is one day, and a turn around the sun, one year. The earth's axis is tilted about 23° from the perpendicular (relative to the plane of its orbit around the sun). Thus at one point in the earth's movement around the sun, called the **northern solstice,** the sun's light falls more directly on the Northern Hemisphere. At the opposite point in its trip around the sun, the **southern solstice,** light falls more directly on the Southern Hemisphere. In both cases the sun's rays come in more nearly vertically near the equator than they do near the poles. In the Northern Hemisphere, the

THE EARTH, THE MOON, THE SUN, AND THE STARS 99

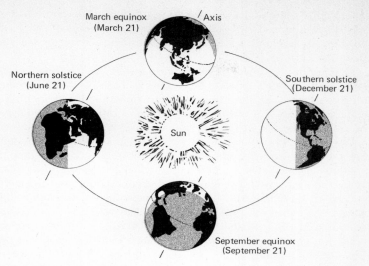

FIG. 4-13 Revolution of the earth around the sun and the tilting of the earth on its axis are responsible for the seasons.

northern solstice is the beginning of summer and the southern solstice is the beginning of winter. The two equinoxes, in March and September, are times when day and night are of approximately equal duration and are the beginning of spring and autumn.

It is the angle at which the sun's rays hit different parts of the earth that primarily determines the differential heating. The closer to 90°, the more energy will be absorbed (Fig. 4-14) rather than reflected. Thus more energy is always absorbed near the equator than near the poles. And more energy is absorbed by the hemisphere tilted *toward* the sun than by the hemisphere pointed *away* from it. It is the different angles at which the sun's light hits the Northern and Southern Hemispheres as the earth circles the sun that creates the seasons and makes them opposite in the two hemispheres. The importance of angle rather than distance from the sun is clear. The earth's path around the sun is not a circle, but an ellipse; winter in the Northern Hemisphere occurs when the earth is *closer* to the sun than it is in the northern summer!

Circulation of the air Warm air is lighter than cool air and rises above it. The circulation of the atmosphere, winds, pressure systems, and all the other aspects of weather trace back to the rising of hot air, especially over equatorial areas, and the descent of cold air, especially over the poles (Fig. 4-15). Prevailing *westerly* winds occur in the regions above about 30°N and S. Easterly winds, called *trades*, occur below 30°N and S. Between the trades and the equator is a region of calm called the *doldrums*.

Local weather patterns are determined by a variety of things. Large bodies of water add moisture to moving air and have a moderating effect on temperature (Fig. 4-16). The windward side of high mountain ranges is usually well watered because rising air is cooled and cool

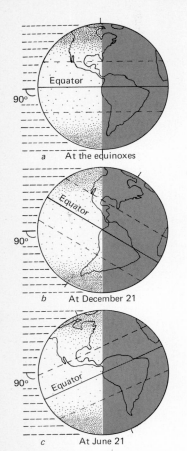

FIG. 4-14 The heating of the surface of the earth is greatest where the sun's rays arrive nearly perpendicular to the surface (*a*). As the angle changes to other than perpendicular (*b* and *c*), the rays must cross a thicker layer of atmosphere and lose heat by absorption.

air can hold less water than warm air. Thus it rains on the windward slopes, and the sheltered side may have a *rain shadow* with little moisture. Forests, cities, farms, and other surface features also have effects on local weather.

Meteorologists understand many aspects of the weather and can now, especially with the help of satellite pictures (Fig. 4-17), produce fairly good forecasts. But air behaves as a fluid, and the exact physics of fluid flow are not entirely understood. Until they are, many of the complex details of climate and weather will remain unknown.

When you think about the weather, you should always remember that the heat from the sun is its main driving force and that the amounts of clouds, dust, and other materials in the atmosphere change its properties of reflection and precipitation. Remember that changes in the amount of CO_2 and water vapor in the atmosphere change the greenhouse effect and that clearing forests and paving land affect the reflectivity of the earth's surface. With these facts in mind, you can understand why scientists are concerned about people's activities changing the weather! In Chap. 17, we shall discuss some scientists' interpretations of weather trends and what the prospects for the future may be as people continue to add both CO_2 from burning fossil fuels and dust from agriculture to the atmosphere, as well as continuing to pave the earth's surface.

The seasons Because the earth is tilted on its plane of orbit around the sun, revolution around the sun causes seasonal changes in the weather. On the day of northern solstice, it is the first day of summer in the Northern Hemisphere and the North Pole has 24 hours of light. In the Southern Hemisphere on the same day, the South Pole has a 24-hour-long night, and the southern temperate zone begins winter. Meanwhile at the equator the days and nights are of equal length, 12 hours each, as they always are.

In the same way, $182\frac{1}{2}$ days later, the reverse occurs. On the southern solstice the North Pole has 24 hours of darkness and the northern temperate zone goes into winter. Intermediate positions of the earth produce midspring and midautumn.

4-8 The rhythms of the earth affect living things

It is not surprising that organisms have adapted to these seasonal changes. For example, many plants and animals become inactive during the winter months. Animals that become inactive during the winter months are said to **hibernate** (from the Latin word for "winter"). In deserts some animals become inactive during the summer; they are said to **aestivate** (from the Latin word for "summer"). Many animals, such as migratory birds, move to a more comfortable climate when winter arrives. Plants may be **annual** or **perennial.** An annual plant produces resistant seeds at the end of the growing season. These seeds survive the winter or dry season and sprout into a new generation of plants when conditions become favorable. Perennial plants

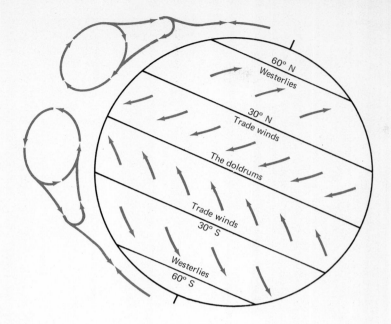

FIG. 4-15 Unequal heating and the rotation of the earth set up the major atmospheric movements and the pattern of winds on the earth's surface.

live for more than one year. When conditions are not favorable for growth, they may lose their leaves and become dormant as do deciduous trees, such as maples, or they may just slow their life processes, as do evergreen trees in the winter. Some perennial plants even die back to underground stems or roots which remain alive until the next growing season. Other organisms have solved the problem of surviving through unfavorable seasons by producing resistant eggs or specialized reproductive structures (see Chap. 6).

FIG. 4-16 The winds of the Pacific Ocean bring moist air which rises, cools, and drops its moisture as rain or snow on the windward side of the Coast Ranges and the Sierra Nevada. The result is what is called a rain shadow on the leeward side of the mountain. The Central Valley of California and the deserts of southern California and Nevada occur in these rain shadows. In the eastern United States, moist air from the Atlantic Ocean and the Gulf of Mexico rises over the less steep Appalachian Mountains, drops some of its precipitation there, and continues to provide sufficient rain for forests as far as the Missouri River.

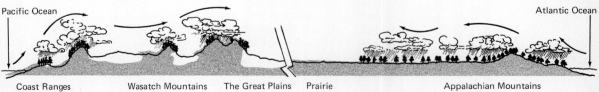

FIG. 4-17 Photograph of the earth from a NASA satellite stationed at 35,786 kilometers (22,235 miles) above the equator off the coast of South America. On this day, September 11, 1974, hurricane Carmen (arrow) was beginning to break up as it moved across the Gulf Coast of the United States. Note South America (*left*) and Africa and Spain (*right*). (*NASA.*)

4-9 The revolution of the moon around the earth produces rhythms on earth

The revolution of the moon around the earth is diagrammed in Fig. 4-18. The moon revolves about the earth in 29½ days. The period of rotation of the moon on its axis is also 29½ days. Therefore, the moon always presents the same face to the earth, and no one had seen the back side of the moon until we were able to orbit it.

The earth and the moon exert a gravitational pull on one another. The gravitational pull of the earth on the moon causes its crust to bulge in the direction of the earth. In the same way, the pull of the moon causes the earth's crust to bulge. Much more conspicuous, however, are the bulges in the earth's oceans. These we call the **tides**, which are periodic changes in water level. The way in which the moon affects tides is shown in Fig. 4-19.

Since the plane of the orbit of the moon changes its relationship to the plane of the equator of the earth, the heights of the tides at different places vary from time to time. The time of occurrence of the tides changes each day because the moon rises about 45 minutes later each night. Thus each of the two daily high tides occurs about 20 minutes later each day. There are some exceptionally high tides occurring about every 50 and 100 years. These occur when runoff of storm water is combined with the lineup of the moon, earth, sun, and other planets. Many people have unknowingly built their houses

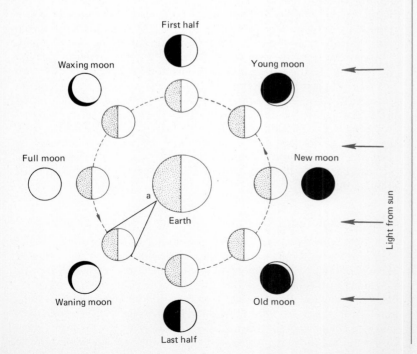

FIG. 4-18 The revolution of the moon around the earth causes the changes in the appearance of the moon. For example, to an observer at a, the moon in its actual position (inner circle) relative to earth and sun would appear as the moon shown in the outer circle.

104 | CHAPTER FOUR

on sites destined to be under water every 50 or 100 years because of this combination of events.

Effect of tides on organisms of the seashore Tides in the ocean are of great importance to animals and plants living on the edge of the sea. In some parts of the world, the difference between high tide and low tide may be 16 meters (52 feet) or more. When the tidal differences are this great and the configuration of the land is suitable, tides could be an important local source of electric power for people nearby. Organisms that live in the tidal zones face the problem that twice a day they may not be covered by water when the tide is out. Algae, such as seaweeds, which live in the intertidal zones must be able to endure some drying out. Barnacles are not only exposed to drying when the tide is out; their food supply is also cut off since they feed by filtering out small organisms from the water surrounding them. Barnacles and many other intertidal organisms have shells which keep them from drying out when they are out of water. The force of waves can wash away organisms which are not fastened down, and barnacles attach themselves to rocks with a "glue" so efficient that dentists have been experimenting with it for holding fillings in teeth. Intertidal plants often have clinging organs called *holdfasts,* and some intertidal animals have suction-cup "feet" which keep them in place.

Effect of lunar cycles on some organisms in the seas Some species of marine worms lay their eggs only during certain quarters of the moon of specific months. The palolo worm, a bottom dweller of the South Pacific, releases sections of its body containing eggs or sperm depending on whether it is female or male. The sections float to the surface of the sea where they burst and the sperm fertilize the eggs.

Along the coast of southern California there is a small fish called the grunion. In the spring and summer months, people come to the beach in thousands on certain nights to collect these fishes, for, on these nights as the tide goes out, hundreds of thousands of grunion are left wriggling on the sands. Grunion lay their eggs in the sand at periods of high tide, and this is when they are found in such numbers. At the next high tide, the eggs are washed to sea where they hatch. If the grunion do not lay their eggs at the highest tide, succeeding tides will wash the eggs away before they have developed sufficiently.

4-10 The rotation of the earth produces the daily cycle

The lengths of days and nights are different at different places and times on earth. At the equator, days and nights *are* of equal length throughout the year. On only two days of the year—once in the spring and once in the fall—is this true for other parts of the planet.

Effects of day length and night length on plants and animals Plants can carry on photosynthesis only during periods of light. Some animals are specialized for **diurnal** (daytime) activity. For example, some desert animals have the lenses of their eyes colored yellow—a built-in

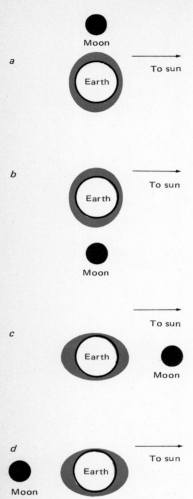

FIG. 4-19 Very low tides occur when the pulls of the sun and moon are at right angles to the earth (*a, b*). Very high tides called *spring tides* are caused by the additional gravitational pull of the sun when the sun, earth, and moon are lined up with each other (*c, d*).

pair of sunglasses. Other animals are **nocturnal** (active at night). Some of these have eyes so sensitive to dim light that they are nearly blinded by daylight. Many mammals that are wholly or partly nocturnal have the back of their eyes lined with a layer which contains light-reflecting crystals. This causes the "eye-shine" which you have probably seen when a cat or other animal faced into your flashlight beam or automobile headlights. Some nocturnal animals rely not on sight but on hearing. Bats and owls, which hunt at night, use their sense of hearing to find their prey. Even parasites may be nocturnal or diurnal. In certain areas, the tiny worms that cause elephantiasis in people remain in the blood deep in the body of the victim during the day and migrate into the blood of the outer layer at night. Elephantiasis is a disease which causes parts of the body to swell to large proportions, and it just happens that it is a nocturnal mosquito that transmits from person to person the worm that causes it. Finally, some animals are crepuscular, active at dawn or dusk, when diurnal and nocturnal species are inactive.

4-11 Plants, animals, and other organisms have periodic behavior

During their evolution, organisms have been exposed to seasonal, lunar, and daily cycles in the environment, and they have evolved many responses to these cycles. Animals and plants often show a cycle of changes in behavior during every 24 hours. Some crabs, for example, change their color, becoming lighter or darker following a roughly 24-hour cycle (Fig. 4-20). Although the movement is much too slow for us to see, some plants move their leaves in regular patterns. Again, they follow a period of about 24 hours.

Rhythmic behavior that occurs regularly in a period of about 24

FIG. 4-20 Certain crabs have a rhythm of color change.

hours is said to follow a **circadian rhythm.** The word *circadian* comes from Latin and means "about one day." Organisms showing certain circadian rhythms can be placed in constant conditions. That is, a crab or a plant can be put under constant light or kept in the dark continuously for several days. If this is done, they continue to show circadian rhythms, just as if the environment were changing. There must be some internal mechanism that regulates circadian behavior, since it continues in a constant environment.

Internal clocks It has been found that most organisms tested have internal timing mechanisms that regulate their activities. These internal mechanisms act as if they were **biological clocks.** Biological clocks regulate such different processes as the chemical activities of organisms and their behavior in response to light and dark. The details of how biological clocks operate are being intensively studied by many scientists. In plants there is a pigment called *phytochrome* which can act as a time-measuring device.

Having a biological clock enables many organisms to accomplish rather surprising feats. Some navigate; some arrive at a food source exactly when the food is available; some arrive at their nesting grounds simultaneously with others of their species even though they have come from different distances. Bees, birds, and some other animals are able to find their way using the sun as a reference point. This feat is *time-compensated sun-compass orientation.* As you can see in Fig. 4-21, a bee can identify the direction of its hive with respect to the sun's position in the sky. As the earth rotates, however, the position of the sun changes. Also, at different times of the year the sun has a different position. In spite of the constantly changing position of the sun, the bee is able to use its biological clock to anticipate the change in the sun's position and adjust its flight pattern properly. If you are ever lost in the woods and know the direction of home you can use your watch or a twig to find north (Fig. 4-22), but this is considerably simpler than what a bee does.

Migratory birds also have been shown to have biological clocks. They are also capable of time-compensated sun-compass orientation; however, the long-distance migratory flights of birds would require true navigation, the kind of navigation you must do if you fly a plane or sail a boat and do not use electronic methods of guidance. You need a map, an accurate timepiece, a compass, and tables and charts to determine where you are in relation to where you are going. Some biologists believe migrating birds use the sun or star patterns along with their biological clocks and some form of built-in map to perform true navigation. This hypothesis, however, has not been confirmed. It does appear that some migratory birds use the stars as a compass. In the Northern Hemisphere when the group of stars around Polaris, the North Star, is visible, you too can easily "navigate" north at night. If you also knew the season and the time, and had some navigational instruments and tables, you could tell where you are and which direction you should take to get to some distant goal.

Before the invention of accurate timepieces it was usual for ships

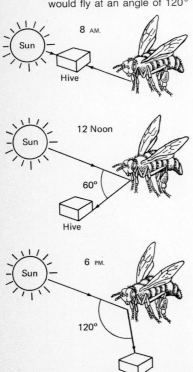

FIG. 4-21 Bees are able to determine the direction in which to fly to their hive by use of the position of the sun. The diagram shows a bee returning from a nectar source to its hive at 8 AM by flying directly toward the sun. By noon, the sun has moved an apparent 60° in the sky. The bee's internal clock *tells* it that it must fly 60° to the left of the sun in order to return to the hive from the same nectar source. At 6 PM, the bee would fly at an angle of 120°

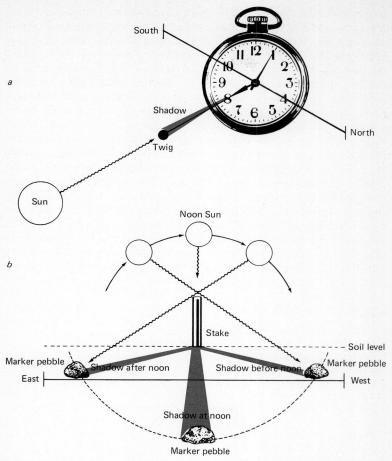

FIG. 4-22 You can find true south (*a*) in the United States and Canada if you have a watch set to local standard time and if the sun is shining brightly enough to throw a shadow. Lay the watch face up with the hour hand pointing directly toward the sun, using a twig so that the shadow falls along the hour hand. South is then midway along the shorter distance between the hour hand and 12 o'clock. An east-west line and north-south line can be determined (*b*) by driving a stake into the ground and marking the tip of its shadow as it shortens and lengthens before and after 12 noon. A line drawn between two marks which are the same distance from the stake will be oriented east-west.

to make a landfall several hundred kilometers from their destination. For thousands of years, however, the golden plover has been flying from the Aleutian Islands to Hawaii, a mere speck in the vast Pacific Ocean. Many other birds have carried out even more remarkable flights generation after generation. Much research remains to be done in order to find out how birds find their way.

Biological clocks of human beings A long-distance, east-to-west or west-to-east flight in a jet airplane can place you far from home while your biological clock is still running on your home time. Under these circumstances, most people experience "jet lag"—they find themselves wide awake in the middle of the night and dropping off to sleep in the middle of the day until their clocks reset in a day or so. For this reason many corporations do not allow their executives to conduct important business within 24 hours of their arrival after a long east-to-west or west-to-east jet flight.

Biologists have kept volunteers deep underground in conditions of constant light or conditions in which they choose their night-day cycle. Usually, their cycles change and are slightly longer or shorter than 24 hours when they are isolated from the surface environment. The biological clocks of human beings and other organisms can be reset by a changed day-night cycle. Nevertheless night workers are somewhat less efficient than day workers, possibly because their biological clocks are exposed to conflicting information. They work under lights at night but then they go home to sleep during the light, and they get up and go to work in the dark.

QUESTIONS FOR REVIEW

1 How does astrology differ from astronomy?
2 Why, since astronomy is such an ancient science and since galaxies are so huge, was the existence and nature of galaxies other than our own not confirmed until the 1920s?
3 What do these terms mean: diurnal, light year, annual plant, aestivate, perennial plant, hibernate, nocturnal, heat balance?
4 Can you name the different motions of the earth that are described in the text?
5 Winter in the Northern Hemisphere occurs when the earth is *closer* to the sun than it is during the northern summer. What, then, is the major factor that determines seasonal temperatures?
6 The old farmer squints skyward and says "Gonna be colder tonight than 'twas last night." What is probably the difference between last night's weather and tonight's, and why would that difference be likely to make a marked difference in temperature?
7 The Martian day (Mar's period of rotation on its axis) is 24.6 hours—almost exactly the same as ours. Yet during its night the temperature plummets an estimated 100° or more. What does this suggest about the Martian atmosphere?
8 What happens in the sun that can be described by Einstein's famous equation $E = mc^2$?
9 What are circadian rhythms, and why do we know that they are controlled by internal mechanisms even if we do not know what those mechanisms are?

READINGS

Aerospace Yearbooks, American Aviation Publications, Washington, D.C.
Bergamini, D., et al.: *The Universe,* Time-Life Books, New York, 1966.
Bok, B. J.: "The Birth of Stars," *Scientific American* **227**(2):48–61, 1972.

Bracewell, R. N.: *The Galactic Club: Intelligent Life in Outer Space*, Freeman, San Francisco, 1975.

Edson, L.: *Worlds around the Sun: The Emerging Portrait of the Solar System*, American Heritage, New York, 1969.

Frontiers in Astronomy, readings from *Scientific American*, Freeman, San Francisco, 1970.

Gingerich, O.: "Copernicus and Tycho," *Scientific American*, **229**(6):86–101, 1973.

Keeton, W. T.: "The Mystery of Pigeon Homing," *Scientific American*, **231**(6):96–107, 1974.

Murray, B. C.: "Mars from Mariner 9," *Scientific American*, **228**(1):48–69, 1973.

Shklovskii, I. S., and C. Sagan: *Intelligent Life in the Universe*, Holden-Day, San Francisco, 1966.

The Solar System, Scientific American, **233**(3), 1975 (entire issue).

Starr, V. P.: "The General Circulation of the Atmosphere," *Scientific American*, **196**(6):40–45, offprint 841, 1956.

Strong, J.: *Search the Solar System: The Role of Unmanned Interplanetary Probes*, Crane, Russak, New York, 1973.

Whipple, F. L.: "The Nature of Comets," *Scientific American*, **230**(2):48–57, 1974.

5
THE DISTRIBUTION OF LIFE

SOME LEARNING OBJECTIVES

After you have studied this chapter, you should be able to

1 State what the biosphere is and describe its extent in terms of the three major environmental gradients discussed in this chapter.
2 Demonstrate some things you have learned about oceans as habitats by
   a Naming one important contribution of ocean-current upwellings to marine organisms.
   b Naming three different ocean habitats and stating which of them contains the intertidal and estuarine subtypes.
   c Saying which ocean habitat is most life-productive and why.
   d Naming a problem faced by intertidal organisms and two problems faced by estuarine organisms.
   e Indicating two *solutions* which estuarine organisms have *invented* (by evolutionary adaptation)—one for each of the two problems referred to in d above.
3 Demonstrate your knowledge of major biome types by
   a Listing the six major biomes discussed in this chapter.
   b Describing each biome in terms of its major plant and animal types.
   c Arranging the six biomes in order of their ecological complexity from most complex to least

complex, and stating how each of several environmental conditions (altitude, growing season, etc.) helps determine the complexity of each.
   d Briefly describing human impact on each biome.
4 Show some of what you have learned about freshwater ecology by
   a Outlining the effects of seasonal temperature changes on temperate-zone freshwater lakes and ponds, and on their organisms.
   b Listing some causes and effects of ecological succession in freshwater lakes and ponds.
   c Outlining the human species' contributions to ecological succession in freshwater lakes and ponds.
5 Demonstrate your knowledge of biogeography by
   a Defining biogeography and stating how a biogeographic region differs from a biome.
   b Naming the six major biogeographic regions discussed in the chapter.
   c Naming the biogeographic region which has the most distinct fauna and giving the reason it does.
   d Stating which biogeographic regions' faunas are similar, explaining why they are, and naming some animals that are found in both of two biographically similar regions.
   e Naming at least three types of natural barriers to the dispersal (migration) of organisms, and naming several kinds of organisms that each type of barrier blocks or retards.
   f Stating how each barrier you named above can also serve as a dispersal route for organisms, and naming some kinds of organisms that employ each as a dispersal route.
   g Naming three continuous dispersal routes (specific geographic areas) and one type of discontinuous dispersal route.
6 Show some of what you have learned about human impact on biogeography by
   a Naming at least two kinds of artificial structures which can act as barriers to the dispersal of organisms or serve as dispersal routes.

b Naming at least two kinds of organisms that were intentionally, and two that were unintentionally, moved to new regions by people.
c Explaining why the introduction of organisms into new biogeographic regions can cause negative impacts ranging from the annoying to the disastrous, and describing the environmental impact of two organisms—the chestnut-blight fungus and the starling—in North America.
d Listing at least two other examples of introduced species which have had negative environmental repercussions, and one whose impact seems, so far, to have been only positive.
e Explaining what biological control is, and briefly describe two effective biological controls that are discussed in the text.

In the last chapter we explored the possible distribution of life in the universe; in this chapter we shall explore something much better known, the distribution of life on earth. Living things are found virtually everywhere on earth. The broad patterns of the distribution of organisms of various kinds on earth can be explained by the distribution of environmental conditions and by past events. Just as the solar system has evolved and is evolving so have the earth and the living things on it changed over time. This process of **evolution** is discussed in more detail in Chap. 11; however, it is impossible to consider the distribution of life without also mentioning its evolution.

Although life appears to be rare in our solar system, it is amazingly common on our planet. The most extreme and severe type of environment you can imagine on earth will probably have something living there. The hottest deserts and the deepest ocean trenches have living things. Even more extreme habitats have life; it is common to find pinkish patches on melting snow in the mountains; if you looked at a sample of such a patch with a microscope, you would find it to be made up of cells (see Chap. 6) of a single-celled alga. This alga survives, photosynthesizes, and grows at temperatures near the freezing point of water. Other kinds of algae as well as some bacteria occur in and around hot springs. For example, there are microorganisms living in some of the hot springs of Yellowstone National Park which have temperatures about 85°C (185°F), very near the boiling point of water, which boils at less than 100°C (212°F) at the high altitude of the park.

It is very nearly true that, on earth, some organism has evolved a way of dealing with every environmental condition. Microorganisms are known that can use pure sulfur or pure iron for energy; other microorganisms have been found growing in petroleum lakes and on the surface of glass. Organisms that live in extreme environments are not abundant; the majority of living things live in the conditions familiar to us, conditions that occur in a relatively thin layer spread over the surface of the earth.

HIGH AND LOW ON THE EARTH

The relatively thin layer of water, soil, and air that contains the living things of earth is called the **biosphere.** Over 75 percent of the earth's surface is covered with saltwater. Although there are some ocean trenches deeper than 9,000 meters (29,800 feet), the average depth of the oceans is about 4,000 meters (13,000 feet). Animals and microorganisms are found at the greatest depth yet explored, over 9,000 meters. The atmosphere extends indefinitely out into space, but only a relatively few feet of it is used by living things. The tallest trees are perhaps 61 to 91 meters (200 to 300 feet) high (although individual exceptions occur). Animals live in the tops of these trees. Flying animals obviously use the air and may fly several thousand meters above the surface. They do not live there indefinitely, however. Bacterial and mold spores have been found drifting in the atmosphere at great heights. It has been suggested that there are microorganisms (bacteria?) which float permanently in the atmosphere. Perhaps they play a role in the cycling of chemical elements in the global ecosystem; however, the importance of these permanently aloft organisms is not known, if, in fact, they do exist.

What about the tops of the highest mountains? The highest peaks are those of the Himalaya Mountains of Asia. British climbers found jumping spiders at 6,700 meters (22,000 feet) on Mount Everest in the Himalayas. At these high elevations spiders were eating insects called springtails and glacier fleas, which in turn fed on windblown debris. And above the tops of the Himalayas, the bar-headed goose has been observed flying with strong wing beat *higher* than the 8,800-meter (29,000-foot) top of Mount Everest. The narrow layer from 9,000 meters below sea level to about 8,800 meters above sea level is the biosphere (Fig. 5-1). Since the biosphere is only about 18 kilometers (11.2 miles) thick at its thickest and the diameter of the earth is 12,740 kilometers (7,918 miles), the vertical distribution of life occupies a crust that is only a little more than 0.1 percent of its diameter. This is comparatively thinner than the skin of an apple!

5-1 The oceans provide a variety of different habitats

People seem to think of the world's oceans (Fig. 5-2) as very uniform; in fact, temperature and oxygen content are different in different ocean regions and even the concentration of salt in the water varies from place to place. The oceans may be divided into a number of different habitats (Fig. 5-3).

The oceans are only very slowly and incompletely mixed by currents. Ocean currents are the result of unequal heating of the seas by the sun, just as atmospheric currents are produced by unequal heating of the land and air. In general, colder waters sink and warmer waters rise. These upwellings bring essential mineral nutrients to the surface layers of the ocean. They are crucial in the recycling of elements. Phosphorus, as we have noted, is washed from the land into the sea, where it forms deep sediments. Upwellings bring some phosphorus and other minerals up into zones where they can be used by living systems. As you might expect, areas of upwellings are rich in living things. The upwellings found off the coasts of California, Peru, Ecuador, Morocco, and southwest Africa are some of our best fishing areas.

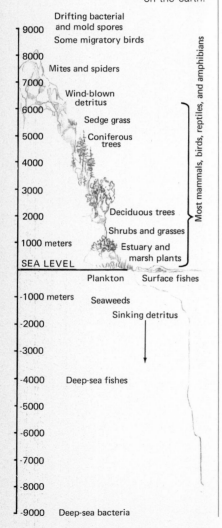

FIG. 5-1 The vertical distribution of life on the earth.

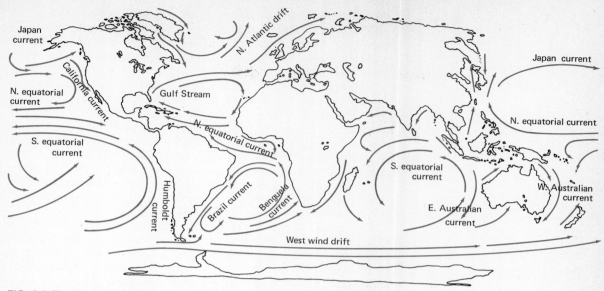

FIG. 5-2 The seas and the major oceanic currents. Warm currents include the Japan Current, the Gulf Stream, and the Equatorial currents. Cold currents include the California Current, the Humboldt Current, and the West Wind Drift.

The open oceans Many strong-swimming, wide-ranging animals such as tuna and whales live in the open ocean; however, the open ocean produces only a small fraction of the world's fish catch. The upper layers of the open sea where there is light enough for photosynthesis are deficient in nutrients. The phytoplankton are very small, resulting in long food chains and waste of energy. Close to shore and in the areas of upwellings rich in nutrients, food chains are

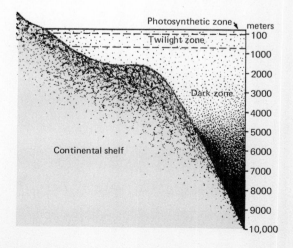

FIG. 5-3 The oceans may be divided into habitats based upon depth and light penetration.

shorter and more of the energy stored by photosynthesis is available for fish which people can harvest.

The ocean depths Organisms in the dark zones of the ocean depend indirectly on photosynthesis. Below the photosynthetic zone many organisms exist on a detritus-based food chain; in the depths, detritus constantly "rains" down from above as surface organisms produce wastes and die. The animals of the depths are often very different from those with which we are familiar (Fig. 5-4). Some have lost their eyes; others have very much enlarged eyes; many have light-producing organs. The light-producing organs, which may produce light of different colors, probably serve to identify species to one another; in one kind of angler fish, they serve as a lure for prey. Some deep-sea squid spray out a cloud of brightly glowing liquid at predators, temporarily diverting the predators until the squid escapes in the darkness.

The organisms of the depths have other problems than the lack of light to contend with, for example, pressure. In the deepest parts of the ocean, water pressure may be as high as 1,000 times that at the surface (1,000 kilograms per square centimeter or 14,700 pounds per square inch). It has been shown that bacteria are able to carry out their functions as decomposers at these pressures; however, organisms adapted to great pressures have difficulties in moving to areas with other pressures. They may literally explode when brought to the surface in the dredges of deep-sea biologists. Since the pressure inside has to be the same as the pressure outside at any depth, biologists have built special high-pressure chambers to study the organisms brought up from great depths.

The edges of the ocean The edges of the ocean provide a severe habitat but one with a variety of subtypes. The exact nature of that habitat depends upon whether the shore is rocky, sandy, or muddy. Organisms must adopt different strategies to live in such diverse habitats (Fig. 5-5). All organisms at the ocean's edge face the problem

FIG. 5-4 Some fishes with light-producing organs.

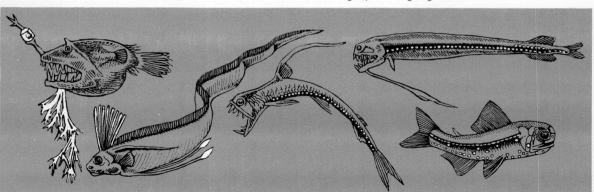

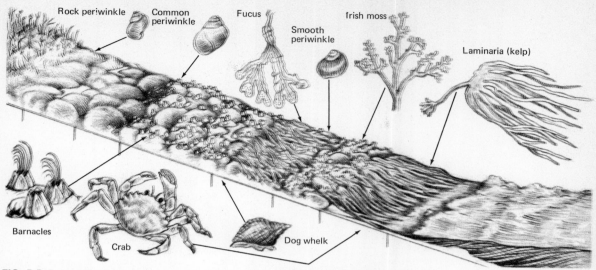

FIG. 5-5 Some inhabitants of a rocky ocean shore. The kelp in the lower part of the habitat is exposed only at low tide. Higher up organisms are exposed for longer periods of time, but are subjected to greater wave action.

of the tides; however, there will always be an area not covered by sea water but exposed to ocean spray. There also will be an area permanently covered with water. Between these is the **intertidal zone.** There, animals and plants will be alternately covered and uncovered by the sea. As has been discussed, they must have specializations which prevent them from drying out at low tide. Animals must be prepared to feed when the tide is high. Some of the animals and plants of a rocky intertidal zone are shown in Fig. 5-5.

Estuaries are bodies of water partially cut off from the sea where fresh water from the land mixes with saltwater from the sea. Many estuarine organisms maintain their position in the estuary in spite of tides from the sea and river currents from the land. Many of these organisms are attached to the bottom or to rocks or pilings or are buried in the mud. Estuarine organisms are faced with another problem, that of changing salt concentration, or **salinity.** Molecules behave in such a way that they tend to equalize their concentration in a process called **diffusion.** When this equalization process occurs across a membrane, such as that surrounding a cell, it is called **osmosis** (Fig. 5-6). A cell in fresh water has more salts and less water proportionately than its environment. Water molecules enter the cell and must be pumped out. A cell in saltwater may have less salt and more water than its environment. Such a cell must continuously adjust for this difference or it will lose water and become dehydrated.

In order to appreciate the problem estuarine organisms face, you need to understand these two important processes, diffusion and osmosis. Ions and molecules dissolved in water are in continual rapid

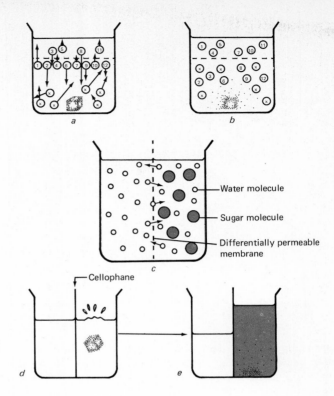

FIG. 5-6 Diffusion (*a* and *b*). Ions and molecules move from regions of high concentration across an imaginary plane to regions of low concentration until they are evenly distributed. Osmosis (*c*, *d* and *e*). If a solution of sugar is placed on one side of a differentially permeable membrane and pure water is placed on the other side, water molecules will cross the membrane until the concentrations are the same on both sides or until some counter force stops them. See text for explanation.

motion, bouncing off each other as well as off the molecules of water. Eventually they become distributed equally throughout the solution. It is very much like a cup of coffee with a sugar lump in it; eventually the coffee becomes sweet even if you don't stir it. This is an example of diffusion—ions and molecules moving from regions of high concentration to regions of low concentration until they are evenly distributed. Figure 5-6*a* and *b* show diffusion of sugar molecules from a sugar cube in a container of water. There are eight sugar molecules directly below an imaginary plane in Fig. 5-6*a* and only four above. An instant later, in *b*, of the original twelve molecules, there are six above and six below, but the concentration below remains higher because of the arrival of more sugar molecules (*x*'s) from the sugar cube. Until the cube is completely dissolved, this random process of diffusion will continue to increase the number of molecules above the imaginary plane. When the cube is dissolved, diffusion will have resulted in a uniform distribution of molecules throughout the container. Figure 5-6*c* shows a diagram of a differentially permeable membrane, through which water molecules can pass but sugar molecules cannot. The result of this process is shown in Figure 5-6*d* and *e*. Water moves through the differentially permeable membrane toward the side of the container with the dissolved sugar, raising the level of the water on that side. This is an example of osmosis, a process of basic importance to life.

Estuarine organisms are often able to withstand the salinity of full seawater at times of high tide and also able to tolerate dilute seawater when storms on the land produce heavy runoff of fresh water. There are limits, however, to the ability of organisms to respond to changing salinity, and there is a gradual change in the kinds of organisms that parallels the gradient in salinity in an estuary.

Estuaries serve as nurseries for many marine organisms. They are extremely important to our commercial fisheries; yet we are rapidly dredging, polluting, or filling estuaries. Usually there are tidal marshes around the edges of estuaries. When the tide is low, you can see little crooked creeks called *tidal meanders* which are caused by the inflow and outflow of the tides in relation to the runoff from the land. Marshes produce an excess of organic material; marsh plants such as cord grass and pickle weed (*Salicornia*) are efficient photosynthesizers. Invertebrates and bacteria feed on the plants as they die, and the excess of organic material washes down the tidal meanders and becomes detritus available as both food and fertilizer for estuarine organisms. Tidal marshes also keep estuaries from being excessively muddy by giving space for eroded soil from the land to settle out. Too much mud in the water of an estuary prevents the sunlight from penetrating and impairs photosynthesis by the phytoplankton. Tidal marshes also provide feeding and resting places for migratory waterfowl and other migratory birds.

In spite of the importance of marshes to oceanic food chains and to migratory birds, people have long regarded them as wastelands and have drained them or filled them, often with garbage. In California over 67 percent of the original marsh land has been destroyed by these means.

5-2 Mountains show a vertical gradient of life forms

Altitudinal gradients are **environmental gradients** that change as you go higher or lower on earth. For example, the higher you go up a mountain, the lower the average temperature becomes, and the daily and annual extremes of temperature become greater, too. Environmental gradients are indicated by gradients in the **life forms** of organisms that live along the gradient. The life form of an organism is its overall shape, size, and way of life. The exact nature of the plants and animals along altitudinal gradients of mountains varies from place to place. Different kinds of organisms will be present on temperate mountains and mountains in the tropics, but they will show an altitudinal gradient regardless of what mountains they inhabit.

Let us examine the vertical, or altitudinal, gradients of a mountain in the tropics (Fig. 5-7). Kilimanjaro, in Africa, or one of the higher peaks in the Andes of South America provides a good example. On the lower slopes of the mountain is the vegetation referred to as tropical forest. Rainfall is high and temperatures are warm to hot.

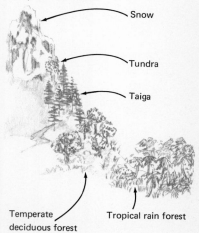

FIG. 5-7 The gradient of vegetation on a tropical mountain.

Snow

Tundra

Taiga

Temperate deciduous forest

Tropical rain forest

Very large trees and a great many smaller ones are found there. A great variety of animals (including many insects and birds) dwell in the tropical forest.

Above the zone of the tropical forest, precipitation (rainfall) is less and temperatures are lower. The tropical forest is replaced by deciduous forest. The trees there lose their leaves during the cold or dry season. Along with the change in vegetation as one goes higher, there is a change in the animals associated with the plants. For example, animals that live in the deciduous forest have to cope with the absence of leaves on the trees during part of the year.

As one goes still higher up the mountain, the deciduous forest gives way to evergreen trees. There will be different kinds of evergreen trees depending on where in the world the mountain is located. Evergreen trees often are able to carry on photosynthesis and grow where water is not abundant enough and temperatures are too low for deciduous trees.

As one goes still higher, eventually the environment is too severe to support trees. The elevation of the line above which trees cannot survive varies in different parts of the world. This **timberline** depends on a great many factors such as exposure to the sun, amount of moisture, temperature, and wind velocities. Mount Washington in New Hampshire is only 1,917 meters (6,288 feet) high, but is treeless at its top because of the very severe climatic conditions there. The tallest equatorial mountains, such as Kilimanjaro (5,894 meters or 19,340 feet), also have treeless tops, but on such mountains the timberline is more than twice as high as the peak of Mount Washington.

Above the timberline, much of the vegetation consists of mosses and grasses. There are also other kinds of low plants of relatively few species as well as a few woody plants that grow low and flat. These plants, and the animals found in this community, are able to survive periodic freezing and intense radiation from the sun. Finally, the tops of some mountains are rocks, ice, and snow. The only organisms able to survive here have an extreme ability to withstand harsh environmental conditions.

NORTH AND SOUTH ON EARTH

Just as there are altitudinal gradients of environmental variables on a mountain, so there are gradients on the lowlands from the equator to the poles. These **latitudinal gradients** parallel in many respects the altitudinal gradients, but on a global scale. The result of latitudinal gradients, again, is a series of communities, each with its characteristic life forms. These also parallel the communities you would see going up a mountain. These global communities based on climatic differences are called **biomes**. A biome is a community which covers a large geographic area, where the life forms are different from those of other climatic areas. The environmental gradient is the same going from the equator to the North Pole or from the equator to the South Pole; thus you would expect to find the same biomes going south from the equator as you would going north. If you look at a map of the world (Fig. 5-8), you will find that there is less land at higher latitudes in the Southern Hemisphere than in the Northern Hemisphere. The Southern Hemisphere thus runs out of land for the polar-most biomes. Also, lands in the middle latitudes most favorable to modern Western agricultural techniques are lacking in most of the Southern Hemisphere.

5-3 The tropical rain forest biome is found near the equator

The **tropical rain forest** biome occurs on either side of the equator in a band around the earth. This biome is thought to be the oldest of the biomes geologically. The most conspicuous feature of the rain forest is the trees (Fig. 5-9), which are commonly very tall and unbranched until near the top. There, their widespreading branches meet and their dense foliage forms a continuous canopy over the forest. In a temperate deciduous forest, you might find 10 to 12 different species of trees; in a tropical rain forest, there may be more than a hundred kinds. Below the tallest trees, there are usually one or more layers, or stories, of shorter trees and perhaps a layer of shrubs.

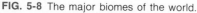Storied vegetation The heavy leaf cover of the tropical forest and

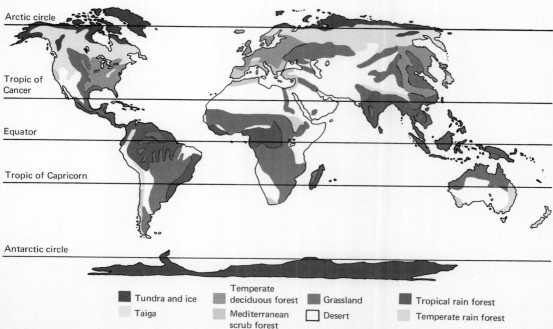
FIG. 5-8 The major biomes of the world.

FIG. 5-9 This tropical rain forest in Costa Rica has a canopy of leaves of the tallest trees, under which are layers or stories of the canopies of shorter trees. (*P. R. Ehrlich.*)

the occurrence of several stories of vegetation cut down on the light reaching the forest floor. There the vegetation consists of relatively few shade-loving plants. Competition for light is extreme, and plants called **epiphytes** have taken to climbing or living perched upon other plants (Fig. 5-10). The rich diversity of plant life determines in large part the diversity of animal life; the greatest number of animals are also found in and around the forest canopy.

Many specialized and curious habitats are found in the rain forest. For example, some bromeliads (the pineapple is a land bromeliad) grow on the branches of trees, not connected with the soil. Their large leaves with overlapping bases form cups in which accumulated water harbors mosquitos, other insects, and even tree frogs, perhaps many meters above the ground.

Common animals of the tropical forests are those at home in the trees. Many different kinds of birds occur here, as do snakes and lizards. Monkeys are usually abundant. Relatively fewer animals are found in the lower levels of the forest.

Most of the trees in a rain forest are evergreen, but this does not mean they keep their leaves forever. In a deciduous forest, all the leaves of all the trees are shed at one time. Evergreen plants shed their leaves individually after the leaves are two or three years old. Therefore, in an evergreen forest there is a continual falling of leaves

FIG. 5-10 A branch of a tree from a tropical forest showing numerous plants, called epiphytes, growing on it. (*Karl Weidmann, National Audubon Society.*)

which have reached that age. Tropical rain forests occur in areas where the rainfall is high, the temperature is high, and the light intensity is high. Rainfall, temperature, and light intensity are uniform throughout the year. In such conditions, photosynthetic production is high and the plants have a high demand for mineral nutrients.

Nutrients in plant tissue Rain forests have the same cycles of nutrients as do other ecosystems. Fallen leaves and the bodies and waste products of other organisms are broken down by decomposer microorganisms. In the warm, moist tropics, decomposition occurs much more rapidly than in other biomes. There is also rapid growth of new vegetation. Therefore, there is *very little accumulation of organic material* or humus. Mineral and other nutrients are released by decomposers and taken up again very rapidly by the plants, which have widespread shallow root systems.

The soil of tropical forests therefore has a different character from that of most other biomes. It does not contain much humus, and it is not particularly rich in nutrients. The rain forest biome survives because it carries on its cycling of elements very rapidly and continuously. Most of the nutrients are in plant tissue.

As we have noted (Chap. 2), rain forest soil is not suitable for

FIG. 5-11 Angkor Wat. These vast ruins are of a city and temples in Cambodia. They were built in part of blocks of laterite. (*P. R. Ehrlich.*)

agriculture in the way we practice it in temperate zones. Once the forest is removed, the nutrients are removed and the cycling of elements stops. Because of its chemical composition, the soil often undergoes **laterization** and becomes very tightly compacted bricklike laterite. Indeed, lateritic soils do make fine construction materials and have been used to build such structures as the great temples of Angkor Wat in Cambodia (Fig. 5-11). Clearing the tropical rain forest would not greatly increase the amount of food available to people because of the unsuitability of the rain forest soil.

5-4 Northward and southward of the tropical forest is the grassland biome

North and south of the tropical zone, rainfall is less, and it is also less evenly distributed throughout the year. Temperature is lower and less uniform during the day and throughout the year. Tropical rain forest trees cannot grow in these conditions, and as we move north or south, we find smaller, often deciduous trees. These trees, making up a *scrub forest,* often have specializations to reduce water loss and spines or thorns to reduce damage by animals. Understories of smaller trees and shrubs are not present in this transitional region, and as we move still further north or south, grasses and other nonwoody plants replace the plants of the forest floor. Eventually we reach the **grassland biome;** Fig. 5-8 shows the distribution of the grassland biome.

Grasses are very distinctive and important plants (Fig. 5-12); rice, wheat, and corn, the three main food crops of humanity, are domesticated grasses. Grasses tend to spread into disturbed areas of other biomes. A grassland community not only has many species of grasses but also many other kinds of plants, all specialized in some way to resist the persistent grazing and trampling of hooves of large grassland herbivores. The evolution of the grasses has paralleled the evolution of the grass-eating animals; there are substances in the urine of grazing animals that act as plant growth hormones and enhance the growth of grasses. Buffalo, antelopes, cattle, and other grazers developed special grinding teeth necessary to eat grasses while grasses were developing special abrasive cells that wear away the teeth of herbivores.

Impact of people Early in its evolution, the human species learned to use fire. There is much evidence that people used fire to kill grazing animals for their meat and skins as well as for their bones and teeth, which were used for tools. Once people had begun to domesticate herds of cattle, they further extended the grassland biome by burning down shrubs and trees. People have played a major role in maintaining this biome, and in many places in the world today people still use fire (and recently, herbicides) to maintain and extend grassland for their herds. People have also destroyed much of the grassland biome. By removing the natural windbreaks of trees, they have allowed winds to erode the soil. By overgrazing, people have allowed animals to destroy the interlocking roots and stems of grasses which hold the soil of grasslands together; then the soil washes or blows away, leaving desert.

Aside from the large grazing animals, many other animals once inhabited the grasslands of the United States. Many of these lived in burrows under the soil. In some areas many acres were covered by animal burrows, as in the so-called towns of prairie dogs. The predators in these grasslands were such animals as snakes, owls, hawks, black-tailed ferrets, and coyotes. Ranchers who raise cattle and sheep have now shot or poisoned almost all these other animals that once shared the grassland with their herds. The United States government helped and is still helping to kill the last remaining burrowing animals and predators in the mistaken belief that the burrowing animals competed with the herds for food and that the predators killed calves and lambs. Recent studies show that the burrowers enhanced soil fertility and are controlled by the predators, if the predators are not eliminated. Predators that kill sheep kill mainly those that are sick because they are improperly cared for. Recent studies show that coyotes can learn not to attack lambs if they are given lamb meat treated with a substance that makes them sick but does not kill them. If this method of predator control were used, coyotes might go back to controlling the burrowers, and the use of poisons which concentrate in food chains might be eliminated.

The grassland biome is in a rather delicate balance with the climate and other environmental factors. With reduced water, increased grazing, or other disturbances, it may change into desert. With increased water and less grazing, grasslands may become scrub forests. The deserts and scrub forests of the world are adjacent to the grasslands. As climate has changed and as human activities have changed, there must have been many shifts of the margins of the biomes.

FIG. 5-12 Grasses have a dense, much-branched, fibrous root system. (*Jack Dermid, National Audubon Society.*)

5-5 The desert biome usually adjoins grasslands

At the latitudes of about 30° north and south, the air is generally descending and dry (see Fig. 4-15). These are the areas of the world's great **desert biomes** (see Fig. 5-8) where rainfall is too low to support the growth of a solid cover of grasses and other plants. Only highly specialized plants, such as cacti and certain other forms, are able to survive (Fig. 5-13). Without abundant plants to modify the environment, heat is absorbed during the day and is lost rapidly at night. Desert temperatures (in air in the shade) may reach 57°C (135°F) at midday, yet nights may be cool or cold in the desert.

Animals that live in the desert usually avoid the heat of the day; most are nocturnal (active at night) or crepuscular (active around twilight) and spend their days in underground burrows or under stones. Desert animals can survive with little or no drinking water; often dew that forms during the night is an important source. Some desert animals obtain all the water they need from the plants they eat. Desert tortoises eat juicy cacti, and some desert rodents can extract water chemically from organic compounds in the seeds they eat.

Impact of people Deserts have been increasing since the early days of recorded history as people's activities added more lands to them. Our clearing of forests, improper agricultural techniques, and ecolog-

FIG. 5-13 Highly specialized desert plants, such as this Joshua tree, successfully survive the extremes of dryness, heat, and cold found in the deserts of southern California. (*FPG*.)

ically unsound irrigation methods have converted many parts of the world into deserts. The once lush Tigris and Euphrates Valley of the Middle East was made into desert by the faulty use of irrigation. The Sahara Desert is presently advancing southward, in some places by as much as 50 kilometers a year, due in part to overgrazing and in part to several years of drought.

Once converted into desert, land is very difficult to restore to agricultural use. Reclamation has been successful in Israel, at least temporarily until water or other resources run out. Half of Israel is desert, yet part of this desert is now being made to support forage and other crops by a combination of hard work and technical skills which have, at least for the moment, overcome the misuse of the last 1,300 years. Israel has a number of outstanding authorities among its own citizens, and it also drew upon the technical skills offered by specialized agencies of the United Nations and the United States.

5-6 The temperate forest biome is found in both Eastern and Western Hemispheres

Temperate deciduous forests are found at those latitudes where rainfall is sufficient for the growth of trees, but the growing season for plants is only six to eight months of the year. During the winter, precipitation may be high; however, it is in the form of snow, and temperatures are too low for growth. With the onset of winter, then, the majority of the trees and shrubs make final stores of food, lose their leaves, and become dormant. Many of the specializations of other plants and animals of this biome are based not only upon the pronounced seasons but also upon the nature of the dominant trees.

The seasonal accumulation of leaves and their gradual breakdown lead to a considerable development of leaf mold, which will eventually become humus, teeming with all sorts of life. The nutrients from the leaves support a rich development of many species of herbaceous plants on the forest floor. These usually bloom more or less simultaneously in the spring and are well on their way to producing seeds before the trees leaf out.

Browsing herbivores, such as deer, are still common in some deciduous forests. Rarely, wild pigs are found, which root about in the leaf mold for food. Omnivores such as raccoons and bears once were abundant, as were smaller carnivores.

Impact of people Deciduous forest at one time covered most of the earth's land surface in the temperate zone. After the human species' destruction of the forests of the Mediterranean region, Western people began to clear the deciduous forest of Europe and North America. Central and Western Europe were mostly forested until neolithic farmers cut and burned the trees of many areas. Neolithic people belonged to the most recent of the stone ages; they made complex stone tools but they did not use metals. Their plows were wooden, and forests remained in many places where the soil was too heavy to till with neolithic plows. During succeeding periods some land returned to forest, but the overall trend on the Continent and in England was the cutting of the forest. As technology became more complex, the wood was used for building ships and for smelting iron ore. In the thousand years between A.D. 900 and A.D. 1900, the face of Central Europe was completely changed by the destruction of many forests. Similar deforestation occurred in Asia and in North America after it was colonized by Europeans. Since deciduous forest can, with proper management, be converted into agricultural land, we tend to assume the same must be true of other kinds of forest as well. Unfortunately, no other forest biome contains the necessary features of climate and nutrient- and organic-matter accumulation. None is successfully made into farmland for more than a short period of time.

5-7 In the Northern Hemisphere there is a biome of evergreen trees

North of the temperate deciduous forest is the **taiga**, a biome characterized by evergreen coniferous trees (see Fig. 5-8). **Coniferous trees** are trees such as spruce, fir, and pine which bear cones (Fig. 5-14). There is little land area at the required latitude in the Southern Hemisphere, and this biome is scarcely represented there.

In the zone of the taiga, winter temperatures often are severe; although there is abundant precipitation, much of it is in the form of snow. The prevailing mood of the forest is monotony, and in fact, this biome is less diverse than other forest biomes. Besides the conifers, deciduous trees such as alders, birches, and aspen may be found. Herbs and shrubs are found in more open parts of the taiga.

FIG. 5-14 A lake surrounded by taiga, showing the coniferous trees characteristic of this biome. (*U.S. Forest Service.*)

Many animals utilize the taiga only during the summer months and migrate southward during the winter. Most of the smaller animals, including the numerous insects, are dormant during the long winters. The common omnivorous bears also hibernate. Predators such as bobcats and wolves are winter-active. The most conspicuous of the herbivores is the moose, which is common in open, marshy areas of the taiga; however, rabbits are also prominent herbivores in the taiga biome.

Impact of people For many years the taiga was utilized by the human species chiefly as a source of fur-bearing animals. Beavers, martins, sables, ermine, and fox are some of the animals sought after by trappers. The early history of northern United States and Canada and their relationship to Great Britain is based upon trappers and the taiga. Today the taiga is being exploited more and more for wood for paper pulp, mainly for newsprint. It is also being exploited for any other resources which can be found there.

5-8 The tundra biome is north of the taiga

Between the taiga and the polar ice cap in the Northern Hemisphere is the biome known as **tundra** (see Fig. 5-8). It is hard to imagine a less suitable place for a community. The climate is very cold, and the growing season may be only a month or so in length. The ground is permanently frozen, with ice (*permafrost*) only one-quarter of a meter or less below the surface, even in summer. Most of the vegetation consists of mosses, lichens, and flat, low-growing shrubs only a few centimeters to a half meter or so high (Fig. 5-15). Most of the animals of the tundra are insects that, like the plants, are dormant most of the year. Larger animals such as caribou and migratory birds move into the tundra during the short growing season and migrate southward for the long winter.

FIG. 5-15 A boggy area of tundra, showing the flat, low-growing shrubs characteristic of this biome. (*U.S. Fish and Wildlife Service.*)

Impact of people The tundra is the least diverse and the least stable of all biomes. Human activities in the past have left permanent scars on this fragile community. For instance, the tracks of a vehicle traveling over the tundra may last for decades or more. We may expect that, with improved technology, we will inflict even greater damage, especially since oil discoveries have been made in tundra areas such as the Arctic slope of Alaska and gigantic pipelines are being built for getting the oil to industrial areas.

FRESHWATER HABITATS OF THE BIOMES

The nature of the freshwater habitat is different in different biomes. Obviously, freshwater communities will be very different in tropical rain forest than in taiga or tundra, where they would be frozen more than half the year. Even within the temperate zone, the community varies widely depending upon whether the habitat is a small pond or a large lake or a small, swiftly flowing stream or a large, lazy river.

5-9 There are many freshwater habitats in the temperate zones

Streams Animals and plants of fast streams often have strong attachments to the bottom keeping them from being washed away (Fig. 5-16). Seed plants have roots, and algae often have specialized cells attaching them to the rocks. Aquatic insects are often streamlined in shape

FIG. 5-16 These organisms use very different strategies to keep from being washed away in fast-flowing streams. The blackfly larva attaches itself with a silken thread to the downstream side of rocks. Caddis fly larvae construct cases around their bodies which protect them from the current and weigh them down, or they spin nets. Algae and mosses are permanently attached to the substrate. Animals that move about, such as fishes, have streamlined bodies. Some immature mayflies and blackflies are flattened and cling to the bottom of rocks. The water penny beetle is so flat it appears to be a bump on a stone.

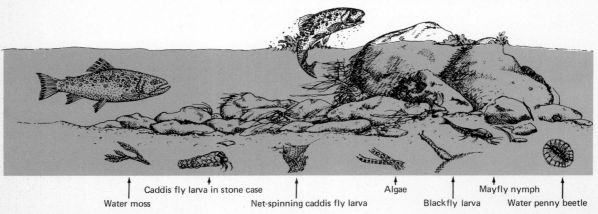

FIG. 5-17 The effects of seasonal temperature change in a temperate-zone pond.

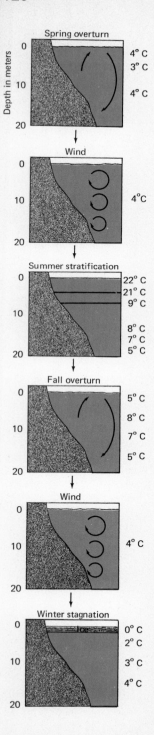

and have strong claws for holding on, and many live under stones where they are protected from the current. Tadpoles in such streams have suckerlike mouths that enable them to hold on to rocks.

Lakes and ponds Lake and ponds in the temperate zone, if they are sufficiently large, have certain features in common with the oceans. There are habitats around the margin and on the bottom; there is a community of phytoplankton and zooplankton; and there may be a zone of little or no light in which photosynthesis cannot occur.

Freshwater lakes and ponds in the temperate zone usually are frozen at least part of the year (Fig. 5-17). When the cold temperatures of winter arrive, the upper layers of water are cooled. Water is most dense at about 4°C (39°F). Both colder and warmer waters are lighter. When the surface waters are cooled to 4°C, they sink and warmer water rises, producing an *overturn*. There is, therefore, a great stirring of the waters as what was formerly on the bottom rises to the top. Wind also contributes to this mixing. The water forms layers, with cold water at the bottom and warm water at the top. Eventually all the water is cooled to 4°C (39°F), and finally the surface layer reaches 0°C (32°F) and freezes.

Spring thawing produces a layer of water from melted ice which warms to 4°C and then sinks, also causing an overturn. Once again there is a stirring of the water, and this stirring of the water is a very important factor in the community. During the summer, when the lake is layered, oxygen production occurs at the surface where there is abundant light. Oxygen is used up by the animals and microorganisms of the lower depths. The fall and spring overturns replenish oxygen in the depths of the lake. They also distribute nutrients more evenly. While they are using oxygen, microorganisms are breaking down organic molecules into simpler ones. Stirring of the lakes makes these nutrients available to the photosynthetic and other organisms living near the surface. In addition, of course, such pollutants as methyl mercury are distributed by the stirring that occurs in lakes and ponds.

5-10 Lakes and ponds undergo succession

Many lakes and ponds in the temperate zone will not remain bodies of water forever. Freshwater lakes begin life relatively low in nutrients and organic matter, and then organisms tend gradually to increase in numbers and the number of species also increases. Around the edges of the lake, plants may grow half in and half out of the water and accumulate organic matter around their stems and roots. This makes the edge less suitable for them, and they tend to grow inward toward the center of the lake. At the same time, the soil they have begun to build up is suitable for plants of marshes

FIG. 5-18 *a* through *d* show a cross section through a lake undergoing succession. A hollow scraped by a glacier fills with water and, over thousands of years, water plants grow and die forming a bog near its edge. The bog is invaded by trees and shrubs and also extends further into the lake. Eventually the lake is entirely filled in and may become a meadow surrounded by forest.

or bogs. Thus there is often a gradual filling in of the lake from its edges. Such changes in vegetation over time are another example of ecological succession. Succession in a lake is diagrammed in Fig. 5-18.

Succession also involves an increase in nutrients. As more and more organisms live and die in the lake, more and more nutrients are made available by microorganisms, creating a demand for oxygen. Oxygen is eventually restored if it can be replaced by photosynthesis in the upper layers. Succession in freshwater lakes is a perfectly natural ecological process. So also is the aging of lakes—the change from nutrient-deficient to nutrient-rich conditions. As you learned in Sec. 2-10, this enrichment of lakes is called eutrophication.

Impact of people People often add excess organic matter or mineral nutrients to lakes which speeds up eutrophication, that is, speeds up the aging process. Not all the causes of eutrophication are what you would think of as impure, dirty water. Crystal clear, drinkable water may cause eutrophication if it contains dissolved nutrients, but sewage, canning wastes, industrial wastes, detergents, and the fertilizer runoff from farms also speed eutrophication. Before industrialization many of these wastes did not exist; only around a few large cities were sewage wastes a problem. The Great Lakes of the United States and Canada are now undergoing rapid eutrophication, as are large lakes in Russia. Rapid eutrophication of fresh waters is a problem in most of the world.

Lake Erie has shown very clearly the effects of eutrophication (Fig. 5-19). Since about 1910, the process has been greatly speeded up as more people and more industry have moved into the area. Large quantities of minerals flow into the lake, coming from sewage, both human and industrial. Phosphates have increased as a result of the use of household detergents and the use of fertilizers in agricultural areas. The phytoplankton and zooplankton communities of the lake have changed in composition. Large numbers of bacteria are found, including the common bacteria found in the human intestine. Swimming beaches have been closed by health services because of the pollution.

The biological oxygen demand (BOD) of the microorganisms has increased also. There are parts of the lake in which there is *no dissolved oxygen during the summer*. Some aquatic insects have been brought to near extinction. The species of fish that were formerly of commercial value because they were considered good eating have been replaced by less desirable species. Eutrophication that occurs naturally in lakes and ponds takes place so slowly it is often difficult to see any change in 50 to 100 years. The kind of rapid eutrophication that people have produced in about 50 years in Lake Erie is clearly a very serious matter. Unfortunately, it is not easy to reverse eutrophication.

EAST AND WEST ON THE EARTH

The same general sorts of biomes are found in both the Western and Eastern Hemispheres. This is because the same latitudinal gradients occur in both hemispheres. Altitudinal gradients on mountains, ecological succession, and eutrophication occur in all parts of the earth because they are responses to environmental factors occurring everywhere.

5-11 The study of the distribution of the kinds of organisms is called biogeography

If we look at the specific kinds of organisms in the same biome in different parts of the world, an interesting fact comes to light. Although the organisms may at first sight seem to be the same, *in fact they are very different*. Monkeys are found in tropical rain forests in both the New and Old Worlds. Monkeys in South America have long prehensile tails which can be used like a third hand to grasp branches. In Africa, however, the monkeys lack such prehensile tails and are in other ways very different from New World monkeys.

We may think of life in biomes as a sort of ecological play with many different roles or parts. All biomes have photosynthesizers, herbivores, carnivores, and decomposers. *These ecological roles are usually played by different actors in different parts of the world.* If we study areas by the kinds of organisms present instead of their life forms or ecological roles we can identify biogeographic regions. Biogeographic regions Biologists who study **biogeography** have found that land plants and land animals, as well as animals of the sea, are arranged in broad patterns called **biogeographic regions** (Fig. 5-20). The environmental factors in the region and the *past history* of the region determine what kinds of organisms live there. In this book only those regions based upon the study of land animals will

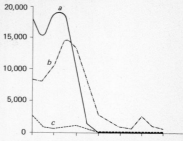

FIG. 5-19 Eutrophication of Lake Erie had resulted in about 70 percent of its bottom waters being depleted of oxygen by 1959. The decline in the catch of commercially valuable fish from Lake Erie is shown in *a*, blue pike; *b*, wall eye; *c*, whitefish. (*After Charles F. Powers and Andrew Robertson, "The Aging Great Lakes," Scientific American,* **215**(5): 95–100, 1966.)

FIG. 5-20 The major biogeographic regions.

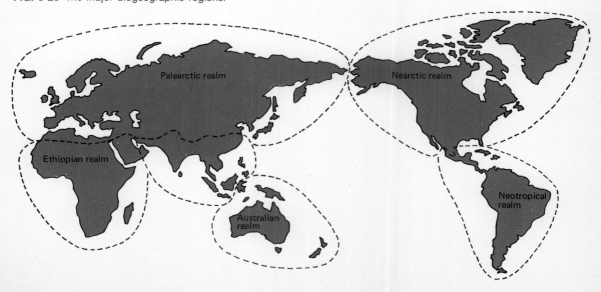

be considered. The animals found in a particular region are called the **fauna** of that region; therefore, it is correct to say that different biogeographic regions have different faunas.

The amount of difference between the faunas of two biogeographic regions depends upon the past history of those regions. Some regions have been isolated from other regions for very long periods of time. Such regions have very distinctive faunas. Other regions have been connected with one another more recently. Animals have had the opportunity to move from one region to the other. Such regions have more in common, and their faunas are less distinct.

5-12 The faunas of the biogeographic regions are very different

Australia The island continent of Australia (Australian region) has been cut off from connection with all other continents for about 200 million years. The placental mammals (see Chap. 6), such as dogs and cats, in which the female carries the young in her body for a considerable period of time before birth, reached the continent only very recently. Most of the mammals found in Australia are pouched animals. Like the opposum of the United States, these animals, called *marsupials,* give birth to tiny young which then develop in a pouch on the underside of the mother.

Australia has many of the biomes that we have discussed: grassland, desert, rain forest. All the ecological roles in these biomes are played by distinctive kinds of organisms in Australia. The opposum of the United States is a relatively unspecialized omnivorous marsupial; it eats many kinds of food and climbs trees. But in Australia, marsupials have become specialized for a great many roles (Figs. 5-21 to 5-23). There are ground-dwelling herbivores (kangaroos), tree-dwelling herbivores (some wallabies and koala bears), and carnivores of various types (wombat, Tasmanian devil). There are also scavenger marsupials, burrowing marsupials, and squirrellike marsupials. Compare these marsupials with the placental mammals shown in Chap. 6.

People came to Australia relatively recently—perhaps 18,000 years ago. The only other nonmarsupial mammals (until Western people arrived 200 years ago) were a kind of dog, the dingo, and bats. People may be perhaps responsible for bringing the dingo to Australia, but the bats presumably flew there. Western people brought their favorite domesticated animals, together with pest species. Horses, cattle, sheep, goats, rabbits, and cats were deliberately introduced; however, rats and mice were accidentally introduced, probably in boxes of cargo from ships. Rats, mice, and rabbits became pests; and in this new region, the dung from cattle is a problem because Australia lacks the appropriate dung beetles to decompose it.

South America The South American biogeographic region (also called the Neotropical region) includes Central America and southern Mexico, in addition to the continent of South America (see Fig. 5-20). This region was not connected by land to other regions for long

FIG. 5-21 Kangaroos are marsupials specialized for grazing. This female has a young kangaroo in her pouch. (*Gabor Csaky, National Audubon Society.*)

periods of time. Its fauna therefore evolved in isolation and grew to be quite distinctive; both pouched and placental mammals diversified into a variety of ecological roles. Judging from fossils, members of the camel family were prominent. Strange, now-extinct, giant relatives of the present-day armadillo have been found as fossils (Fig. 5-24).

From time to time in the past, the surface of the land of South America was connected to North America by a land bridge or the distance between them was such that some animals could get across by swimming or drifting on natural rafts. This permitted occasional exchange between the two regions. For the past few million years, however, the two continents have been continuously connected. Many kinds of animals moved both northward and southward. As the two faunas came together, many species became extinct, possibly because two "actors" were attempting to play the same ecological role, but the southern fauna lost the most species. Although members of the North American biogeographic region invaded the South American region and vice versa, the latter remains a very distinctive region.

North America and Eurasia A land bridge in the region of the Bering Straits has connected North America and Eurasia throughout much

FIG. 5-22 The koala bear is not a true bear, but a marsupial specialized for climbing. Its diet consists exclusively of the leaves of eucalyptus trees. This photo shows a young koala on the back of its mother. (*R. VanNostrand, National Audubon Society.*)

FIG. 5-23 The Tasmanian devil is a marsupial specialized for a carnivorous life. (*R. VanNostrand, National Audubon Society.*)

of their history (Fig. 5-25). During a great deal of this period of connection the climate of Alaska and Siberia was much more moderate than it is today. Extensive plant and animal fossils of species that occur elsewhere in subtropical and temperate zones have been found in Alaska. There is abundant evidence that for many millions of years animals and plants have moved back and forth across this land bridge; indeed, people came to the New World by this route. It is not surprising that the faunas of North America and Eurasia are very similar, but they are usually given the names Nearctic and Palearctic, respectively.

Africa and South Asia As you can see from the map (Fig. 5-20), Africa and South Asia are connected at the eastern end of the Mediterranean Sea. Over the ages there has been much interchange between these regions. The African region (called the Ethiopian region) is cut off from Eurasia to the north by the Sahara Desert. The South Asian region (the Oriental region) is isolated from Eurasia by the Himalayan Mountains. Exchange between the African and South Asian regions is responsible for there being lions, leopards, elephants, and rhinoceroses in both. Some animals of these two regions are shown in Fig. 5-26.

DISPERSAL

In the course of time organisms have increased in numbers and gradually spread over the earth wherever they have found a way. New forms evolved as animals and plants encountered new habitats and changed environmental conditions. The earth's surface has not always been the same as it is now. The extent of the seas has varied. Continents have become connected and disconnected.

FIG. 5-24 *Glyptodon,* a giant armadillo which lived in South America about seven million years ago.

FIG. 5-25 A land bridge (color) connected North America and Eurasia in the region of the present Bering Sea. Today most of the bridge is below sea level, and only the Aleutian Islands are visible evidence of its existence.

FIG. 5-26 The Asian one-horned rhinoceros (*Rhinoceros unicornis*), *a*, differs significantly from the African black rhinoceros (*Diceros bicornis*), *b*. There are two other kinds of rhinoceros, not pictured, alive today; however, there are only two kinds of elephants. The Asiatic elephant (*Elephas maximus*), *c*, and the African elephant (*Loxodonta africana*), *d*, are also quite different.

5-13 Continental drift rearranges the earth's surface

There is abundant evidence that, until some 200 million years ago, all the continents formed one land mass called Pangaea (a word meaning "all earth"). Gradually, throughout millions of years, this land mass broke apart. The continents drifted apart, collided with each other, and eventually came to occupy their present positions. If you look at a map of the continents (Fig. 5-27e), you will see that the continents do fit roughly together, like parts of a badly made jigsaw puzzle. The stages of separation called **continental drift** are shown in Fig. 5-27. Recent studies of certain types of rocks found on now widely separated coasts of continents seem to confirm that continental drift actually did occur. The types and magnetic orientation of the rock particles confirm evidence also presented by fossils of microorganisms. Moreover, it appears that the continents are still drifting.

Earthquakes It is now thought that the earth's crust is composed of huge plates, called **tectonic plates,** on which the continents ride (Fig. 5-28). On the West Coast of the United States, the Pacific plate is sliding northward along the American plate, carrying the coast of southern California, including Los Angeles, and Baja California, with it. The earthquake zone that results is called the San Andreas fault, and it was responsible for the earthquake and fire that devastated San Francisco in 1906. Studies of the earth's crust suggest that these plates slide by or across one another, or one may slide under another, at deep trenches on the ocean floor where there is a great deal of volcanic activity. Where the two plates slide past one another, there may be zones of frequent earthquakes. Geologists predict that in 50 million years, Africa will be split in two and a slice of California will be an island slowly heading north. It is unlikely that our species will be around, however, to witness this result of millions of years of slipping and sliding of the tectonic plates. As long as we are around we *are* apt to be inconvenienced by the earthquakes caused by this process.

Separation of today's five major continents occurred somewhat more recently than 180 million years ago. Although continental drift explains many puzzling features of the distribution of ancient fossils, it does not explain the distribution of more familiar and recent forms. The latter have dispersed over the surface of an earth with continents in positions very similar to those of today.

5-14 Many species of organisms have moved from continent to continent

Barriers to dispersal The movement or migration of organisms is referred to as **dispersal.** If you look at a map of the world, you will see many **barriers to dispersal,** which prevent migration.

For organisms that live in the water, land masses are barriers to dispersal. The same area that is a dispersal route for some organisms

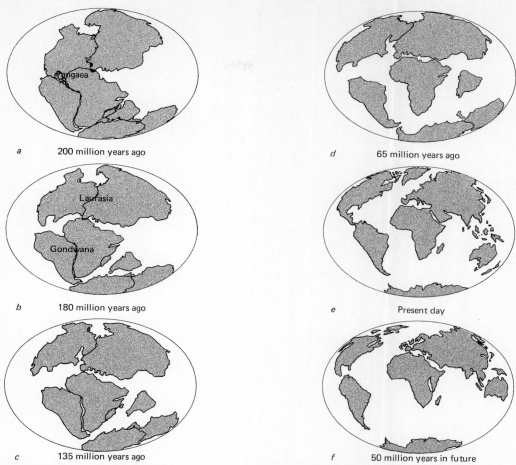

FIG. 5-27 Postulated stages of continental drift. Not all details are agreed upon. (*Data from R. S. Dietz and John C. Holden,* Scientific American, **223**(4): 30–41, 1970.)

may be a barrier to dispersal of others. Central America made it possible for organisms to move between North and South America. Once the Central American connection was established, however, it became a barrier to the migration of fishes and other marine life from the Atlantic to the Pacific Ocean and vice versa. As a result, organisms in the ocean on either side of this land barrier have diverged evolutionarily (Chap. 10) from each other somewhat during the millions of years they have been kept apart.

Barriers as filters Barriers to dispersal need not be quite so obvious, however. In order to cross any dispersal route, organisms must be able to live in the environmental conditions found there. We have seen that not all the animals of South America were able to migrate

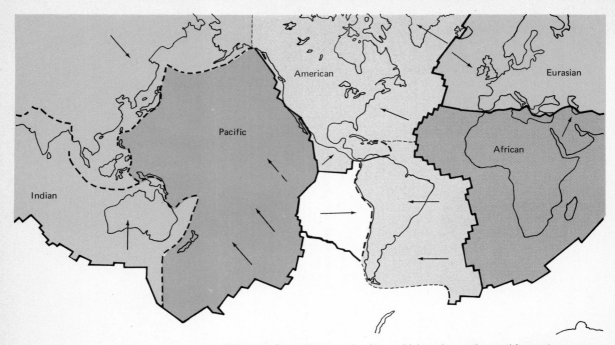

FIG. 5-28 Six major tectonic plates which make up the earth's crust—arrows indicate direction of movement (assuming the African plate were stationary). Small unnamed plates are also shown. Note that plates include both continents and oceans. Heavy lines indicate edges of plates; dashed lines indicate uncertainty about boundaries. (*Data from Edward Bullard,* Scientific American, **221**(3): 66–75, 1969.)

northward. Conditions in Central America were not suitable for them. In the same way, not all the animals of Africa were able to move into South Asia.

In a sense, dispersal routes act as a kind of filter or sieve; that is, only organisms of particular kinds are able to pass through the filter. In order to pass through, an organism must be able to live and *reproduce* along the dispersal route. Long-distance dispersal is not a rapid process; it occurs over thousands of years. Species advance a short distance and reproduce themselves, and their offspring carry on the process. The filtering nature of dispersal routes is primarily responsible for maintaining the distinctiveness of two interconnected biogeographic regions with respect to the kinds of plants and animals found in them.

We have already discussed some routes of dispersal: Central America, the Bering Strait area, and the Middle East. A dispersal route can be rather like a highway connecting two continents; a highway, however, on which traffic moves very slowly and on which not all traffic is accepted. We can imagine checkpoints where passengers lacking the proper papers are turned back. Another kind of dispersal route, however, is a chain of islands; plants and animals can move long distances by island hopping, if the distance between any two islands is not too great.

Dispersal by chance From time to time, things happen in nature by chance which permit organisms to move to a new area. In many parts of the world, the vegetation along the edges of both freshwater and saltwater habitats extends out over the surface of the water. The margins of the vegetation often are floating, and the roots and stems accumulate organic matter in which other organisms can grow. If a portion of this mat of vegetation should break off, it may form a natural raft on which a portion of an *entire community* can be carried.

In somewhat the same way, airborne rafts may be formed. During severe tropical storms, the entire tops of trees may be blown away. All the plants which grew attached to the tree will be carried along, together with the animals and microorganisms associated with them. As with rafting on the water, organisms may travel great distances. Dispersal, however, *does not* result in the organisms finding a new home unless they are able to live and reproduce in their new areas.

PEOPLE AND BIOGEOGRAPHY

We have seen that the biogeographic regions have developed because their faunas were isolated, at least to some degree. This isolation results from the filtering nature of dispersal routes and the very slow rate at which dispersal takes place. The biogeographic regions also have distinctive groups of plants called **floras.** People have had a major impact on the floras and faunas of most of the biogeographic regions.

Each biogeographic region is a large-scale ecosystem in which many different ecological roles are played. Within such an ecosystem, energy is transferred from sunlight to plants to herbivores to carnivores, and minerals are continually cycled between organisms and environment. There is a built-in stability in such a whole region. The numbers of individuals in most species do not suddenly increase or decrease. When people began to take on the role of an ecologically dominant species, other species *were* drastically affected. As people became dominant in various biogeographic regions, energy flow, cycling of minerals, and the stability of whole ecosystems were changed in various complex ways.

5-15 People have had major impacts on the floras and faunas of the biogeographic regions

Extinction of animals As we noted in Sec. 5-4, one of the early ways in which people modified nature was by the use of fire. One of their discoveries was that fire could be used to frighten or even kill large animals. Many animals that were potential sources of food, hides, bones, and teeth were very large indeed (see Fig. 11-32). One does not run up to a mammoth or a giant ground sloth and kill it readily with a primitive spear; however, fire could be used to confuse and terrify a lame animal. It also could be used to drive a herd of animals over a cliff; then the wounded animals could be killed more easily.

At about the time that people were becoming more numerous during the period of great glaciations some 8,000 years ago, a curious thing happened. The fossil record shows that in this period, which is called the Pleistocene, 70 percent of the large land mammals of North America became extinct. In Africa 30 percent of such animals became extinct. Specialized predators of these animals, such as the saber-toothed cats, also became extinct. In fact, the extinction of various large land mammals in different parts of the world coincided with the time of entrance of early people into their biogeographic regions (Fig. 5-29).

Scientists used to believe that the rapid climatic changes which occurred during the Pleistocene glaciations caused these extinctions. However, smaller animals and plants did not die off at the time their large neighbors became extinct. Therefore, many scientists now think that the hunting activities of human beings were causes of these extinctions and that they already had begun to transform the ecosystems of the earth. Because these hunters could not possibly have eaten all the animals they killed, their hunting techniques have been referred to as *Pleistocene overkill*.

Increase of deserts and grasslands People's early use of fire probably also was responsible for the increase in area of grasslands (Sec. 5-4) and possibly deserts. Grasses and some other grassland plants have

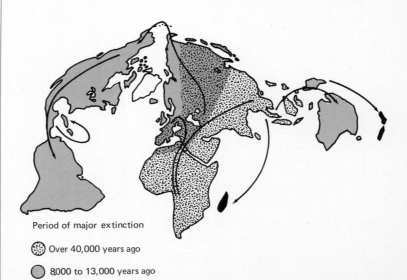

FIG. 5-29 The migrations of human beings (arrows) coincided with periods of extinctions of prehistoric animals. (*Data from Paul S. Martin*, Natural History, **76**(10):32–38, 1967.)

Period of major extinction
- Over 40,000 years ago
- 8,000 to 13,000 years ago
- 400 to 4,000 years ago

evolved underground fire-resistant stems and roots, and they can grow back quickly after a fire. Most trees and shrubs lack fire-resisting specializations; however, some trees, such as sequoia and sugar pine, have evolved resistance to fire. In fact, fire seems to be a necessary event for the sprouting of seeds of some trees. The U.S. Forest Service has stopped controlling fires in forests containing these species. In general, however, human beings can easily turn a forest into a grassland with fire. In many parts of the world, people are still using fire for this purpose; they also burn the nonharvestable portions of crops to prevent the spread of plant diseases. Agricultural burning contributes a great deal of particulate matter to the earth's atmosphere each year.

Deserts, as noted in Sec. 5-5, may be increased in size by people's activities. Some deserts were apparently of human origin; we know that improper agricultural techniques eventually transform fertile land into desert. We have already noted how erosion can remove fertile topsoil; certain types of plowing or irrigation lead to the formation of **hardpan,** an impenetrable layer of soil below the surface. Irrigation can also cause the buildup of salts in the soil (rain, of course, is essentially distilled water and does not carry salts). As salts build up, the soil becomes more and more alkaline and only very highly specialized plants such as salt bush and mesquite can survive. These plants are presently of no use for agriculture; however, some attempts are being made to breed salt-resistant crops. If there is plenty of water, it is sometimes possible to leach out the salt using underground drain tiles. In spite of people's efforts, however, as we have seen, many acres of land are lost to desert every year.

When an environmental problem, such as a shortage of water, is recognized by people, it is often relatively easy to find some technological solution. The ancient engineers of the Middle East must have been very proud of their irrigation systems that brought water to their fields. Unfortunately, a simple technological solution often creates other problems. The Babylonians did not have the necessary ecological knowledge to foresee that their irrigation techniques and other agricultural practices would eventually make deserts of their fertile farmlands.

Impact of structures on dispersal The Isthmus of Panama is a barrier to dispersal for organisms in the Atlantic and Pacific oceans. Isolated from one another, populations along the western and eastern coasts of Central America have become differentiated from one another. They are stable ecosystems with built-in checks and balances.

The present Panama Canal does not significantly affect this situation. It is a freshwater canal with many locks and changes of water level; however, it is too small for many large modern ships to go through. Therefore, a sea-level canal connecting the two oceans has been proposed. This would make it possible for dispersal of organisms to occur in both directions. Unfortunately, we do not have the necessary ecological information to know what the long-range effects of a sea-level canal will be; if organisms of these two isolated ecosystems intermingle, the normal checks and balances of each ecosystem may be upset. For example, predator-prey and parasite-host relationships may be disturbed. Invading species may compete with species already present and cause their extinction. In this, as in many other situations, we do not know *exactly* what will happen in detail. It is not unreasonable, however, to predict that drastic changes will take place, and it is not likely they will be for the better.

People also create barriers to the dispersal of organisms. Dams and breakwaters obviously prevent the movement of aquatic organisms, and similarly, canals on the land may serve as barriers to dispersal of land organisms. It is important to realize that a barrier need not be effective against all organisms; nor need it be a solid structure, such as a dam. *Any* change in the environment may serve as a barrier for some species or many. If a river is polluted by wastes from a chemical plant, the polluted area is a barrier to many organisms, even though it may not be visibly different to us. To a small animal, such as an insect, a country road may be as effective a barrier as a river. To a forest-dwelling bird, a plowed field may be as significant as a mountain range.

People as agents of dispersal   A key feature of dispersal and migration is the *slowness* with which it occurs in nature. For many organisms, the human species has speeded up the rate of dispersal immensely, and this is by no means a purely recent phenomenon. When hides were first used for clothes, and baskets as vessels for storing things, movable habitats were created for many kinds of organisms. Parasites, living in clothes, moved with people as they dispersed. Grain- and fruit-eating insects moved along with people in their baskets and pots of food.

When boats were developed, an important means of dispersal was invented. Early seagoing peoples in the South Pacific carried their food staples—pigs, taro roots, coconuts, etc.—from island to island in their canoes. Eventually, in the course of human history, domesticated plants and animals spread far from the places in which they had originated (Fig. 5-30).

In recent times, people have chosen to move animals from one region to another for various reasons. The starling (Fig. 5-31) was brought to this country in 1890 because it was thought to be an attractive bird and because the introducer wanted to bring all the birds of Shakespeare into the United States. First introduced into Central Park in New York City, it has since spread to the West Coast of the United States (Fig. 5-32). The bird is such a serious pest in so many ways that it is difficult to calculate the damage it has caused. In the cities, it is noisy and its droppings deface the buildings and kill the trees upon which it roosts in the hundreds. In agricultural areas, it does great damage also, not least by consuming and contaminating food for cattle, chickens, and other domesticated animals. Researchers are hard at work trying devices to frighten it away with loud noises, poisons to kill it, and birth control to inhibit its reproduction. Nothing seems to work; in the United States there seem to

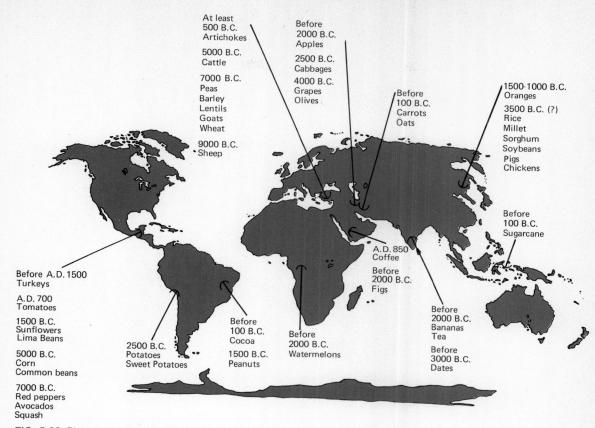

FIG. 5-30 Places of domestication of plants and animals.

be very few checks on the population size of this bird and it is competing seriously with other hole-nesting birds, such as the acorn woodpecker in California.

The introduction of other kinds of birds has not led to such population explosions. Many species of game birds have been introduced into the United States for the "sporting" pleasure of shooting them once they have become established, but most have not flourished. Only two of the introductions have been notably successful: the Hungarian partridge and the Oriental ring-necked pheasant. These species have found a way of life in a new biogeographic region.

Introduction of a new organism may be inadvertent and disastrous. For centuries different species of chestnut trees have been found in eastern United States and in China; each had its associated species of parasites to which it had become adapted in the course of time. Specimens of the Chinese chestnut were brought to this country; growing on these young trees was a fungus called the chestnut blight which causes little serious damage to the Chinese chestnut;

FIG. 5-31 Starlings in America occur in huge flocks which do heavy damage by eating seeds and by roosting on trees and buildings. This flock was photographed near Bloomington, Indiana. (*Dave Repp, National Audubon Society.*)

however, it proved to be a fatal disease for the American chestnut tree. First introduced by accident in the early 1900s, the chestnut blight had killed virtually every American chestnut by 1950. Resistant chestnut trees are now available which were produced by hybridizing American trees with resistant Asian trees.

Sometimes a second purposeful introduction can be used to combat an introduced pest. Because of its edible fruits and its possibility as cattle food, a cactus called prickly pear was transplanted from the Western Hemisphere to Australia. There it became an enormous pest; it grew in huge stands covering millions of acres and prevented cattle grazing or other use of the land. Biologists discovered a small moth in South America whose caterpillars ate the cactus and they deliberately moved some to Australia. The caterpillars happily munched on the cactus and reduced it to the scattered, harmless clumps it exists as today. The moth was an effective **biological control** of the cactus. Normally biological control does not exterminate the unwanted species; it and its controlling species simply live in a dynamic balance. The rabbits that were released in Australia and became pests overgrazed huge areas of rangeland; eventually a South American virus that attacks rabbits was released in Australia and effectively controlled the rabbits.

FIG. 5-32 The spread of the starling (*Sturnus vulgaris*) across the United States from the place of its release in New York in 1890. The dots represent the first publication of a sighting in an area. The lines indicate approximate *front* that birds had reached at the indicated years. (*Data from R. W. DeHaven and P. J. DeHaven, "A contribution toward a bibliography on the starling (Sturnus vulgaris)," unpublished report in files of Denver Wildlife Research Center, U.S. Fish and Wildlife Service, 1972.*)

Unfortunately, we have not found a good biological control for starlings or for imported grasses, such as cheat grass and wild oats, which have replaced the native perennial grasses on much of our Western land. There are many other serious imported pests for which we have not found biological control. Very little money has gone into biological control since it is not profitable for companies to sell a product that reproduces itself and spreads across the land, solving the problem as it goes. Insecticides and herbicides are profitable, however, since they have to be applied again and again in larger and larger quantities as the pests become resistant to them and as they kill off any local predators and parasites which may attack the pest. Some farmers have banded together to pay for biological control, and the city of Berkeley, California, recently funded the importation of a parasite which controlled insects which were harming its shade trees.

There are many other examples of introductions of species, some with good consequences, some with bad consequences. We generally consider the introduction of horses and cattle into the Western Hemisphere beneficial. Some introductions thought at first to be beneficial turned sour; carp were introduced to the United States from Germany as a food fish. They turned out *not* to be popular with fishers and to interfere with populations of sportfish and waterfowl. The mongoose was introduced from India to Jamaica to control rats and ended up eating almost everything, including chickens, eggs, lambs, kittens, and puppies. Large African grazing mammals such as the kudu have recently been introduced to the United States and Mexico; only time will tell whether they turn out to be beneficial in their new environment. Organisms are often thoughtlessly introduced by people who bring them back from abroad or who release pets they are tired of. The giant African land snail has been introduced into Hawaii, Ceylon, and other Pacific areas, where it devours crops with abandon. The walking catfish from Asia escaped from a pet dealer in Florida; it can move across dry land when its ponds dry up, is reported to be dangerous to dogs and cats, and has invaded swimming pools. The gypsy moth was accidentally released in Massachusetts in 1867; since, it has ravaged vegetation and has been a major cause of DDT use in the United States.

Honeybees are native to Europe, Asia Minor, and Africa, and were probably first brought to the New World in 1622, by English settlers, to Virginia. An African strain of honeybees was brought to Brazil in 1956. These bees produce more honey, but they are considerably more aggressive than European and American bees. Fifty stings will kill a person, and the bees chase their victims ten times farther than ordinary bees. It was hoped that, by hybridization, less-aggressive strains producing more honey could be developed. In 1957, two dozen bees escaped from experimental hives in Brazil and interbred with local bees to form very aggressive bees producing 50 percent more honey. Since the escape, the hybrid bees have spread widely covering most of east, central, and southern South America. They continue to spread and, it is estimated, could reach the United States in 10 to 18 years. Barriers to their dispersal may exist, for example, deserts in Mexico. However, they remain a potential threat to people and to the economics of honey production.

Before an animal or plant is introduced into a new biogeographic region, a very great deal should be known about the total ecology of the situation; otherwise, serious problems may result. The United States now has strict regulations about the introduction of any animal or plant and severe penalties for violations; nevertheless, newspapers and magazines often have stories of new organisms which have been brought in anyway.

QUESTIONS FOR REVIEW

1 What causes ocean currents?
2 What is the direct source of energy and nutrients for oceanic organisms below the photosynthetic zone? What is the indirect source? Why are most organisms which live at great oceanic depths *prisoners* of their deep habitats?
3 What do these terms mean: flora, laterization, dispersal route, fauna, humus, Pleistocene overkill, taiga, storied vegetation, salinity, estuarine?
4 The following types of vegetation are found on the slopes of a tall mountain. Can you arrange them in order, from highest to lowest, along the altitudinal gradient: tropical rain-forest trees and shrubs, meadow grasses, coniferous trees, lichens, deciduous trees?
5 Which biome is: the oldest? usually nearest the equator? the most diverse? the wettest? the most uniform in terms of rainfall, temperature, and sunlight? the most stable? the least diverse? dominated by conifers?

Which biome has: the richest "standing crop" (the most plant tissue)? the fastest plant growth? the greatest temperature extremes? the longest growing season? the shortest growing season?
6 The crystalline, or solid, form of water—ice—is less dense (heavy) than its liquid form. Its liquid form is most dense at 4°C, which is somewhat above its freezing point. What do you suppose would be the consequences to the creatures which live and breed in freshwater lakes and ponds, and to those terrestrial and aerial creatures that feed on them, if water did *not* have these two peculiar properties?
7 What type of marine habitat may be called *the ocean's nursery*? Why? How are these *nurseries* being damaged and even destroyed in many areas?
8 Deciduous-forest regions can be converted into productive farmlands, but rain-forest regions cannot be. Why?
9 Why are sheep and cattle ranchers only making things worse for themselves when they try to exterminate the burrowing and predator animals on their property?
10 "The tundra is the least diverse of all biomes." Even without being told anything else about the tundra, what can you say about its ecological stability?
11 What is continental drift, and what effect has it had on the biosphere's flora and fauna? Is it still having this effect?
12 What do you suppose causes continental drift? (Hint: What does the text say is going on in the deep-ocean trenches where the earth's tectonic plates are converging?)
13 "Dispersal routes act as a kind of filter or sieve." Can you explain this statement and give some examples?

READINGS

Alexander, T.: "A Revolution Called Plate Tectonics Has Given Us a Whole New Earth," *Smithsonian*, **5**(10):30–40, 1975.
⸻.: "Plate Tectonics Has a Lot to Tell Us About the Earth as It Is—and as It Will Be." *Smithsonian*, **5**(11):38–47, 1975.

Denison, W. C.: "Life in Tall Trees," *Scientific American,* **228**(6):74–80, offprint 1274, 1973.

Ehrlich, P. R., and R. W. Holm: "Evolution," *Biocore,* unit XXII, McGraw-Hill, New York, 1974.

Elton, C. S.: *The Ecology of Invasions by Animals and Plants,* Methuen, London, 1958.

Farb, P., et al.: *Ecology,* Life Nature Library, Time Inc., New York, 1963.

Findley, R.: *Great American Deserts,* National Geographic Society, Washington, D.C., 1972.

Gregg, M. C.: "The Microstructure of the Ocean," *Scientific American,* **228**(2):64–77, offprint 905, 1973.

Hylander, C. J.: *Wildlife Communities from the Tundra to the Tropics in North America,* Houghton Mifflin, Boston, 1966.

Laycock, G.: *The Alien Animals—The Story of Imported Wildlife,* Natural History Press, Doubleday, Garden City, N.Y., 1966.

Martin, P. S.: "Pleistocene Overkill," *Natural History,* **76**(10):32–38, 1967.

McNeil, M.: "Lateritic Soils," *Scientific American,* **211**(5):96–102, offprint 870, 1964.

Richards, P. W.: "The Tropical Rain Forest," *Scientific American,* **229**(6):58–67, offprint 1286, 1973.

Teal, J., and M. Teal: *Life and Death of the Salt Marsh,* Little-Brown, Boston, 1969.

Thomas, W. L. (ed.): *Man's Role in Changing the Face of the Earth,* University of Chicago Press, Chicago, 1956.

Wooster, W. S.: "The Ocean and Man," *Scientific American,* **221**(3):218–234, offprint 888, 1969.

THE KINDS OF LIVING THINGS

SOME LEARNING OBJECTIVES

After you have studied this chapter, you should be able to

1 Demonstrate your knowledge of some basic aspects of cell biology by
  a Naming the two basic kinds of cells and stating how they differ.
  b Explaining the major differences between plant cells and animal cells, and stating whether each of these two types is prokaryotic or eukaryotic.
  c Explaining what cell differentiation and cell coordination mean, and why each is important to multicellular organisms.
  d Defining tissue, organ, and organ system, and giving an example of each.

2 Show what you have learned about the nature and purpose of the classification of organisms by
  a Explaining why classification systems are so important in the study of organisms.
  b Naming the seven main hierarchical levels of organization used in the formal classification of plants and animals, and listing them in order from narrowest (most specific) to broadest.
  c Stating the criterion (basis for deciding) which biologists generally use to determine whether two organisms should be assigned to the same or to different species.
  d Giving the full, formal classification of a familiar animal—yourself.

3 Demonstrate your knowledge of the four-kingdom classification by
  a Giving a reason why most biologists no longer use the older, simpler two-kingdom classification of *plants* and *animals*.
  b Naming the four kingdoms, the basic cell type (prokaryotic or eukaryotic) of each, and the body type(s) (unicellular and/or multicellular) found in each.
  c Saying which of the four kingdoms each of these kinds of organisms is assigned to: plants, bees, fungi, earthworms, blue-green algae, all other algae, animals, protozoa, vertebrates, clams.

4 Show some of what you have learned about the kingdom Metazoa by doing the following:
  a Name the phylum that all vertebrates belong to, and name the five major groups (classes) of vertebrates.
  b Name at least three characteristics shared by all mammals.
  c Explain the main difference between placental and nonplacental mammals, and name at least two examples of each type.
  d Say which major vertebrate group (class) is most closely related to mammals, and name that group's four major subgroups.
  e Name the phylum with the most species.
  f Name at least four different major kinds (subgroups) of arthropods, and state two characteristics all arthropods have in common.
  g Name one structural feature shared by all mollusks and one structural feature common to most but not all of them; name at least three kinds (subgroups) of mollusks—including one that is carnivorous, one that is a filter feeder, and one whose subtypes are mostly herbivores.
  h Name a kind of invertebrate which is a disease-causing parasite in humans, and describe its effects on its human host.

5 Demonstrate your knowledge of the kingdom Metaphyta by
  a Explaining how reproduction in angiosperms differs from reproduction in conifers.
  b Naming two plant structures which are evolutionary adaptations to life on land and stating the function of each.

   c Stating two important functions performed by each of these plant structures: leaves, roots, stems.
   d Naming the two major kinds of plant vascular tissues and stating the major roles of each.
   e Naming the type of nonvascular land plant discussed in the text, stating one reason why it is able to survive in severe habitats like tundras and deserts, and stating one way in which its reproductive process is similar to that of a fern.
6 Demonstrate your knowledge of the kingdom Protista by
   a Giving the text's definition of Protista and naming the three most common types.
   b Naming two common aquatic types of Protista and stating the critically important role of one of these types in aquatic food chains.
   c Naming one kind of unicellular and two kinds of multicellular fungi, stating two ways in which fungi are used by people, and describing one direct and one indirect way in which fungi are injurious to people.
   d Stating how most protozoa obtain their food and naming a major human disease caused by a parasitic protozoan and the alternate host of this parasite.
7 Demonstrate your knowledge of the kingdom Monera by
   a Giving the common names of its two major subgroups (phyla).
   b Stating how the blue-green algae obtain their food and why some of them play an important role in agriculture.
   c Saying where in the biosphere bacteria may be found, and naming two ways in which they obtain their food.
   d Describing three useful (beneficial) ecological roles played by decomposer bacteria.
   e Listing at least four serious human diseases caused by bacteria and stating how each is transmitted.
8. Show some of what you have learned about viruses by
   a Explaining why all viruses can grow and reproduce only within living cells.
   b Naming at least four human diseases caused by viruses, including three whose modes of transmission differ.
   c Naming at least one way in which viruses are useful to people.

Nobody knows for sure how many species of organisms there are alive today, but biologists estimate there are between 2 million and 10 million. It is very probable that we will never know how many species of organisms there are. There are two reasons for this; first, people will probably destroy many habitats such as tropical rain forests before biologists have time to classify the abundance of species there, and, second, even biologists cannot agree on how to define a species. Nevertheless, it is important for you to know that you share this planet with a vast number of other species all evolving together in the biosphere.

   Our one species, *Homo sapiens,* is dominant today in the impact it has on other species and on the biosphere as a whole. For the last 200 years, people have brought to extinction, on the average, about one major species of large animal per year; yet, we have not succeeded in eliminating one single pest or disease species except in local areas. About 5 percent of all photosynthesis now occurs in our agricultural ecosystems. For one species out of 2 million or more to corner 5 percent of the basic food-making process does indeed indicate its dominance.

   Of the various characteristics used to classify organisms into broad groups, the kind of cells they have and how these cells are arranged at critical life periods are the most important. Therefore, we begin this chapter with a brief description of these structural units of organisms and the variety of arrangements they take in living things.

THE KINDS OF LIVING THINGS

CELLS, THE BASIC UNITS OF LIFE

Cells are the tiny, immensely complex structural units of living things. Most cells are between 1 and 20 micrometers in size (a micrometer is one-millionth of a meter or about 1/25,000 of an inch). All organisms are composed of cells; they may be **unicellular,** made up of just one cell, or they may be **multicellular,** made up of many cells put together in complex ways. Viruses are not cells; they can only reproduce inside living cells, using the energy and materials of their host cell to make more viruses. They seem to represent the borderline between living and nonliving things, and there does not seem to be much point in arguments as to whether they are organisms or just very elaborate molecules; even though they are not cells, they interrelate with cells to play important ecological roles.

6-1 There are two basic kinds of cells

Cells are either **prokaryotic** or **eukaryotic.** Under an electron microscope, many cells from your own body would look something like the cell shown in Fig. 6-1. Cells like this are called eukaryotic cells; they all have a well-defined central body called a **nucleus** (except when they are dividing to make two new cells) and many other **organelles.** Chloroplasts, which were mentioned when we studied photosynthesis in Chap. 2, are good examples of organelles, structures inside cells that are specialized to do different jobs. Most familiar organisms, both plant and animal, have eukaryotic cells. Cells from a bacterial culture, on the other hand, would look like the cell in Fig. 6-2 under an electron microscope; cells like this are called prokaryotic cells. Prokaryotic cells lack a nucleus and do *not* have the same kind of well-defined organelles. Certain algae, the blue-green algae, are also prokaryotic.

There is a theory, **serial symbiosis** (*symbiosis* means "living together"), that eukaryotic cells got at least some of their organelles by engulfing prokaryotic cells, in early evolution. A photosynthetic prokaryote which could live inside another cell could serve the function of a chloroplast.

The cells of both animals and plants are eukaryotic and, except for the presence of chloroplasts in plant cells, have basically the same kinds of organelles. Both plant and animal cells are surrounded by a cell membrane, and all material entering or leaving a cell must pass through this membrane. Most plant cells, however, have another structure, a cell wall outside the cell membrane (Fig. 6-3). This cell wall is rigid and full of tiny openings called *pits* through which

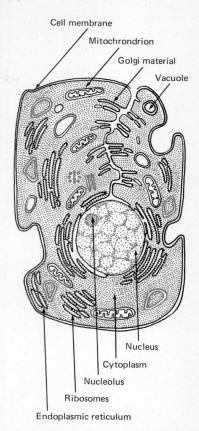

FIG. 6-1 A typical eukaryotic cell from an animal showing the various organelles. Organelles are discussed further in Chaps. 12 and 13.

FIG. 6-2 A cell of the bacterium *Escherischia coli* as seen with the electron microscope in a thin section. Bacterial cells are prokaryotic. Lighter areas are regions of genetic material. Magnification about ×30,000. (*E. Kellenberger.*)

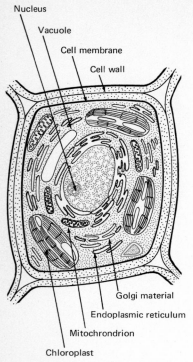

FIG. 6-3 A typical eukaryotic cell from a plant.

materials can pass. In a group of plant cells, the walls of adjacent cells are held together by an intercellular cement.

6-2 Organisms may be either unicellular or multicellular

Unicellular organisms Unicellular organisms may be free-living or parasitic. Most unicellular organisms live in water or, if they are parasitic, in the body fluids of other organisms. Many of the phytoplankton and zooplankton which we have discussed are unicellular organisms; many soil organisms, decomposers, and disease organisms are unicellular. Unicellular organisms may remain permanently in one place or they may have some means of moving about. Some unicellular organisms, such as amoebas (Fig. 6-4), move by changing their body shape and flowing in one direction or another. When they eat, they flow around a food particle and engulf it. Unicellular organisms may also move by means of threadlike extensions of their **cytoplasm** (cytoplasm consists of all the material inside a cell's membrane except its nucleus). These threadlike extensions may be of two kinds: (1) **flagella** (singular, *flagellum*), which are few and long in comparison with the size of the cell, or (2) **cilia** (singular, *cilium*) which are many and short. Organisms with flagella use these long movable cytoplasmic extensions as propellers to move them along (Fig. 6-5). Organisms with cilia may have their entire cell covered with cilia (Fig. 6-6) which whip them along at a fast pace by wavelike movements.

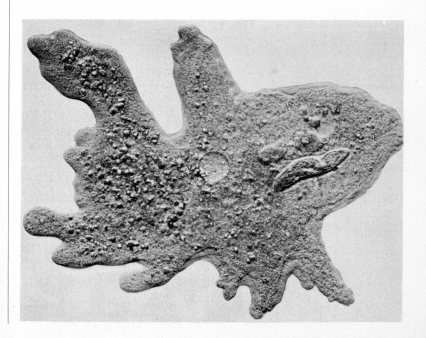

FIG. 6-4 An amoeba is a unicellular eukaryote which moves by changing the shape of its body. This amoeba has taken in two paramecia (see Fig. 6-6) as food (*right*). (*Carolina Biological Supply Co.*)

THE KINDS OF LIVING THINGS | 153

Colonies There is no sharp dividing line between unicellular organisms and multicellular organisms. Some of the so-called unicellular organisms actually live in colonies, such as the alga shown in Fig. 6-7. These colonies are composed of thousands of cells. Within a colony such as this, there is coordination of the swimming of cells and division of labor since some cells are destined to reproduce while others are not. In such cases, it is difficult to decide whether we are dealing with a colony of single-celled individuals or a multicellular individual. It is not worth arguing about; there simply is no sharp dividing line between some colonies of cells and a multicellular individual.

Furthermore, some cells of multicellular organisms behave very much like unicellular organisms. The white blood cells in your body may behave very much like a free-living amoeba when they engulf and remove bacteria and foreign materials. Human sperm cells are propelled by flagella, and many cells in the bodies of multicellular organisms have cilia. Your respiratory tract is lined with ciliated cells which use the movement of their cilia to move particles out of your lungs.

Multicellular organisms Most of the familiar organisms we see around us are made up of billions of cells. They are extremely complex, and it is obvious that a dog or a tree, for example, is not just a colony of cells. Within such a multicellular organism, individual cells become specialized for a variety of different functions during development. This process is called **differentiation.** Coordination of the functioning of millions of cells must be very precise if the organism is to survive. Basically, coordination is achieved by cells with similar functions being grouped into **tissues.** Various kinds of liver cells are shown in Fig. 6-8. The many functions of the liver cannot be carried out by only one kind of tissue. For example, blood must be carried to and from the liver. In this figure you can also see ducts that carry the digestive fluid, bile, from the liver to the gall bladder, where it is stored. Several tissues that function together make up an **organ.** The liver is an example of an organ in your body. So is a salivary gland, which secretes saliva containing substances necessary for digestion of food. Saliva is secreted by some of the cells that make up the glandular tissue. Other tissues, such as blood vessels and nerves, are also part of a salivary gland.

The liver is part of the digestive system. So is a salivary gland. The digestive system also includes many other organs: the stomach and the intestines, to mention only two. You can see that carrying on one major function—in this case digestion—may require more than one organ. Animals, such as human beings, usually have organs that work together as **organ systems** (Fig. 6-9).

CLASSIFICATION OF ORGANISMS

There is an immense diversity of single-celled and multicellular organisms. In order to discuss and understand this great diversity, scien-

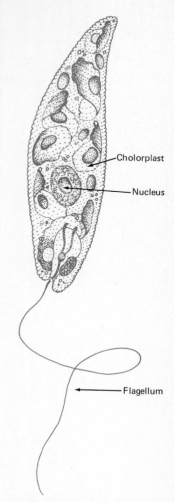

FIG. 6-5 *Euglena,* a flagellate, free-living, unicellular organism.

tists have developed systems of **classification** of organisms. In fact, all peoples classify organisms; that is, they place them into categories for convenience in talking about them. Classification seems to be a necessary part of dealing with the world around us. Most classifications are based upon **utility.** The kind of classification depends upon the use to which it is put. Farmers may classify insects as harmful or useful; a hunter may classify plants as poisonous or edible.

6-3 Informal systems of classification are widely used

Classification based on similarity People tend to lump together organisms that look alike or have other physical properties in common. Thus when you say that roses and pine trees are plants and that butterflies and cockroaches are animals, you are using a classification based on **similarity.** A rose has more features in common with a pine tree than it does with any animal. At another level, just to call an individual organism a rose, a pine tree, a butterfly, or a cockroach is to classify it by similarity. Since there are, for example, many different kinds of plants, any individual rose plant is classified into

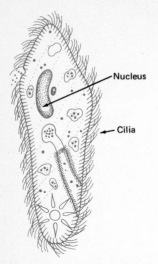

FIG. 6-6 *Paramecium,* a ciliate organism.

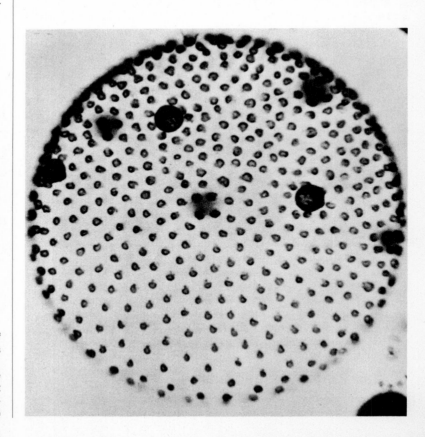

FIG. 6-7 *Volvox* is a colony formed of thousands of single-celled individuals arranged in the form of a hollow sphere. Offspring colonies can be seen inside the larger sphere. Magnification about ×100. (*Ward's Natural Science Establishment, Inc.*)

a subgroup of plants as soon as it is called a rose. It has been given the same name as a group of similar organisms and thus has been classified by similarity.

Classification based on ecological relationships Grasshoppers, parrots, and cows are herbivores. Is "herbivore" a category based on similarity? Praying mantises, hawks, and lions are carnivores. But hawks are not structurally more similar to lions than they are to parrots (Fig. 6-10). Classification based on ecological relationships often differs from that based on similarity of structure. The most commonly used ecological classifications are those of communities. When you talk about a "meadow" or a "forest" or a "desert," you are using an ecological classification of communities of organisms.

6-4 The formal system of classification

Ecological relationships are often difficult to see, and may require a great deal of study of organisms under natural conditions to discover. In contrast, overall similarities and differences in structure are apparent in dead museum specimens and even in pictures. It is not surprising, then, that biologists have used such similarities as the foundation of their system of classification.

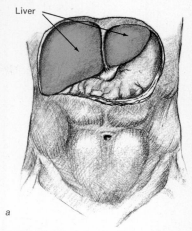

FIG. 6-8 (*a*) Position of liver in abdomen. (*b*) Section of a portion of liver tissue showing liver cells, blood vessels, and bile ducts.

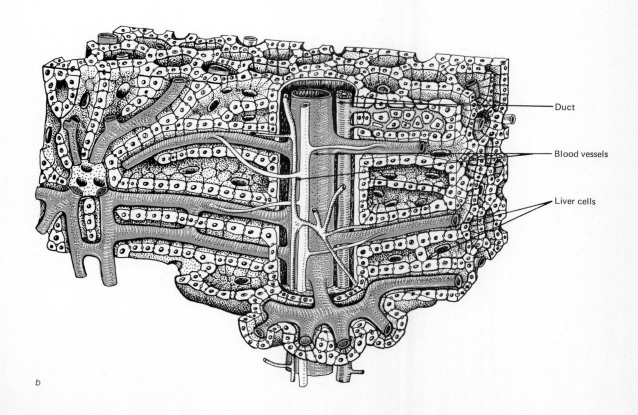

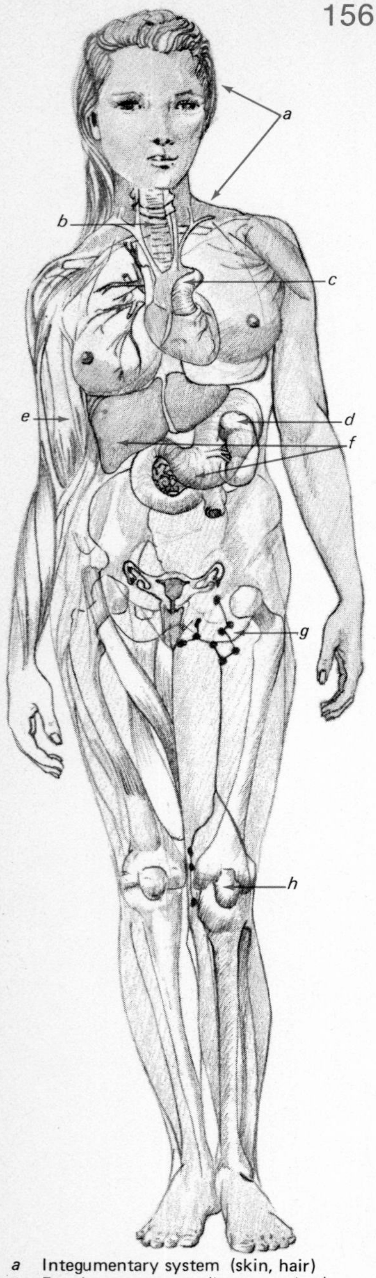

CHAPTER SIX

FIG. 6-9 The major organ systems and some of their organs (only parts of some systems are shown).

a Integumentary system (skin, hair)
b Respiratory system (lungs, trachea)
c Cardiovascular system
        (heart, blood vessels, blood)
d Excretory system   (kidneys, bladder)
e Muscular system (muscles, tendons)
f Digestive system (stomach, liver, etc.)
g Nervous system
        (brain, nerves, sense organs)
h Skeletal system (bones)

Biologists whose primary concern is classifying organisms will ordinarily specialize in studying some particular group. They then learn all they can about that group so that they can base their estimates of similarity on the most detailed and extensive possible comparison.

Biologists use a classification based on *many characteristics* of organisms because such a classification forms a convenient basis for communication. As you may know, although whales are more or less fish-shaped, they are not classified as fishes, but as mammals. We are mammals also. That means whales are more similar to us than they are to fishes. Look at Fig. 6-11. Notice how it is *only* in shape that whale and fish are similar.

Hierarchy An army has a definite arrangement of units and ranks (Fig. 6-12). Each soldier belongs to a *squad*. Each squad is part of a *platoon*. Each platoon is part of a *company,* and several companies make up a *regiment*. Regiments make up *divisions,* and divisions

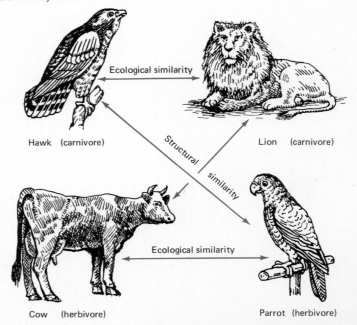

FIG. 6-10 How an organism is classified depends upon the point of view of the observers and the purpose of the classification. A hawk can be classified as a bird or as a carnivore. If we are interested in its structure, it is much more like a parrot than a lion. But, if we are interested in the hawk's way of life, we would classify it as a carnivore.

Hawk (carnivore)   Lion (carnivore)

Cow (herbivore)   Parrot (herbivore)

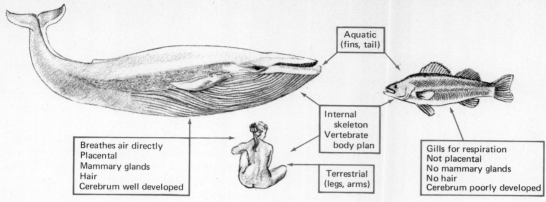

FIG. 6-11 Whales are more similar to human beings than they are to fish.

FIG. 6-12 The hierarchical organization of military units.

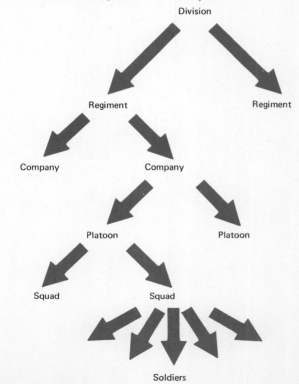

make up *armies*. This kind of organization found in an army is called a **hierarchy.** Each soldier belongs to one squad, one platoon, one company, one regiment, one division, and one army. In a hierarchy there are different levels of category (platoon, regiment, etc.), and any item classified (in this case, a soldier) can belong to only one unit at that level.

The formal biological system of classification is also hierarchic. Every kind of organism, called a **species,** belongs to a **genus.** Each genus belongs to a **family.** Each family belongs to an **order,** and each order to a **class.** Every class is in a **phylum,** and every phylum is in a **kingdom.** Each organism belongs to one genus, one family, one order, one class, one phylum, and one kingdom (Fig. 6-13).

Nomenclature Each species of organism is given a two-part name, and the system of giving names is called nomenclature. The first part is the name of the *genus* (plural, *genera*) to which the species belongs; the second part distinguishes it from other members of the same genus. Names of species are always two words, the first of which is capitalized, the second not. They are always printed in italics or

FIG. 6-13 The hierarchical organization of animal classification.

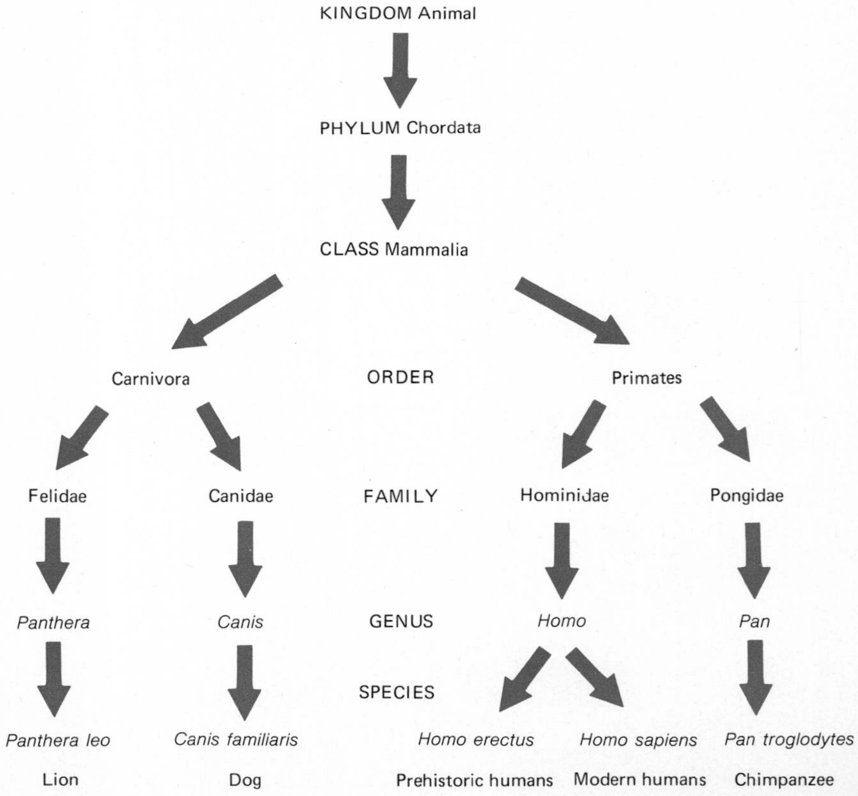

```
                        KINGDOM Animal
                              ↓
                       PHYLUM Chordata
                              ↓
                       CLASS Mammalia
                        ↙              ↘
                  Carnivora    ORDER    Primates
                   ↙    ↘                ↙    ↘
              Felidae  Canidae  FAMILY  Hominidae  Pongidae
                 ↓       ↓                 ↓         ↓
              Panthera  Canis   GENUS    Homo       Pan
                 ↓       ↓              ↙    ↘       ↓
                                SPECIES
            Panthera leo  Canis familiaris  Homo erectus  Homo sapiens  Pan troglodytes
               Lion           Dog          Prehistoric humans  Modern humans  Chimpanzee
```

underlined in typing and handwriting. Thus a dog has the zoological name *Canis familiaris,* which in Latin means, roughly, "familiar dog." Biologists use a Latin form in making up scientific names because when the system was started, Latin was the language of all educated Western Europeans. It is still convenient that a scientific report about dogs, in any language, written anywhere in the world, will refer to them as *Canis.* If the second part of the name is not known, *sp.* is substituted, for example, *Canis* sp.

There are other "dogs" besides *Canis familiaris. Canis latrans* is the coyote (Fig. 6-14); *Canis lupus* is the wolf (Fig. 6-15). The common horse is *Equus caballus.* There are several species of wild "horses," such as *Equus przewalskii* (Mongolian wild horse), *Equus burchelli* (zebra), and *Equus asinus* (wild ass) (Fig. 6-16). There are also many species in the oak genus *Quercus.* Figure 6-17 shows the leaves and acorns of several of these species.

The scientific name of modern people is *Homo sapiens.* There is only one living species of the genus *Homo.* There is an extinct species, however, *Homo erectus* (Fig. 6-18). As far as we know, there are no other genera of people alive today either. No doubt you have read accounts in newspapers and magazines of huge humanlike creatures purported to be roaming around remote parts of Tibet and North America. Satisfactory scientific proof of their existence does not exist because there are no specimens of such creatures anywhere and no clear photographs exist.

The genus *Canis,* along with a number of other genera, such as *Urocyon* (gray foxes) and *Lycaon* (cape hunting dogs), make up the family Canidae, the dog family. The genus *Felis* contains the house cat *Felis catus.* Some of the big cats, like cougars, are in the genus *Felis,* and some (lion, tiger, panther, leopard) are in the genus *Panthera. Felis* and *Panthera* and several other genera make up the family Felidae, the cat family. *Homo* and at least one other genus of people, the extinct *Australopithecus,* make up the family Hominidae. Family names of animals are made by taking the first part of the name (sometimes with slight changes) of one of the genera in the family and adding the ending *-idae.*

The oaks, *Quercus,* are in the family Fagaceae. The family is named for the genus *Fagus,* the genus of beech trees. Plant family names generally end in *-aceae.*

Complete classifications of several familiar organisms are shown in Table 6-1. As you can see, there may be no particular rules for the making of names for categories above the level of family.

Definition of species So far our definition of "different kind" has been rather vague. After all, men and women are "different kinds." Should they be considered different species? Red, white, and pink roses are "different kinds." Are they different species?

The answer is that biologists usually consider kinds to be sufficiently distinct to call them different species if they are thought unlikely to *interbreed* (breed with each other) under natural conditions. Thus horses and asses, living in the wild, do not ordinarily

FIG. 6-14 A coyote. Coyotes are smaller than wolves and usually do not form social groups larger than a mated pair. The coyote is the wolf's closest wild relative, and the original ranges of the two overlapped in North America. Unlike wolves, coyotes feed mostly on small mammals and only occasionally on large mammals such as deer. In some places coyotes have successfully mated with dogs forming *coydogs*.

FIG. 6-15 A wolf. Wolves have wider nose pads than coyotes and are larger and heavier. When running, a wolf usually carries its tail high, while a coyote usually carries its tail below the level of its back. Wolves are believed to mate for life and form complex social groups. Originally, wolves inhabited all Northern Hemisphere biomes, except tropical rain forest and desert. The range of wolves is presently much reduced because people have shot and poisoned them.

interbreed; however, in captivity they can be interbred to produce hybrids called mules. But because they do not interbreed in nature, they are considered separate species. All human populations may interbreed when they come in contact with each other; thus, tall people and short people, or Europeans and Eskimos, may be different kinds, but they are not different species. Similarly, all the different colored garden roses freely interbreed—they all belong to the same species.

The four kingdoms   If you were to play "twenty questions" with a biologist, he or she could quite honestly answer "no" to all three questions: "Animal?" "Vegetable?" "Mineral?" This biologist might have in mind a bacterium, which in the biological way of classifying things fits in the kingdom Monera. In fact, most biologists no longer classify living things into just two kingdoms, plants and animals. Some classify them into four kingdoms, some into five or more kingdoms. We prefer the four-kingdom classification, which is shown in Table 6-2. You will notice that organisms with prokaryotic cells have a kingdom all their own, the Monera. Thus the biologist, or anybody else, no longer has to worry about whether bacteria are plants or animals. The kingdom Protista consists of organisms with eukaryotic cells, which may be either unicellular or multicellular as long as their reproductive structures are single-celled. The kingdom Metaphyta is made up of all those organisms commonly called plants, and the kingdom Metazoa includes all those organisms commonly called animals. The Metaphyta and the Metazoa are all multicellular organisms with multicellular reproductive structures; the difference between them is that the vast majority of the Metaphyta can perform photosynthesis while the Metazoa must eat other organisms in order to obtain energy for their life processes.

Table 6-3 is a simplified version of a classification of all living things so that you can get some idea of the great diversity of living things in the world. We do not expect you to memorize lists of organisms, especially ones you will probably never see in your entire life. In fact, most biologists have never seen a live brachiopod or chaetognath. In the rest of this chapter, we discuss mainly those organisms that are familiar or important to you. You should not forget, however, that we are only one of over 2 million species of organisms, most of which are very different from us.

FIG. 6-16 Some members of the genus *Equus:* (a) *Equus burchelli,* (b) *Equus asinus,* (c) mule, the hybrid offspring of a male ass and a female horse.

SOME MAJOR KINDS OF ANIMALS

Animals originated in the sea over 500 million years ago and did not move onto land away from the ocean's edge until there were land plants for them to feed on. There are still more *major kinds* (phyla and subphyla) of animals in the sea than on land, but probably 95 percent of the millions of *species* of animals live on the land today. In Chap. 10, we discuss the evolutionary processes that led to the development of land animals.

FIG. 6-17 Leaves and acorns of some species of the genus *Quercus*. These oaks are all native to California. (*a*) *Quercus agrifolia*, (*b*) *Q. wislizenii*, (*c*) *Q. chrysolepis*, (*d*) *Q. engelmannii*, (*e*) *Q. douglasii*.

6-5 The most familiar animals are vertebrates

Vertebrates, animals with backbones, are what most people mean when they say "animal." Vertebrates are the major group of large land animals as well as the dominant animals of the sea. There are five major groups of vertebrates. These are the mammals, birds, reptiles, amphibians, and fishes (Fig. 6-19).

Mammals **Mammals** are vertebrates that have hair and females that nurse their young with milk produced by mammary glands. You are a mammal; dogs, cats, rabbits, bats, and whales are mammals. Mam-

FIG. 6-18 One representation of *Homo erectus*.

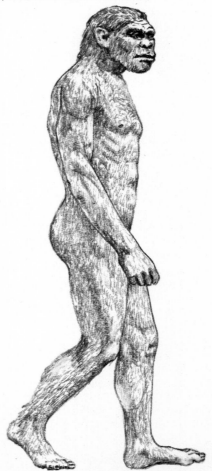

TABLE 6-1 CLASSIFICATION OF SOME FAMILIAR ORGANISMS

Animal Kingdom

Division	Chordata	Chordata	Chordata	Arthropoda	Arthropoda	Coelenterata
Class	Mammalia	Mammalia	Mammalia	Insecta	Crustacea	Hydrozoa
Order	Primates	Carnivora	Carnivora	Lepidoptera	Decapoda	Hydroida
Family	Hominidae	Canidae	Felidae	Nymphalidae	Homaridae	Hydridae
Genus	*Homo*	*Canis*	*Panthera*	*Danais*	*Homaris*	*Chlorohydra*
Species	*Homo sapiens*	*Canis familiaris*	*Panthera leo*	*Danais plexippus*	*Homaris americanus*	*Chlorohydra viridissima*
Common name	Human being	Dog	Lion	Monarch butterfly	American lobster	Hydra

Plant Kingdom

Division	Eumycota	Bryophyta	Coniferophyta	Anthophyta
Class	Basidiomycetes	Musci	Coniferinae	Monocotyledonae
Order	Agaricales	Eubrya	Coniferales	Graminales
Family	Agaricaceae	Funariaceae	Pinaceae	Gramineae
Genus	*Agaricus*	*Funaria*	*Pinus*	*Zea*
Species	*Agaricus campestris*	*Funaria hygrometrica*	*Pinus strobus*	*Zea mays*
Common name	Mushroom	Moss	White pine	Corn

mals evolved from a group of reptiles—our hairs are actually modified scales. If you look in a mirror and open your mouth, you will see another characteristic of mammals—differentiated teeth. Your teeth are not all alike, as in most reptiles. Instead they are differentiated into incisors, canines, and molars (Fig. 6-20).

A very small group of mammals, which includes the duck-billed platypus of Australia, lays eggs. All other mammals give birth to their young. There are two major groups of mammals that give birth to their young. As we noted previously one group, the marsupials, gives

TABLE 6-2 FOUR-KINGDOM CLASSIFICATION

KINGDOM	CELL TYPE	BODY TYPE	REPRODUCTIVE STRUCTURES	NUTRITION	EXAMPLE
Monera	Prokaryotic	Unicellular	Unicellular	Absorption Photosynthesis Chemosynthesis	Bacteria Blue-green algae
Protista	Eukaryotic	Unicellular or multicellular	Unicellular	Absorption Photosynthesis Ingestion	Fungi Algae Protozoa
Metaphyta	Eukaryotic	Multicellular	Multicellular	Photosynthesis (rarely saprophytism or parasitism)	Plants
Metazoa	Eukaryotic	Multicellular	Multicellular	Ingestion	Animals

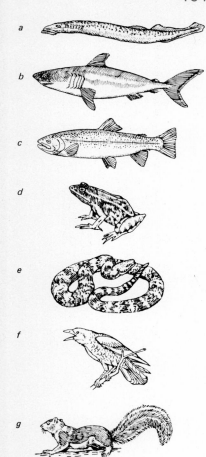

FIG. 6-19 Some examples of the classes of vertebrates. (*a, b, c*) the three classes of fishes, (*d*) amphibia, (*e*) reptiles, (*f*) birds, (*g*) mammals.

TABLE 6-3 CLASSIFICATION OF LIVING SPECIES

CLASSIFICATION	EXAMPLES	ESTIMATED NUMBER OF LIVING SPECIES ALREADY NAMED
Kingdom Monera		
Phylum Cyanophyta	Blue-green algae	200
Phylum Schizophyta	Bacteria	1,600
Kingdom Protista		
Phylum Protozoa	Microscopic, unicellular, or simple colonial "animals"	30,000
Phylum Chrysophyta	Golden algae and diatoms	8,000
Phylum Pyrrophyta	Golden-brown algae	1,100
Phylum Euglenophyta	Euglenalike organisms	450
Phylum Gymnomycota	Slime molds	490
Phylum Mycota	Fungi (molds, mildews, and mushrooms) and lichens (fungi plus algae)	100,000
Phylum Rhodophyta	Red algae and seaweeds	4,000
Phylum Phaeophyta	Brown algae and seaweeds	1,100
Phylum Chlorophyta	Green algae	7,000
Kingdom Metaphyta		
Phylum Bryophyta	Mosses, hornworts, and liverworts	23,500
Phylum Tracheophyta	Vascular plants	
Subphylum Lycophytina	Lycophytes	1,000
Subphylum Sphenophyta	Horsetails	24
Subphylum Pterophytina	Ferns, gymnosperms, and flowering plants	
Class Filicineae	Ferns	11,000
Class Coniferinae	Conifers	550
Class Cycadinae	Cycads	100
Class Ginkgoinae	Ginkgo	1
Class Angiospermae	Flowering plants	275,000
Kingdom Metazoa		
Phylum Coelenterata	Sea anemones, hydra, corals, etc.	9,600
Phylum Ctenophora	Combjellies and sea walnuts	80
Phylum Platyhelminthes	Flatworms	15,000
Phylum Nemertea	Ribbon worms	550

THE KINDS OF LIVING THINGS | 165

TABLE 6-3 (Continued)

CLASSIFICATION	EXAMPLES	ESTIMATED NUMBER OF LIVING SPECIES ALREADY NAMED
Kingdom Metazoa (*continued*)		
Phylum Nematoda	Roundworms	80,000
Phylum Acanthocephala	Spiny-headed worms	300
Phylum Chaetognatha	Arrow worms	50
Phylum Nematomorpha	Horsehair worms	250
Phylum Rotifera	"Wheel animalcules"	1,500
Phylum Gastrotricha	Microscopic wormlike animals	140
Phylum Bryozoa	"Moss" animals	4,000
Phylum Brachiopoda	Lamp shells	260
Phylum Phoronidea	Wormlike animals that live in leathery tubes	15
Phylum Porifera	Sponges	4,800
Phylum Annelida	Earthworms and other segmented worms	8,000
Phylum Mollusca	Snails, clams, octopus, etc.	100,000
Phylum Arthropoda	Insects, spiders, lobsters, etc.	1,250,000
Phylum Echinodermata	Starfish, seaurchins, etc.	6,000
Phylum Chordata	Chordates	
Subphylum Hemichordata	Acorn worms	90
Subphylum Tunicata	Tunicates	1,600
Subphylum Cephalochordata	Amphioxus	20
Subphylum Vertebrata	Vertebrates	
Class Agnatha	Lampreys and hagfish	50
Class Chondrichthyes	Sharks and rays	550
Class Osteichthyes	Bony fish	20,000
Class Amphibia	Salamanders, frogs, and toads	2,000
Class Reptilia	Snakes, lizards, turtles, and crocodiles	6,000
Class Aves	Birds	8,200
Class Mammalia	Mammals	4,500

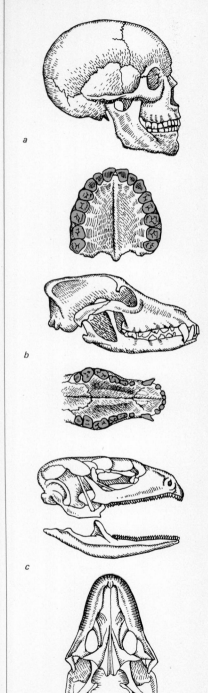

FIG. 6-20 The jaws and teeth of (*a*) *Homo sapiens,* (*b*) a dog, (*c*) a lizard.

birth to a relatively poorly developed new individual after it has spent a short time developing inside the mother. The fragile and almost helpless new individual is born and crawls into a pouch (called a *marsupium*) on the mother's belly, where it continues its development, nursing from a nipple in the pouch. As you will remember (Sec. 5-12) most marsupials live in Australia, where they play a wide range of ecological roles.

The females of all mammals that do not lay eggs have an internal organ called an **uterus** in which the young begin their development. Most mammals do not have a pouch and their young spend a relatively long time in the uterus. These are the **placental mammals.** The developing mammal is nourished in the uterus by a complex structure, the **placenta.** In the placenta, the blood of the mother is brought close to the blood of the developing young, and nourishment is transferred to it while wastes are transferred to the mother.

Placental mammals include the primates (monkeys, apes, and human beings), carnivores (cats, dogs, bears), seals and walruses, whales, elephants, rabbits, rodents (rats, mice), moles, bats, and all the hoofed animals—just to name a few (Fig. 6-21). Roughly 450 genera of placental mammals are known to science, including some 200 in North America.

Reptiles **Reptiles** have scales but not hair, and they do not have mammary glands to supply milk for their young. Most reptiles lay eggs; some give birth to their young, but they lack a true uterus such as mammals have. Most reptiles are able to fend for themselves when they are born or hatched. Even though there is no period of nursing, some reptiles show parental care. Pythons coil themselves around their eggs, and alligators guard their nest of eggs and the young hatchlings (see Fig. 8-9).

Mammals and birds have internal mechanisms to keep their body temperatures constant and at a level above that of their surroundings in cold weather. We say that they are "warm-blooded." Reptiles are generally considered "cold-blooded," because they are unable to control their temperatures internally. They generally use behavioral strategies to regulate their temperatures (see Fig. 14-1). On cold days they bask in the sun; on hot days they hide in the shade. When there is no sun and the air temperature drops, they become sluggish and their blood gets "cold."

In spite of all their differences, reptiles are the nearest relatives of mammals. Indeed, our distant mammalian ancestors were descendants of mammallike reptiles that are now extinct. There are only four orders of living reptiles—turtles, crocodiles, lizards and snakes, and a reptile called the tuatara (which resembles a lizard) (Fig. 6-22). Snakes are legless reptiles that evolved from lizards; the remnants of their legs can be seen in some species (Fig. 6-23). Some snakes have modified salivary glands that produce poison instead of saliva. The poison is injected into prey or attackers through modified teeth called fangs.

Snakes produce two main types of poisons: one affects the nervous

FIG. 6-21 Some placental mammals: (*a*) a whitenosed monkey from Africa, (*b*) a sea lion, (*c*) a bat, (*d*) an antelope. [(*a*) *American Museum of Natural History;* (*b*) *Ira Rosenberg/Image;* (*c*) *courtesy of Don Carroll;* (*d*) *Tom Ackerson/Image.*]

FIG. 6-22 Some examples of reptiles: (a) a tuatara, (b) a turtle, (c) an Asian cobra, (d) a grass anole, (e) a crocodile. [(a) M. F. Soper, National Audubon Society; (b) Carolina Biological Supply Co.; (c) American Museum of Natural History; (d) Courtesy of J. Roughgarden; (e) Leonard Lee Rue, National Audubon Society.]

system and is common in cobras and coral snakes, the other affects primarily the blood and is predominant in vipers and pit vipers. Cobras and their relatives have short, permanently erect fangs, while vipers and pit vipers have very long fangs attached to a modified jawbone in such a way that they can be folded back when not in use. Pit vipers, which include rattlesnakes, copperheads, and water moccasins, are so named because they have a deep pit in either side of their heads between the nose and the eye. These pits are detectors of heat (infrared waves) and enable the snake to aim accurately at warm-blooded prey in total darkness.

Many snakes lack poison and kill their prey by crushing it in their coils. Some of these snakes, pythons and boas, for example, have heat detectors on their lips. Snakes cannot chew and thus have to swallow their prey whole; most can unhinge their jaws to swallow prey bigger around than they are. There are about 6,000 known species of reptiles.

Birds Birds are "feathered reptiles." Both birds and mammals are descended from reptiles; but they are descended from different groups of extinct reptiles. In fact, birds are descended from the earliest dinosaurs! Just as the hair of mammals is thought to have evolved from scales, the feathers of birds are thought to have evolved from scales. Birds still have scales on their legs. Most birds can be distinguished from reptiles because they are warm-blooded and can fly; however, some birds, like the ostrich (Fig. 6-24) cannot fly, and some lizards (Fig. 6-25) and snakes can glide. Some birds which seem to be flightless, the penguins, for example, "fly" underwater. Birds, like most reptiles, lay eggs; unlike most reptiles, however, they usually take care of their eggs and young. Most birds keep their eggs warm by sitting on them, and many carry food to the young after they have hatched. There are about 8,200 species of birds, of which about 650 live in North America.

Amphibians The mammals, reptiles, and birds that lay eggs usually do not lay them in water. Tough shells keep the eggs from drying out and protect them from other environmental hazards. Amphibians lay eggs, without tough shells, in water or in damp places. Young amphibians develop in water, undergoing metamorphosis to become adults (Fig. 6-26). Frogs, toads, and salamanders are the best-known amphibians (Fig. 6-27). Adult amphibians absorb oxygen and give off carbon dioxide in part through their skin, which must remain continually moist. Therefore, amphibian adults normally live in or near water, and very often are active only at night because they are less likely to dry out at night than in the daytime. There are about 2,000 known species of amphibians.

Fishes The largest group of vertebrates is the bony fishes, with 20,000 known species. Fishes live in both salt and fresh water, but the most diverse kinds live in the sea (Fig. 6-28). The majority of fishes have bony skeletons; however, sharks and their relatives have skeletons made of cartilage—a tough, flexible material that you may call "gristle" when you find it in the meat you eat. (Cartilage is the stuff that gives your ears and the end of your nose their shape.)

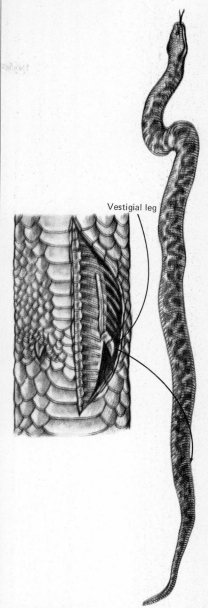

FIG. 6-23 Vestigial legs of a python. The skin is cut back on the left to show the skeletal arrangement. Male pythons use these movable spurs to scratch the female during courtship.

FIG. 6-24 The ostrich has lost the ability to fly and is specialized for terrestrial life. (*FPG/Allan Rose.*)

Fishes either lay eggs or bear their young alive. Some fishes ignore their young after they are born or hatched; some may even eat them if they can catch them. Others give considerable maternal care, herding the newly hatched young and defending them. In some species, the male scoops up the eggs after they are laid and keeps them in his mouth until they are hatched. After they are hatched, his mouth serves as a convenient portable cave into which the young can flee if they are threatened (Fig. 6-29). What keeps the male from accidentally (or intentionally) swallowing his own babies? During the time he is brooding his throat is plugged with mucus; not only can he not eat his offspring, he cannot eat anything at all! Still other fishes make nests of plants or of bubbles for their eggs and young (Fig. 6-30).

6-6 Invertebrates do not have backbones

Arthropods Vertebrates have their skeletons inside (Fig. 6-31) the muscles that work them. Insects and their relatives have exactly the opposite body plan. They have their skeletons on the *outside* and the muscles that work them inside the skeleton. Insects and related animals, such as mites, spiders, crabs, and lobsters (Fig. 6-32), have many-jointed legs or appendages. This is the source of the zoological name of the phylum Arthropoda, which means "jointed foot." (Some familiar expressions have the same root referring to joints. For instance, *arthritis* means "inflamed joints.") Insects are smaller than vertebrates for the most part because a very large external skeleton leads to mechanical difficulties in walking, flying, and in simply holding the body up. A housefly the size of an elephant would collapse of its own weight.

FIG. 6-25 This lizard (*Draco volens*) is able to glide by flattening its body, extending its ribs outward.

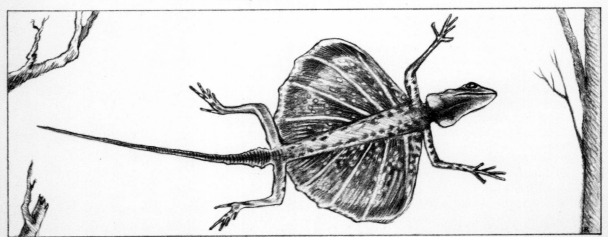

FIG. 6-26 Stages in the metamorphosis of a tadpole into a frog. (a) Male fertilizes eggs produced by female (usually in shallow water), (b) eggs, (c, d) tadpoles which live in water and feed on algae, (e, f) tadpoles losing tails and developing legs, (g) adult frog which lives on land and in water and feeds on insects.

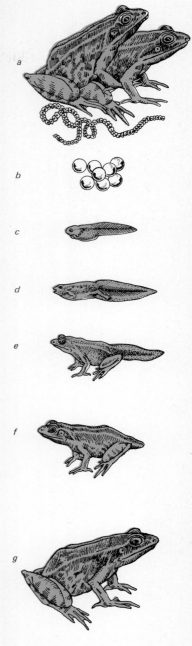

Insects make up for their small size by the diversity of their species and by the vast numbers of individuals in many of the species. Insects are the most abundant and varied of the land animals. Table 6-3 estimates that over 1,250,000 arthropods have been formally named. Experts working in insect identification believe that if all living species could be identified, there would be between 2 million and 10 million species of insects. One order of insects alone, the beetles, contains roughly 280,000 named species. A famous biologist, when asked to name an outstanding characteristic of God, said, "He has an inordinate fondness for beetles."

As large a group as the beetles are, the three orders containing the butterflies and moths, the flies, and the bees, ants, and wasps are not far behind. These orders contain species that usually have conspicuous wings as adults. In fact, with minor exceptions, all adult insects have functional wings. Some, like beetles, carry their wings concealed and may not fly readily, but most of them *are* winged. Once they become adults, insects do not grow any more—and, with a single minor exception, all winged insects are adults. Thus you know if you see two similar flies whizzing about, one large and one small, they are adults of different species. Most insects lay eggs and go through a series of distinct developmental stages (Fig. 6-33). This process is called **metamorphosis** (see Sec. 7-6). Some, like termites, bees, ants, and wasps, live a complicated social life in large colonies (Fig. 6-34).

There are many other arthropods besides insects. A great many other arthropods such as mites, ticks, and so forth have yet to be identified; there may be as many species of them as of insects. In terms of numbers of individuals and numbers of species, the present era should really be called "the age of arthropods."

FIG. 6-27 This salamander is a terrestrial amphibian which must return to the water to breed. (*Carolina Biological Supply Co.*)

CHAPTER SIX

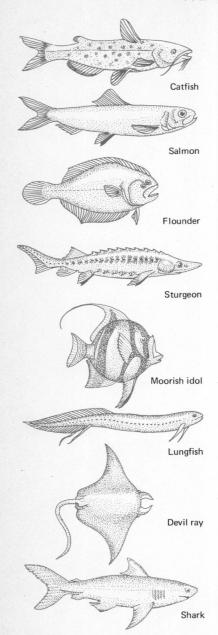

FIG. 6-28 A selection of fishes to show the great diversity in this class of vertebrates.

Many arthropods live in the sea. Although no group of aquatic arthropods rivals the insects and mites in numbers of species, aquatic arthropods show an amazing variety of shapes and life styles. The most common are crustaceans (arthropods with a very hard outer shell). Many different kinds of crustaceans, shrimps, lobsters, crabs, barnacles, and others inhabit the sea (Fig. 6-35). Crustaceans include both herbivores and carnivores. Barnacles (Fig. 6-36), which superficially look like shellfish, are actually crustaceans that have settled in one place and use their legs to strain food from sea water. Some of the zooplankton consists of tiny crustacea of many different species that feed on each other, on phytoplankton, and on detritus. Many of these crustaceans are microscopic but multicellular. This is true of many other organisms so you should not get the idea that only unicellular organisms are microscopic.

Mollusks Animals of the phylum Mollusca (Fig. 6-37) have a body made up in part of a large muscular "foot." Snails, clams, oysters, and octopuses are perhaps the best known mollusks. Snails and their relatives have a single "foot" with which they crawl by rhythmic movements of its muscles. They are called *gastropods,* which means "stomach foot." Octopuses and squid have their foot divided into eight or ten tentacles. They are called *cephalopods,* which means "head foot."

Most mollusks have thick shells that serve to protect them from enemies and, in land-dwelling snails, keep them from drying out. The shells of mollusks may be extremely beautiful. If you are lucky, some

FIG. 6-29 In some fishes, the eggs are held in the parent's mouth until they hatch. While the young are still small, they may return to the parental mouth when danger threatens. (*T.F.H.*)

day at the beach you may see a mollusk shell with legs sticking out of it (Fig. 6-38). One kind of crab, the hermit crab, uses the shells of dead mollusks to protect its own soft body.

Snaillike mollusks usually feed on plants. Some, such as the beautiful tropical cone shells, have poison darts with which they catch fishes and other small animals. These darts are stabbed into the body of the prey and paralyze it. They are also used for defense and are very dangerous to unwary divers. Octopuses and squid (Fig. 6-39) are carnivorous. They have eyes amazingly like our own in construction. They are also able to swim rapidly by a kind of "jet propulsion," often leaving a "smokescreen" of ink behind to confuse their enemies.

Another group of mollusks are called *bivalves* because they have paired shells. This group includes clams, oysters, and mussels. Most of this group feed by filtering food out of the water. Water is pumped into a tube past the mouth. There, food particles become stuck to mucus and are pushed by cilia toward the mouth. The water and nonfood particles are rejected, and the water exits via another tube.

Other invertebrates Table 6-3 lists some of the other phyla of inver-

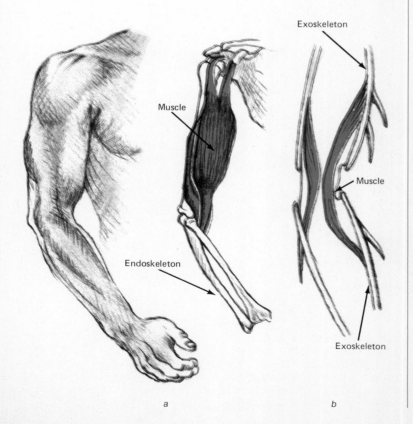

FIG. 6-30 Fish nests: (*a*) The male *Betta* (Siamese fighting fish) makes a nest of floating mucus bubbles to which he attaches the eggs after fetching them from the bottom where the female has dropped them; here the male, wrapped around the female, is fertilizing the eggs. (*b*) Male stickleback fish make nests from aquatic plants, binding them together with a sticky secretion from their kidneys and then enticing a series of females into the nest to lay eggs; here a male is tickling a female in his nest in order to stimulate her to lay eggs.

FIG. 6-31 Movement in a vertebrate (human being) (*a*) and an arthropod (insect) (*b*). In the insect, the muscles attach to an exoskeleton; in the human being, the muscles attach to an internal skeleton or endoskeleton.

FIG. 6-32 The diversity of Arthropoda includes scorpions, bees, flies, butterflies, lobsters, mites, and spiders (not drawn to scale).

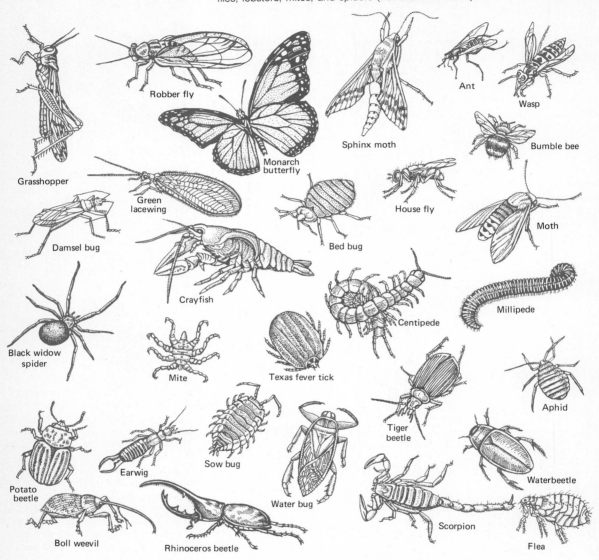

FIG. 6-33 Stages in the development of a checkerspot butterfly (*Euphydryas chalcedona*): (*a*) mating pair, (*b*) female laying eggs on leaf of larval foodplant, (*c*) eggs, (*d*) larvae in web of silk which they spin, (*e*) closeup of small larvae, (*f*) larva just before diapause (a period of rest which in this species lasts most of the year), (*g*) diapausing larvae in a hollow stem, (*h*) postdiapause larva, (*i*) pupa, (*j*) adult female and male.

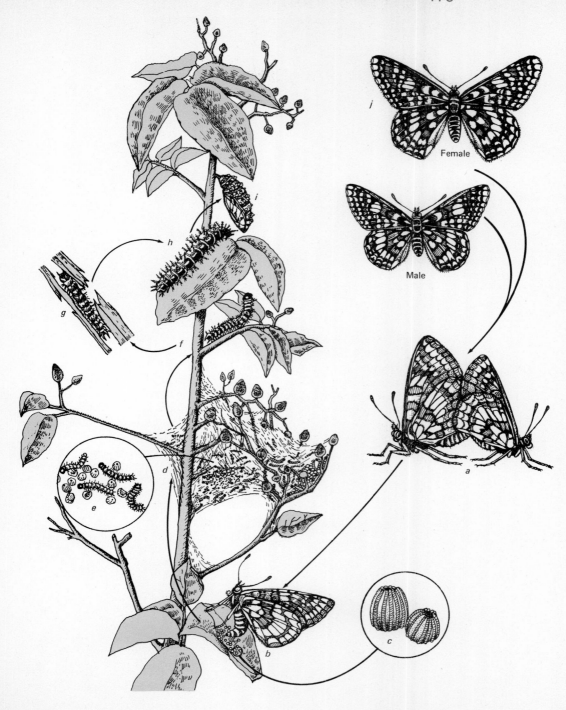

FIG. 6-34 This colony of wasps lives in a nest constructed by them from chewed-up plant tissue held together by secretions from glands in their mouths. The walls of the nest resemble paper. Hexagonal compartments called cells contain eggs and developing larvae. (*Karl H. Maslowski, National Audubon Society.*)

tebrates that are recognized by biologists. You have noticed by now no doubt that even though Vertebrata is a subphylum, it is common practice to refer to all other phyla of animals as invertebrates. In other words, if an animal is not vertebrate, it is an invertebrate. So far we have discussed only two of the invertebrate phyla, Arthropoda and Mollusca. There are three other subphyla of the phylum Chordata and 16 other invertebrate phyla. The other chordates are rare and mostly of interest to people studying the evolution of the vertebrates. The other invertebrate phyla, however, have many organisms that play important ecological roles. Even though you may not have many opportunities to see them, you should be aware that there are

FIG. 6-35 A male fiddler crab. The greatly enlarged claw is waved at the female during courtship. (*New York Zoological Society.*)

FIG. 6-36 A group of barnacles on a rock. Until a little over a century ago, because of their appearance, barnacles were classified as mollusks instead of as arthropods, the phylum they best fit, based on their internal anatomy and mode of development. (*Gordon S. Smith, National Audubon Society.*)

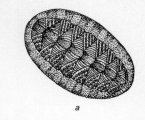

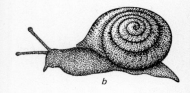

FIG. 6-37 Some representative members of the phylum Mollusca. (*a*) A chiton, one of the most primitive mollusks which have a shell consisting of eight plates, (*b*) a snail, (*c*) a clam.

FIG. 6-38 Hermit crabs have soft bodies and use the abandoned shells of mollusks as protection. As they grow, they have to select a new shell and move out of the old one and into the new one. This hermit crab is living in a conch shell. (*American Museum of Natural History.*)

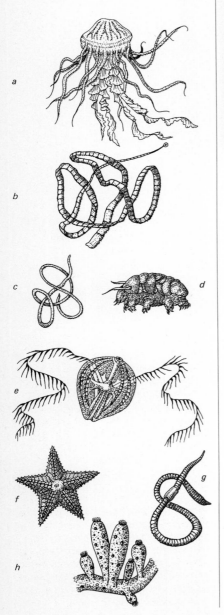

FIG. 6-39 A squid is a cephalopod mollusk with ten tentacles. (*George Lower, National Audubon Society.*)

a great many individuals of these phyla sharing this planet with us. Fig. 6-40 shows some of the other invertebrates.

Some of the "other" invertebrates are beneficial to people. A great many of the lesser-known invertebrates play important roles in food chains in the oceans and estuaries. Others, such as earthworms (phylum Annelida), are important to land ecosystems. Earthworms play a vital role in keeping soils loose and fertile; without them people would not be able to grow as much food. In spite of this, certain pesticides used over many years have been allowed to deplete the soil of its earthworms. There are orchards in southern England in which repeated spraying with a fungicide called Bordeaux mixture has seriously affected the earthworm populations.

Members of the phylum Nematoda (roundworms) are almost everywhere, and there are many thousands of species, most microscopic. Some are found in polar seas, in boiling hot springs, deep in the oceans, in the most arid deserts, and in all soil. Some are also parasitic in the bodies of birds, fish, and mammals, including people. Some

FIG. 6-40 Some other kinds of invertebrates. (*a*) a jellyfish—most are marine, many have stinging tentacles; (*b*) a tapeworm—all are parasitic, reproduction is accomplished by dropping off of segments which appear in the feces; (*c*) a roundworm—common parasites of dogs and cats; (*d*) a tardigrade or "water bear"—these tiny organisms withstand drying for very long periods; (*e*) a comb jelly—marine jellyfishlike invertebrates; (*f*) a starfish—starfish are predators on bivalves and coral; (*g*) an earthworm—extremely important in maintaining soil fertility; (*h*) a sponge—most are marine, they feed on particles of detritus and small organisms they filter from the sea.

prefer to eat living organisms, and those that live in the soil often attack plant roots.

Members of the phylum Platyhelminthes are often parasitic in other animals; flatworms called *flukes* cause several diseases including schistosomiasis, one of the most serious diseases of people in wet, tropical areas. The organisms (*Schistosoma*, formerly *Bilharzia*) penetrate the skin when any part of the body is immersed in infested water. In the body they spread to many organs over a period of years and cause fever, weakness, weight loss, and diarrhea. The liver, intestines, lungs, and other organs may be invaded, producing a variety of serious symptoms which may be prolonged for years until the patient eventually dies of pneumonia or other infections. The complex life cycle of one of these flukes is shown in Fig. 6-41. Notice that both people and snails are alternate hosts for this parasite. One of the serious consequences of the building of the Aswan Dam in Egypt has been the spread of *Schistosoma*. Snails thrive in irrigation canals. In fact, many irrigation projects, for the purpose of increasing food production in Africa, have led to an increase of schistosomiasis, which one would expect to lead to many deaths. One scientist has argued that it is less humiliating to die of schistosomiasis than to starve to death. Death from malnutrition often takes about 20 years, which is about the time it takes for death from schistosomiasis. The

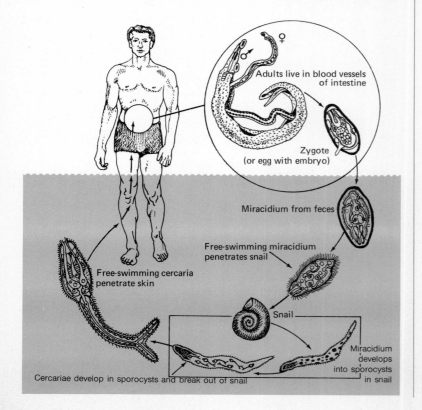

FIG. 6-41 The life cycle of *Schistosoma mansoni*, a parasitic fluke that causes schistosomiasis. The adult flukes live in the blood vessels of the intestine, obstructing blood flow to the liver and intestine; the obstruction may seriously damage liver and intestinal cells. Eggs leave the human body in the feces; if they get into water, they hatch into ciliated swimming miracidia, which must quickly find a particular kind of snail or die. After going through certain developmental stages within the snail, the flukes emerge as free-swimming cercariae, which penetrate human skin and travel in the blood to the vessels of the liver and intestine, where they mature and live for many years. Keeping fecal material away from fresh water will break the life cycle and prevent infestation.

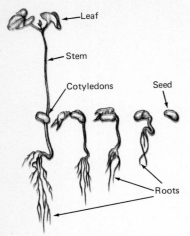

FIG. 6-42 Sprouting seeds: Bean plants store food in the two fleshy cotyledons (first leaves) of the seed. The cotyledons are succeeded by photosynthetic foliage leaves.

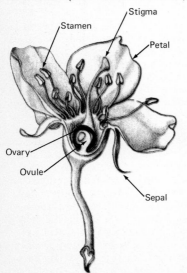

FIG. 6-43 A longitudinal section through a flower. (*After Raven and Curtis.*)

catch is that most of the food the local people grow with the water from the snail-laden irrigation canals does not go to them but to people in the cities. Thus, people in the villages now may have the opportunity to die of both schistosomiasis *and* malnutrition.

SOME MAJOR KINDS OF PLANTS

Plants, like animals, undoubtedly originated in the sea. For millions of years photosynthetic Protista, the algae, evolved and diversified in the oceans. Eventually, and probably over millions of years, descendants of the green algae developed more complex multicellular structures and were able to survive on land. The descendants of these pioneers became the Metaphyta. In the land environment, they formed a number of new evolutionary lines. Today there are more species of plants on the land than in the sea.

6-7 The most familiar plants are seed plants

Your mental image of a typical land plant is no doubt that of a **seed plant.** The conifers, cycads, ginkgo, and angiosperms of Table 6-3 are seed plants; included in these groups are all "trees," "grasses," and "flowers." Seed plants evolved most recently and are the dominant plants on land. The evolution of seed plants is intimately connected with the evolution of land animals. In part, this is because of the evolution of the seed. As we shall see in Sec. 7-6, the seeds of plants contain an embryo in which development has been halted. The seed is a little packet of high-energy food stored for the time when the seed germinates and the embryo grows into a young plant (Fig. 6-42). Land animals were able to evolve means of utilizing this energy. Many different kinds of animals have evolved into seedeaters.

Fruit-bearing plants Most seed plants produce seeds inside fruits. A **fruit** is a specialized container for the seeds of angiosperms (flowering plants). Angiosperms are the most abundant of the seed plants, with over 275,000 species (Table 6-3). They have a special structure called a flower (Fig. 6-43), which includes the reproductive parts of the plant. In most flowers there are both male and female reproductive structures (Fig. 6-44). Other kinds of flowers may be either male or female; for example, a corn tassel is made up of male flowers and the corn cob of female flowers. The daisy is actually a group of many flowers called an *inflorescence,* as is a corn tassel or cob. Both corn tassels and corn cobs occur on the same plant. Other plants form male and female flowers on separate plants. Marijuana and date palms are examples of species which have male and female flowers on separate plants.

The female structure is an **ovary,** which later develops into a fruit. The male structures are **stamens,** which produce pollen. Pollen consists of tiny grains that produce the male cells (Fig. 6-45). Many flowers have brightly colored petals, a sweet nectar, and scent glands producing a pleasant fragrance.

Reproduction in the angiosperms requires two processes: (1) polli-

THE KINDS OF LIVING THINGS 181

FIG. 6-44 (*a*) Magnolia, (*b*) sweet pea, (*c*) daisy. These three species have both male and female reproductive structures in the same flower. [(*a*) *Courtesy of G. R. Roberts;* (*b*) *W. Atlee Burpee Co.;* (*c*) *I. L. Brown.*]

nation and (2) fertilization. **Pollination** consists of getting the pollen grains to a specialized portion of the ovary called the **stigma**. **Fertilization** is the process by which the nuclei of male and female cells fuse. It is discussed more fully in Sec. 7-6. The first problem faced by an angiosperm is how to accomplish pollination. Some flowers have both male and female parts and are self-pollinating. Other plants, for example, grasses and oak trees, depend on the wind blowing pollen from flower to flower. These have flowers with nothing which properly can be called petals. Flowers with brightly colored petals, nectar, and fragrance are pollinated by insects almost exclusively, although some attract bats and birds. Patterns on petals serve as guides on many flowers to where the nectar is located (Fig. 6-46). Although we cannot see some of these patterns without special cameras, the flowers evolved patterns using ultraviolet light that are seen by insects, for insects see different wavelengths (colors) of light than we do (Fig. 6-47). Insects such as honeybees use both nectar and pollen for food. Honeybees make honey from the nectar. In moving from flower to flower, gathering what they need, the insects carry pollen to the flowers, accomplishing pollination.

After being brought to the flower, the pollen grain grows a tube that extends from the stigma to the **ovules,** which contain the female cells. The male cells move through the tube and fertilize the ovule. Stored food is laid down, and a seed is formed.

Cone-bearing plants   Not all seed plants produce flowers. Pines, spruces, firs, and redwoods are examples of the seed plants called conifers. These trees have cones, not flowers (Fig. 6-48). Large woody cones are made up of cone scales that bear the ovules on their upper surface. Smaller cones produce the pollen. In the conifers, pollination always takes place by wind. Pollen grains from the male cones are blown to the female cones. The male cells move into the ovule through a pollen tube, and fertilization of the egg takes place. Once again pollination and fertilization are separate events. The seeds of conifers are also rich in stored energy. Some people eat the seeds of conifers, for example, pine nuts; and seeds of many conifers are eaten by herbivores. Conifers form the basis of many food chains, and people use conifers for wood for heating, for building, and as pulpwood for paper.

The pollen tubes of both flowering plants and conifers are a specialization for life on land. In organisms living in the water, the delicate male cell can move through the water to an egg. This is, obviously, impossible for land plants unless there is a film of water over the reproductive organs. The pollen tube provides a sort of canal through which the sperm can move to the eggs.

Vascular tissue   Life on land poses many problems for plants besides pollination and fertilization. All cells, including those of plants, must be provided with water and nutrients. Cells must be protected from drying out. Photosynthetic cells must be exposed to light and carbon dioxide. Water and nutrients are found down in the soil. Carbon dioxide and light are found up in the air. A land plant thus has the problem that it must be in two places at once and yet cannot move as do animals.

Seed plants have solved this problem by developing roots, stems, and leaves (Fig. 6-49). Roots are the underground, absorbing, and anchoring organs of the plant. Water and dissolved nutrients are

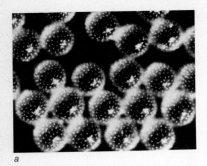

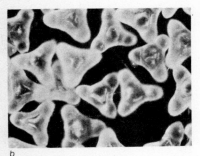

FIG. 6-45 Pollens grains of (a) squash (×75), (b) evening primrose (×75), and (c) bergamot (×150). (*Hugh Spencer, National Audubon Society.*)

taken up from the soil by fine hairs on the roots of plants. Photosynthesis is carried on primarily in leaves. Leaves have pores called **stomata** (singular, **stoma,** meaning "mouth") on their surfaces through which gases, such as carbon dioxide, oxygen, and water vapor, can pass. There is a large system of air spaces among the photosynthetic cells of a leaf, and all cells are exposed to carbon dioxide. Unfortunately, when the stomata are open, letting in carbon dioxide, water vapor can also escape. Plants have to balance the need for carbon dioxide against the need to protect their cells from drying out. When the stomata are open, pollutants in the air can also enter.

FIG. 6-46 Patterns on the petals of this flower, *Penstemon,* help guide insects to the nectar and pollen. (*I. L. Brown.*)

FIG. 6-47 Many flowers have patterns which reflect wavelengths in the ultraviolet, which people cannot see but which insects can. (*a*) This daisy from Colorado was photographed in light visible to human beings; no special markings are apparent, (*b*) the same daisy, photographed in such a way as to show what an insect (that can see ultraviolet) sees when approaching the flower, and to show conspicuous dark nectar guides toward the tiny flowers in the center of this inflorescence. (*Courtesy of Ward B. Watt.*)

FIG. 6-48 A branch of a pine tree showing numerous small male cones and a larger female cone. (*William Harlow, National Audubon Society.*)

Leaves and roots are connected by stems; but that is not all that stems do. They also hold the leaves up to the light and transport materials up and down from roots to leaves and vice versa. In seed plants and ferns, the tissues that transport materials are called **vascular tissues.** For this reason seed plants, ferns, and relatives of ferns are called collectively "vascular plants." Vascular tissues are found in all parts of vascular plants: leaves, stems, roots, even flowers. There are two kinds of plant vascular tissues, each with a different function. The tissue called **xylem** serves to transport water and dissolved nutrients upward and laterally from the roots. The tissue called **phloem** conducts the products of photosynthesis from the leaves to nonphotosynthetic parts of the plant.

Xylem is composed of thin-walled living cells and thick-walled cells that become functional after they die at maturity (Fig. 6-50). The thick-walled cells are of two kinds: fibers and conducting cells. Fibers give mechanical strength to roots, stems, and leaves. Conducting cells of the xylem die and lose their contents when they become functional. They are similar to miniature pipes or a series of pipes largely made up of a carbohydrate, cellulose. The exact details of how water movement takes place in xylem are very complex. Water molecules tend to cohere or stick together very tightly. The result is that there is a *continuous column of water* from root to stem to leaf. The water which evaporates from the leaves when the stomata are open must be replaced by water entering from the roots. Water is taken up by roots and moves to the xylem and through it to other parts of the plant. Leaves give off large quantities of water to the air. This process is called transpiration and its importance was discussed in Sec. 2-7.

In the center of the stems of trees and shrubs, the xylem cells stop conducting water. Their cell walls become harder and often they are filled with water-resistant compounds. The greatest mass of the trunks of trees and the stems of shrubs is made up of this nonfunctional xylem. Such xylem is called *heartwood*. Wood can be used directly for making many different things from furniture to pencils. The wood of angiosperm trees such as oak, maple, and walnut is called hardwood; it makes especially fine furniture. The wood of conifers is usually less dense and is called softwood. Wood products are used in many industries: fibers in wood can be separated out and used to make paper; resins are used in paints. Wood already is becoming scarce in many parts of the world. In the United States, wood is still used as a building material and in some places even as a fuel. Wood is potentially a renewable resource and, if forests are carefully managed, we need not run out in this country.

Phloem is composed of thin-walled cells associated with specialized conducting cells (Fig. 6-51). The latter may lose their nucleus when they are mature; they do not lose their contents, however. Conducting cells of phloem transport the products of photosynthesis —sugars—from the leaves to nonphotosynthetic parts of the plant. Nonphotosynthetic cells, for example, in a root or stem, use sugar from the leaves in carrying out their functions. Photosynthetic cells

THE KINDS OF LIVING THINGS   185

make amounts of sugars beyond what they need. These sugars move in complex ways through the conducting cells of the phloem. Apparently the tendency of molecules to move from regions of greater concentration to regions of lower concentration, which we talked about earlier (Sec. 5-1), is involved in this process.

6-8 Not all plants produce seeds

Ferns and their relatives Ferns and their relatives (Fig. 6-52) are vascular land plants that do not reproduce by means of seeds. Modern seed plants seem to have had ancestors resembling today's ferns. In the ferns and their relatives, pollen tubes are not produced. Therefore they are not independent of water for reproduction as are the seed plants. The male cells must have a thin film of water to swim to the egg. Many ferns and their relatives do grow in dry places, but they can form new individuals by combining male and female cells only when water is present. Not very many animals eat ferns. In damp climates, the fern called bracken often invades grazing lands, and attempts are made to get rid of it by burning. One kind of bracken fern has often been associated with fatal poisoning of cattle, and

FIG. 6-50 A longitudinal section of a portion of wood (xylem) of a silver maple (*Acer saccharinum*) showing the narrow fibers and the larger water-conducting cells. Some conducting cells are sectioned so that the pits on their walls are visible; some show the absence of end walls. The elliptical groups are composed of ray cells which conduct horizontally.

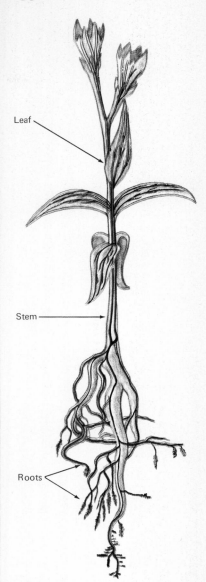

FIG. 6-49 The basic structure of a seed plant, showing photosynthetic leaves, transporting stem, and anchoring and absorbing roots.

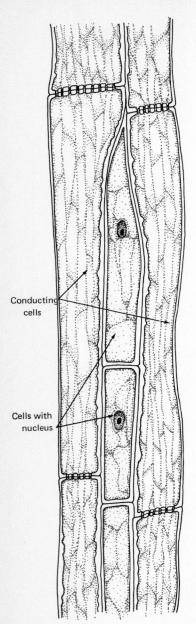

FIG. 6-51 A longitudinal section through phloem tissue showing the large conducting cells and the smaller cells associated with them, which contain nuclei.

mutagens, substances that affect the hereditary material, have been found in other bracken ferns. Some peoples eat the uncurled tips of other ferns, called fiddleheads, which are said to taste rather like asparagus.

Mosses Not all land plants are vascular plants. Mosses are relatively small plants that are able to survive without highly specialized conducting cells. Mosses (Fig. 6-53) appear to be thin-leaved, delicate plants, and they often occur in moist, shady places. In the course of reproduction, moss plants produce organs that contain either male or female cells at the tips of their stems. After fertilization, club-shaped spore cases begin to develop, attached to the moss plant. When mature, these open and discharge the spores. Spores are single-celled reproductive bodies. If they land in a suitable place, they germinate and produce a new moss plant.

Despite their fragile appearance, many mosses occur in deserts. Often they are the first plants to appear in ecological succession. This means they are able to live in very severe habitats not suitable for other plants. Many species of mosses are able to withstand drying out for long periods of time. When water finally reappears, they rapidly absorb it and become active. All species of mosses require at least a film of external moisture through which the sperm swim to the eggs during sexual reproduction.

Mosses are found at high elevations on mountains and at high latitudes. They are abundant in the tundra and taiga biomes. In arctic and subarctic areas, a species of moss called peat may form huge marshy bogs. Peat mosses grow for hundreds of years, and it has been estimated that the mosses in some deep peat bogs may be many meters in length. Peat moss produces acid substances that slow or stop growth of bacteria and other microorganisms. A peat bog is therefore a good place for the preservation of fossils. The remains of over 700 men and women have been found in the peat bogs of northern Europe. Several were in almost lifelike condition after as long as 2,000 years in the bog. Most of these people appear to have been killed as part of the ritual sacrifices of some ancient cult. When people out digging peat to burn on their hearths have dug up these remains, they have often been so shocked that they reburied the find before scientists could examine it. More of these peat burials are being revealed every year since peat is still used for burning even though it is smoky compared with coal and other fossil fuels.

SOME PROTISTA

Protista are organisms with eukaryotic cells and unicellular reproductive structures. The most common Protista are algae, fungi, and protozoa.

6-9 Most algae are aquatic Protista

All those organisms commonly called algae, except the blue-green algae, are classified (Table 6-3) in the kingdom Protista. This means

FIG. 6-52 Vascular plants which do not produce seeds: (a) ferns, (b) a club moss, (c) stem tips of a horsetail. [(a, c) American Museum of Natural History; (b) courtesy of G. R. Roberts.]

FIG. 6-53 A group of moss plants. The thin, leafy appendages at the stem tips contain reproductive structures. (*Carolina Biological Supply Co.*)

that they have eukaryotic cells and unicellular reproductive structures. Their body type may be either unicellular or multicellular. All the algae are able to photosynthesize. Although a few species occur on land, algae as a group are aquatic. Some of the different kinds of algae are illustrated (Fig. 6-54). They may be classified in a general way by color because this represents basic chemical differences among them. Almost all algae contain chlorophyll. Therefore you would expect them to be green. However, many of them contain other pigments as well that give them a red, brown, or yellow appearance.

A great many of the "plants" called seaweeds are red algae or brown algae. Brown algae are multicellular and are found as rockweeds in the intertidal zone (see Fig. 6-54). They also form great masses of floating or attached **kelp**. Individual kelp "plants" may be as long as 60 meters (200 feet). The substance algin, extracted from kelp, is used for stiffening ice cream and puddings. Kelp is also a source of iodine and is sometimes used for fertilizer. Red algae also occur along the ocean's edge. They are able to carry on photosynthesis at greater depths than brown algae; therefore, they often extend deeper along the bottom. Some red algae are used for food, especially in Oriental countries. One red alga is also the source of the substance called agar. Agar is a jellylike material that has many uses; for example, it is used to make the capsules that hold vitamins, etc. It is also used as a medium on which to grow bacteria. Bacterial cultures on agar have contributed a great deal to our understanding of basic genetic mechanisms (Chap. 9). Not only that, it is also used in quick-set desserts and as a base for cosmetics. Even though it is used in food products, agar cannot be digested by human beings.

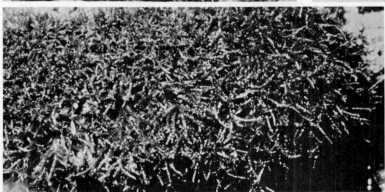
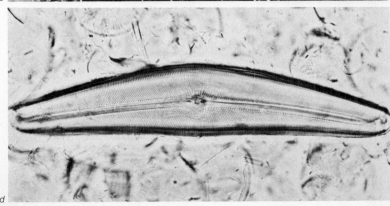

FIG. 6-54 Some different kinds of algae: (a) a freshwater green alga, (b) an intertidal brown alga called a rockweed, (c) an intertidal red alga, (d) a diatom (diatoms have a two-part outer layer, glassy in nature; fossil diatoms are used in toothpaste as an abrasive and in swimming pool filters). [(a) *F.P.G./H. Spencer;* (b) *courtesy of Don Carroll;* (c) *F.P.G.;* (d) *Carolina Biological Supply Co.*]

FIG. 6-55 Spore-producing structures of three kinds of fungi: (a) a mushroom, (b) bracket fungi growing on and in a tree, (c) mold growing on ravioli which was left unrefrigerated. [(a) Carolina Biological Supply Co.; (b) F.P.G./ Charles R. Potter; (c) courtesy of Don Carroll.]

It is important to remember that microscopic algae are the plankton that form the base of oceanic food chains. They are also the major primary producers in many freshwater lakes.

6-10 Fungi are Protista that get their food by absorption

There are many different kinds of fungi. The fungi are very highly specialized organisms that we have classified in the kingdom Protista (Fig. 6-55); however, some other biologists prefer to give the fungi a kingdom of their own. Fungi obtain their living by breaking down organic substances produced by other organisms; most wait until the organism dies and are thus decomposers; others invade living organisms and are parasites. Some fungi consist of only one or a few cells; other fungi are large-bodied and multicellular. Yeasts are fungi with only one or a few cells; mushrooms and molds are multicellular. Usually the filaments that make up most of the body of a multicellular fungus are imbedded in the soil, a tree trunk, or whatever organic material it is using for food. The only parts you are apt to see are the parts (containing reproductive structures) that produce spores, which will make new individuals. Mushrooms, bracket fungi, and mold on foods (such as the mysterious growths found on leftovers after they have been forgotten for months in the back of the refrigerator) are all reproductive structures of fungi. Spores of fungi are very tiny and blow about extensively; if they land in a suitable spot, they form new fungal filaments and eventually produce new reproductive structures. The spores of fungi have thick walls and can survive for a long time waiting to find conditions that are suitable for them to grow.

Uses of fungi Some biologists have referred to yeast as the human species' oldest domesticated organism. Very early in our history, people found that sweet solutions of fruit juices or grains become intoxicating to drink if they are left for a time under the right conditions. Beerlike drinks or wine may have been our species' earliest manufactured beverages. The discoverers of beer had no idea what caused the intoxicating effect; we now attribute it to alcohol produced by the biochemical activities of yeast cells (Fig. 6-56). Alcohol is a poison to yeast cells (as it is to us if we drink too much), and eventually they are killed by their own waste product, alcohol. This is the reason that beer and wine contain relatively low percentages of alcohol compared with distilled spirits such as whisky or brandy. The yeasts have given "their all" for the cause by the time the beverage is 15 to 20 percent alcohol. We must use other methods such as distillation to get higher concentrations. Yeasts are also important in making leavened (raised) breads. Bread yeasts make bread rise by producing bubbles of carbon dioxide gas as they respire. Bread is kneaded in order to get oxygen into the dough which the yeasts use in respiration; without sufficient oxygen bread yeasts will produce alcohol. Yeasts have the ability to grow either with or without oxygen, depending on their surroundings; this made yeast ideal for the study of biochemical pathways. Many of the metabolic pathways we

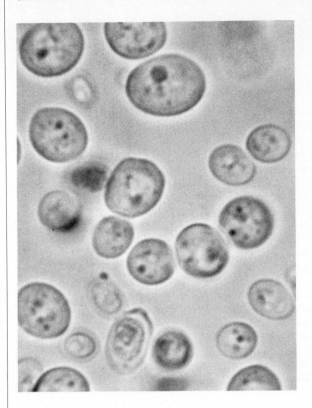

FIG. 6-56 Cells of brewer's yeast in the process of budding (center). Brewer's yeast and bread yeast have been domesticated by people for making beer and bread, respectively. Wild yeasts are used in wine production. Bread yeasts make bread rise. (*Courtesy of Richard H. Gross.*)

shall present in Chap. 12 were first worked out in test-tube colonies of yeast.

Some fungi are good to eat, mushrooms and truffles, for example. Mushrooms are often cultivated for food; however, some people insist on picking their own, often with fatal results. It is extremely difficult to be sure that mushrooms picked in the woods are not poisonous. An expert in mushroom poisoning says: "There are old mushroom hunters, there are bold mushroom hunters, but there are no old, bold mushroom hunters. The best place to pick mushrooms is in the grocery store."

Some mushrooms are used because of the psychoactive substances they contain. During the long winter nights the primitive peoples of northeastern Asia used to relieve their boredom with bouts of drunkenness induced by the fly agaric, a mushroom that in prolonged usage and large doses leads to raving madness.

People also use fungi to produce **antibiotics,** substances produced by one living organism that injure other living organisms. Penicillin, effective against a wide variety of bacterial diseases, is produced by the mold *Penicillium.* Although it was accidentally discovered in 1928, it was not widely used until the Second World War; now there are many hundreds of antibiotics.

Plant diseases caused by fungi Fungi are the single most important cause of plant disease; every year diseases caused by fungi destroy enough food to feed 300 million people. A fungus was responsible for the Irish potato famine when it wiped out most of the potato crop of Ireland in just about a week in the summer of 1846. In the 1870s, downy mildew of grapes brought in from America nearly destroyed the entire French wine industry. In fact, all of our agricultural crops are vulnerable to some fungus; two of the most important are wheat rust and corn blight. These parasites may weaken the plants and seriously reduce the size of the crop. In 1970, a new genetic strain of the fungus *Helminthosporium maydis* attacked the United States corn crop; about 17 percent of the crop was lost. Over 80 percent of the corn planted that year was of a variety susceptible to this new strain. The corn-blight epidemic illustrates well the dangers of planting large acreages with a crop of one variety which may prove susceptible to some new disease. Such fungi are a major problem in the massive monocultures which are our agricultural fields; diseases like these can spread like wildfire through these fields where all the plants are identical.

People who eat plants or plant products which have been attacked by fungi may be poisoned. In Europe there is a fungus called ergot that infests food grain plants such as rye, wheat, millet, and corn. The ovary of the grain flower is replaced by the ergot. Ergot possesses highly toxic compounds, including the parent compounds of the famous LSD-25. When grain contains more than 1 percent of ergot, the food produced from it can cause a disease called ergotism. In medieval Europe, periodically tens of thousands of people were stricken with delusions, nervous disorders, gangrene of the limbs, agonizing burning sensations, and convulsions. Increased appetite was often a symptom so that, not knowing the cause of their plight, victims often ate more bread made from the poisonous grain. By 1735 the cause of these outbreaks was discovered and grain was inspected for ergot. As late as 1951, however, an outbreak occurred in France. Officially, it was attributed to fungicides containing mercury; but reports of victims and physicians strongly suggest ergot poisoning.

Fungi growing on stored food such as peanuts have been shown to produce poisonous materials called *aflatoxins*. These substances have caused liver damage in poultry and have been shown to be able to cause cancer in people.

Human diseases caused by fungi Fungal diseases in people are far more common in tropical areas than in temperate zones. Several fungal skin diseases, a couple of lung diseases, and several diseases of the body openings are found in temperate zones. Athlete's foot and ringworm are fungal diseases of the skin. After treatment with antibiotics a patient often experiences a secondary infection of fungi. Apparently, the harmless bacteria which normally live in parts of our bodies serve the useful function of keeping fungi in check; when the antibiotic kills all the bacteria, the fungi take over. Women often get a "yeast infection" of the vagina; and both men and women may

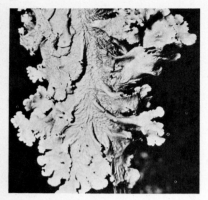

FIG. 6-57 This lichen is growing on a tree branch. (*Carolina Biological Supply Co.*)

get fungus infections of their mouths and disturbances of the intestines and rectum.

Lichens Some fungi form an interesting association with certain algae. The cells of the alga are enclosed within the filaments of the fungus that keep the algal cells from drying out and otherwise protect them. The algal cells carry on photosynthesis, and the fungus obtains a portion of the products for its own use. Such a combination of a fungus and an alga is known as a **lichen.** Lichens occur in several different forms (Fig. 6-57). Their shapes and colors depend more on the kind of fungus than the kind of alga. Lichens are extremely resistant to cold, heat, drying, and even being stepped on or nibbled by a caribou. Lichens, however, have a fatal weakness that threatens their survival in the modern world; they are extremely sensitive to pollutants put into the air by people. In England the forests downwind from the heavy industrial coal-burning areas have no lichens growing on trees or fenceposts. In fact, some scientists have suggested that we should use lichens as pollution monitors.

6-11 Protozoa usually get their food by ingestion

Protozoa are for the most part unicellular organisms that "eat" very much like animals. Some manage to flow right around their food and engulf it; others have special regions of their cell bodies that serve as "mouths." Often the "mouth" is surrounded by cilia that create water currents bringing in food. Their food consists of smaller phytoplankton, bacteria, and particles of larger organisms that have been decomposed and are floating in the water. Protozoa make up an important part of the zooplankton, the first level of herbivores depending on the primary producers. They themselves in turn are eaten by other slightly larger zooplankton. The aquatic protozoa are thus important in food chains.

Parasitic protozoa Just as some fungi have made their homes in or on living organisms, so have some protozoa. The most notorious is probably the malarial parasite. Several species of the protozoan genus *Plasmodium* cause malaria, the most important disease of mankind (it has killed more people than any other disease). The parasite lives part of its life cycle in a mosquito and the other part in a human being (Fig. 6-58). In a human being, the parasite has a stage it spends *inside* the red blood cells of its host. At periods of 48 to 72 hours, depending on the type of malaria, the host undergoes chills followed by a high fever. Malaria is most prevalent in tropical regions, where it causes untold suffering and death. In the United States, in 1935 there were 900,000 cases of malaria; by 1945 this was down to 62,763. Draining of the swamps in which mosquitos breed and treatment of hosts were largely responsible. After 1945 DDT was used with abandon in mosquito abatement programs. By 1960 there were only 72 cases of malaria in the entire country. Mosquitos were resistant to DDT in many places by that time, however, and people returning from abroad brought new infestations of *Plasmodium*. The

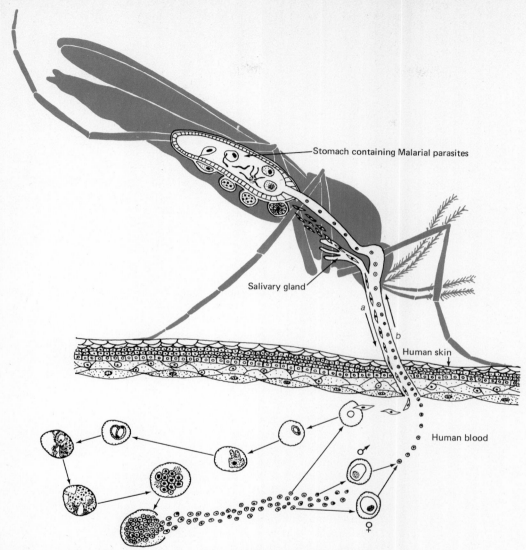

FIG. 6-58 Life cycle of a malarial parasite. (a) A female mosquito which has previously bitten an infected person carries elongated cells of malaria in her salivary glands and injects them into each new person she bites. In a bitten person, the malarial parasites enter the red blood cells and multiply. After a period of 48 to 72 hours, depending on the kind of malaria, the red blood cells burst, releasing many more malarial parasites into the blood stream. At this time, the afflicted person experiences fever and chills. The new parasites invade other red blood cells, some of which repeat the cycle above, while others form cells destined to fuse to become a new generation. They do this (b) not in the human, but when they are taken into a mosquito.

FIG. 6-59 Example of a free-living protozoan. *Vorticella* is a ciliated freshwater protozoan. (*Carolina Biological Supply Co.*)

frequency of malaria has been rising in this country since 1960, almost entirely because of cases arriving from abroad. In California there are now mosquitos resistant to all known insecticides. The origin of such resistance is discussed in Chap. 10. Malaria is just one of a great many diseases caused by protozoa. Most of them are best known in tropical regions and in the undeveloped countries. Protozoa also cause diseases in wild and domestic animals; in some cases, wild animals of a region are disease reservoirs for protozoan parasites of domestic animals and man.

Harmless protozoa Protozoa may be free-living as single cells or colonies in moist soil or, more commonly, fresh or seawater (Fig. 6-59). They can form resistant spores which may be blown around by the wind. If you go out to a park or vacant lot and get a handful of dried grass and put it in a jar of water, within a few days the water will contain a rich culture of protozoans that have grown from spores which had fallen on the grass. You will need a microscope to see the individuals of this beautiful little ecosystem. This culture is an ecosystem that undergoes succession like any other. There will also be algae present, and with sunlight as the only input, species will replace each other, and the water will evaporate as time goes on. As the water becomes stagnant and dries up, some of the organisms will form resistant resting stages and will be able to repeat the cycle again whenever water is available.

SOME MONERA

The Monera are all microscopic as individuals; however, colonies of these individuals may be visible to the unaided eye. They are different from all other organisms in that they have prokaryotic cells.

6-12 Blue-green algae are photosynthetic Monera

Blue-green algae are found almost everywhere on earth. Blue-green algae live in hot springs, melting glaciers, as plankton in the oceans, in soil, and many other places. Blue-green algae were the first organisms to invade the Indonesian island of Krakatoa after it was denuded of all life by a volcanic explosion in 1883. One species has been found in wet soil or fresh or brackish water all the way from Greenland to Antarctica and vertically from the floor of Death Valley to the top of Pike's Peak. Many blue-green algae live inside other organisms such as amoebae, fungi, plants, and even other blue-green algae. Inside other organisms, they usually lack a cell wall and function as do chloroplasts. Because of their varied habitats, and because many of them are not in fact blue-green in color, blue-green algae are difficult to identify. The periodic algal "blooms" which give the Red Sea its name are caused by a *red* blue-green alga.

Some blue-green algae can use nitrogen from the air to make compounds of nitrogen which are necessary for plant life. This ability of blue-green algae allows them to be early invaders of bare rock; it also makes them useful in agriculture. In Southeast Asia rice can be grown year after year in the same rice paddies without addition of fertilizer because there is a rich growth of nitrogen-fixing blue-green algae in the paddies.

6-13 There are many different kinds of bacteria

The bacteria are a very diverse group of organisms. While they have only three main growth types—rods, spheres, and spirals (Fig. 6-60) —they have developed enormous complexity in their metabolic patterns, their habitats, and their habits. Bacteria occur everywhere on the earth's surface and well up into its atmosphere; they are able to reproduce very rapidly when conditions are right. Sometimes as a result of this rapid reproduction they occur in colonies. A few species are photosynthetic. If circumstances are unfavorable, many species produce thick-walled spores. These are highly resistant to drying out and may be blown great distances. If you were to prepare an agar plate or a beef broth solution and then expose it only briefly to the air in a typical classroom, you would find a great many different kinds of bacteria growing there in a few days (Fig. 6-61), especially if you incubated the agar at a temperature favorable to the growth of bacteria.

Ecological roles of bacteria Many are decomposers, breaking down wastes and the bodies of dead organisms so that they can be recycled. Bacteria play roles in most of the cycles of the elements. Sometimes these roles are such key ones that an entire cycle would be disrupted if a key type of bacterium were to be destroyed. An example is the group of bacteria that convert ammonia into less poisonous products in the nitrogen cycle. Ammonia is a by-product of the breakdown of protein; without this group of organisms our atmosphere would soon become unpleasant and eventually toxic. Many bacteria have

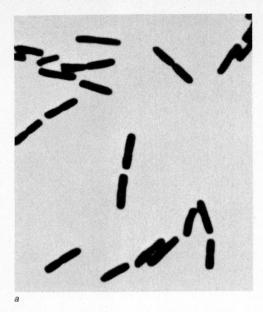
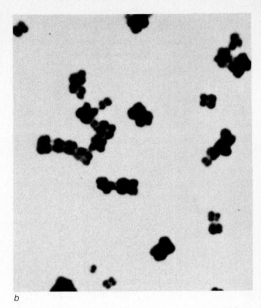
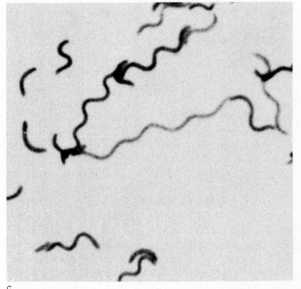

FIG. 6-60 The three main types of bacteria. (a) Rod-shaped or bacillus-type bacteria include the microorganisms known to cause lockjaw (tetanus), diphtheria, and tuberculosis. (b) Round or coccus bacteria include bacteria important in the nitrogen cycle as well as those known to cause bacterial pneumonia. (c) Coiled- or spirillum-type bacteria include the microorganism causing syphilis. (*Ward's Natural Science Establishment, Inc.*)

the ability to remove pollutants from the air. Soil bacteria that can take up carbon monoxide and nitrous oxide have been found. For many pollutants, such as some herbicides, fungicides, and insecticides, we do not know whether bacteria can break them down. One Environmental Protection Agency report suggests that we should breed bacteria for this purpose. We do know that bacteria have great

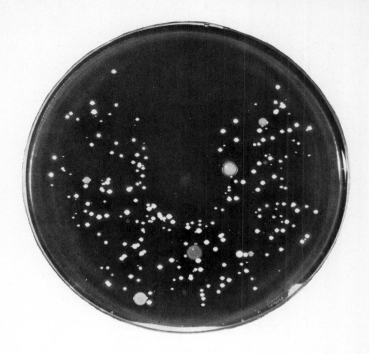

FIG. 6-61 A glass dish called a petri dish with colonies of several kinds of bacteria growing on a solid culture medium. (*Environmental Services Branch, National Institutes of Health, Public Health Service.*)

difficulty breaking carbon-chlorine bonds such as are found in DDT. Sometimes they may turn an inert pollutant such as mercury into an active one such as methyl mercury, which concentrates in food chains (see Sec. 1-11). Bacteria are also helpful in the roles they play in helping break down cellulose in the guts of herbivores and providing vitamins and nutrients to other organisms.

Diseases caused by bacteria Bacteria often cause disease in plants, animals, and microorganisms. These diseases are usually produced by some product of the bacteria, commonly a **toxin** (poisonous substance). Some diseases are **contagious;** that is, they may be "caught" from another person by the transfer of bacteria or their spores. Others may be contracted by eating contaminated food, drinking contaminated water, or even from touching contaminated towels or doorknobs. Fortunately, most of the contagious bacterial diseases are under control in our country. Diphtheria and scarlet fever are no longer the common threats of childhood that they once were. However, there remain many other disease-causing bacteria that are not under control. Table 6-4 lists some of the most frequently reported bacterial infections in the United States in 1969 and 1972. Strep throat (caused by *Streptococcus*) is distressingly common, but it can be treated with penicillin. Unfortunately, many bacteria, especially many strains of *Staphylococcus* ("staph"), are now resistant to such antibiotics. Some bacteria are harmless to us directly, but may grow in our food and make it poisonous. Many foods, such as cream pies,

TABLE 6-4 SOME MAJOR REPORTABLE BACTERIAL AND RICKETTSIAL DISEASES IN THE UNITED STATES IN 1969 AND 1972

DISEASE	BACTERIA OR RICKETTSIA*	MODE OF TRANSMISSION	Number of cases 1969	Number of cases 1972
Gonorrhea	Neisseria gonorrhoeae	Sexual relations with infected person	534,872	767,215
Strep throat and scarlet fever	Streptococcus pyogenes	Air and objects contaminated by infected person	450,008	n.a.
Syphilis	Treponema pallidum	Sexual relations with infected person	92,162	91,149
Tuberculosis	Myobacterium tuberculosis	Air and objects contaminated by infected person; also contaminated food	39,120	32,932
Salmonella type food poisoning	Salmonella sp.	Contaminated food	18,419	22,151
Bacillary dysentery	Shigella sp.	Contaminated food and water	11,946	20,207
Typhoid fever	Salmonella typhosa	Contaminated food and water	364	398
Diphtheria	Corynebacterium diphtheriae	Nasal secretions of infected person; contaminated food and milk	241	152
Tetanus	Clostridium tetani	Present in many soils; becomes pathogenic in deep puncture wounds	116	128
Rocky Mountain spotted fever (tick-borne typhus)	Rickettsia rickettsia	Bite of tick from infected animal	498	532

* All are bacteria-caused except Rocky Mountain spotted fever.

potato salad, canned mushrooms, and soups, are good media for bacterial growth under certain conditions. Botulism, a disease that is often fatal, is caused by eating food containing toxin produced by botulism bacteria. This toxin, botulin, is one of the most poisonous substances known. It has been calculated that as little as one cup is enough to kill every person in the world! This bacterium lives only in conditions where there is *no oxygen,* and so it is usually found in improperly canned foods. The toxin it makes, however, is not harmed by oxygen, and it is this toxin accumulated in the canned food that can kill you. Recently there have been recalls of many commercial products thought to contain botulism toxin, and the increase in home canning has resulted in more cases from this source. Other bacteria may enter the body with food. *Salmonella* often grows in foods left out of the refrigerator for too long a time. It produces an illness often mistaken for stomach flu. Typhoid fever is classified in the same genus as this mild food poisoning. Typhoid, too, can be transmitted via food if you are so unlucky as to eat food prepared by some persons having typhoid bacteria. Such persons, called *carriers*, are relatively unharmed by the bacteria but are able to spread them wherever they go.

Bacteria as a group are indispensable to the continuation of life on this planet. Some of them, however, give people a good deal of trouble. In fact, in some cases, exposure to some of them may be fatal. This is one of the reasons the government has agencies that inspect food canning and preparation areas in canneries and restaurants, and also keep statistics on communicable diseases.

Rickettsia Rickettsia are smaller than most bacteria but larger than most viruses. Rickettsia resemble bacteria, but they are intracellular parasites like viruses. They attack both people and animals. Rickettsial diseases have killed more people than any other diseases except malaria. Typhus, for example, is carried from rats to people by fleas and from person to person by lice, especially under crowded conditions. It has played a major role in war; spreading through crowded, unsanitary military camps, it has often killed more soldiers in an army than were killed in battle.

VIRUSES

Viruses are noncellular; they are at the borderline between living and nonliving. Viruses can grow and reproduce only inside living cells; they take over the cellular mechanisms of their hosts and use them to make more virus particles. The cell dies, and the viruses are released to infect other cells. Some viruses have the peculiar property of "living" inside cells for long periods of time without damaging them. Viruses usually consist only of protein and RNA or DNA (Chap. 9).

6-14 Viruses cause diseases

Plant diseases Tobacco mosaic virus (Fig. 6-62) was the first virus to be isolated and crystallized. Crystallization is a property of compounds, not of organisms; therefore, it was quite surprising when crystals from infected tobacco plants could be dissolved and reapplied to healthy plants and cause the characteristic spotted leaves which are a symptom of the disease. Plant viruses are commonly transferred from plant to plant by sucking insects such as aphids. It is said, however, that people who smoke tobacco should not smoke around tomato plants because they can spread tobacco mosaic virus to the tomato plants.

Human diseases Table 6-5 lists some of the human diseases caused

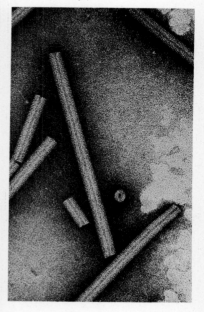

FIG. 6-62 Tobacco mosaic virus (TMV) as seen with the electron microscope. TMV was the first virus to be extracted from an infected plant, purified, crystallized, and then found to be capable of infecting another plant. This experiment, done in 1933, indicated that viruses are not simply smaller bacteria but are composed of only two compounds instead of the vast array of organic compounds found in even the smallest bacterium. Tobacco plants infected with TMV show a characteristic mosaic pattern on their leaves as the virus kills leaf cells in patches. Magnification about × 100,000. (*Virus Laboratory, University of California, Berkeley.*)

CHAPTER SIX

TABLE 6-5 SOME DISEASES CAUSED BY VIRUSES

DISEASE	TYPE OF VIRUS	MODE OF TRANSMISSION	Number of cases reported in U.S. 1969	1972
Infectious hepatitis	Not known	Contaminated water or shellfish, blood	48,416	54,074
Measles	Myxovirus (RNA)	Air contaminated by infected person	25,826	32,275
Rubella (German measles)	Myxovirus (RNA)	Air contaminated by infected person	57,686	25,507
Common cold	Rhinovirus (RNA)	Air and objects contaminated by infected person	Not reportable	
Influenza	Myxovirus (RNA)	Air and objects contaminated by infected person	Not reportable	
Mumps	Myxovirus (RNA)	Air and objects contaminated by infected person	Not available	
Cold sores	Herpesvirus (DNA)	Probably present in skin cells of everyone	Not available	
Poliomyelitis	Picornavirus (RNA)	Air and objects contaminated by infected person	20	31
Smallpox	Poxvirus (DNA)	Air and objects contaminated by infected person	Not available	
Yellow fever	Arbovirus (RNA)	Mosquito bite (*Aedes aegypti* or others if transmission is from monkey)	Not available	
Rabies	Rhabdovirus (RNA)	Bite or scratch of infected mammal especially dogs, cats, skunks, and bats	3,490 (in animals)	4,369 (in animals)

by viruses. Many of these diseases are mild in healthy adults, for example, the common cold; others have severe effects and high death rates, for example, yellow fever. Polio is another well-known, severe virus disease (Fig. 6-63). There are a number of different strains of virus that cause influenza. Influenza viruses very readily mutate (change) to more or less infective forms; often a new strain will spread widely and rapidly enough to cause an epidemic. Worldwide, during 1918 and 1919 there were an estimated 200 million cases of "the 1918 flu" and 10 million deaths resulting from the influenza or the secondary bacterial infections that accompanied it. No one knows why the epidemic ended or why the many epidemics of influenza which have occurred since have all been much milder, for example the Asian flu and the London flu.

There are many viruses which are poorly known and extremely dangerous. The Marburg virus discussed in Chap. 17 is an example of a virus making a sudden and deadly appearance. Another example is Lassa virus: a missionary nurse was flown from Africa to New York via commercial airline for emergency treatment of an unknown disease. Several research workers were infected while attempting to identify and treat the disease. Finally the research was transferred to the maximum-security virus laboratory in Atlanta, Georgia; meanwhile an unknown number of people in Africa died of the disease. In this age of jet travel it is possible to spread diseases from continent to continent in a matter of hours; it is indeed fortunate that none of the other passengers on the plane with the nurse with Lassa fever was infected.

Some viruses persist in their hosts for long periods of time without causing symptoms. In people, the herpesvirus which causes cold sores remains harmlessly in cells for long periods until stress (such as sunburn or fever) triggers it into action. Apparently, measles virus can become inactive in some people who have had the disease and then reappear months or years later as an acute fatal brain disease. This reappearance is exceedingly rare; however, slow, inapparent, and

FIG. 6-63 The viruses causing the diseases (a) polio and (b) influenza, as seen with the electron microscope. Magnification about ×300,000. (*Virus Laboratory, University of California, Berkeley.*)

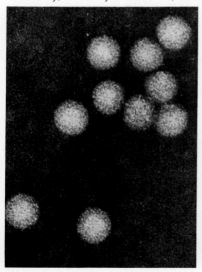
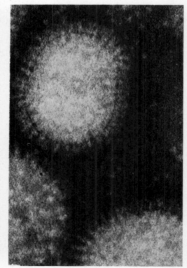

a
b

recurrent viruses are suspected of playing a role in triggering a number of chronic degenerative diseases in people. Viruses are known to cause tumors in plants and cancers in laboratory animals. They are suspected of playing some role in human cancer.

6-15 Viruses can be studied in the laboratory

Virus laboratories must take many precautions to avoid the escape of viruses. Since viruses are not alive and can survive as crystals, they are difficult to destroy. Usually materials infected with viruses are burned; and at least once, a chamber designed to isolate astronauts returning from the moon was used to isolate a suspected Lassa fever case. In order to study viruses, it is necessary to grow them in living cells, either in the bodies of live animals or plants or in tissue cultures or eggs. An enormous amount of skill is needed to work safely in the laboratory with highly infectious viruses such as Marburg or Lassa virus.

Uses of viruses   Viruses which attack bacteria are called **bacteriophages;** it is these viruses which have led to our understanding of the nature of the genetic material (Chap. 9). Unfortunately bacteriophages have not proved to be useful in treating bacterial diseases; however, they are useful in identifying the strains of bacteria in typhoid outbreaks.

Viruses which attack pest organisms may be used in the biological control of the pest. The disease which was introduced into Australia to control the rabbit population was a virus disease. Presently research is underway on virus diseases of insects which might be used to kill insect pests. Many insects have population explosions which are then controlled naturally by virus diseases.

In the laboratory, viruses can be used to transfer bits of genetic material from one organism to another. It is not known how important this role is in nature. If genetic engineering (Sec. 17-4) ever becomes a reality, it will probably make use of harmless viruses to transfer genetic material.

QUESTIONS FOR REVIEW

1 Can you arrange these classification categories in order from the broadest (most general) to the most specific: order, kingdom, species, family, class, phylum, genus—and say which of them are used in the scientific name of every living thing?

2 To what overall "class," broader than a kingdom, do all unicellular and multicellular organisms belong? (Define this class or group in such a way as to exclude viruses.)

3 Why are viruses and rickettsias not included in Table 6-3 (the four-kingdom classification) or Table 6-2 (classification of living species)?

4 Various kinds of marine invertebrates are commonly called *shellfish*. Can you name at least four different kinds—including two in the same phylum and two in different phyla?

5 What do these terms mean: eukaryotic, placenta, pollination, cilia, organelle, prokaryotic, lichen, uterus, vertebrate?
6 Why must amphibians normally live in or near water? What is one reason they are so often nocturnal?
7 What criterion (measure) does the text use in asserting the dominance of *Homo sapiens*? Can you think of one criterion by which arthropods could be called the dominant organism of the biosphere? Can you give a criterion by which the nematodes (roundworms), the bacteria, or the blue-green algae have a better claim to this title than does the whole vertebrate phylum, including *Homo sapiens*?
8 Where in the biosphere did animals originate? Where did plants originate? Which came first—land animals or land plants?
9 Which kingdom has the most species? Which phylum has the most species? Which kingdom do you suppose has the most individual organisms at any given time?
10 What is another name for the nonfunctional xylem of plants?
11 What are rickettsias? What is one deadly rickettsial disease of humans, and how is it spread?
12 What feature of the skeletal design of all insects ensures that huge, people-sized insects will continue to be a potential threat only in science-fiction stories?

READINGS

Berg, C. O.: "The Fly That Eats the Snail That Spreads Disease," *Smithsonian,* **2**(6):9–17, 1971.
Biddulph, S., and O. Biddulph: "The Circulatory System of Plants," *Scientific American,* **200**(2):40–49, offprint 53, 1959.
Burnet, M.: "Viruses," *Scientific American,* **184**(5):43–51, offprint 2, 1951.
Carefoot, G. L., and E. R. Sprott: *Famine on the Wind—Man's Battle against Plant Disease,* Rand McNally, Chicago, 1967.
Fuller, J. G.: *Fever! The Hunt for a New Killer Virus,* Reader's Digest Press, New York, 1974.
Glob, P. V.: *The Bog People: Iron-Age Man Preserved,* Ballantine Books, New York, 1969.
Grant, V.: "The Fertilization of Flowers," *Scientific American,* **184**(6):52–56, offprint 12, 1951.
Hedgpeth, J. W.: "Animal Diversity: Organisms," *Biocore* unit XIV, McGraw-Hill, New York, 1974.
Hegner, R. W., and K. A. Stiles: *College Zoology,* 7th ed., Macmillan, New York, 1959.
Holland, J. J.: "Slow, Inapparent and Recurrent Viruses," *Scientific American,* **230**(2):32–40, 1974.
Lamb, I. M.: "Lichens," *Scientific American,* **201**(4):144–146, offprint 111, 1959.
Pfeiffer, J., et al.: *The Cell,* Life Science Library, Time Inc., New York, 1964.
Postgate, J.: *Microbes and Man,* Penguin, Baltimore, 1969.
Raven, P. H., and H. Curtis: *Biology of Plants,* Worth, New York, 1970.
Rickson, F. R.: "Plant Diversity: Organisms," *Biocore* unit XV, McGraw-Hill, New York, 1974.

7

BEING YOUNG AND GROWING OLD

SOME LEARNING OBJECTIVES

After you have studied this chapter, you should be able to

1. Distinguish between genotype and phenotype by defining each, stating which is largely determined by environmental conditions during development, and stating what is included in your own genotype.
2. Name and briefly describe four types of asexual reproduction, name some unicellular and multicellular organisms that reproduce asexually, and state which type of asexual reproduction is employed by each organism you named.
3. Demonstrate your understanding of some basic aspects of sexual reproduction by
   a. Stating how the outcome of mitosis differs from that of meiosis.
   b. Stating why genetic recombination is important, saying whether it takes place during mitosis or meiosis, and naming the three different processes by which it takes place.
   c. Describing how gametes are produced.
   d. Explaining how *identical* twins originate, what is *identical* about them, and why that term correctly describes them only during the earliest stage of their embryonic development.
4. Demonstrate your knowledge of certain developmental processes by doing the following:

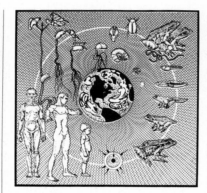

   a. State the text's definition of development and name the two kingdoms, all of whose organisms may be properly said to develop, and the kingdom some of whose organisms may be said to develop.
   b. Say whether development involves meiotic or mitotic cell division.
   c. Briefly describe the course of embryonic development from zygote to three-layered embryo, and name the organ systems mentioned in the text which develop from each of the three embryonic layers in vertebrates.
   d. Distinguish between the following types of development—complete metamorphosis, incomplete metamorphosis, gradual development—by naming a kind of animal whose growth is characterized by each type, and by describing the developmental stages of each of the animals you have named.
   e. Explain, with examples, what it means to say that a plant's development is less determinate than an animal's.
5. List some examples of negative environmental impacts on human development.
6. Demonstrate some of what you have learned about aging and death by
   a. Naming at least one type of human cell in each of these categories: Is continually dying and being replaced, even in young people; Is replaced only slowly during and after maturity; Dies in increasing numbers, and is not replaced, as the person grows older.
   b. Listing at least three bodily changes other than the death of cells which contribute to the aging process.
   c. Stating one positive and one negative consequence of the prolongation of human life well past the reproductive age.
   d. Briefly explaining how the cessation of life—that is, death—is essential to continuation of life, in two senses: the replacement of individuals of existing species, and the evolution of new species.

One of the distinguishing characteristics of organisms is that they reproduce themselves. Unicellular organisms usually divide into two, and each new cell, called a **daughter cell,** then enlarges. Multicellular organisms usually begin life as a single cell which divides many, many times to produce a multicellular organism. As cell division proceeds, the cells differentiate and become organized into tissues and organs. Unlike unicellular organisms, multicellular organisms usually have a long period of development before they are fully functional as adults. Also, unlike unicellular organisms, which are potentially immortal, multicellular organisms almost always undergo a process of aging and eventually die.

LIFE CYCLES OF ORGANISMS
7-1 Some organisms reproduce asexually

Division   Asexual reproduction does not involve two individuals of different sexes. One type of asexual reproduction is that which occurs when an unicellular organism such as an amoeba or a bacterium divides into two daughter cells. When this happens, each daughter cell must receive the same genetic information. Genetic material is composed of a chemical compound called DNA (see Sec. 9-9) and contains the information needed if the cell is to function and reproduce. In prokaryotic cells, the genetic material is in the form of a long molecule of DNA, which makes a copy of itself before the cell divides. The original goes to one daughter cell, the copy to the other.

Eukaryotic cells have their genetic material in organelles called **chromosomes,** which are visible during the process of cell division. When eukaryotic cells divide in asexual reproduction, they undergo a process called **mitosis.** In mitosis each chromosome is duplicated and the two daughter organisms have the *same number of chromosomes as the parent and the same genetic information.*

Mitosis can be divided into four stages: prophase, metaphase, anaphase, and telophase (Fig. 7-1). In *prophase,* the chromosomes gradually become visible as they shorten by coiling up. As they shorten, each can be seen to consist of two strands held together by a structure called a *centromere.* As prophase merges into metaphase, the nuclear envelope breaks down and proteins in the cytoplasm become arranged into a spindle-shaped mass with pompomlike asters at either end. (These proteins make up tiny tubules which are called *fibers* because of their appearance.) During *metaphase,* the maximally shortened chromosomes are oriented with their centromeres along the equator of the spindle. The beginning of *anaphase* is marked by division of each centromere, and the daughter chromosomes move to opposite ends of the spindle. *Telophase* is the stage during which nuclear envelopes are formed around each group of daughter chromosomes and the chromosomes uncoil, lengthening and losing their visibility.

Usually both daughter cells also have the same kinds of organelles

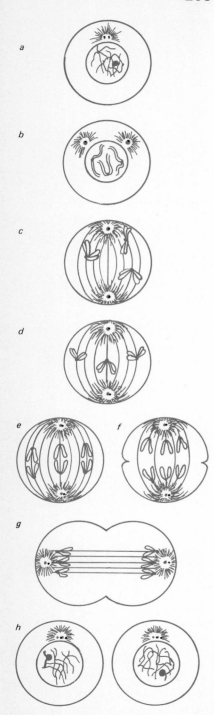

FIG. 7-1 The stages of mitosis. Mitosis has been divided into four stages: (*a* to *c*) prophase, (*d*) metaphase, (*e* and *f*) anaphase, (*g*) telophase. The two daughter cells are shown in *h*.

as the parent cell although they may not be equally divided. Generally, cytoplasmic organelles need not be divided exactly between the two daughters since each cell can manufacture more organelles. Some organelles contain their own genetic material and are able to divide and reproduce themselves; others may be made by the daughter cells.

Differentiation after division Unicellular organisms undergo differentiation after division. In some of the Protista, such as protozoa, each daughter cell grows in size and completes its set of organelles. It becomes structured the same way as its parental cell and is then specialized to carry on the way of life of its species and divide again when conditions are suitable. The genetic material which the organism received from its parent cell contains information called the **genotype** of the organism. The structure and functioning of the organism make up its **phenotype.** The phenotype is controlled both by the genotype and by the environment in which it develops. The phenotype of the protozoan shown in Fig. 6-6 includes all its specialized organelles, its size, its shape, and all the details of its cellular chemistry.

In nature, the same species of unicellular organism is sometimes found in slightly different environments; species may also be grown in different environments in the laboratory. Often the process of differentiation is different in different environments. Because mitosis duplicates the chromosome sets exactly, the genetic material is the same, but the phenotypes that result from differentiation are different. Studies of this situation have shown that the phenotype is the result of the *interaction* of the genetic material and the environment. The genotype sets the capability or the potential for the phenotype. The environment determines what the actual phenotype will be. This is true also for multicellular organisms.

Budding, cloning, and parthenogenesis Many kinds of multicellular organisms can reproduce *asexually* as well as sexually. Members of the phylum Coelenterata often reproduce by developing a small new individual on the outside of their bodies. When the new individual is sufficiently large, it can break away and start life on its own. This process of **budding** is also common in plants. Plants often send out underground shoots which develop into new individuals; often a whole patch of such plants as wild strawberry is produced this way. Such a group of genetically identical individuals is called a **clone.** Plant breeders often exploit the ability of plants to replace missing parts. Branches of particularly desirable plants usually can be made to produce roots and eventually a new plant if they are broken off and placed in water or moist sand (see Fig. 9-20). Some animals can replace missing parts **(regeneration),** and some worms can divide in two and each half will replace its missing parts. Starfish can also regenerate lost parts, much to the chagrin of oyster fishermen who used to cut starfish in half when they found them preying on their oyster beds. Of course the more starfish the fishermen "killed" this way, the more there were.

Some insects and even some vertebrates can reproduce asexually. Aphid females may have daughters all summer long without mating, and some species of lizards and fishes are all females. In such animals asexual reproduction involves development of eggs without true fertilization; this is called **parthenogenesis** or "virgin birth."

Exchange of genetic material Many asexual organisms have some kind of exchange of genetic material with others of their species from time to time. In single-celled organisms there are often **mating types**, sometimes more than two. Cultures of a pure strain often will die off unless members of the other mating type are introduced. In bacteria a "male" transfers genetic material to a "female" through a tube called a pilus. In members of the phylum Protozoa, nuclei may be exchanged between individuals in a process called *conjugation*. The genetic exchanges of bacteria are particularly important because harmless bacteria may pass on antibiotic resistance to disease-causing types, or mild strains may become virulent after genetic exchanges. Furthermore, bacteria are not particular about whom they exchange genetic material with; they may exchange material between species or even genera.

7-2 Sexual reproduction begins with production of gametes

In organisms with sexual reproduction, adult individuals produce sex cells. These sex cells are called **gametes.** Some organisms produce gametes that are all the same size; other organisms produce gametes of very different sizes. In these organisms, male gametes are called **sperm** and are small in comparison with female gametes, which are called **eggs** or **ova** (singular, **ovum**). Sometimes, as in snails, one individual may produce both sperm and eggs; when this occurs in animals, these species are called **hermaphroditic** species. In higher plants (where male gametes are carried by pollen grains) individuals which bear both male and female flowers on the same individual plant are said to be **monoecious.** Plants which have flowers of different sexes on different plants are called **dioecious.** In the vertebrates, excluding the Agnatha and certain fishes, one individual can produce only one kind of gamete, either sperm or eggs, and is classified as male or female accordingly.

Most animals have motile sperm (Fig. 7-2); that is, they swim using a flagellum. This "tail" pushes the sperm along like the propeller of a motorboat. An ovum is usually a relatively large cell with stored food reserves (Fig. 7-3). A sperm fertilizes an egg when it penetrates the cell membrane of the egg. The body of the sperm, which is mostly genetic material, enters the egg (Fig. 7-4), and a fertilized egg, or **zygote,** is formed, containing genetic material from both parents. The zygote begins to divide by mitosis, and eventually a young animal

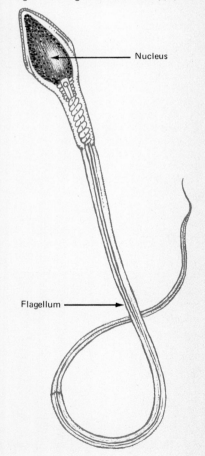

FIG. 7-2 A human sperm showing flagellum. Magnification about ×4,000.

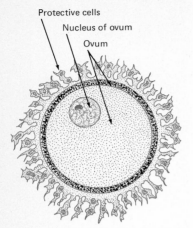

FIG. 7-3 A human egg surrounded by protective cells. Magnification about ×250.

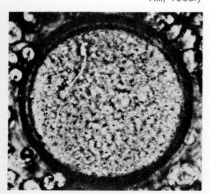

FIG. 7-4 Fertilization in a mouse. A sperm is seen on the surface of the egg. Fertilization of human eggs outside the body is a controversial subject. The zygote produced cannot survive at present because safe techniques for reemplanting the zygote in a human uterus have not been devised, nor has a satisfactory artificial uterus. Implantation of eggs fertilized outside the body has been accomplished in cattle and some other animals, and there are reports from time to time that it has been accomplished in human beings. (*From J. Smiles, in* Biology of Laboratory Mouse, *E. Green (ed.), McGraw-Hill, 1966.*)

or plant called an **embryo** is formed (Fig. 7-5). This embryo goes on to develop into an **adult** which produces either sperm or eggs depending upon its sex, and the life cycle continues.

Halving the genetic material The formation of eggs and sperm involves halving the genetic material. When an egg and sperm fuse in fertilization, their nuclei fuse also, and the resulting zygote contains two sets of chromosomes carrying two sets of genetic material. A cell with two complete sets of chromosomes is called a **diploid cell.** Thus the zygote and all the cells produced by mitotic divisions from it are diploid. If such a diploid individual produced eggs or sperm with the same amount of genetic material it has in each cell, the fertilized egg of the next generation would have four sets of genetic material and the generation after that eight sets and so on. Except for rare instances this does not happen. A special process of reduction division called **meiosis** reduces the genetic material to one set of chromosomes for each gamete. Cells with one set of chromosomes are said to be **haploid;** thus gametes are haploid, and meiosis is the process which changes diploid cells into haploid cells. Each gamete—egg or sperm—contains one set of chromosomes and therefore one set of genetic information.

The set of chromosomes that would be found in a human sperm and the set in a human egg are shown in Figs. 7-6 and 7-7. There are 23 chromosomes in each set. When the chromosomes differ from one another in size, shape, and chemical staining, individual chromosomes can be identified. In the past this was not an easy procedure, especially in organisms such as *Homo sapiens* that have a large number of small chromosomes. But modern methods of chemically staining chromosomes have greatly improved our ability to distinguish individual chromosomes. By photographing and carefully measuring them, the chromosomes in the set can be recognized individually and classified into groups.

If you compare the set of chromosomes from the male parent with those from the female parent, you will notice that, with one exception, there is one chromosome of each type in each set. In other words, *each chromosome in the male set has a duplicate in the female set.* The exception is that one of the pairs does *not* contain similar chromosomes. This pair of chromosomes, known as the **sex chromosomes** because of the role it plays in sex determination, will be discussed in greater detail later, in Chap. 9.

Obviously, when a sperm fertilizes an egg, the resulting cell has *two* sets of chromosomes. One came from the father and one from the mother. The individual which develops from the fertilized egg produces gametes that have only *one* set of chromosomes because the organs producing gametes—**testes** (singular, **testis**) producing sperm, and **ovaries** (singular, **ovary**) producing eggs—are able to carry out meiosis. Meiosis occurs in a testis when sperm are produced or in an ovary when eggs are produced. In the course of two *cell* divisions—but only one division of the chromosomes—meiosis accomplishes a number of important things. (Each of the two divisions has

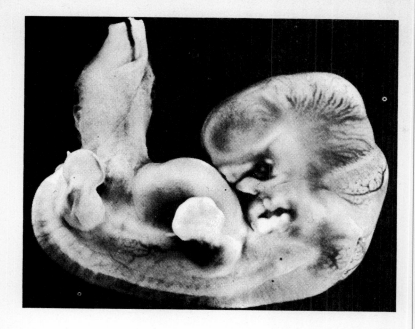

FIG. 7-5 A very young human embryo. (*Carnegie Institute of Washington.*)

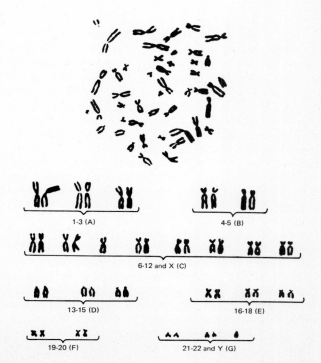

FIG. 7-6 Chromosomes of a normal human male. *Above,* as seen under the microscope; *below,* numbered and arranged in groups, indicated by capital letters, according to size and shape. The X chromosome is in group C; the Y chromosome is in group G. (*Jeroboam, N.Y.*)

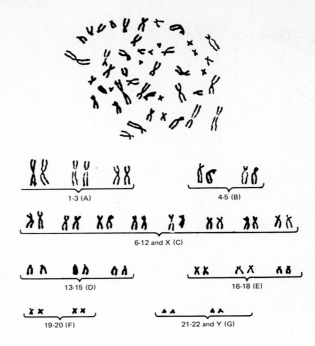

FIG. 7-7 Chromosomes of a normal human female. The X chromosomes are in group C. (*Courtesy of Park S. Gerald, M.D.*)

the same four stages as mitosis.) First, the *number* of chromosomes in each cell is reduced by half. In human beings this means the chromosome number of 46 in each of the parental cells is reduced to 23 in their descendant cells, the sperm and the egg (Fig. 7-8).

Independent behavior of chromosome pairs Second, meiosis assures that not just any 23 chromosomes wind up in an egg or sperm. Each gamete receives *one complete set of chromosomes*. The set of chromosomes that an egg or sperm receives, however, is *not* one of the two original sets received in the egg or sperm from the previous generation. During meiosis each pair of chromosomes behaves independently of all the other pairs (Fig. 7-9). Let us make this clearer by numbering the chromosomes and labeling the members of each pair M if they come from the male parent and F if they come from the female parent.

The original sperm would have the following set of chromosomes: $1M, 2M, 3M, \ldots, 23M$. The original egg would have: $1F, 2F, 3F, \ldots, 23F$. When the sperm fertilized the egg, the zygote would have: $1M, 1F; 2M, 2F; 3M, 3F; \ldots; 23M, 23F$, or a total of 46 chromosomes. The individual developing from that fertilized egg would have 46 chromosomes in every cell of its body. When it began to produce gametes, meiosis would assure that each gamete would have one set of chromosomes. Because of the independent behavior of members of pairs,

FIG. 7-8 The stages of meiosis: (*a* and *b*) first prophase, (*c*) first anaphase, (*d*) second prophases, (*e*) second anaphases, (*f*) four daughter cells with the haploid number of chromosomes. Note that the stages are given the same names as in mitosis, but meiosis consists of two divisions of the parental cell and only one division of the chromosomes.

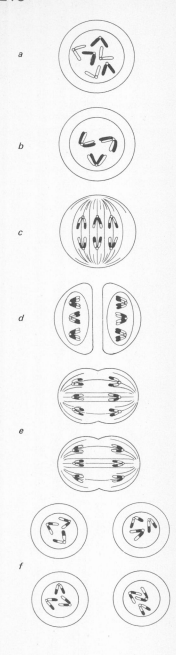

the chances that any one gamete would have the same set as one of the original sets are extremely small; this behavior is called **independent assortment** (see Fig. 7-9). All possible combinations of chromosomes from the male and female parents can occur. A sperm could contain 1M, 2F, 3M, etc., to 23F. Another sperm might contain 1F, 2F, 3M, etc., to 23M. In fact, there are 2^{23} different chromosome sets possible in either sperm or eggs.

Crossing over A third very important thing happens in meiosis. During meiosis chromosomes from the male and female parent come together in pairs or **synapse.** While they are associated (during **synapsis**), breakage of the strands of the chromosomes usually occurs. At this stage each chromosome is composed of two longitudinal strands called **chromatids.** Chromatids of the same chromosome are called **sister chromatids.** When the chromatids break during synapsis, much of the time broken sister chromatids reunite. Sometimes, however, nonsister chromatids become joined (Fig. 7-10). This is called **crossing over.** If a particular location on a chromosome can be shown to control a genetic trait (such as eye color), that location is called a **locus** (plural, **loci**) and the genetic material at that place a **gene.** Figure 7-10 shows two different loci, *A* and *B*. A chromosome that was formerly completely *M* may now contain portions of *F* chromosomes. Its partner, formerly completely *F*, now contains portions of *M*.

We can symbolize a chromosome from the male parent with a series of *M*'s—*MMMMMMMMMMMM*. In the same way, a chromosome from the original mother would be *FFFFFFFFFFFF*. During meiosis, breakage and rejoining might produce a pair of chromosomes such as this:

MMMFMMFMMMM . . . and *FFFMFFMFFFF* . . .

These processes leading to new combinations of the genetic material of the mother and father are called **recombination.** Recombination is the most important source of genetic variability in populations and, as we will see, is essential to the evolutionary process of most organisms. The genetic material is a very important determinant of the appearance and functioning of an individual. Recombination of parental genetic material assures that although offspring will be similar to their parents in many respects, they will be different from them in others. Because of independent assortment of chromosomes, crossing over, and the determination by chance of which sperm will fertilize which egg, *it is very unlikely* that two individuals of a species with sexual reproduction will look exactly alike or function in ex-

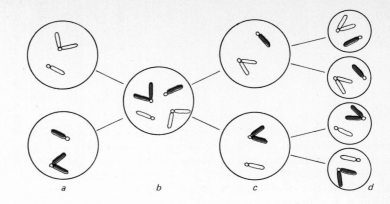

FIG. 7-9 Independent assortment. Paternal (*white*) and maternal (*color*) chromosomes behave independently in meiosis. (*a*) Male and female gametes, (*b*) zygote, (*c*) first meiotic products of zygote, (*d*) new gametes showing new combinations of the original chromosomes.

actly the same way. Exceptions are identical twins (or triplets, quadruplets, etc.). Identical twins are produced by mitotic division of a single fertilized egg resulting in two separate individual cells. This means that each twin will have precisely the same genetic information as the other. Thus any difference between such twins will be due only to effects of differing environments. Fraternal twins are not genetically the same.

7-3 Cells vary in the number of chromosome sets they contain

Many organisms have a life cycle similar to that of the human species (Fig. 7-11). Such a life cycle begins with a fertilized egg and is completed by gamete formation. Most of this life cycle is spent in the diploid condition; the only haploid cells are the gametes, which must quickly unite or die. A few of your body (somatic) cells such as some in your liver have more than the diploid number of chromosomes; those cells are said to be **polyploid** (three or more sets). Many plants have life cycles in which the cells in the mature plant are polyploid. Those with four haploid sets are said to be tetraploid, those with six, hexaploid, etc. Many of our important agricultural and crop plants are polyploid, for example, bananas, wheat, and apples.

Some organisms spend most of their lives as haploids. Many Protista, as well as a few Metaphyta, spend the greater part of their life cycle as haploids (Fig. 7-12). The mosses and certain algae and fungi are the only multicellular organisms for which this is true. In all these organisms meiosis occurs shortly after fertilization, and the only division the fertilized egg undergoes is meiosis. The daughter cells

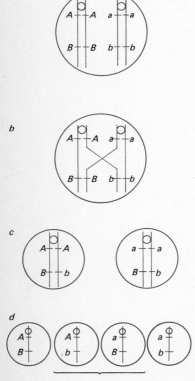

FIG. 7-10 Crossing over of strands of chromosomes occurs during synapsis (pairing) of homologous chromosomes in the first division of meiosis. A, a, B, b represent loci on the chromosomes. (*a*) Pairing (synapsis of chromosomes), (*b*) crossing over, (*c*) products of first division of meiosis, (*d*) gametes showing two with recombination and two without. (Small circles indicate centromeres.)

of this division become new haploid individuals of the next generation, which may divide by mitosis; the only diploid cell in this sort of life cycle is the fertilized egg. Notice that mitosis can take place in either haploid or diploid cells; the haploid individuals in this sort of life cycle are produced by meiosis or by mitotic divisions of haploid cells.

DEVELOPMENT

Multicellular organisms undergo a period of development before they are sexually mature adults. Development is the sequence of events in which the fertilized egg divides repeatedly by mitosis and its daughter cells undergo specialization and become organized into tissues, organs, and systems of organs.

7-4 Organisms grow by mitotic divisions of their cells

From zygote to embryo As we noted previously, the egg is a large nonmotile cell; often it is loaded with food reserves called **yolk**.

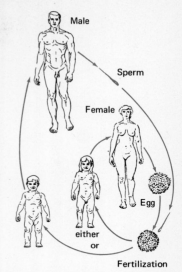

FIG. 7-11 Life cycle of human beings.

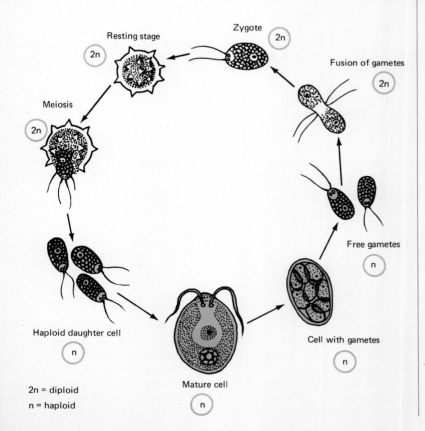

FIG. 7-12 Life cycle of the green alga *Chlamydomonas*. After fusion of gametes, the zygote forms a thick wall, becoming a resting stage. This resting stage later undergoes meiosis producing four haploid daughters which may produce gametes by mitosis.

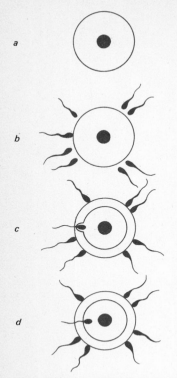

FIG. 7-13 The fertilization of a sea urchin egg. (*a*) Egg, (*b*) sperm cluster around egg in sea water, (*c*) one sperm penetrates egg and fertilization membrane appears, (*d*) remaining sperm cannot penetrate fertilization membrane.

During fertilization many sperm swarm around it, but only one sperm fertilizes the egg. The first sperm to enter the egg triggers the formation of a *fertilization membrane* (Fig. 7-13) which prevents other sperm from entering. In organisms predominantly diploid, after the sperm and egg nucleus fuse, the fertilized egg, now a zygote, divides mitotically. First there are two cells, and then each of these divide and there are four. These divide and the cells continue to increase in number—8, 16, 32, 64. . . . These cells may form a hollow multicellular ball (Fig. 7-14) or, in eggs with a great deal of yolk, a flattened hollow plate at one end of the egg (Fig. 7-15). Cell division continues until the embryo is made up of hundreds or thousands of cells.

The three-layered embryo What comes next is a process converting a hollow ball or flat plate of cells into an embryo with three cell layers. The details of this process vary from species to species. Basically, movement of cells and folding of the hollow ball occur so that some of the outer cells are inside (Fig. 7-16). You can understand this process by imagining poking your finger into a balloon which is not blown up completely. The part of the balloon around your finger represents the inner layers of cells. In various ways, in the different major groups of animals, a middle layer of cells develops between the inner and the outer layers of either the hollow ball or the flat plate.

In the vertebrates, each of these layers gives rise in the course of development to different organs and organ systems of the embryo. The inner layer becomes the digestive tract and the lungs. The outer layer gives rise to the entire nervous system and the outer layers of skin. The middle layer produces the skeleton, the muscles, and the sex and excretory organs as well as the inner layers of skin. Some organs combine cells of different layers, however.

7-5 Movement of cells and tissues results in organ formation

As the embryo develops, cell movement occurs in all parts. Individual cells become specialized for particular functions. In other words, they undergo differentiation. Differentiation is affected by a number of things. Some cells and tissues may cause other cells to differentiate in a particular way. This process is called **induction.** Some cells produce substances called **inducers** which affect the development of other cells.

Interaction of cells A good example of the interaction of cells in development occurs in the formation of the eye (Fig. 7-17). The major portion of the eye arises as a swelling on each side of the head and the nervous system. This is the part of the nervous system that will become the brain. The tip of the bulge from the nervous system becomes cup-shaped as it pushes out toward the skin. When it reaches the skin layer, it is deeply cup-shaped. The presence of this extension of the brain causes the skin to fold inward. Eventually a hollow, flattened mass of cells separates from the skin and occupies

FIG. 7-14 Formation of the hollow-ball stage after division of the zygote in a frog. (*a*) Fertilized egg, (*b*) two-celled stage, (*c*) four-celled stage, (*d*) eight-celled stage, (*e*) many-celled stage is hollow.

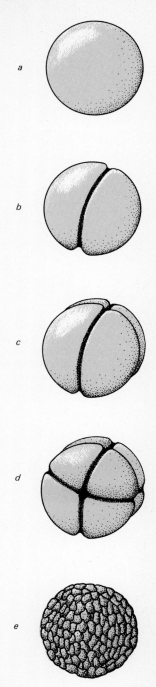

the opening of the cup. This group of cells will become the lens of the eye.

Biologists have found, in experiments with salamander embryos, that if the eye cup is removed on one side, no lens tissue develops. The other side develops normally. It is possible to remove the eye cup from the head and place it under the skin somewhere else in the body of another embryo. The skin will form a lens (Fig. 7-18) but the resulting third eye will not be functional because it will not be connected to the brain. A final functional eye is the result of the interaction of different kinds of cells, and also contains blood vessels which are from cells of the middle layer of a three-layered embryo.

Interaction of genotype and environment The millions of cells of which an embryo is formed are derived by mitosis from a fertilized egg. Mitosis produces daughter cells all of which have the same genetic material or genotype. All cells of an embryo therefore have the same genotype. But in the course of development, different cells differentiate for different functions. Some become liver cells, some become eye cells, some become bone cells, and some become blood cells, etc. (Fig. 7-19). If all cells of the embryo have the same genetic information, how can there be so many different cell phenotypes?

Earlier it was mentioned that the same kind of unicellular organism grown in different environments may produce different phenotypes. The phenotype, then, is the result of the interaction of the genotype and the environment of the differentiating cell. This appears to be part of the answer to how differentiation occurs in multicellular organisms. If you think about it for a moment, you will see that different cells of a multicellular embryo are in different environments. Some are on the inside; some are on the outside. For example, some of those on the outside are very close to a developing eye cup, while most are not. The cellular environment plays a role, then, in determining the phenotype of cells in a multicellular organism. Inducers appear to be substances which turn parts of the genotype on and off. A cell in the skin of your big toe once had the potential of being an eye lens cell; but since it was nowhere near a developing eye cup, this potential was never realized.

7-6 Different kinds of organisms have different kinds of development

Gradual development The process of development in some animals is a gradual one from early embryo, to the young individual, to the adult. It is not possible in these organisms to divide growth into a series of distinct stages. For example, society determines the ages at which we classify a human being as an infant, a child, an adolescent,

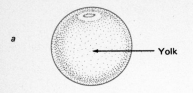

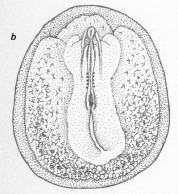

FIG. 7-15 Development of a chicken embryo. (a) After division of a group of cells atop the massive yolk, the cells become arranged in two and then three layers; (b) shows the developing embryo of (a) in greater detail.

or an adult. In people, as in some other organisms, growth slows as an individual becomes sexually mature, and eventually it stops. Very old people even become slightly smaller. In many fishes and reptiles, growth slows as sexual maturity is reached, but an individual continues to increase in size until its death. In plants, growth continues until the individual reaches the maximum size for its species in a particular environment.

Distinct stages of development Tadpoles are immature stages in the development of a frog or toad (see Fig. 6-26). The fertilized egg of a frog develops into an embryo, which gradually develops into a tadpole. Tadpoles have no legs, but they have well-developed tails with which they swim. Like fishes, they have gills for oxygen uptake. They eat algae and plant materials, living as tadpoles for a month to several years, depending upon the species.

Eventually a tadpole begins to change (see Fig. 6-26). Hind legs start to grow on each side of the body at the base of the tail. Front legs appear at the time the tail is gradually being absorbed and growing smaller and smaller. The tadpole then begins to breathe air as the gills stop functioning. It changes into an adult frog that spends part of its life on land and, as a carnivore, eats insects and other small animals. This remarkable change in structure, function, and way of life is called metamorphosis. It requires several weeks to several months in most frogs and is controlled by hormones from the thyroid gland.

Metamorphosis occurs in a great many other kinds of animals. Insects have several kinds of metamorphosis. **Incomplete metamorphosis** is found in such species as cockroaches and grasshoppers (Fig. 7-20). In this kind of development, the egg hatches into an immature stage called a **nymph.** The nymph looks much like the adult but lacks wings and its sex organs are not functional. Nymphs feed and grow larger and molt, shedding their rigid "skins" (which are their external skeletons) several times in the process. This shedding is necessary to permit growth, since the external skeleton does not grow between molts. The final shedding reveals an adult that has wings and is sexually mature. All insects with fully developed wings are adults and will grow no more—a small winged bug is *not* a baby that will grow into a big bug.

The second type of metamorphosis is called **complete metamorphosis** (see Fig. 6-33). This is the type occurring in bees, houseflies, and butterflies, for example. In this type of development, the egg hatches into a **larva.** The larva can be thought of as the stage of life specialized for eating and growing; it looks nothing like the adult. It consumes enormous quantities of food and increases in size, shedding its skin several times. With the final shedding of skin, the larva enters into its second stage of development, called the **pupa.**

The pupa is a stage during which the "feeding and growing machine" larva is transformed into the adult. After a period of time, the pupa's skin is shed and the adult emerges. The adult is winged and has mature sex organs. You may think of it as a "reproducing and dispersing machine." In the life cycle of an insect with complete metamorphosis, each stage is specialized for a particular role. Hormones (chemical messengers) control each stage. In the future, artificial insect hormones may play an important role in controlling pest insects by upsetting their normal development.

Plant development Plants develop differently from animals. As we saw in Sec. 6-7, when a pollen grain falls on the stigma of a flower, a pollen tube grows down to a ovule in the ovary. One of the two sperm nuclei in the pollen tube fuses with the egg and the other fuses with two other nuclei in the ovule to form the triploid (three haploid sets of chromosomes) **endosperm,** which becomes stored food for the embryo (Fig. 7-21). The wall of the ovule hardens and becomes the coat of a seed. Thus a seed contains an embryo plant and stored food. Sometimes the wall of an ovary develops into a fleshy fruit—for example, an apple or a peach. Sometimes it develops into a dry covering around a seed, as in an acorn or a peanut.

When a seed sprouts, cells in the embryo divide and become differentiated into various kinds of tissues. A root, stem, and leaves form; soon the new plant has used up the stored food provided by the endosperm and provides its own food by photosynthesis. At the tip (apex) of the root and at the tip of the stem, there are growing regions called **apical meristems.** These consist of permanently embryonic cells which continue to divide for long periods of time, increasing the length of the root or stem. The development of plants is less determinate than that of animals; for example, a horse always has a head, four legs, and a tail attached to the body in the same place. A forest of oak trees will probably not have two trees exactly alike in their branching pattern. You can prune part of a branch from a tree and some of the permanently embryonic tissue in a bud on the stem behind the cut will grow to form a strong, new branch. The apical meristem produces a hormone called **auxin** which prevents buds lower on the branch from growing. When it is removed, usually the bud near the cut develops a branch with a new apical meristem at its growing tip.

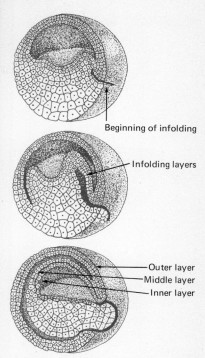

FIG 7-16 Formation of the three-layered stage in a frog.

7-7 Development may be upset by an unfavorable environment

We have seen that the phenotype of cells and individuals is determined by their genetic material interacting with the environment. Our discussion so far has assumed that the environment will support the normal growth of the organism. It has been found by biologists that unusual environmental situations may cause abnormalities in development. If frog eggs are raised at too high or too low a temperature, they may never hatch or the resulting tadpoles may be so abnormal that they die before metamorphosis.

Environmental conditions affecting human development Foreign substances in the environment of the developing human embryo may cause abnormal development. If a pregnant woman smokes or uses

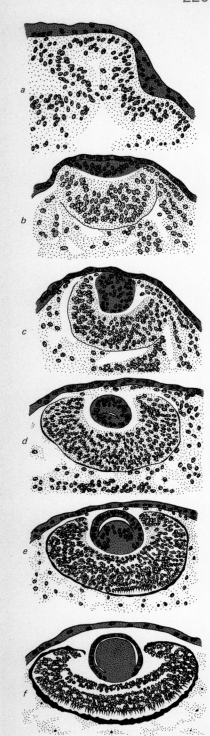

FIG. 7-17 Stages in the formation of the eye in a frog. (a) Outgrowth of brain approaches skin, (b) skin thickens opposite brain, (c) brain outgrowth becomes cup-shaped and skin folds inward, (d, e, and f) brain outgrowth becomes eye-cup and infolded skin becomes lens. (*Adapted from D. Bodenstein.*)

certain kinds of drugs, her child may be affected. Such effects may be very mild. For instance, there is some evidence that the children of mothers who smoke heavily are smaller and less healthy at birth than those of mothers who do not smoke at all. Unfortunately, other effects may be much more serious, for example, the effects of thalidomide. Thalidomide is a tranquilizer which had been shown to reduce nervousness apparently without any known serious side effects. It was prescribed for men and women, including pregnant women. Babies born to women who had used thalidomide during the early months of pregnancy were grossly malformed. Their arms were small and distorted in shape, often little more than flipperlike appendages. Thalidomide was prescribed by doctors until this side effect became known. Needless to say, it is no longer used for pregnant women. Such substances which cause defects in developing embryos are called **teratogens**.

The sad point of the thalidomide story is that the drug was thought to be unusually safe—until a great many babies were born who are growing up without normal arms. Thalidomide would have been widely used in the United States except for one doctor of the Public Health Service who insisted that it needed more testing before it was approved. She later received the President's Medal for her caution. Suppose thalidomide had caused only a minor deformity or one that occurs normally with low frequency, such as cleft palate or slight mental retardation. It would probably be widely used now and would still be considered safe, because detecting these effects would require an extensive study of large groups of children. Even if such a study were to be done, it would be almost impossible to point to any specific drug as causing an increase in frequency of a defect that normally occurs at low frequency. Think of all the drugs that can be bought in any supermarket or drugstore. Even without a prescription, a pregnant woman can expose herself to any number of things that may affect her unborn child. Many prescription drugs that were once widely used are no longer recommended for pregnant women. It is not wise for a pregnant woman to use large quantities of aspirin, cough medicines, antihistamines, sleeping pills, etc. without her physician's guidance.

Medicines are not the only chemicals that have been shown to affect development of unborn infants. People are putting many chemicals into their environment which may have an effect on developing babies. The substance 2-4,5T is used as an herbicide (weed killer). It has been used to clear roadsides of weeds and to remove plants from the vicinity of high-voltage electric lines and the towers that carry them. It has also been used to create grassland for grazing. In

FIG. 7-18 If an eye at a very early stage of development in a salamander (*color*) is transplanted to another salamander, it will cause the formation of a lens and a third eye will be produced. (*a*) Section through early embryo showing removal of the area of the brain destined to be eyecup, (*b*) arrow shows site of transplant in section of second embryo, (*c*, *d*) sections of second embryo showing eye developing on flank, (*e*) young salamander with third (nonfunctional) eye.

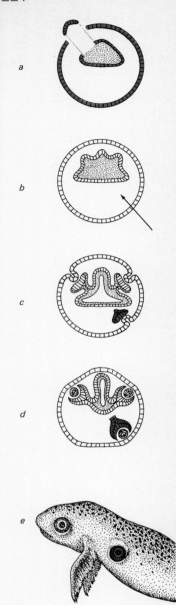

Southeast Asia it was used to remove leaves from trees and kill brush by American forces wishing to deny "cover" to the Viet Cong. It is now known that animals exposed to 2-4,5T often produce deformed offspring, and the use of this substance is banned in many places. In New Zealand two babies with the same malformations were born of mothers whose water supplies were collected from rooftops sprayed with 2-4,5T. The closely related 2-4,D has now been shown to be teratogenic in rats also. You can go down to your local garden store and buy as much of *it* as you want. If your college has lawns, they have probably been sprayed with 2-4,D to kill broad-leaved plants among the grasses.

Medical authorities are becoming increasingly wary about exposing pregnant women to chemical hazards. The testing of chemicals for teratogenicity can be done only on small laboratory animals. Unfortunately, species differ greatly in their response to chemicals. Thalidomide was tested on laboratory animals. The human embryo turned out to be 60 times more sensitive to the teratogenic effects of thalidomide than the most sensitive animal tested, the mouse embryo. In some states, health authorities now advise pregnant women not to eat tuna or swordfish. These fishes contain the element mercury in their meat at a level which is not thought to be high enough to affect adults. It may, however, be high enough to damage a developing embryo. Presumably, these fishes are at the top of a food chain which is concentrating mercury from both natural and industrial sources by biological amplification. Unfortunately, medical advice is not available for many people in the world who may be exposed to larger doses of even more potent chemicals. Even in Japan, which has well-organized medical facilities, a number of babies were born with defects because their mothers ate fish contaminated with methyl mercury (Sec. 1-11).

Treatment of such congenital abnormalities (birth defects) is possible in some cases if they are not too severe. In fact, about 1 in 20 human beings is born with some congenital malformation; most of these, such as moles, harelips, or slightly misplaced internal organs, are not serious enough to impair life processes. Others, for example, anencephaly (no brain) or severe spina bifida (open spinal column), are incompatible with life. Recently, some very severe congenital heart malformations in infants have been treated by a new kind of surgery. Surgeons are using deep hypothermia (cooling of the body) to put these tiny infants into a state in which essentially no blood is flowing through their bodies. They can then perform open heart

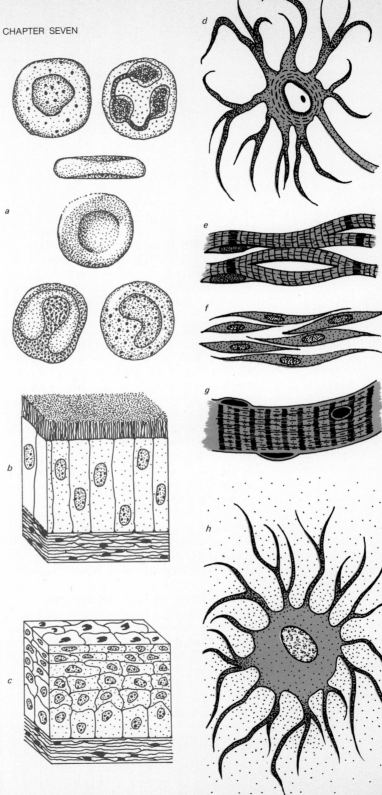

FIG. 7-19 Examples of differentiated cells in animals. (a) Various blood-cell types of human beings, (b) cells found on membranes on parts of the respiratory tract have cilia, (c) skin cells, (d) cell body of nerve cell, (e) muscle fibers of heart, (f) smooth muscle cells, (g) skeletal muscle fiber, (h) bone-forming cell.

surgery. Some of these operations have lasted as long as 90 minutes, and so far these infants appear normal after they recover from surgery. As of 1974 none of these children was yet old enough to tell for certain whether the period of prolonged arrest of their blood flow had damaged their brains. Certainly, it is far better to avoid birth defects by seeking out their causes, including environmental chemicals, than to use such difficult and dangerous procedures to repair them.

AGING AND DEATH

Although single-cell organisms are potentially immortal, many of them are killed and eaten by other organisms and others are killed by unfavorable environmental conditions. For most animals and plants, death from disease or attack by a predator is the inevitable end. For those organisms that are not eaten from without or from within (by disease or parasites), aging and death is their usual end. There is great variation in the length of time different species live before they age and die. Some plants and animals have a short lifetime of a few months or years and appear to be "programmed" to die fairly suddenly. Annual plants often live only a few weeks; some adult insects live only a day or two. On the other hand, some species grow as long as they live and appear to be programmed to live to great ages. The oldest known living things are trees—the bristle cone pines in the White Mountains of California. Some of these trees are over 4,600 years old. Nevertheless, if organisms did not die (assuming cycles of elements did not first break down), eventually all available matter would be tied up in old organisms and no new organisms could be formed; thus death is a natural and essential part of the processes of the biosphere.

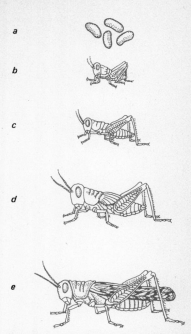

FIG. 7-20 Life cycle of a grasshopper, showing incomplete metamorphosis. (*a*) Eggs, (*b*, *c*, and *d*) nymphs, (*e*) adult grasshopper.

7-8 Aging is related to specialization of cells

Some cells die during the process of tissue and organ formation. During the development of a multicellular individual, cells die and are replaced with others. For example, certain kinds of bones are built on a foundation of cartilage, and as the cartilage is replaced with bone, the cartilage cells die and disappear. Cells of many embryonic tissues are resorbed and replaced with different, more developmentally advanced tissues.

Cells are dying and being replaced continuously in a multicellular individual. Special cells and organs take care of removing dead cells. For example, your red blood cells live only about 120 days; when they are first released into your bloodstream, they already lack a nucleus and should thus really be called corpuscles, not cells. Every second, your liver and spleen are busy destroying 2 million overaged red blood cells. Fortunately, new red blood cells are being produced at about the same rate. Meanwhile, the cells of your skin are renewed about every week, and even your skeleton is renewed about every two years.

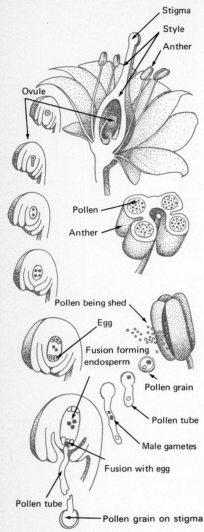

FIG. 7-21 Life cycle of a flowering plant, showing double fertilization and the formation of endosperm. Pollen grains on the stigma produce pollen tubes which grow down the style to the ovule. In the ovule an egg and seven other nuclei are formed. Two male gametes from a pollen tube enter the ovule through an opening at its tip, and one fuses with the egg, while the other fuses with two nuclei in the center of the ovule. The nutritive endosperm tissue develops from the latter fusion. The other nuclei disintegrate.

In plants, some cells are not even functional until they are dead, for example, the water-conducting cells of the xylem. Each year, trees produce new xylem cells which take over the function of old ones which remain in the interior of the tree and become heartwood. Cells in the heartwood can no longer conduct; they do, however, serve as structural support for the tree. This seasonal replacement of xylem cells is the source of the rings we see when we cut down a tree which has grown in a climate with pronounced seasons.

In multicellular organisms some cells are not replaced after maturity. Many cells become highly specialized for a particular job when they differentiate. When such cells die from aging or injury, they are not replaced; nerve and muscle cells in human beings fall into this category. If the long extensions of nerve cells called axons are cut, new axons will grow out again; however, if an entire nerve cell is killed, others nearby cannot divide and replace it. As people age, nerve cells die and the brain decreases in weight. Loss of these irreplaceable cells contributes to the overall aging process.

7-9 There are other aspects of aging in people

Slowing of body processes Aging is related to slowing of body processes. As we grow older, changes occur in the production of the chemical messengers of the body, the hormones; these changes may affect the functioning of cells. The thymus gland, which is partly responsible for our ability to resist disease, begins to atrophy (waste away) as soon as we reach sexual maturity. Much research is being done on cellular aspects of aging. One theory suggests that our cells are programmed to divide just so many times and no more; body processes slow as our cellular replacement slows.

Structural changes Aging is also related to structural changes. Sometimes cells are not able to get rid of all their waste products; this reduces their ability to function. Materials such as calcium may accumulate. Muscles and their attachments lose their elasticity. Walls of arteries may also lose their elasticity and become prone to breakage or to having blood clots form. Bones often become brittle and joints stiffen. Similar processes also occur in animals other than people. As aging occurs, an organism becomes more prone to accident, to attack by a predator, or to disease. It is unlikely that any wild animal will die of old age, and it is unlikely that very many of our early ancestors died of old age. In fact, until quite recently in the history of our species, very few people lived past the age of 30.

7-10 Human attitudes to death have varied greatly from culture to culture

How people react to death is very much a cultural matter. Some cultures, for example, those mentioned earlier that left bodies in bogs, practiced human sacrifice. In other cultures, death was celebrated as a time of rejoicing for an individual freed of the hardships of life. In early Western European culture, death was accepted calmly by both the dying individual and his or her relatives. Dying individuals departed easily, as if they were simply moving into a new house. Usually family members gathered about the deathbed; in fact, until the eighteenth century no portrayal of a deathbed scene failed to include children. Now, in the United States we exclude children from deathbed scenes and we try to hide death by putting the dying person in a hospital. Undertakers and funeral parlors were first licensed in the United States and do not exist in much of the world where the dead are buried or cremated by their relatives.

Efforts to prolong life People have tried many ways to prolong life. Our very earliest ancestors tried to protect themselves from predators. Today we have succeeded almost entirely in this endeavor. We have also succeeded in protecting ourselves from many diseases so that now infectious disease is no longer a major cause of death in much of the world. This leaves us with the problems of middle and old age: heart disease, cancer, stroke, as well as failing eyesight and hearing, which may lead to accidents as causes of death. We have devised special diets, operations, heart pacers, hearing aids, and many other techniques and devices for dealing with these problems.

Attempts are also being made to understand more thoroughly the biochemistry of aging. If we could understand the changing hormone balance that accompanies aging, we might be able to restore the youthful balance artificially. As yet, however, no simple technique of injecting hormones can restore youth to an aging individual. Some success has been had by doctors who have given hormones to women approaching menopause (the time when menstruation ceases and women can no longer bear children). These treatments seem to slow the aging process and prevent the loss of calcium from bones which causes "dowager's" hump and other skeletal abnormalities in older women. Women are the only female primates who live long beyond the reproductive years and thus suffer this severe hormone imbalance.

7-11 Prolonging human life has created changed social conditions

Prolonging human life has increased the size of the human population. Many people alive today would have died of childhood diseases if they had been born 100 years ago. Because more people live longer, there are more people around at any given time. In fact, it is a decrease in death rates, not an increase in birthrates, that has led to the population explosion, which we discuss in Chap. 8.

Dependency of aged Prolonging human life has increased the dependency load. In all societies, people who are disabled or too young or too old to work are dependent on the rest of society to provide for them. In hunting and gathering cultures, old people who could not keep up might be left behind to die. In times of famine, infants might be allowed to die because they could not survive if their parents starved, whereas if the parents survived they could have another child. In most contemporary societies, people feel a moral obligation to keep people alive whether they can work or not. We have a great many people today who live past the age at which they want to work or are able to work; we also have rules which require people to retire at a certain age. Unless these people were able to save money for their retirement, somebody else must support them. In the United States many retired people live on social security checks which are so little that they must live in near poverty. Older people have more illness than young or middle-aged people; unless they have wealth, or private or government insurance, they must often "go on welfare" if they have a serious illness.

When older people become senile or too weak and ill to care for themselves, they create grave problems for their families. In the past and in some traditional cultures, they would be cared for at home until they died. Today, with most members of a household working or in school, there is often no one at home who can care for a sick or weak person. To meet this need, a great many nursing homes and convalescent hospitals have been built. These are often profit-making organizations, although some are sponsored by religious and other nonprofit groups. While some of these institutions are good, far too many are simply "dumping grounds" for the dying in which "care" is given by poorly paid, overworked, and underskilled personnel. Several books, one with the intriguing title *Tender Loving Greed,* have recently documented the excessive profits made by unscrupulous operators of shabby nursing homes full of suffering old people. It does not seem to be a good idea to prolong life without making provision that it can be healthy and rewarding.

Cultural advances Prolonging human life has allowed many cultural advances. Postreproductive individuals do not make any significant contribution to the evolution of most species; a rare exception is sheep. Among sheep, the oldest females are the leaders of the flock. The human species, because it evolves not only biologically but also culturally (Chap. 15), is another exception; postreproductive individuals often make significant contributions to future generations. In traditional societies, postreproductive individuals were given the task of remembering and imparting the cultural wisdom of the tribe to new generations. In modern societies such as ours, postreproductive individuals often make significant contributions in science, art, music, philosophy, etc., and in making sure that our immensely complex cultures are transmitted to the coming generations.

7-12 Death makes way for new life

Death is a necessary event if evolution is to proceed. Death and reproduction make possible the replacement of existing individuals with new ones. The environment in all parts of the world is continuously changing, and new kinds of phenotypes and genotypes eventually must replace those that went before. In the course of sexual reproduction, the genetic material is continually being recombined and new genotypes which can develop into new phenotypes are constantly being produced. If the environment has changed, some of these new phenotypes may be superior to the existing individuals, but most will be inferior. Death, like reproduction, is important in the gradual replacement of less fit organisms with more fit ones.

Death is necessary to return elements to natural cycles so that they can be used again. If no living things ever died, no new living things could come into being.

QUESTIONS FOR REVIEW

1 What are two advantages of sexual reproduction?
2 Why is it so very unlikely that there is, anywhere in the world, another person with exactly the same genotype as yours (unless he or she came from the same fertilized egg)?
3 Can two organisms ever have the same phenotype? If so, when?
4 What do these terms mean: chromosome, parthenogenesis, gamete, meristem, hermaphrodite, embryo, teratogen, nymph, diploid cell, haploid cell, pupa?
5 "Culture, not biology, determines the developmental status of human individuals." What does this mean? Explain, using another species for comparison.
6 A zygote has two sets of genetic information—one from the sperm cell and one from the egg cell which joined to produce it. Why, then, don't fertilized eggs of the next generation have four sets, the next generation after that eight, the next sixteen, and so on?
7 To biologists, *the life cycle* of the human being (and of many other kinds of organisms) can mean something other than the round which begins with birth and ends with death. With what does another meaning of *life cycle*, discussed in the text, begin and end?
8 More is known today about the causes, prevention, and treatment of heart disease and cancer than was known 100 years ago. Why, then, do these diseases kill proportionately more people today than they did a century ago?

READINGS

Aries, P.: *Western Attitudes toward Death from the Middle Ages to the Present,* Johns Hopkins, Baltimore, 1974.
Balinsky, B. I.: *An Introduction to Embryology,* 2d ed., Saunders, Philadelphia, 1965.
Ebert, J. D., and I. M. Sussex: *Interacting Systems in Development,* 2d ed., Holt, New York, 1970.
Etkin, W.: "How a Tadpole Becomes a Frog," *Scientific American,* **214**(5):76–88, offprint 1042, 1966.
Fogg, G. E.: *The Growth of Plants,* Penguin, Baltimore, 1963.
Gifford, E. M.: "Plant Diversity: Cells and Tissues," *Biocore,* unit XI, McGraw-Hill, New York, 1974.
Keller, D. E.: *Sex and the Single Cell,* Pegasus, Bobbs-Merrill, New York, 1972.
Leaf, A.: "Getting Old," *Scientific American,* **229**(3):44–52, 1973.
Mannes, M.: *Last Rights—A Case for the Good Death,* William Morrow, New York, 1974.
Mendelson, M. A.: *Tender Loving Greed,* Knopf, New York, 1974.
Tanner, J. M.: "Growing Up," *Scientific American,* **229**(3):34–43, 1973.
Wessels, N. K., and W. J. Rutter: "Phases in Cell Differentiation," *Scientific American,* **220**(3):36–44, offprint 1136, 1969.

THE MARCH OF GENERATIONS

SOME LEARNING OBJECTIVES

After you have studied this chapter, you should be able to

1. State the two main strategies by which living things allocate their reproductive efforts; say which one is apparently limited to the organisms of one phylum in one kingdom, and name the phylum; state at least one advantage and one disadvantage of each strategy; and name two kinds of animals that employ each.
2. Demonstrate your knowledge of the human reproductive process by
   a. Explaining how fertilization is achieved in human beings, and naming the structures and substances involved in the process as discussed in the text.
   b. Outlining the events which take place in the female reproductive system during the menstrual cycle, stating the cycle's normal period (length), and listing at least four things which can disrupt the cycle.
   c. Outlining the process of embryonic and fetal development from fertilization to birth.
   d. Describing the course of birth from the beginning of labor to the physical separation of mother and newborn child.
3. List several ways in which *sex* (sexuality and sexual activity) shapes human relationships, including some ways which greatly increase the chances of maximizing the basic reproductive strategy employed by our species.

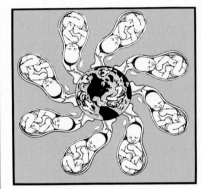

4. State the difference between contraception and the more general term, birth control; list three male and three female methods of birth control including two female methods which are not, strictly speaking, contraceptive techniques; and rearrange in order of the most effective to the least effective the methods you have named.
5. Explain what the gestation period is, and compare the gestation process in placental and nonplacental mammals.
6. Briefly describe the reproductive strategies developed by each of these groups—placental mammals, nonplacental mammals, birds, most reptiles—for overcoming the following problems of reproduction on land: (*a*) fertilization, (*b*) protection of the embryo, (*c*) nutrition of the embryo, (*d*) care of the young before hatching (by egg-laying animals), (*e*) care of the young after birth.
7. State some ways in which the reproductive strategies of land plants are similar to those of land animals, and name one strategy employed by many land animals that is *not* employed by any plant.
8. Demonstrate your knowledge of some basic principles of population dynamics by
   a. Stating the three factors which determine changes in the size of a population, and saying which of them are the most important.
   b. Stating how each of the following measures—birthrate, death rate, growth rate—is expressed.
   c. Stating what the relationship of a population's birthrate to its death rate must be for the population to (1) grow, (2) decline, or (3) remain stable.
   d. Calculating the birth, death, and growth rates of a population if you are given the number of individuals who were born into the population in a given year, the number of individuals in that population who died that same year, and the size of the population at midyear.
   e. Explaining what exponential population growth means and giving two reasons why populations grow exponentially.
   f. Describing the causes and results of an outbreak-crash population cycle, explaining why the human population has experienced a population outbreak over the last 8,000 years, and giving some reasons why humanity might be headed for a population crash.

228

Reproduction and death are responsible for the march of generations. Asexual reproduction results in offspring genetically the same as their parents; sexual reproduction results in offspring who do not have exactly the same genetic information as their parents. Both kinds of reproduction lead to the growth of populations of organisms. Species must have successful reproductive strategies or they cease to exist; that is, a species must produce enough offspring so that the parents are replaced when they die. As we shall see in Chap. 11, extinction has been the usual fate of species over the history of life. Some scientists estimate that 99 percent of all the species that ever existed are now extinct. Many of these, however, left descendants better able to survive than they were. In this chapter we shall review some of the successful reproductive strategies of present-day organisms and consider particularly human reproduction and the present status of the human population.

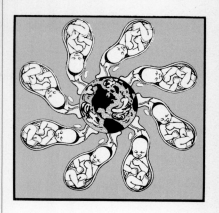

REPRODUCTIVE STRATEGIES

Reproductive strategies constitute a series of "game plans" by which a species is able to produce new individuals to replace those that die off. These game plans involve the anatomy (structure), the physiology (function), and the behavior of the individuals of the species in such a way as to make sure that there is an unbroken transmission of genetic material from generation to generation. Each species must have reliable methods of passing on its genetic material; if it is a sexual species, it must have ways of getting the sexes together so that their gametes can fuse. Each species must have ways to survive during fragile developmental periods or harsh environmental conditions; it must allocate some of its life energy for reproductive effort. Finally, in a world of changing environment such as ours, it must find ways either to be adaptable or to evolve adaptable offspring; that is, it must find ways to ensure variability.

8-1 There are two main strategies by which species allocate their reproductive effort

Large numbers of offspring Some species produce large numbers of offspring and provide little or no parental care. Many marine invertebrates cast huge numbers of eggs and sperm into the sea. Some freshwater mollusks are known to produce as many as 3 million larvae in a season. Parasitic species also produce enormous numbers of offspring: apparently they must do so to survive since the chance of encountering a suitable host is so slight. The female *Ascaris*, a roundworm parasitic in human beings, can produce 26 million eggs in her lifetime at the rate of 200,000 per day. These are expelled in the feces of the infected individual and must be eaten by another human being if they are to develop. Obviously, lack of cleanliness or the use of untreated human excrement for fertilizer is necessary for the success of this reproductive strategy. Many terrestrial insects also produce large numbers of offspring. In most insects, the parents die shortly after reproduction. Female insects often lay 1,000 fertile eggs; in some there are new generations every few weeks; in others there are new generations annually.

Parental care Other species produce few offspring and give parental care. Most mammals, birds, and a few reptiles provide parental care for their offspring. When there are overlapping generations, elaborate social structures may result. In many primates, the young remain with the parents, or the mother, for more than one year. A family group is then set up that includes the mother and several offspring, often of different ages. The father may or may not remain with the group. Frequently, a number of families become associated into a larger social group. Here several males may be accompanied by many females and their offspring of varying ages. In the primates other than *Homo sapiens* this is not strictly speaking an association of "families," since the males may mate with any female who is reproductively ready.

8-2 Animals may have either internal or external fertilization

The production of eggs and sperm The male parent produces sperm in organs called **testes**. Sperm are very specialized cells which cannot survive drying out and must also have a suitable chemical environment or the sperm may be killed or immobilized. Eggs are produced by the female parent in organs called **ovaries**. They are usually thousands of times larger than a sperm.

Since neither eggs nor sperm can survive without moisture, land animals are faced with the problem of getting sperm to eggs without either drying out. Since the chemical environment must be proper also, many aquatic animals have the problem of getting sperm to eggs without exposing them to a chemical environment (for example, fresh water) which they cannot endure. Some aquatic animals, for example, frogs and sea urchins, have developed very hardy sperm and eggs. No one knows how this process of **external fertilization** is being affected by the many pollutants in most bodies of water today.

Copulatory organs Many animals have solved the problem of getting sperm to eggs by developing special organs for the transfer of sperm from the male to the female called **copulatory organs.** These transfer sperm to a special opening in the body of the female. This process of **internal fertilization** by transferring sperm is called **copulation.**

In mammals (and in some reptiles), the male organ of copulation is called a **penis**. Associated with the penis are glands which secrete fluids assuring the proper chemical environment for sperm. When the penis is not being used in copulation, it is relatively small and may be enclosed in a protective sheath, as in a dog or cat.

REPRODUCTION IN HUMAN BEINGS

Reproduction in mammals is always sexual. The anatomy and physiology of reproduction in human beings is an example of reproduction in other mammals. There are some significant differences, however, which will be noted later. In Sections 8-3 to 8-7 we discuss sexual reproduction primarily in an anatomical and physiological sense. For human beings, the word *sex* connotes these things and much more. In later parts of this chapter and in Chap. 15 we consider the social and behavioral aspects of sexual reproduction in greater detail.

8-3 Internal fertilization in human beings is accomplished by copulation

Erection of the penis Human sperm are produced continuously in the testes (Fig. 8-1), and there is no significant monthly or yearly cycle of sexual activity in the human male. As in most other mammals, when a human male is sexually excited, the penis becomes enlarged and erect because of increased blood flow into its spongy tissues. The penis is then capable of being inserted into the vagina (Fig. 8-2), a muscular tube in which the female receives the penis during the process of copulation. (The human female has an organ homologous to the penis, the **clitoris,** which is made up of erectile tissue.) Surrounding the opening of the vagina is the **vulva,** consisting of the *labia majora* and *labia minora,* which together with the vagina produce lubricating secretions.

Copulation Friction between the vagina and penis during copulation stimulates nerve endings, inducing a state of tension that causes muscular contractions within the glands and ducts associated with the penis as well as the muscular tissue at its base. The sperm from the testes, which have been stored in the epididymis and the **vas deferens,** are mixed with glandular secretions. This mixture, called **semen,** is forcibly ejected in a process called **ejaculation.** The fluids in semen are produced by the prostate gland, Cowper's glands, and seminal vesicles (see Fig. 8-1). The fluids carrying the sperm serve as a source of energy for the sperm cells and also protect them from the usually acid environment of the vagina, which would kill them. The fluids also activate the sperm to begin their swimming activity. About 350 million sperm are released at each ejaculation. Substances in the seminal fluid encourage the contractions of the muscular wall of the female tract, which are important in helping the sperm arrive at their destination, the **fallopian tubes,** where fertilization may occur. It takes about 70 minutes for the sperm to travel to the fallopian tubes, and it is likely that only about 1 in 50,000 makes it. Studies suggest that the fertilizing capacity of sperm once they are in the female reproductive tract does not last more than a day or two.

Orgasm In males, the release of tension built up during sexual stimulation by ejaculation is called an **orgasm.** Women also undergo a similar build-up of tension during sexual intercourse and its preliminaries. They also experience an orgasm with release of tension, usually accompanied by contractions of various regions of the reproductive tract. Many women are able to experience a series of closely spaced orgasms, whereas men have a period following orgasm during

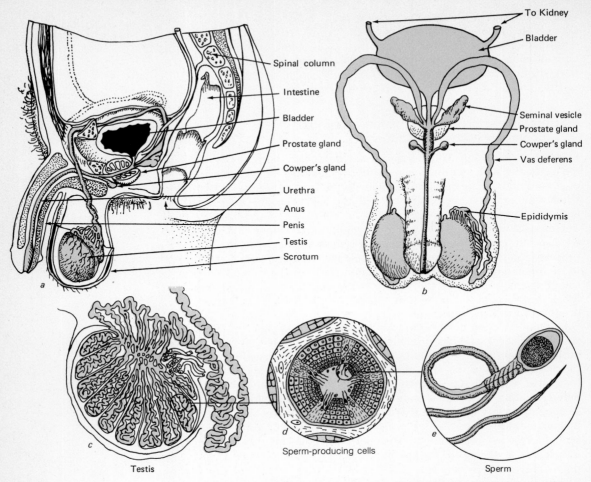

FIG. 8-1 Sex organs of a human male: (*a*) a longitudinal section through the male pelvic region, (*b*) frontal view showing the accessory glands, (*c*) testis in section showing epididymis, (*d*) sperm-producing cells, (*e*) a sperm.

which they are not able to experience a second orgasm. Both sexes usually feel extremely relaxed after orgasm. In the female, this helps ensure fertilization since the semen is less likely to run out of the vagina if she is lying down. Almost all people find sexual behavior extremely pleasurable.

8-4 The reproductive system of human females undergoes a monthly menstrual cycle

Ovulation Usually only one of the two ovaries produces an egg each month. When a human egg completes development in the ovary, it

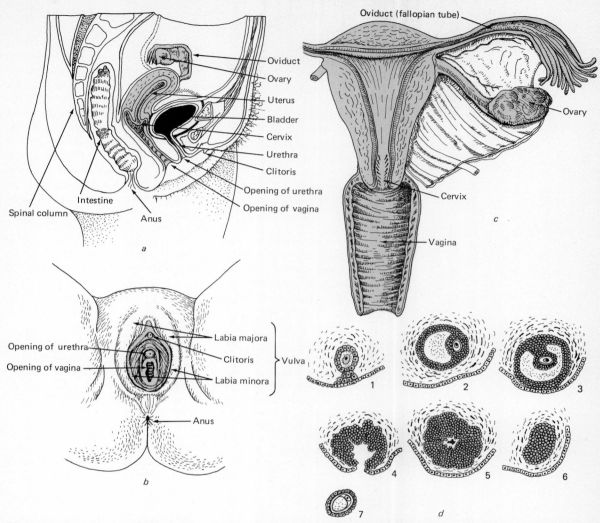

FIG. 8-2 Sex organs of a human female: (*a*) a longitudinal section through the female pelvic region; (*b*) external view; (*c*) longitudinal section of reproductive organs; (*d*) stages in the development (1 to 3) and release (4) of an egg from a follicle, and production (5) and regression (6) of a corpus luteum.

is released and enters the funnellike mouth of the **fallopian tube** (also called an oviduct, the tube through which the eggs pass). Egg production is called **ovulation**. If fertilization is to occur, the sperm must meet the egg somewhere in the upper regions of the fallopian tubes. Muscular contractions of the uterus and the fallopian tubes aid the sperm on their way. The sperm also use their flagella to push themselves along and one may penetrate the membranes of the egg at the point of fertilization in one of the tubes. Once fertilization has occurred, the fertilized egg begins to divide as it is moved from the tube to the uterus by action of cilia on the tube walls. Dividing fertilized eggs may implant (Sec. 8-5) in the wall of the uterus and undergo development. Unfertilized eggs disintegrate and pass out of the body through the **cervix** (muscular neck of the uterus) and vagina.

Preparation for implantation In human females, before ovulation the lining of the uterus, the **endometrium,** thickens and becomes richly supplied with blood. This prepares it for the **implantation** of the fertilized egg. If fertilization and attachment do not occur, part of the thickened uterine lining is eventually expelled by the uterus and passes out the vagina accompanied by some blood **(menses).** Then the whole process of egg development and thickening of the lining begins again. This repeated series of events is known as the **menstrual** (from the Latin word for "month") **cycle** because it takes 28 days or so on the average. The onset of bleeding (which usually lasts about five days) marks the end of one cycle. The entire cycle is coordinated by a series of hormones released by the ovaries and the pituitary gland at the base of the brain. A particular balance of hormones is responsible for the period of vaginal bleeding called **menstruation.**

There are four key hormones active in the human menstrual cycle (Fig. 8-3). The pituitary gland secretes *follicle stimulating hormone*

FIG. 8-3 The human menstrual cycle: (*a*) (1) FSH secreted by the pituitary gland; (2) estrogen produced by the maturing follicle inhibits release of FSH, stimulates the lining of the uterus to thicken, and stimulates LH release; (3) LH causes ovulation and induces corpus luteum; (4) corpus luteum begins secreting less estrogen and more progesterone which conditions uterus lining and prevents maturation of another follicle by inhibiting secretion of FSH and LH; lower levels of LH cause corpus luteum to regress and to secrete less progesterone; (5) menstruation with low progesterone levels; (6) FSH inhibition is released because of low levels of ovarian hormones; return to (1). (*b*) The pathways of hormonal control of menstruation, same sequence as in (*a*).

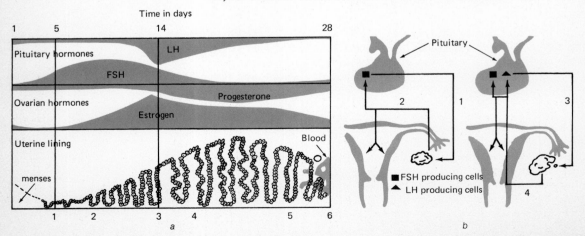

(FSH), which stimulates the growth and maturation of a follicle and its egg. As the follicle matures, it produces estrogen, which has three effects: (1) inhibition of release of FSH, (2) stimulation of endometrium to thicken, and (3) stimulation of release of luteinizing hormone (LH) by the pituitary. LH causes ovulation and the formation in the follicle of a mass of yellow cells called a *corpus luteum*. The corpus luteum begins to secrete progesterone and less estrogen, and the progesterone conditions the endometrium to receive an early embryo. Progesterone also inhibits secretion of FSH and LH by the pituitary, thus preventing maturation of another follicle. Lower levels of LH cause the corpus luteum to dwindle and to secrete less progesterone. With lower levels of progesterone, the endometrium sloughs off (menstruation). With only low levels of ovarian hormones in the blood, the pituitary begins secreting FSH and the cycle begins anew.

The menstrual cycle The menstrual cycle in the human female usually does not begin until between the ages of eleven and fourteen in temperate climates, and somewhat earlier in the tropics. It is often very irregular in timing when it first begins; several months may occur between menstrual periods. In some women it never stabilizes at the 28 days considered average, or at any other definite number of days. For these women, methods of birth control that depend on regular cycles are useless. Furthermore, such women have great difficulty deciding whether they are pregnant or their period is late again. The menstrual cycle also is disrupted by emotional stress, illness, drugs, starvation, and the approach of **menopause.** The menopause is the time when menstruation ceases and a woman is no longer able to bear children, because she no longer ovulates. This may occur any time between the ages of forty and fifty-five.

Some women experience feelings of depression and emotional instability during the week before menstruation. This experience is called **premenstrual tension** and appears to be caused by accumulation of excess water in the tissues. Some women also experience considerable pain during menstruation. Occasionally menstrual difficulties are the result of attitudes instilled in the child during upbringing. Both premenstrual tension and menstrual discomfort can be treated successfully by a physician or by nonprescription remedies. The period of menopause also may be accompanied by emotional problems such as depressions in some women. These problems, however, can usually be treated successfully by physicians. Most women have no problems with their menstrual cycles; some do, however, and men should be aware of this. Many needless arguments could very easily be avoided by men who understand that women can sometimes be temporarily depressed because of a slight menstrual malfunction.

Most women ovulate 14 days after the onset of the menstrual period. Human eggs are capable of being fertilized for only somewhere between 12 and 24 hours after ovulation. After that time the egg cell disintegrates. However, as we have seen, many women have irregular cycles, and fertilization has been known to occur at any time in the cycle, even during menstruation. Unlike other mammals and even other primates, human females if properly courted may be willing to engage in sexual intercourse at anytime in the menstrual cycle, not just during the period when they are fertile. This behavior appears to have played an important part in the evolution of the human family (Chap. 11).

8-5 Human embryos develop within the mother's uterus

Implantation is the attachment of the dividing fertilized egg (early embryo) to the endometrium. Once an early embryo is implanted in the wall of the uterus, it continues to divide. In human females it may take a little less than a week for the early embryo to travel to the uterus and implant itself. The egg lacks a true yolk mass; however, there is enough stored food in the early embryo to keep it alive until implantation. After implantation the embryo begins to form the **placenta,** a nutritive organ connecting the embryo with the uterus. Infoldings of the surface of the placenta become very closely associated with the endometrium of the uterus (Fig. 8-4). The bloodstream of the mother and that of the growing fetus are very close together. Waste materials and carbon dioxide from the embryo pass into the mother's blood. Nutrients and oxygen pass into the placental bloodstream from the mother. Each embryo is enclosed within a pair of membranes called the **amnion** and the **chorion,** which are filled with fluid. Thus the embryo is protected from injury not only by the body of the mother but by its own fluid-filled membranes. The embryo is connected to the placenta by the umbilical cord, through which the blood vessels run.

FIG. 8-4 The development of a human embryo and placenta: (a) implanted embryo, (b) section of embryo at about the fifth week, (c) full-term embryo and placenta.

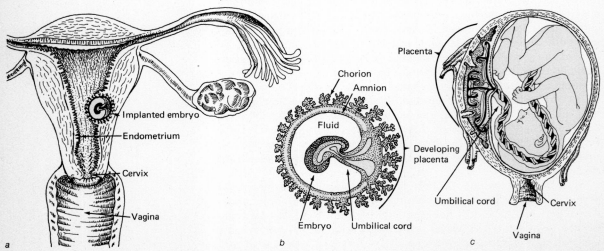

8-6 Birth and infancy are difficult times in the human life cycle

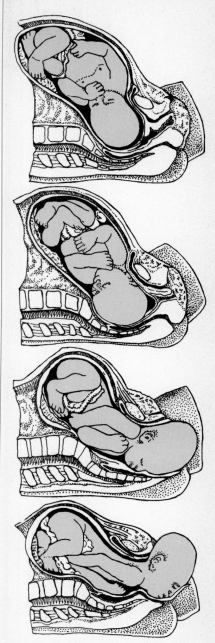

FIG. 8-5 Stages in a human birth.

Labor Labor is the period of uterine contractions leading up to birth. Birth of a baby (Fig. 8-5) begins with **contractions** of the muscular wall of the uterus; these contractions often cause pains called labor pains. As soon as strong contractions begin, the fetal membranes break, releasing the fluids they contain through the vagina. As contractions continue, they occur more frequently and the openings of the uterus and vagina become relaxed and stretched. Hormones produced during the pregnancy have made the birth canal (cervix, vagina, and vulva) more elastic to accomodate the passage of the baby. Eventually, the **fetus** (a human embryo is called a fetus from the end of its third month until birth) is pushed by muscular contractions from the uterus, through the vagina, and to the outside of the mother. The head of the infant arrives first in a normal birth. A newborn baby is connected by its **umbilical cord** to the placenta, which is still in the uterus. After the blood vessels in the cord stop pulsating, it is cut and tied. The membranes which form the placenta and protect the baby during its stay in the uterus are expelled later and are called the afterbirth.

Nursing and weaning human infants Following birth, the mother's mammary glands begin to **lactate,** that is, to secrete. The first substance secreted is not milk but a substance called **colostrum,** which helps to protect the infant from disease. The secretion of colostrum is followed by the secretion of highly nutritive milk—the period of **lactation.**

In American and Western European countries, a human infant is nursed by the mother, usually for a few months, perhaps as long as half a year. It is then switched to bottled milk, usually cow's milk, for up to two years. Sometimes other nutritive substances are added to the cow's milk. The child is gradually changed from a milk diet to other foods. This process is called **weaning.** Sometimes porridge or canned baby foods are given to the infant when it is still quite young. Monosodium glutamate (MSG) was formerly added to many baby foods as a flavor enhancer. MSG was added to baby food not because it mattered to the infant how the food tasted, but so that the food would taste better to the *mother* when she sampled it herself before feeding it to her baby. MSG is now believed to be harmful to babies. Excessive amounts of salts and spices were also sometimes added to baby foods to make them taste better to mothers.

While we in the United States are putting needless and possibly harmful additives in the foods of our babies, many human societies do not have any baby foods other than mother's milk. In many cultures, children are still breast-fed for three years or longer. After that, they eat whatever is available where they live. Many societies now have diets deficient in protein and the first symptoms of protein-deficiency diseases appear when the three- or four-year-old child is weaned from nursing. Because they are growing rapidly, children need much more protein than adults. Lack of sufficient protein during a child's developmental years may be fatal because resistance is lowered to disease and parasites. A protein-deficient child may survive to be a stunted adult, often with irreversible impairment of mental ability. The disease *kwashiorkor* is especially common in West Africa. The word means "the sickness that comes when another child is born" because the mother often stops nursing an older child when another is born, and protein starvation follows weaning. The symptoms of kwashiorkor include spindly legs and swollen bellies. The disease may be fatal if not checked by adequate diet.

Mother's milk is the ideal food for human infants. The trend away from nursing babies that developed in Western countries in the early part of the twentieth century fortunately seems to be on the wane. Sadly, a similar trend has begun recently in underdeveloped countries, where the nutritional results will be even more catastrophic. Whenever possible, women should nurse their babies, especially in the critical early months. Most physicians feel that the nutritional and psychological benefits of nursing outweigh the potential damage to the infant from such substances as DDT which contaminate mother's milk because they concentrate in food chains.

8-7 In human beings sexual maturity is called puberty

The human male The production of viable sperm does not occur in the human male before he reaches **puberty,** the stage of development when an individual first becomes capable of sexual reproduction. Often a young man's first ejaculation occurs when he is sleeping, in a so-called wet dream. Many young men are capable of copulation before they are able to produce many sperm. Sexual maturity occurs more gradually in young men than it does in young women. Although a girl of age eleven may often be capable of getting pregnant, it is unlikely that a boy of age eleven could be a father. The appearance of a beard, if he belongs to a group with abundant facial hair, and a deepened voice generally indicate that a young male has reached sexual maturity. The ability to fertilize a female often extends into old age in men, and males in their eighties have fathered children.
The human female In girls, when puberty is reached, menstruation and the production of eggs begin. As a young girl approaches puberty, her breasts develop, her hips become broader, her uterus enlarges, and she develops sexual feelings. Among girls there is a great deal of difference in the timing of these events. Some girls may be physically mature but mentally and emotionally still children. They may innocently behave in a sexually enticing manner toward older males without realizing the significance of their actions. Other girls may develop an interest in boys long before they are sexually mature. In the last 20 years in the United States, the average age at the time of the first menstruation has dropped from 13.5 years to 11.5 years. Many junior high school girls are now capable of becoming pregnant.

8-8 "Sex" and reproduction are not the same thing

Reproduction can occur without sex, and sex can occur without reproduction. As you already know, many plants and some animals reproduce asexually—that is, without any sex whatsoever. Furthermore, many mammals copulate at only one season of the year and yet they seem to be able to reproduce quite satisfactorily. Human couples often copulate when the woman is known to be pregnant already. Finally, couples in their seventies and eighties, long beyond the reproductive years, continue to enjoy sex. Therefore, there is quite clearly more to what we call "sex" than producing babies.

The many roles of sex One of the most important roles of sex in human society seems to be the bonding of human relationships. Human infants are born quite helpless and need a long period of parental care and protection. Mutual sexual enjoyment tends to keep couples together and provides a proper environment for the raising of children. The result is a family in which, ideally, the parents obtain mutual satisfaction from each other and from sharing in the care of their offspring. In many cultures, fathers and other relatives as well as mothers care for children, and a new baby is not the devisive influence that it often is where the new mother is so busy with the baby that she neglects her husband. Long pregnancy and nursing results in a close bond between mother and child. Women who nurse their children usually report that it gives them great pleasure. (Nursing also has been shown to stimulate the uterus to shrink back to its normal size after childbirth.)

Sex is extremely important in many facets of our lives seemingly very far removed from human reproduction. It is the basis of much of our humor, and it is thought to influence not just whom we marry but what cars we drive and for whom we vote for President! Because of its widespread influence, it is not surprising that all societies have developed sets of rules concerning sexual behavior. In our society, as well as others, there is often considerable anxiety, especially among young people, concerning what is "right" sexually.

Sexual activities and attitudes Homosexual bonds are those between members of the same sex and also often lead to close and enduring human relationships. Homosexual bonds played an important role in ancient Greece, along with bonds to family and children. Some of our species' most famous artists, poets, politicians, and military men are reported to have been either homosexual or bisexual (sexually attracted to members of both sexes). Immense books have been written on the so-called sexual perversions. It is probably safe to say that anything physically possible in the way of sexual activity has been tried by someone. Most types of sexual activities commonly found among consenting adults are considered "normal" by biologists. Certain religious and government authorities have spent much time trying to decide whether people should be allowed to engage in or read about activities of a sexual nature which do not fit their idea of what is proper. If one looks at these attempts in the perspective of history, they appear futile. American and European medical texts of the last century said that women did not enjoy sex and, furthermore, that those who did were evil. Medical texts of the present consider women who do not enjoy sex to be "frigid" and suggest ways to cure them. Western medicine once thought masturbation (sexual self-stimulation) could lead to anything from acne to insanity. Now it is not considered unhealthy and studies show that almost everybody—male or female—practices it from time to time.

If you have taken an anthropology course, you probably have discovered that sexual practices vary from culture to culture. Practices such as kissing on the mouth, which we consider normal, are considered "perverse" in other cultures. The only common themes running through human cultural attitudes toward sex seem to be (1) that there is some kind of family structure oriented around sexual bonds and (2) that penalties are usually imposed on people who take advantage of children sexually or use force and violence to satisfy their desires.

8-9 Sex is nothing to panic about

There is no reason people should *worry* about having sexual feelings. Most people, young and old, spend a great deal of time thinking about sex, and almost everyone at some time or other has doubts about the adequacy or acceptability of his or her sexual behavior. As an individual, however, *you are responsible for the consequences of your sexual behavior*. These consequences could include unhappy people, unwanted pregnancies, and even venereal disease. If people have common sense, however, and are considerate of others, they should find sex a source of joy and satisfaction throughout their lives, even into old age.

Premarital copulation A common question among young people is whether or not premarital copulation is proper. Various religious groups have differing views on this, but from the point of view of biology you should remember that most often copulation involves strengthening the bond between a male and a female, in part to provide a proper environment for *rearing offspring*. At one time it was common to think of sex as something a woman "gave" to a man or something a man "took" from a women. Most people now feel that a sexual relationship should be a relationship entered into *mutually*, with great attention being paid to the wishes and feelings of the other partner.

8-10 Everyone should be familiar with methods of birth control

It is very important that couples be able to control whether or not copulation leads to pregnancy. Unwanted pregnancies can lead to many problems. **Contraception** is preventing sperm from reaching an egg, and there are a number of devices, drugs, and methods which do just this. In practice, the meaning of contraception is commonly extended to include prevention of a fertilized egg from developing or an embryo from being born. Contraception may be used if a couple desires no children, already have the number of children they want, or if pregnancy would endanger the female. Couples may wish to delay parenthood or to space children to reduce the burden of care and expense.

Condoms and diaphragms A contraceptive used by males is the **condom,** or rubber—a sheath that is placed over the penis before copulation. The condom contains the semen after ejaculation and prevents sperm from reaching the uterus. Condoms are now advertised in national magazines and can be bought by mail. Many drug stores no longer hide them under the counter; they even display them, often near the cash register. Another rubber device, a **diaphragm,** can be used by females to keep sperm from entering the uterus. The diaphragm is a round, concave disk inserted over the **cervix,** the muscular opening of the uterus (Fig. 8-6). A diaphragm must be obtained from a medical doctor and is normally used with a sperm-killing jelly.

Hormones A different kind of contraception is commonly known as "the pill." There are several kinds of contraceptive pills which contain an oral dose of synthetic hormones similar to those that control the menstrual cycle. They function by stopping ovulation. Pills must be obtained from a medical doctor, and not all women can safely take them. Pills give some women headaches, cause some to gain weight, and affect the emotional state of others. A study of women in England showed a correlation between the use of pills and formation of blood clots; however, this effect has not been found in American women. Research is now being carried out to produce a pill for men which would prevent viable sperm from being formed. If such a pill can be produced, it will appeal to men who do not like to bother with condoms but who feel that men should take some responsibility for birth control.

Intrauterine devices Birth control also can be accomplished in several other ways. For instance, a plastic coil, called an **intrauterine device (IUD),** can be inserted by a medical doctor (Fig. 8-7). It remains semipermanently in the uterus. Such devices apparently prevent fertilized eggs from implanting in the uterus. Some women are unable to use present IUDs; however, improved IUDs are being developed which may make them acceptable to more women. Metal IUDs and IUDs containing hormones are also being developed. Several other methods of birth control are listed in Table 8-1, which gives the failure rates of various methods. As you can see, the only method which never fails (besides sterilization or total abstinence, which are not listed) is abortion.

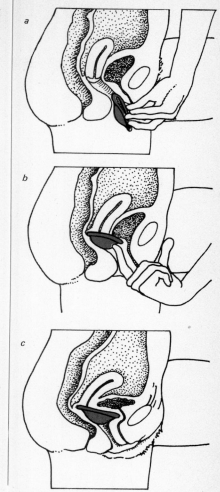

FIG. 8-6 Insertion of a diaphragm: (*a*) the diaphragm is compressed and inserted into the vagina, (*b*) the diaphragm is pushed up toward the cervix, (*c*) the diaphragm covers the cervix completely.

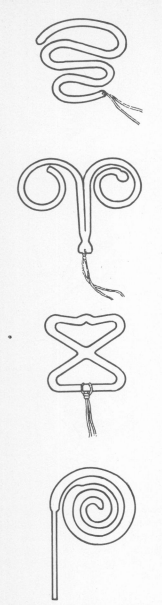

FIG. 8-7 Types of intrauterine devices. All are flexible and can be inserted through the cervix into the uterus. IUDs usually have a taillike thread by which they can be removed.

TABLE 8-1 FAILURE RATES OF BIRTH-CONTROL METHODS

METHOD	Pregnancy Rates for 100 Woman—Years of Use	
	HIGH	LOW
No contraceptive	80	80
Aerosol foam	—	29
Foam tablets	43	12
Suppositories	42	4
Jelly or cream	38	4
Douche	41	21
Diaphragm and jelly	35	4
Sponge and foam powder	35	28
Condom	28	7
Withdrawal	38	10
Rhythm	38	0
Lactation	26	24
Steroids (the "pill")	2.7	0
Abortion	0	0
Intrauterine devices (averages) Lippes loop (large)		
0–12 months		2.4
12–24 months		1.4

Based on Berelson et al., *Family Planning and Population Programs,* University of Chicago Press, Chicago, 1966.

8-11 Abortion is an artificial end to pregnancy

An **abortion** is a simple operation by which a physician can end a pregnancy after implantation and after development has begun. If it is performed by a physician in the first few months of pregnancy, it is statistically safer for the mother than carrying the embryo to term and bearing the child. Some people, however, believe that abortion is immoral and claim that the developing embryo is a human being. Other people feel that calling an embryo a human being is like confusing a set of blueprints with a building which might be constructed using them. They point out that a human being results only when the hereditary "plan" starts to interact with a cultural environment—that is, after birth. In recent years, abortion was illegal in some states and legal in others. A 1973 Supreme Court ruling, however, made it legal in all the United States. There are still some local laws restricting the conditions under which abortions can be obtained, and some physicians and hospitals are unwilling to participate in abortions.

In nations where it is illegal—for example, Italy—a great many women have abortions anyway. These abortions are often performed by incompetent friends, relatives, or professional abortionists. All too frequently, death results from bungled illegal abortions. Since rich people can easily travel to a country where abortion is legal, it has been argued that laws against abortion discriminate against the poor.

The availability of birth control varies throughout the world. Many nations have free or very inexpensive birth-control clinics and advertise birth control widely. In other nations, it is even illegal to dispense birth-control information. In some of the most overpopulated countries, people do not want to use birth control even if it is available because of religious or cultural taboos against it. Because of the unavailability of other birth-control methods in many countries, abortion is the most widely used birth-control method throughout much of the world. Obviously, foolproof and safe ways of keeping fertilization or implantation from occurring are preferable to abortion. Abortion is morally objectionable to more people than other methods and involves some risk to the woman, especially when it is done illegally by unskilled persons. Abortion will remain as a "back-up" as long as an acceptable and perfectly foolproof means of preventing fertilization or implantation has not been devised. If birth-control methods and information were made widely available throughout the world, the number of abortions would certainly decrease.

8-12 Sterilization is a method of birth control

Some people, faced with the uncertainties of contraception and abortion, decide to have themselves sterilized once they know they do not want more children. In women, the operation consists of cutting the fallopian tubes **(tubal ligation),** a procedure which now usually can be done without an overnight stay in the hospital. The operation for men is even simpler; it can be done in less than an hour in a doctor's office. It consists of cutting each **vas deferens** (see Fig. 8-1), the tubes which lead from the testes to the urethra. The operation, known as a **vasectomy,** includes cutting and tying off each tube before the junction with the accessory glands. There is a period after a vasectomy during which sperm stored in the male tract are cleared out. During this period the couple should continue with their usual contraceptive method. Usually, after about 12 ejaculations, there are no sperm left in the semen; however, the physician may want to test the semen to advise when it is safe to stop using contraceptives.

Both vasectomy and tubal ligation are theoretically reversible; but, in practice, fertility may not be restored. Neither operation changes the sex drive (although men who do not know this or who are persuaded to have the operation by others may have some psychological problems about loss of virility). A male ejaculates exactly as he did before the operation, but his semen no longer contains sperm. Vasectomy cannot be shown to be totally without risk (no operation can), but no side effects as serious as those associated with the pill have yet been demonstrated.

Vasectomy is becoming quite a popular operation among American

TABLE 8-2 INCIDENCE OF VENEREAL DISEASE IN THE UNITED STATES*

YEAR	GONOR-RHEA	SYPHILIS	OTHER
1945	313,363	351,767	10,261
1950	286,746	217,558	8,187
1955	236,197	122,392	3,913
1960	258,933	122,033	3,811
1965	324,925	112,842	2,015
1967	404,836	102,581	1,309
1968	464,543	96,271	1,486
1969	534,872	92,162	1,778
1970	600,072	91,382	2,152
1971	670,268	95,268	2,101
1972	767,215	91,149	2,251

*Cases reported, civilian only.
SOURCE: *Statistical Abstract of the United States,* 1974, p. 86.

men. Such contraceptive devices as diaphragms, IUDs, and especially the pill, put the bother and risk of contraception on the woman. Many men look on vasectomy as a way they can assume responsibility in human reproduction which is, after all, a partnership. About 70 percent of the 6 million Americans who have had themselves voluntarily sterilized by 1974 were men who had vasectomies. Men who worry that something may happen to their children may have a store of their sperm frozen before the vasectomy. Experience with domestic animals indicates that semen survives freezing quite well; in fact there is a thriving international trade in frozen bull semen, which is used to introduce desirable characteristics into herds of cattle around the world. With present techniques human sperm seem to be able to survive freezing for two or three years. Semen from human donors are often used when a couple wants children and the man is sterile.

8-13 Venereal disease is a serious problem

As was noted in Table 6-4, some human diseases are transmitted from individual to individual during sexual relations. These are known as **venereal diseases** and are often referred to by the initials **VD.** The changing incidence of these diseases in the United States is shown in Table 8-2. The two that are most widespread are **syphilis** and **gonorrhea,** caused by different kinds of bacteria. The first sign of syphilis is an open sore, usually on the reproductive organs, and a rash. These symptoms then disappear and the disease "goes underground." It often reappears later in life, attacking the brain and causing insanity or death, or perhaps attacking the arteries leading from the heart, also with fatal results. Gonorrhea is somewhat less serious. In males the major symptom is pain when urinating; in females there may be no noticeable symptoms. Gonorrhea can, however, lead to sterility, especially in females, because of formation of scar tissue.

A person who suspects he or she has a venereal disease should see a doctor at once. Catching VD is no more shameful than catching a cold or the measles, even though ignorant people may say so, but the disease is apt to be a lot more serious. The doctor should also be given the names of all sex partners involved so that they may be contacted for treatment. Doctors are not required to discuss venereal disease with parents of minors in some states.

REPRODUCTION IN OTHER ANIMALS AND PLANTS

Reproduction in organisms other than human beings involves a great number of different strategies. The other mammals are similar to human beings in much of the anatomy and physiology of their reproductive structures; however, they differ among themselves and from human beings in many important respects.

8-14 There are some important differences between human reproduction and that of most other animals

Nonprimate mammals Mammals other than human beings vary in their egg production. Some produce only one egg at a time; others produce dozens. The uterus is often two-branched in nonhuman mammals, and multiple births are common in many small mammals such as rodents. As we have seen, ovaries produce eggs periodically according to a cycle regulated by hormones produced by the pituitary gland. In most mammals, the female is unwilling to accept the male except when eggs can be produced. This period during which the female is willing to copulate is called **estrus** or "heat." The receptive period is part of a complex cycle of events in the female called the **estrus cycle.** Most wild mammals have only one such period a year; thus their estrus cycle is a year in length. In dogs and cats, it may be from four to six months long. In domestic mice and rats, the females are fertile and accept copulation about every five days.

Primates Most female primates have menstrual cycles. Female chimpanzees are receptive to the male only during a 10-day period of the month. In chimpanzees as in most mammals as a conclusion to courtship behavior, the female assumes a receptive position indicating she is willing for the male to copulate.

Gestation The **gestation period** is the period of development in the uterus. The gestation period varies in length depending upon the animal. In the pouched animals, such as the opossum or kangaroo, the gestation period is very short—only 10 days in the opossum. Then the fetus crawls out of the uterus of the female and into the pouch on her abdomen. Usually the female helps it along. There it attaches itself to one of her mammary glands. Subsequent development takes place in the pouch. Pouched mammals, or marsupials, do not develop a placenta. The marsupial fetus takes in nourishment from the mammary glands in the pouch, where it spends much more time than the very brief time it spent in the uterus. Gestation in the placental mammals ranges from some 21 days in a mouse to 280 days in the human female, 360 days in a whale, and 660 days in an African elephant.

Birth Some mammals are born still surrounded by their fetal membranes. In these mammals, the mother tears away the membranes and frees the baby animal. She licks the baby clean of blood and fluids and usually she eats the membranes and the placenta, which is pushed from the vagina by further contractions. In nonhuman mammals, the mother licks the new infant until it is clean and dry. During this process she thoroughly massages the baby animal with her tongue. This massage stimulates the breathing and blood circulation of the baby. The mother also learns very thoroughly the odor of each of her offspring.

Lactation and weaning The milk of different species varies a great deal in the relative concentrations of fats, carbohydrates, proteins,

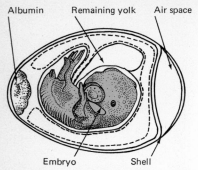

FIG. 8-8 A section of a chicken egg showing the developing chick and the surrounding membranes.

and other substances. In many cases, what is satisfactory for one species may be quite unsatisfactory for another. For example, the milk of fur seals is 50 percent fat and obviously would not be good for human infants.

Different animals nurse their young for different periods of time. Animals such as dogs, cats, and mice that are born relatively helpless nurse for comparatively longer periods than animals such as guinea pigs or horses that are able to see and walk about almost immediately. Often animals that are born well developed start to eat the same food as their parents soon after birth while they are still nursing.

8-15 Reptiles and birds lay eggs which withstand drying

Like the mammals, both reptiles and birds have internal fertilization. They do not require external water for the sperm and egg. Very rarely in reptiles and never in birds do the eggs develop inside the female. The problems of drying out and getting nutrition to the embryo are solved by production of a shelled egg which can withstand drying and which contains stored food. A chicken egg contains food a hen provided for a baby chick. The egg that would have become the baby chick is a tiny spot at one end of the yolk. The yolk is stored food; the white of an egg is both protective and nutritive. Surrounding the embryo are membranes (Fig. 8-8), one of which is filled with fluid in which the embryo grows. The other membrane is near the shell and contains numerous blood vessels which give off carbon dioxide and take up oxygen through the shell. This membrane also receives waste material from the embryo.

In reptiles and birds, fertilization of the eggs occurs in the oviduct. Before the egg is fertilized, it accumulates very large amounts of yolk. The nucleus of the egg and most of its other cellular components, except the yolk, are pushed to the surface. The yolk of a chicken egg may weigh 19 grams (0.7 ounces), while the yolk of an ostrich egg may weigh 460 grams (16 ounces). After fertilization, the fertilized egg moves down the oviduct through a number of specialized regions. There it receives the protective egg white, and then membranes are

FIG. 8-9 An alligator nest. This female alligator is defending her eggs, most of which are buried in the mass of vegetation, although a few may be seen near her tail. (*Dee Jay Nelson, National Audubon Society.*)

laid down around the white. Finally, cells of the oviduct secrete a shell on the outside of these membranes. In reptiles, the shell is usually leathery and flexible. In birds, it is hard and brittle.

Incubation A batch of eggs is called a *clutch*. Most birds do not begin **incubating** (keeping their eggs warm) until their entire clutch is laid. Either the male or the female or both may incubate. Usually when birds are incubating, they have a temporary patch on their chests which has an extensive blood supply. This *brood patch* has no feathers and is pressed close to the eggs to warm them. The eggs of many small birds hatch in 11 or 12 days. However, the royal albatross, the bird with the longest incubation period, has to incubate its eggs for almost 80 days. Reptiles, on the other hand, often bury their eggs or deposit them in rotten logs and never bother with them again, although a few reptiles give parental care (Sec. 6-5). Alligators make a large nest of decaying vegetation in which they lay their eggs (Fig. 8-9). While the decomposing vegetation gives off heat, keeping the eggs warm and speeding their development, the mother alligator guards the nest to make sure it does not dry out. Female pythons lay large numbers of eggs and coil about them to protect them. Even though she is technically "cold-blooded," at this time the female python can raise her body temperature to incubate the eggs.

Birds not only incubate their eggs, many also incubate their young. Most birds build some sort of nest in which the eggs are laid. A nest may range from a scarcely noticeable hollowed-out spot in the ground to a very elaborately woven structure hanging from a tree branch (Figs. 8-10 to 8-12). The high body temperature of birds, their brood patches, and their practice of periodically turning the eggs provide ideal conditions for development of the embryos.

Parental care after hatching All the eggs laid in a clutch usually hatch at about the same time. With a special temporary structure called an *egg tooth* on the tip of its beak or nose, the young bird or reptile cuts or breaks the shell of the egg. Reptiles are usually able to live quite independently soon after they hatch. Most do not receive any parental care. The American alligator cares for its offspring for a time after they hatch. Other reptiles may find their newly hatched offspring irresistible as food.

Many birds are able to walk and feed shortly after hatching. Chickens and ducks fall into this category (Fig. 8-13). In other species of birds, the newly hatched young are quite helpless for some time (Fig. 8-14). They are fed often, usually by both parents, and protected from cooling and from predators. The parents of many hatchlings even carry away their offsprings' droppings, which are produced conveniently in special fecal sacs.

FIG. 8-10 A depression in the gravelly substrate plus a few twigs serves some terns as a nest.

FIG. 8-11 Housefinches make a nest of grass and twigs.

8-16 Many animals have external fertilization

Some animals that live in water or spend part of their life in water have external fertilization. In some invertebrate sea animals—sea urchins, for example—the development of the male and female reproductive organs is synchronized. When the gametes are mature, eggs

FIG. 8-12 Woodpeckers nest in holes they make.

and sperm are released into the water in great quantities. Fertilization occurs if sperm and egg come into contact, and the embryo then develops without any sort of parental care.

In vertebrates, external fertilization may also occur, as in most fishes and amphibians. Frogs and toads, for example, commonly come together in large aggregations in ponds or lakes in the spring. In one area, there may be several different species. Each kind of male frog or toad has its own distinctive call, however. The call serves to bring reproductively mature individuals of the same species together in one place. Males grasp the female with their forelimbs firmly behind her forelimbs (Fig. 8-15). If he grabs another male, the latter has a special croak to indicate a mistake has been made. When the female lays the eggs, the male sprays them with sperm. Frogs and toads do not provide parental care, except in certain highly specialized forms such as the Central European midwife toad in which the male picks up the eggs the female has laid and carries them about with him until they hatch.

8-17 Plants have reproductive strategies similar to those of animals

Fertilization and pollination were explained in Chaps. 6 and 7. Plants have faced problems similar to those presented by the environment

FIG. 8-13 A hen with her chicks. The baby chicks are able to walk and pick up food shortly after they hatch, but they are closely guarded by the hen. (*C. P. Fox, National Audubon Society.*)

FIG. 8-14 This female hummingbird is bringing food to its offspring in the nest. Such birds are unable to feed themselves for several weeks after they hatch. (*G. Ronald Austing, National Audubon Society.*)

to animals, and they have solved them in similar ways. Pollination tubes are similar to copulatory organs in making internal fertilization possible in seed plants. The embryo seed plant develops for a time inside the ovary (developing fruit) of the parental plant. It is then released to the environment inside a tough seed coat and provided with food reserves. Some seed plants produce relatively few large

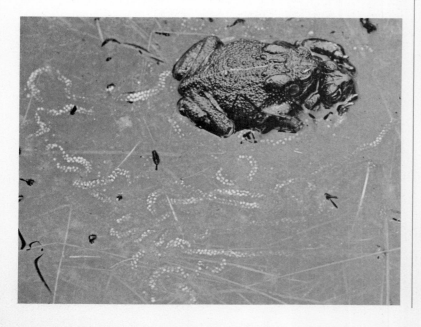

FIG. 8-15 In tailless amphibians such as these toads, the male clasps the female (head barely visible in water) firmly behind her forelimbs and releases sperm as she produces eggs (seen as long strings behind the pair). (*American Museum of Natural History.*)

seeds, with large amounts of stored food. The acorn of an oak is a good example. Other plants, such as orchids, produce hundreds of thousands of very tiny seeds. These have relatively little reserve nutrients. These strategies are analogous to those of animals: producing large numbers of fertilized eggs without parental care or having only a few offspring and caring for them during their early development.

REPRODUCTION AND THE FORMATION OF POPULATIONS

Reproduction of individuals leads to groups or *populations* of organisms (see Sec. 3-6). Population size is the number of individuals in a population. For instance, the size of the entire human population in 1975 was about 4.0 billion people. Populations change in size; the study of size changes in populations is the study of **population dynamics.**

8-18 Changes in the size of populations depend primarily on birth and death rates

Ignoring for the moment the possibility of migration (the movements of individuals from one population to another), we can say that two factors determine changes in the size of a population: *birth* and *death*. Birth and death may be thought of as inputs and outputs to the population. Imagine for the moment that the world is a glass bowl, with a faucet running into it and a drain in its bottom. The faucet is the input, the drain the output, and we start with some water in the bowl. Suppose the faucet and the drain are both flowing at the same rate. What will happen to the water level? Obviously, it will remain constant. What if you turn the faucet on harder? The level will rise. What if you plug the drain a bit? It will rise even faster. What if you open the drain wider? As you can see, changes in the level of the water will depend on the *balance* between input and output, or in the case of populations, between births and deaths.

Population growth is analogous to a bowl of water with more input than output. If you were studying the water system, you would measure the rate of flow of water into and out of the bowl—that is, how much water went in and out *per unit of time*. You would probably measure the flow in liters per hour. In the human population, you could measure the input as births per year and the output as deaths per year. Thus for the year 1970 there were 123 million births in the entire human population and 51 million deaths. In the United States, there were 3.6 million births and 2 million deaths. In Brazil there were 3.6 million births and 1 million deaths.

Which country is growing most rapidly? That question is not easily answered by looking at the inputs and outputs we have just given you. After all, you cannot tell how fast the water will rise in a bowl with an input of 8 liters (2 gallons) per hour and an output of 1 liter (0.3 gallons) per hour unless you know the size of the bowl.

In order to compare birth and death rates conveniently from population to population, the number of births per year and the number of deaths per year are divided by population size at midyear. For the world rate in 1970, this means 123 million births ÷ 3.6 billion midyear population, or 0.034 births per individual. For the United States, it is 3.6 million ÷ 205 million = 0.0176. Note that 0.034 and 0.0176 are equal to 3.4 and 1.76 percent, respectively. By general agreement, however, birth and death rates are not given as rates per individual or rates per hundred (percent) in the population, but as *rates per thousand*. Thus the birthrate in the world in 1970 was 34 per 1,000. That is, for every 1,000 people in the population, 34 babies were born. About 18 babies were born for every 1,000 people in the population of the United States. As you can see, with the world birthrate at 34 per thousand and the United States at 17.6 per thousand, the world birthrate is proportionately almost twice as high as that of the United States. Brazil's birthrate of 38 per 1,000, on the other hand, is somewhat higher than that of the world as a whole.

Whether the water in our bowl rises or falls depends on the relative rates of input and output. Similarly, whether a population increases or decreases in size depends on the relationship of the birth and death rates. If the birthrate is above the death rate, the population is growing. If the birthrate is below the death rate, the population is shrinking. If the birthrate is the same as the death rate, the population is stationary, or as it is often put, there is "zero population growth." In the United States, in the mid-1970s the birthrate was still well above the death rate, and the population was growing. The size of families, however, had fallen to the point where, if no change occurred, the birth and death rates would be in balance around 2030.

8-19 The natural growth rate is the difference between the birth and death rates

Humanity is rapidly increasing in numbers. The birth and death rates for the world population in 1970 were 34 and 14 respectively; 34 people were born and 14 died for each 1,000 people in the population. In the United States, the birthrate was 17.6, the death rate, 9.6. Virtually all human populations were growing in 1970 and continue to grow today. Table 8-3 gives some representative birth and death rates. The difference between the birth and death rates is corrected in some cases for migration (migrants subtracted from or added to population increase). This rate is conventionally not given as a rate per 1,000 but as a percent. Thus, the world growth rate is 2.0 percent, or 20 per 1,000.

Populations grow like compound interest in a savings account. Such growth is called **exponential growth.** Just as the interest itself earns interest, so people added to a population have a way of themselves adding more people to the population. Therefore populations grow with deceptive speed. For instance, if you simply *add* 20 persons per year to a population of 1,000 people, it will take 50 years to double that population (20 × 50 = 1,000). But under the compound interest

TABLE 8-3 BIRTHRATES, DEATH RATES, AND GROWTH RATES OF REPRESENTATIVE COUNTRIES

REGION	COUNTRY*	POPULATION MID-1974 (IN MILLIONS)	ANNUAL BIRTHS PER 1,000	ANNUAL DEATHS PER 1,000	ANNUAL RATE OF GROWTH, PERCENT
World		4,061.1	35	13	2.2
Africa					
North Africa	Egypt	38.0	44	16	2.8
West Africa	Nigeria	79.8	50	25	2.5
East Africa	Ethiopia	27.9	46	25	2.1
Middle Africa	Zaire	24.1	45	23	2.2
South Africa	Republic of South Africa	24.3	40	16	2.4
Asia					
Southwest Asia	Turkey	38.9	40	15	2.5
Middle South Asia	India	619.4	43	17	2.6
Southeast Asia	Indonesia	133.7	48	19	2.9
East Asia	China	917.0	38	14	2.4
North America	U.S.	217.5	14.9	9.4	1.0†
Latin America					
Middle America	Mexico	57.9	45	9.0	2.8
Caribbean	Haiti	5.3	44	20	2.4
Tropical South America	Brazil	110.7	38	10	2.8†
Temperate South America	Argentina	24.5	22	10	1.2
Europe					
Northern Europe	United Kingdom	56.0	13.9	12.0	0.2
Western Europe	Federal Republic of Germany	61.9	10.2	11.8	−0.2
Eastern Europe	Poland	33.7	17.4	8.0	0.9
Southern Europe	Italy	55.2	16.0	9.9	0.6
U.S.S.R.	U.S.S.R.	252.0	23.0	9.7	1.7
Oceania	Australia	13.4	18.9	8.5	1.7†

*The country with the largest population in each area was picked as representative.
†Includes migration
SOURCE: Environmental Fund, based on UN and other data.

TABLE 8-4 THE RELATIONSHIP BETWEEN EXPONENTIAL GROWTH RATE AND DOUBLING TIME

ANNUAL INCREASE, PERCENT	DOUBLING TIME, YEARS
0.5	139
0.8	87
1.0	70
2.0	35
3.0	23
4.0	17

principle, a population increasing at 2 percent per year will actually double in only 35 years. Table 8-4 shows the relationship between exponential growth rate and **doubling time**.

8-20 Fruit flies show an outbreak-crash type of population cycle

Look at Fig. 8-16. It shows a typical pattern of population growth and decline such as you might find if you put a pair of fruit flies in a bottle with some suitable fruit-fly food. The flies would mate, the female would lay eggs, fruit-fly larvae would grow on the food and form pupae, and a new generation of fruit flies would be produced. Slowly at first, and then more rapidly, the population of fruit

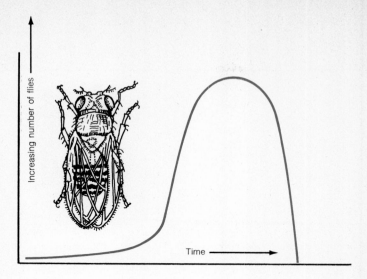

FIG. 8-16 An outbreak-crash population curve as seen in the fruit fly *Drosophila*.

flies would increase. The birthrate would be high; the death rate, low. Eventually the food would become exhausted or fouled with wastes. As a result, the death rate would catch up with or pass the birthrate. Population size would drop, at first slowly and then very rapidly. There would be a **population crash.**

Outbreak-crash population cycles This sequence is known as an **outbreak-crash population cycle.** It is quite typical of fruit-fly populations in nature. A pregnant female fly discovers a suitable place to lay eggs, say a rotting fruit. She lays her eggs, and a large population of flies is produced in a few generations. Then the population runs out of resources (fruit) or so pollutes its environment with its waste products that fruit-fly life can be supported no longer. The population crash then occurs. This is not the end of the fruit flies, though. Some of the flies have long since fled the population and established themselves on other rotting fruits.

8-21 Human population growth can be considered an outbreak

In Fig. 8-17 you can see the history of human population growth. About 8,000 years ago, when people first started to practice agriculture, the human death rate began to drop. The birthrate remained high and the population started to grow. Further improvements in farming, the industrial revolution, and the development of modern medicine and public health produced further lowered death rates, and this speeded growth rates. Does the curve of human population growth look familiar to you? It should; it resembles the first half of an outbreak-crash population curve. We have had our outbreak; will we now have a crash? There are signs that we have used up much

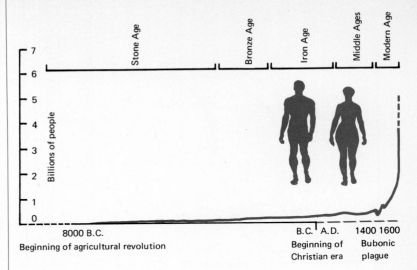

FIG. 8-17 Population curve showing the growth of the human population. (Before the agricultural revolution; time is not drawn to scale. If it were, the base would be another 7 meters to the left.)

of our resources (shortages of food and fuel) and that we have fouled our medium (pollution and extinction of other species). Unfortunately, *we* do not have any other fruit to move to.

Human population crashes Remember the crash of Irish population following the collapse of Ireland's unstable potato monoculture. Population crashes due to famine have been a common occurrence in human history. One study lists 200 famines in Great Britain alone between A.D. 10 and 1846. Another counts 1,828 Chinese famines in the 2,010 years preceding 1911, a rate of nearly one a year. Some of these famines, and similar ones in India, have resulted in many millions of deaths. Even in this century famine has killed millions. For example, perhaps 5 million to 10 million deaths have been attributed to starvation in Russia (1918 to 1922, 1932 to 1934), perhaps as many as 4 million deaths in China (1920 to 1921), and 2 million to 4 million deaths in India (1943). Each year now, some 10 million to 20 million human beings die who would survive if they had adequate nourishment.

8-22 Organisms normally produce more offspring than would be necessary to replace the parents if all survived

A pair of butterflies may have 1,000 or more offspring. Other invertebrates may have millions of offspring. Fish also may produce millions of eggs. Orchid plants produce millions of seeds. Obviously, if all these offspring survived generation after generation, we would soon be up to our necks in butterflies, fish, and orchids! Various environmental factors intervene in most populations to produce a correspondingly high death rate to balance this very high birthrate. Predators eat the offspring; some offspring cannot find enough food (or, in the case of seedlings, light, water, or minerals) and perish. Others succumb to disease or bad weather. When conditions happen to be very favorable (abundant food, fine weather, predators wiped out by an insecticide), the death rate will be low and the population will grow; but sooner or later some factor will bring growth to a halt.

As populations grow and become more dense (have more individuals per unit of area), the competition for resources gets more severe, and crowding makes the population more susceptible to disease and a more inviting target for predators. The death rate increases. When environmental conditions are unfavorable, the death rate may remain above the birthrate and *extinction* of the population results. Understanding the ways in which death rates are determined in animal populations is of great practical importance, since people have a sizable stake in being able to manipulate animal death rates. People often wish to raise death rates in populations of organisms they consider as pests and to lower death rates in populations of organisms they rate as beneficial.

8-23 Control of birthrates by the human species is unique in the animal world

You may have heard an opponent of contraception or abortion complain that such practices "reduce people to the level of animals." Regardless of the moral pros and cons of birth control (and setting aside the question of whether being compared to animals is an insult), the statement is nonsense. Human beings are, in fact, the only animals that have the ability voluntarily to control their birthrate, and they have habitually done so. Indeed, virtually all societies have social measures that affect birthrate. These may involve such things as delay of marriage, a tradition of long nursing (nursing mothers are less likely to get pregnant than nonnursing mothers), extensive segregation of the sexes, abortion, or infanticide. And, of course, they also may involve the use of contraceptive devices discussed in Sec. 8-10 or more primitive ones used in antiquity and by so-called primitive tribes.

Humans are the only organisms that can be aware of the size of their species population and of its rate of growth. They also are in a position to influence both rationally. It remains to be seen if the growth of the human population will be ended the "human way," primarily by limiting births, or the "animal way" by a dramatic rise in the death rate.

QUESTIONS FOR REVIEW

1 Why is the production of tremendous numbers of offspring—often millions of them—essential to the survival of certain human parasites?
2 What do these terms mean: intrauterine device (IUD), weaning, population dynamics, embryo, vasectomy, testes, puberty, exponential growth, estrus, orgasm, placenta, lactation?

3 What are some consequences of protein deficiency in children, and what childrearing trend in some underdeveloped countries threatens to increase the incidence of protein deficiency in the young?

4 The sensible use of any birth-control method demands that other factors besides the method's effectiveness be taken into account. What major consideration in addition to contraceptive effectiveness is especially important in the case of birth-control pills?

5 What are the harmful effects of untreated syphilis? Untreated gonorrhea? Why do you suppose that the incidence (number of reported cases) of gonorrhea in this country has risen steadily since the mid-1950s while the incidence of syphilis has declined?

6 Why was the shelled egg such an important evolutionary development?

7 Work the following problems, and then check your answers against the data in Table 8-3:
  a The reported mid-1974 human population of China was 917,000,000, its annual birthrate was 38, and its death rate was 14. What was its growth rate? (Remember that growth rate is expressed as a percentage.) Did you need all the information provided in this question to answer it?
  b Mexico's mid-1974 population was reported as 57,900,000, and approximately 2,605,000 children were born there in 1974. What was Mexico's birthrate for that year?

8 "It remains to be seen whether the growth of the human population will be ended the 'human way' or the 'animal way.'" What are these two "ways"? Which way do you think would be the better one in terms of minimizing human suffering and social, political, and economic disruption? Do you think it more likely, realistically speaking, that the human population growth will be ended the "human" or the "animal" way? Why?

READINGS

Burns, E.: *The Sex Life of Wild Animals,* Faucett, New York, 1956.

Ehrlich, P. R., and A. H. Ehrlich: *Population Resources Environment: Issues in Human Ecology,* 2d ed., Freeman, San Francisco, 1972.

———, ———, and J. P. Holdren: *Human Ecology—Problems and Solutions,* Freeman, San Francisco, 1973.

Frejka, T.: "The Prospects for a Stationary World Population," *Scientific American,* **228**(3):15–23, 1973.

Griffin, D. R., and A. Novick: *Animal Structure and Function,* 2d ed., Holt, New York, 1970 (see chap. 9, "Reproductive Systems").

Hedgpeth, J. W.: "Animal Structure and Function," *Biocore,* unit XII, McGraw-Hill, New York, 1974.

Jaffe, F. S.: "Public Policy on Fertility Control," *Scientific American,* **229**(1):17–23, 1973.

Katchadourian, H.: *Human Sexuality. Sense and Nonsense,* Freeman, San Francisco, 1975.

Kochert, G.: "Plant Structure and Function," *Biocore,* unit XIII, McGraw-Hill, New York, 1974.

Munger, B. L.: "Animal Diversity: Cells and Tissues," *Biocore,* unit X, McGraw-Hill, New York, 1974.

Rosebury, T.: *Microbes and Morals: The Strange Story of Venereal Disease,* Ballantine, New York, 1973.

Segal, S. J.: "The Physiology of Human Reproduction," *Scientific American,* **231**(3):52–62, 1974.

Shneour, E. A.: *The Malnourished Mind,* Anchor Press/Doubleday, Garden City, N.Y., 1974.

9

THE STUDY OF HEREDITY

SOME LEARNING OBJECTIVES

After you have studied this chapter, you should be able to

1. Demonstrate some of the things you have learned about Mendelian genetics by
   a. Explaining what Mendelian genetics is (what it studies).
   b. Explaining what, in Mendelian-genetic terms, each of the following is: a truebreeding strain, a parental generation, a dihybrid cross.
   c. Explaining, in terms of the dominance and recessiveness of the alleles involved, how a cross of two tall pea plants can result in the production of one or more short plants as well as tall plants.
   d. Giving some examples of traits with discontinuous variation, including a much-studied human disease.
2. State at least four kinds of inheritance of discontinuous traits which involve aspects of heredity other than those studied by Mendel, and state which one of them controls the A-B-O human blood types.
3. State what genetic linkage is; outline the process of inheritance of a sex-linked gene; name two human traits that are sex-linked; state whether each trait you have named is a continuous or a discontinuous trait.
4. State what is the basic genetic difference between continuously and discontinuously varying traits; say

which type of trait is the more common in organisms; give two examples of continuously varying traits in humans; name one environmental factor that may be involved in each of the two continuous human traits you have named.

5. Explain why identical twins are the preferred subjects of those who study the relative importance of *nature* and *nurture* in the determination of human phenotypes; and describe an effective strategy employing identical twins that is used by geneticists to study the relative contributions of heredity and environment to phenotypes.
6. Demonstrate your understanding of some of the basic workings of heredity by
   a. Stating the text's definition of a gene.
   b. Describing DNA's physical shape or configuration, and its chemical structure.
   c. Describing how the genetic information of a chromosome, contained in DNA molecules, is duplicated when a cell is ready to divide.
   d. Naming the four kinds of DNA nucleotides; stating how many of them combine to form a "word" in the genetic code, how many such "words" there are in the code, and what the function of the "words" is.
7. Demonstrate your knowledge of the nature and causes of mutations by
   a. Giving the text's definition of a mutation.
   b. Explaining why mutations are more likely to have harmful rather than helpful or inconsequential effects on the organisms in which they occur.
   c. Saying what a mutagen is, and naming some products that contain mutagens.
   d. Explaining why there is *no* lower limit (safe level) of human exposure to, or intake of, mutagens.
8. Name at least four genetic (hereditary) human diseases, including two that can be treated successfully if detected in time, and at least one that cannot be successfully treated.
9. Distinguish between eugenics, euphenics, and genetic engineering and say which of them is not yet a practical possibility.

CHAPTER NINE

As was discussed in previous chapters, the genetic material contains the information necessary to produce a new individual. The study of **genetics** is the study of the genetic material, its nature, and how it is passed from generation to generation. Genetics also includes the study of how the genetic material interacts with the environment to produce the phenotypes which we observe.

Long before anything at all was known about the genetic material, people knew that organisms inherited characteristics from their parents; they called the pattern in which inheritance took place **heredity**. Although the terms are often used interchangeably, we can say that genetics is the scientific study of heredity.

HEREDITY AND PEDIGREES

Long before the genetic material was discovered, people had been using their knowledge of heredity to breed animals and plants which had desirable qualities. They drew or wrote down a family tree or **pedigree** of the characteristics of the organisms they wanted to breed.

9-1 Pedigrees can show patterns of inheritance

A pedigree is the family tree of an organism; often it is drawn in such a way as to show only one trait or characteristic. Figure 9-1 shows a pedigree for the inheritance of coat color in guinea pigs (Fig. 9-2). In this diagram, a square symbolizes a male and a circle symbolizes a female; color indicates black coat color and white indicates white coat color. In this family of guinea pigs, a white female was mated with a black male. This white female gave birth to a litter of two black males and two females, one black, one white. On the other side of the family tree, a white male was mated to a white female; this female gave birth to a litter of two males and two females, all white. When both litters reached sexual maturity, one of the black males from the first litter was mated with a white female from the second litter. Their first litter consisted of three black males.

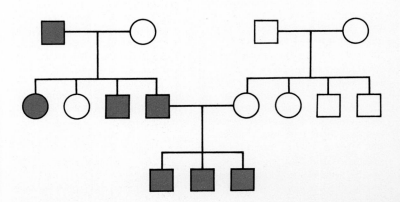

FIG. 9-1 A pedigree showing inheritance of coat color in guinea pigs. Squares indicate males, circles denote females; color indicates black coat, no color indicates white coat.

THE STUDY OF HEREDITY | 259

FIG. 9-2 Guinea pigs, also called cavies, are tailless rodents often kept as pets. Many varieties with different color and types of hair are known. (*American Museum of Natural History.*)

The pattern of inheritance shown by these few matings is that white mated to white produces only white offspring, but that black mated to white sometimes produces some black and some white offspring and sometimes only black offspring.

9-2 Pedigrees can be very puzzling

Although, as we shall soon see, the explanation for this seemingly inconsistent pattern is actually very simple, early geneticists were very puzzled by these kinds of pedigrees. Why should black plus white produce some black and some white sometimes but only black other times? Practical people, however, went right on using pedigrees to make decisions on breeding animals and plants. For the most part, they simply bred desirable types with each other and did not allow the others to reproduce. If they wanted a strain of white guinea pigs, they would exclude from breeding any who were known to have black ancestors. Sometimes this worked and sometimes it did not. Overall, however, it was remarkably successful; all our crop plants and farm animals were domesticated without any scientific knowledge of how heredity actually works. It was only in the 1930s that scientific knowledge of the actual way the characters observed in pedigrees were inherited began to be applied to domesticated plants and animals.

MENDELIAN GENETICS

The laws of genetics were first worked out in the mid-nineteenth century by an Austrian monk, Gregor Mendel. Mendel also knew nothing of the nature of the genetic material and simply crossed ordinary garden peas with varying characters. He not only drew pedigrees, he also counted the number of offspring of each type in each generation and made many crosses of the same type. This procedure gave him data from which he was able to draw conclusions about the mechanism of the inheritance of discontinuously varying traits.

9-3 Mendel studied discontinuous traits

Discontinuously varying traits are those that occur in only a few easily distinguished states, such as white or black coat color in guinea pigs. Mendel chose a number of such traits in garden peas. These plants produce many seeds each generation, and it is possible to mate, or **cross,** particular kinds of individuals by pollinating the plants with pollen for the desired cross.

Mendel found that his plants were either short or tall; none were medium height. Height, then, was a discontinuous trait. Mendel soon

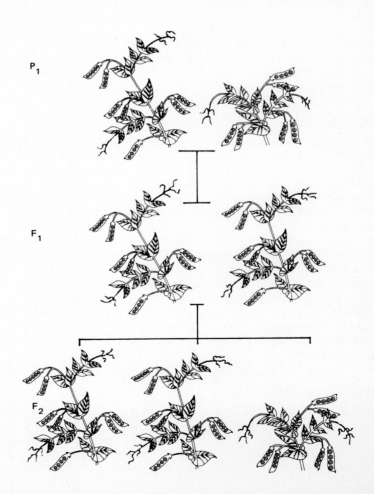

FIG. 9-3 Inheritance of tallness and shortness in pea plants. A tall plant crossed with a short plant has offspring which are all tall. When these tall offspring were crossed with one another, both tall and short offspring resulted.

produced strains of tall plants which always had tall progeny (offspring) and strains of short plants which always had short progeny. Thus he had two strains which bred true for his **parental generation,** symbolized P_1. When he crossed these tall plants with these short plants, he found that the progeny of this cross were all tall (Fig. 9-3). This generation is usually called the first familial or F_1 **generation.** When Mendel crossed the members of this F_1 generation with each other, he found that even though they were all tall, some of their *progeny* were tall and some were short. Mendel counted the numbers of tall and short plants in this, the F_2 **generation.** He found that although the exact number varied from cross to cross when he repeated the experiment, the ratio of tall plants to short plants in the F_2 generation was always the same. There were always about three times as many tall plants as short plants (a ratio of 3:1).

Mendel made several observations from his experiments. In a cross between true-breeding tall plants and true-breeding short plants, the offspring resemble one parent but not the other. Today we say that one trait is **dominant** to the other; tallness is dominant to shortness; and shortness is said to be **recessive** to tallness. Mendel found, however, that the recessive trait did not just disappear, it reappeared in the F_2 generation in one-fourth of the plants. In order to explain this reappearance, he assumed that there were two "factors" (now called genes) in each plant which affect each trait. The parent true-breeding for tallness would contain two factors for tallness; let us call them *TT*; and the parent true-breeding for shortness would contain two factors for shortness; let us call them *tt*. In the first generation cross all the progeny of the F_1 would receive one factor for tallness and one factor for shortness (Fig. 9-4). The next generation, the F_2, has parents both of which are *Tt* individuals. Figure 9-4 shows how when these factors are put together in all possible combinations, there are *TT*, *Tt*, and *tt* offspring. The *tt* offspring are short, of course, and Mendel's factors explain not only the reappearance of the recessive trait in the F_2 generation but also the ratio of three tall plants to one short plant.

Mendel's observations can explain the coat color pedigree in guinea pigs (see Fig. 9-1). The pattern of coat color inheritance in guinea pigs follows the same rules as inheritance of height in garden peas. If black is symbolized by *B* and white by *b*, the pedigree can be redrawn with symbols (Fig. 9-5) which account for the results obtained. The all-black litter of the black male and white female does not contradict the rules. To determine *his* results, Mendel counted *large numbers* of progeny. In any one cross, particularly one with only three offspring, the expected results might not be found. In fact, it is possible to predict how often the observed result is likely to occur. There are laws of probability which allow one to predict the outcome of chance events such as the sorting of genetic factors. In this cross, $Bb \times bb$, there are two possible types of offspring, *Bb* (black) and *bb* (white); thus each offspring has a 50 percent chance of being black. The laws of probability show that the chance of

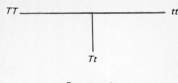

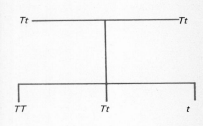

FIG. 9-4 Inheritance of tallness and shortness in pea plants showing behavior of factors controlling height in the F_1 and F_2 generations.

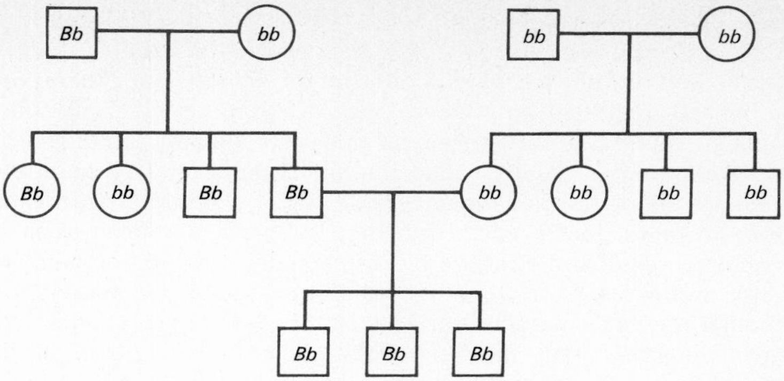

FIG. 9-5 A possible pedigree of coat color inheritance in guinea pigs showing the factors.

getting a group of events together is the product of their separate probabilities. Thus the probability of getting three black offspring is .5 × .5 × .5, or .125. Translating this back into percent, 12.5 percent of the time three black offspring are expected in such a family.

9-4 Mendel studied traits that behaved independently

As you know from Chap. 7 the different chromosomes of a set assort independently at meiosis when the gametes are formed. If the factors controlling two different traits are on different chromosomes, they should behave independently. After Mendel studied the inheritance of seven traits separately, he experimented with their behavior in combinations. Now, garden peas have seven pairs of chromosomes, so it is quite remarkable that the seven characters Mendel studied were all controlled by factors on *different* chromosomes. In Sec. 9-6 we shall see what happens when factors on the same chromosome are tested for inheritance.

One of Mendel's crosses which involved two traits is diagrammed in Fig. 9-6. He crossed a true-breeding tall strain with yellow seeds (*TTYY*) with a true-breeding short strain with green seeds (*ttyy*). The F_1 generation will all have the genotype *TtYy* since each offspring receives a factor for each trait from each of its parents. Phenotypically, the F_1 plants will all appear tall and have yellow seed; both tallness and yellow seed are dominant. The F_1 plants are all **heterozygous** for both traits; that is, they have both possible factors present for each trait. Sometimes organisms heterozygous for a trait are said to be **hybrids**. Thus, the F_2 generation of this series of crosses is often said to result from a **dihybrid cross**. The results are discovered by first listing all the kinds of gametes each parent can produce and then combining them. If these traits are on different chromosomes and thus behave independently, the possible gametes are *TY*, *Ty*, *tY*, and *ty*. Figure 9-7 shows all the genotypes which would result from such a cross. There are four possible phenotypes: tall, yellow; tall, green; short, yellow; and short, green. These occur in the ratio 9:3:3:1.

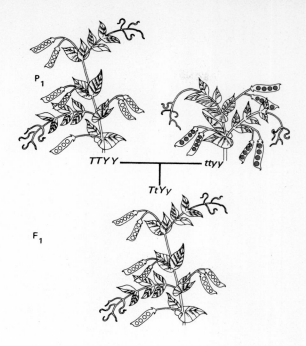

FIG. 9-6 Tall pea plants with yellow seeds produce tall, yellow-seeded plants when crossed with short plants with green seeds.

9-5 Mendel's factors are now called genes and his observations have become laws

Mendel published his report on heredity in 1865. Other biologists of the period were not interested in his results, which did not agree with their theories. In fact, one distinguished scientist urged Mendel not to publish his results because they were "obviously wrong." As a result, Mendel's work was largely ignored until the early 1900s when it was rediscovered. By that time, the behavior of chromosomes at meiosis had been worked out and the time was right for a theory which involved factors and independent assortment. A search was made for organisms in which genetics could be studied in the laboratory. Several organisms, especially the fruit fly *Drosophila* (Fig. 9-8), were found to be excellent for this purpose. *Drosophila* could be raised in jars in the laboratory and produced many offspring very rapidly. All of Mendel's work was confirmed and many new things were discovered as investigations were made in laboratories and gardens and with domesticated animals and plants.

What Mendel called factors affecting the traits of an individual we now call genes. The inheritance of each of Mendel's traits was controlled by a different factor. Today we say that there is a gene controlling each characteristic and that each gene has a particular place, its locus on a chromosome. Each gene can occur in two or more states, for example, the dominant and the recessive. The alternative (two or more different) states of a gene are called **alleles**. The

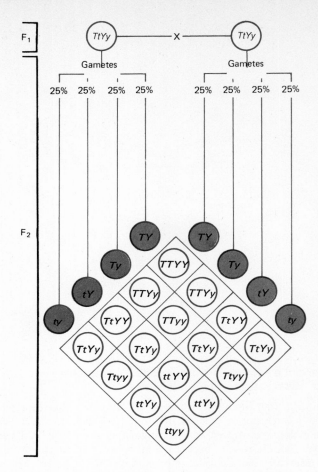

FIG. 9-7 Diagram to show the results of crossing two pea plants that are heterozygous for the genes affecting height and seed color.

FIG. 9-8 The fruit fly *Drosophila melanogaster* is commonly used in studies of heredity because it is easily grown in the laboratory and produces a new generation every 21 days.

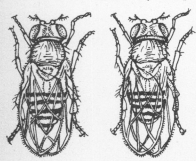

Female   Male

gene for height in peas has two alleles: T and t. T is dominant to t. An individual with the same two alleles of a gene (for example, TT or tt) is said to be **homozygous** at the T locus. An individual with two different alleles of a gene (for example Tt) is said to be heterozygous at the T locus. We now know (Sec. 9-9) that each gene is made up of DNA (deoxyribonucleic acid) and that alleles represent slight changes in a specific region of the DNA.

Mendel's observations are now referred to as **"Mendel's laws,"** and the study of the inheritance of discontinuous traits is called **Mendelian genetics.** Now that chromosomes, genes, and meiosis are known, Mendel's laws are often stated as:

1. *The law of segregation.* When sexual organisms reproduce, the paired hereditary factors (genes) segregate from each other during meiosis in a way assuring that each egg or sperm receives only one member of each pair of genes (for example, a female with a Tt genotype produces T eggs and t eggs).

2. *The law of independent assortment.* The offspring of individuals heterozygous for several traits exhibit the traits independently of each other; that is, each gene pair segregates independently of the other gene pairs of the parents. (For example a female with a *TtYy* genotype produces four kinds of eggs with respect to these loci: *TY*, *Ty*, *tY*, and *ty*.)

SOME COMPLEXITIES OF HEREDITY

Most organisms grow up to look more or less similar to their parents. No two individuals, however, are ever exactly alike. Look at other students the next time you walk across campus. They differ in skin color, hair color, hair type, eye color, size, and many other characteristics. This is true of all organisms (Fig. 9-9). Biologists such as George Schaller, who studies mountain gorillas, and Jane van Lawick-Goodall, who studies chimpanzees and wild dogs, quickly learn to recognize individuals of the animals with which they are working. Of course, the animals themselves recognize individuals within their own species. Even penguins that all look alike to most people are recognized individually by their mates in a breeding colony.

What causes this variation? The fact is, not all aspects of heredity are as simple as those studied by Mendel. Genes commonly have more than two alleles. Genes may be on the same chromosome and *not* assort independently. Some traits are controlled by many alleles, and some alleles control many traits. Dominance is often not complete; alleles can mutate to different alleles. Chromosomes can break and pieces can be rearranged or lost. All these processes combined with the basic processes of crossing over and random assortment of chromosomes at meiosis (Sec. 7-2) assure individual variation *and* complexities in the study of heredity.

9-6 The location of genes on chromosomes can be important in their inheritance

Linkage Soon after the genetics of such organisms as *Drosophila* began to be studied, it was found that some genes do not behave independently; this finding led people to hypothesize that such genes were on the same chromosome. Tests showed that indeed some genes were on the same chromosome and that their position on the chromosome could be mapped in relation to all the other genes on the same chromosome. As noted in Sec. 7-2 and Fig. 7-10, crossing-over of chromatids (half chromosomes) occurs in meiosis. Unless this breakage and reunion of chromatids occurs, genes on the same chromosome are inherited as one unit; they are said to be **linked**. That is, a male with hypothetical genes *ZR* on one chromosome and *zr* on the other of the pair could produce, without crossing over, only sperm which carried *ZR* together or *zr* together. Usually crossing-over occurs only occasionally between two genes on the same chromosome so that a certain percent of the meiosis-produced gametes carry recombined alleles. There is more crossing-over between genes that

FIG. 9-9 Sketches of individual differences in head markings in four individual acorn woodpeckers. Birds of this species are black and white with a red head patch (shown in stipple). *a* and *b* are identified as males because they lack wide black bands in front of the red patches. The black bands, conspicuous in *c* and *d*, indicate females. Male *a* is distinguished from male *b* by the square shape of the black area around its beak in contrast to the rough edge of this black area in male *b*. Female *c* has jagged edges on the black areas around her beak and eye, while in female *d* these black areas are smooth at their edges and pointed. Ornithologists use these patterns (easily seen in the field through a 20-power spotting scope) to study behavior in colonies of this hole-nesting species. (*Sketches courtesy of Ruth Troetschler.*)

266 CHAPTER NINE

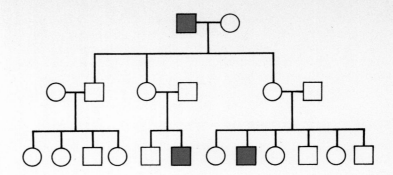

FIG. 9-10 A pedigree showing inheritance of color blindness in humans. Color indicates color blindness.

are far apart on a chromosome than between genes that are close together. The amount of crossing-over is usually expressed as a percent; with data from many genes, these percentages can be used to map the positions of genes on chromosomes without the chromosomes even being seen.

Sex linkage Although human beings are not ideal subjects for genetic studies—they live too long, have too few offspring, and resist being subjects of controlled crosses—human pedigrees have contributed to our knowledge of some aspects of genetics. In fact evidence that genes are on particular chromosomes came from studies of pedigrees of human families with red-green color blindness and hemophilia.

Figure 9-10 is a pedigree for red-green color blindness. Individuals having this characteristic are unable to distinguish red from green. If this pedigree is compared with that for guinea pig coat color (Fig. 9-1), it will be seen that the patterns of inheritance of the two characteristics are different. Red-green color blindness is much more common in males than in females. Red-green color-blind fathers pass their traits on to their daughters but not to their sons. However, the daughter is usually not color-blind. She *does* carry the genetic potential of having a color-blind son. This kind of pedigree is also shown by another human trait, hemophilia (Fig. 9-11). Hemophilia is a disease in which the blood does not clot normally so that individuals with hemophilia may lose great quantities of blood from a simple cut or nose-bleed, or suffer massive bruises or internal bleeding. This pedigree begins with Queen Victoria of England and Prince Albert. It shows the pattern of inheritance of this gene in the marriages among the royal families of Europe. Only the males have the disease and, at first sight, it appears to occur randomly.

These pedigrees are very simply explained if we assume that the troublesome genes are carried on a chromosome present in females but not in males. The photographs of the chromosomes of a human male and a human female in Figs. 7-6 and 7-7 show that one pair of chromosomes is different in the two sexes. Males have one pair of two different-appearing chromosomes, while females have two chromosomes that appear the same. The two sex chromosomes of a human female are called X chromosomes and the two sex chromo-

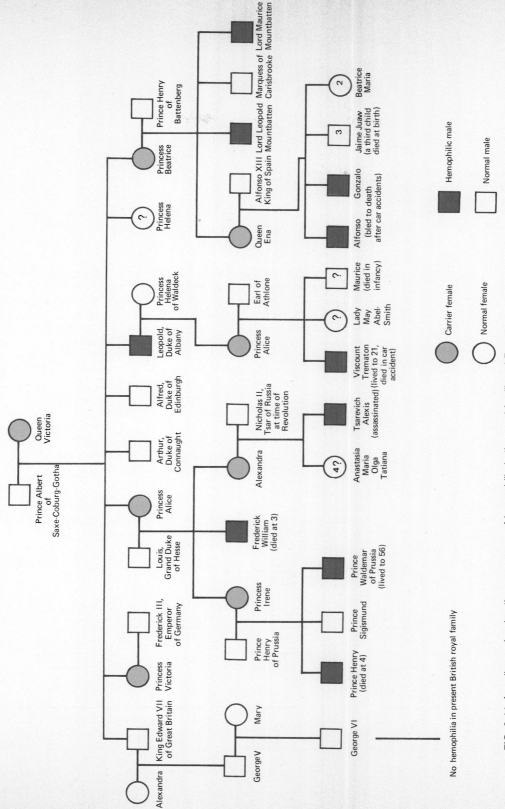

FIG. 9-11 A pedigree showing the occurrence of hemophilia in the royal families of Europe.

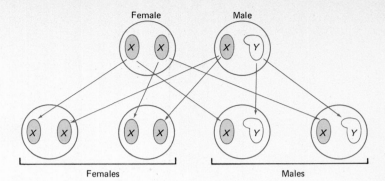

FIG. 9-12 The pattern of X chromosome inheritance. Males give an X chromosome (colored) only to their daughters. Sons receive their X from the mother and their Y from the father.

somes of a male are called an X chromosome and a Y chromosome.

In meiosis in a human female, every egg will receive an X chromosome. In a male, however, meiosis results in half of the sperm getting an X and half a Y chromosome. When a sperm containing an X chromosome fertilizes an egg, which always has an X, the resulting individual will be *female*. If a Y-containing sperm fertilizes an egg, the offspring will be *male* (with one X and one Y chromosome). If the pattern of behavior of the X chromosomes of males (Fig. 9-12) is compared with the pattern of inheritance of red-green color blindness (Fig. 9-10), it will be seen that they are similar. This pattern of inheritance is called **sex-linked inheritance,** and it occurs for every gene located on the X chromosome. If you symbolize the allele for color blindness by X_{cb} and normal color vision by X, you can figure out the genotypes of individuals in the pedigree shown in Fig. 9-10. The correct genotypes are shown in Fig. 9-13.

FIG. 9-13 The color blindness pedigree of Fig. 9-10 showing the behavior of genes on the X chromosome. Light color indicates carriers, darker color indicates afflicted males.

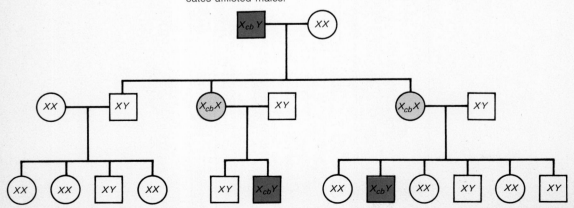

9-7 There are other problems in studying the inheritance of discontinuous traits

Incomplete dominance There are many genes which express their full potential effect on the phenotype only when they are homozygous. When they are heterozygous, the phenotype is intermediate between the homozygous dominant and the homozygous recessive phenotypes. An example is the inheritance of coat color in shorthorn cattle (Fig. 9-14); coat color in these cattle may be white, red, or a mixture of white and red called roan. A pure red strain crossed with a pure white strain gives offspring that are all roan. Red is said to be **incompletely dominant.** Two roan individuals, when crossed, produce calves in the ratio of one red to two roan to one white.

Multiple alleles All the characteristics discussed so far have been controlled by genes with two alleles. One allele may be dominant over the other, or dominance may be incomplete. More complex situations are known, however, in which a gene has more than two

FIG. 9-14 Coat-color inheritance in shorthorn cattle showing incomplete dominance. Dark color indicates red coat, light color indicates roan.

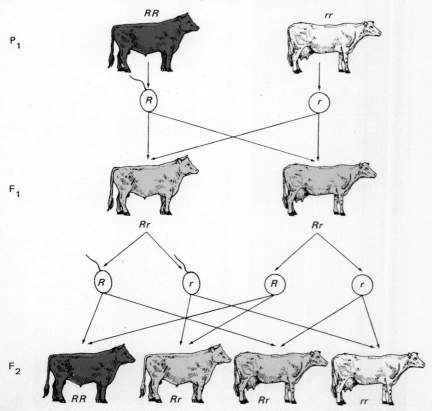

FIG. 9-15 Four types of coat color in domestic rabbits: (a) wild type (agouti), (b) albino, (c) chinchilla, (d) Himalayan. [(a) *Knolan Benfield Jr., Image;* (b) *Mary Maddick, FPG;* (c) *Clyde Keeler and Virginia Cobb,* Journal of Heredity; (d) *From W. E. Castle.*]

alleles. Coat color in many mammals illustrates the effects of **multiple alleles** (Fig. 9-15). Wild rabbits commonly have a grayish brown fur called *agouti* (each hair has stripes of several different colors). When wild rabbits are raised in captivity, rarely an all-white **albino** rabbit with pink eyes is produced. Crosses can be made which show that the wild-type color is dominant to white. Rarely, however, other kinds of coat color are produced. One type is all gray and is called *chinchilla*. Another is all white, except for the ears, nose, and feet, which are black. This type is called *Himalayan*. Crosses have shown that four alleles can explain this situation. Wild-type is dominant to all others, chinchilla is dominant to Himalayan and albino, and Himalayan is dominant to albino. Other mammals have similar patterns of coat-color inheritance.

A-B-O blood types in human beings are also controlled by multiple alleles. Blood type is dependent upon the kind of antigen (see Sec. 14-30) on the surface of the red blood cells (Table 9-1). The antigen on red blood cells which determines blood type is a polysaccharide (a complex sugar) and can be either type A or type B. In any individual, the red blood cells can have type A antigen or type B antigen or both or neither, on their surfaces. Therefore, the blood type can be classified as A, B, AB (both antigens present), or O (neither kind of antigen present). The alleles of the gene affecting the type of antigen are symbolized by I^A, I^B, and i. Genotypes $I^A I^A$ or $I^A i$ result in antigen A. $I^B I^B$ or $I^B i$ result in type B antigen. Individuals with $I^A I^B$ have AB blood, while ii individuals have type O blood. Since the kind of antigen on the cell affects the behavior of the red blood cells in blood tranfusions (Table 9-1) A-B-O blood typing is of great medical importance. It may also be important in identifying the natural father of a child. For example, a man with type O blood (ii) cannot father an offspring with type AB blood ($I^A I^B$). Although in some cases it can be shown by blood type that a given man *was not* the father of a certain child, it can never be shown by blood type that a given man *was* the father of a certain child. Even though the blood types allow for the possibility, there are many other men in a community with the same blood type as the man in question.

Multiple loci Sometimes a single phenotypic trait is controlled by several genes at different loci. In fact, most traits are affected by more than one gene. Often the effects are very subtle so that the pattern of inheritance cannot be easily studied using Mendelian genetics. There are some cases, however, where two loci clearly control one character; an example is feather color in chickens (Fig. 9-16). In this case, allele A inhibits the expression of allele B, which is the allele for colored feathers, and we say that allele A is *epistatic* to allele B.

9-8 The inheritance of continuously varying traits is controlled by many genes

Most characteristics are difficult to put into clear-cut categories. We tend to think, for example, that people have either blue eyes or brown eyes. If you try to classify your friends' eye colors this way, you will

TABLE 9-1 A-B-O BLOOD GROUPS IN HUMAN BEINGS

BLOOD GROUP	GENOTYPE	ANTIGEN ON RED BLOOD CELLS	ANTIBODIES IN SERUM OF BLOOD	BLOOD GROUPS ACCEPTABLE FOR TRANSFUSION
O	ii	None (universal donor)	Anti-A and anti-B	O
A	$I^A I^A$, $I^A i$	A	Anti-B	A, O
B	$I^B I^B$, $I^B i$	B	Anti-A	B, O
AB	$I^A I^B$	A, B	None	All (universal recipient)

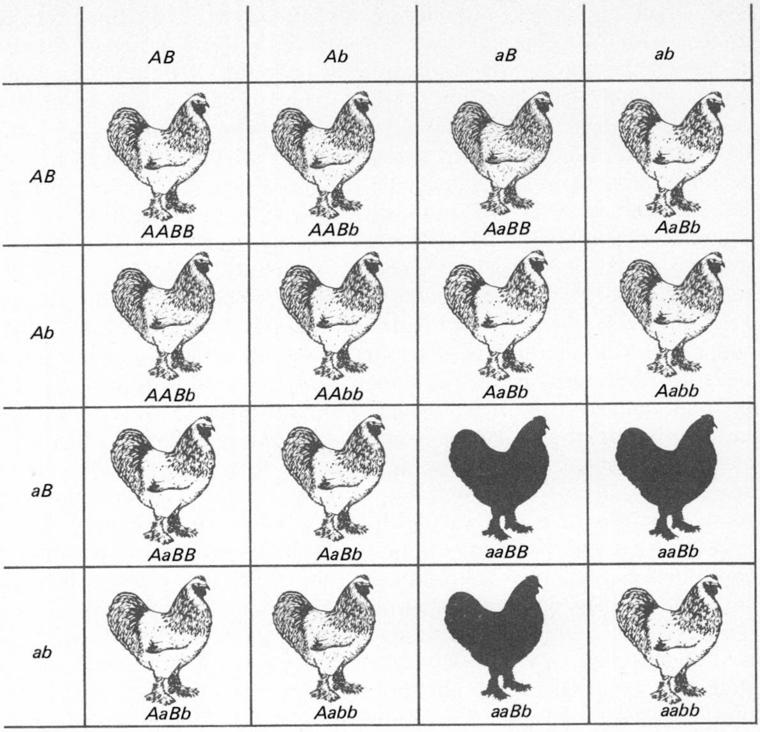

AB
Ab
aB
ab
AB
AABB
AABb
AaBB
AaBb
Ab
AABb
AAbb
AaBb
Aabb
aB
AaBB
AaBb
aaBB
aaBb
ab
AaBb
Aabb
aaBb
aabb

FIG. 9-16 Inheritance of feather color in chickens. Allele *B* is responsible for production of color, but allele *A* inhibits its action. The parents are both *AaBb*. Such interaction of genes is called epistasis.

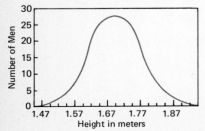

FIG. 9-17 Distribution of heights in a sample of males in the United States showing continuous variation.

30
25
20
15
10
5
0
Number of Men
1.47
1.57
1.67
1.77
1.87
Height in meters

probably find that some have gray eyes and some have greenish or hazel eyes. Hair is not just black or blond; all intermediate colors of brown occur. People do not occur in two height classes, tall or short, nor even in three classes—tall, medium, and short.

Characteristics such as black coat versus white coat in guinea pigs or color-blind versus not color-blind in human beings are relatively rare. Most characteristics of organisms vary *continuously*. If you measure the height of all the men in one of your classes, you could make a graph such as that in Fig. 9-17. This graph shows continuous variation in height; there are short people and tall people, but most fall in between these extremes. Most characteristics in men and women show this kind of continuous variation. Continuous variation was a very difficult problem for the early geneticists. They had little success in understanding the inheritance of continuously varying traits until they had thoroughly studied **discontinuous variation,** such as that of guinea pig coat color.

Characteristics which vary continuously, such as height, are now known to be inherited in the same way as discontinuous traits. The only difference is that they are controlled by a great many different genes. The study of continuously varying characteristics is difficult because one cannot just count the number of individuals of each type.

Often even complex mathematical analysis is inadequate to understand such inheritance.

Let's use a hypothetical example of inheritance of a continuously varying trait. Suppose adult pigs are weighed and a graph like that in Fig. 9-18 is drawn. The inheritance of adult weight in pigs could be explained by assuming that there are 50 different genes at 50 different loci; alleles which affect weight are symbolized by either W or w. The allele W adds 2 kilograms (4.4 pounds), while the allele w adds 1 kilogram (2.2 pounds). If a pig has Ww at each of the 100 loci, it could weigh 150 kilograms. If it has WW at all 100 loci, it could weigh 200 kilograms. The minimum weight is 100 kilograms with ww for all loci, and all intermediate phenotypes between 100 and 200 kilograms can occur. The possible phenotypes for this example are graphed in Fig. 9-19. Notice that, according to this system, a pig could weigh 150 kilograms or 152 kilograms or 154 kilograms, etc., but it could not weigh 139 kilograms or 151 1/2 kilograms. Each allele W adds 2 kilograms, no more, no less.

Obviously when the large number of pigs were weighed to get the data in Fig. 9-19, some pigs did not weigh even numbers of kilograms, and some did have weights which had to be expressed as fractions. The exact weight of a pig depends not only on its genotype but also on its environment. Suppose a pig whose genotype specified 150 kilograms was very well fed. It might easily reach a weight of 167 kilograms. If a pig with the same genotype did not receive enough food, it might weigh only, say, 143 kilograms. Any one allele can affect weight by only 1 or 2 kilograms. The phenotype, with respect to weight of the pig, is affected by the environment, which may act to add or subtract, say, 10 to 20 kilograms. This illustrates that with continuously varying traits the effects of the environment may be greater than the effects of one or several genes.

Nature versus nurture Much debate has occurred over the relative importance of environment and genotype in determining phenotype. This has sometimes been referred to as the *nature-versus-nurture problem*. We have seen that nature (the genotype) and nurture (the environment) are both important. The question is: Which is more important? This is a very difficult question to answer. The ideal way of studying the problem is to have two organisms with exactly the same genotype. These could then be raised in different environments. Obtaining two individuals with the same genotype and carrying out such an experiment is not easy in animals.

Cloning As was discussed in Chap. 7, plants can be reproduced asexually. Branches can be cut off a plant and rooted in sand or soil; the branches each form new plants genetically identical to the original plant (Fig. 9-20). Such clones of plants can be useful in separating the effects of genotype and environment on the phenotype. Suppose that individuals of one species of plant are low-growing and have thicker and grayer leaves when they grow in sandy soil near the edge of the sea. In inland areas, they are tall and have thinner, brighter-green leaves. The seashore genotype plants could be grown in an

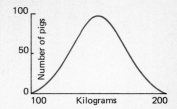

FIG. 9-18 Curve showing the distribution of weights in a hypothetical population of pigs.

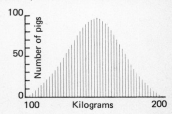

FIG. 9-19 Frequency diagram showing expected distribution of weights of pigs based upon the discussion in the text.

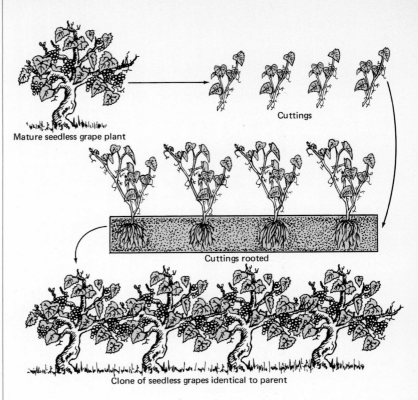

FIG. 9-20 Many plants can be propagated by taking cuttings (branches) which will root and grow. For example, seedless fruits, such as grapes, must be propagated this way. The group of genetically identical individuals is called a clone.

inland area and the inland genotype plants could be grown in a *seashore habitat*. By comparing the performance of the two genotypes in each of the environments, we could decide how much of the difference between the two kinds of plants was due to environment and how much to their genotypes. Such studies (Fig. 9-21) have shown that although the environment can modify many traits in plants somewhat, differences such as those described for seashore and inland plants are largely genetic.

Identical twins One of the difficulties in studying human genetics is the impossibility of making controlled crosses. Another is that nearly all the characteristics in which we are interested are controlled by a great many genes. Furthermore, humans cannot be reproduced asexually to produce many individuals with the same genotype. Therefore it is difficult to answer such questions as: What diseases are inherited? Is intelligence genetically determined?

Nature, however, has provided a way in which these questions can be studied. On the average, once in some 900 births twins are produced. Twins may be fraternal or identical, fraternal being more common. Fraternal twins result when *two* eggs are fertilized and the embryos become implanted in the uterus. Because they result from

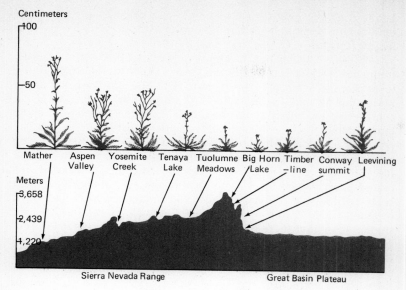

FIG. 9-21 Variation in size of the plant milfoil (*Achillea*) along a transect across the Sierra Nevada Range from California to Nevada. (*Courtesy of Carnegie Institute.*)

two *different* eggs and fertilizations, fraternal twins are no more alike genetically than *any* two **siblings** (offspring of the same parents).

Identical twins result from a separation of *one* fertilized egg into two, after it has divided into two cells. The two cells proceed to develop as if they had been separate eggs. The resulting embryos, however, have the same genotype. When they are born, identical twins are, indeed, genetically identical. By studying the phenotype of identical twins and comparing them with fraternal twins, one can learn something of the effects of heredity and environment. It turns out that identical twins *remain* much more similar physically than do fraternals. Identical twins often get the same diseases at about the same time. This applies to certain mental diseases as well. Therefore there must be some genetic control of various kinds of disease, as well as of bodily characteristics.

But this refers to twins reared in the same environment. How can we determine the relative importance of genotype and environment? Rarely, identical twins are separated at birth and raised in very different environments. Sometimes they are put up for adoption and sent to families in different geographical areas and often of different religious or ethnic backgrounds. Here, then, is the kind of test needed to study the effects of different environments on the same genotype in people. Scientists have tracked down and studied as many cases as possible of identical twins separated very early in life and raised in different homes.

From such studies, it is clear that there is strong genetic influence on such bodily traits as physical appearance. Twins raised in different families may have different weights, depending upon their diet and habits of exercise. But their general look-a-likeness is not changed by different environments, and their scores on IQ and other psychological tests are very similar.

Intelligence, defined as the ability to learn or understand, has been frequently evaluated by IQ tests. IQ tests actually measure the ability to perform academically on certain tests and most are culturally biased toward middle-class values. As we shall see in Chap. 10, great controversy has raged about the use of IQ tests to attempt to show that different races of people are genetically different in IQ. The scores of identical twins on IQ tests tend to be more alike than the scores of other siblings, and the scores of identical twins reared together tend to be even more alike than those of identical twins reared apart. When the various studies are all put together, there is a large range of overlap between the similarities of twins reared together, twins reared apart, and siblings reared together. Recent studies show that there is a strong relationship between birth order and IQ; firstborn children score highest and last-born the lowest. Thus there appear to be both genetic and environmental factors that affect an individual's score on an IQ test. Unfortunately, many of the early twin studies have serious defects which do not permit an accurate estimate of the role played by genetics in determining score on IQ tests. Because of these difficulties, about all we can do is make a tentative model based on our overall knowledge of the inheritance of continuously varying characteristics.

The genetic control of intelligence probably is not as strong as the genetic control of physical features. An oversimplified model would be that the genotype determines a general level of intelligence—a potential range as measured by IQ scores (Fig. 9-22). If a twin is brought up in an intellectually stimulating environment and encouraged to read and learn, most of his or her genetic potential will be realized, and the IQ will be higher than that of the identical twin raised under less stimulating circumstances. We must emphasize that this is an oversimplification, however, since in actuality one cannot think of a rigid, genetically controlled potential independent of environment. For instance, no lower limit to IQ could be set genetically which could not be exceeded environmentally. One of the twins in Fig. 9-22 with the 80 to 120 genetic range, if brain-damaged in an accident, might end up with an IQ of only 20. Similarly, an extraordinarily stimulating environment might boost the IQ to 124—although

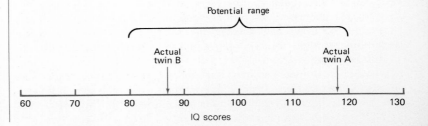

FIG. 9-22 Diagram showing the effects of the environment on the IQ of twins. Twins A and B have identical genetic potential (IQ between 80 and 120). Twin A has had a favorable environment and scores 118 on an IQ test. Twin B has had a less stimulating environment and scores 87.

the ways in which IQ performance can be enhanced environmentally are much less well understood than ways in which it can be environmentally reduced.

HOW HEREDITY WORKS

We have referred to the genetic material as consisting of the chemical substance DNA. Now it is time to consider the structure and functioning of DNA, in other words, how the chemical structure of DNA controls the development of the phenotype of organisms.

9-9 Genes are sections of giant DNA molecules

A gene is the amount of genetic information necessary to control production of one polypeptide. Proteins are made up of sequences of chemical building blocks called **amino acids.** A sequence of amino acids is called a **polypeptide.** Genes contain a "code," which is "decoded" by a cellular mechanism and translated into the specific sequence of amino acid building blocks which characterizes a polypeptide. Each allele (barring mutation, Sec. 9-10) is always translated into the same sequence of amino acid building blocks according to its code. About 20 types of amino acids are commonly found in proteins. It is the different arrangement of these amino acid subunits of the giant protein molecules that gives each protein its special qualities. Some genes carry coded instructions for making **structural proteins**—those which are the main constituents of muscle, hair, and so on. Other genes carry the code for the production of a kind of protein called **enzymes,** the biological catalysts (Sec. 13-1).

Structure of DNA molecules The genetic information of eukaryotic organisms is stored in the chromosomes of the cell nucleus. It is in the form of a code which is a sequence of subunits of molecules of deoxyribonucleic acid. Each DNA molecule consists of two strands (chains) of subunits winding around each other in the form of a **double helix** (Fig. 9-23a). Each subunit consists of a chemical group called a *base* with an attached phosphate group. If you could unwind a DNA molecule, its two-stranded structure would look something like a ladder. Each subunit would make up part of an upright and part of a rung (Fig. 9-23). Where subunits are joined in the rungs, they are held together by relatively weak chemical bonds called **hydrogen bonds** and are said to be paired (Fig. 9-23b and c). Where subunits are joined together in the uprights, they are held together by relatively *strong* chemical bonds. The solidly joined parts of the subunits (the uprights of the ladder) are referred to as the backbone of each of the two strands. The backbones of both strands are made up of the sugar-phosphate parts of the subunits. The backbones of the two strands are the same except that their subunits are oriented in opposite directions.

The DNA of prokaryotic organisms (Monera) is not found in well-defined nuclei. It consists of ring-shaped molecules not incorporated

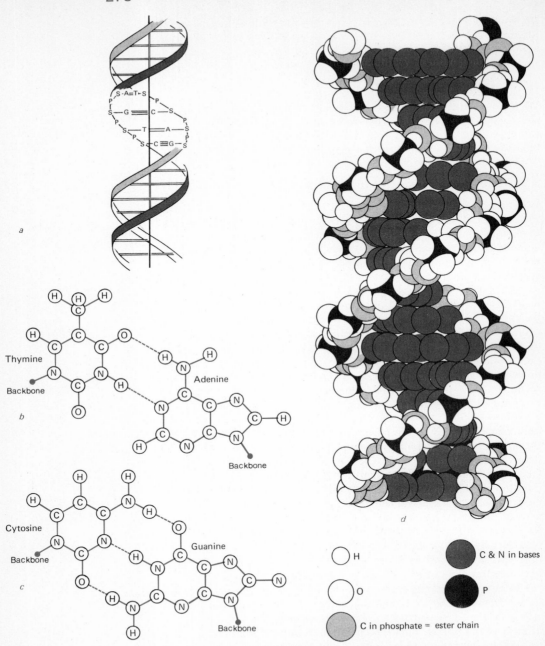

FIG. 9-23 The structure of DNA as proposed by James Watson and Francis Crick. (*a*) Helical structure showing two helical backbones held together by hydrogen bonds (S indicates sugar group; P indicates phosphate group; A, T, G, and C indicate bases); (*b*) the pairing of T and A by two hydrogen bonds (dashed line); (*c*) the pairing of C and G by three hydrogen bonds (dashed lines); (*d*) model showing how all the atoms fit together.

into chromosomes. Nearly all our knowledge of the way DNA is put together and how it functions comes from studies of the simpler systems found in the bacterium *Escherichia coli* and other bacteria. There is evidence that DNA functions in the same way in eukaryotic organisms.

DNA nucleotides DNA subunits differ in the chemical structure of the parts of the molecule which form the rungs of the DNA ladder (see Fig. 9-23). There are four different nucleotides, which we shall call A, C, G, and T, after the names of the bases which make up the rung sections of the nucleotides: adenine, cytosine, guanine, and thymine. A, C, G, and T subunits are able to pair with each other only in the patterns A:T and C:G. This is because the chemical structure of the bases permits hydrogen bonds to be formed only between those combinations.

Thus you can think of a DNA double helix as being composed of two **complementary strands.** Each is made up of subunits, each has the same backbone of sugar phosphates, and each has its bases arranged so that they are able to pair with the bases of the other strand (see Fig. 9-23). Pairing bases are said to be complementary to each other. Thus a sequence AAACGTT on one strand is complemented by TTTGCAA on the other.

This double-helix model of DNA has been confirmed in numerous experiments and provides the basis for an explanation of how the information of a chromosome is duplicated when cells are ready to divide. What apparently happens is that the hydrogen bonds joining the pairing units break and the two strands unwind (Fig. 9-24b). Then, in a chemical reaction, new strands are synthesized against each of the old. In a sense, each old strand serves as a mold, or a **template,** for the construction (Fig. 9-24b) of a new complementary partner for itself. Thus a single DNA double helix is converted into two DNA double helices, each an exact copy of the original (Fig. 9-24c). One daughter double helix then goes into each daughter cell at mitosis—each cell gets a complete copy of the genetic information.

Language of the genetic code If a gene is a section of DNA specifying the sequence of amino acid residues in a protein, then the code language must be able to specify some 20 types of amino acids. Combinations of two nucleotides out of the four could specify only 16—not enough. But combinations of three nucleotides **(triplets)** out of four could specify 64 amino acids—more than enough. Ingenious experimental work, for which several biologists won Nobel prizes, has confirmed that the code does indeed use "words" that are three nucleotide units long. Thus, a dictionary has been assembled that tells which triplets code for which amino acid residue (Table 9-2). Notice that the nucleotide uridine (U) substitutes for thymine (T) in Table 9-2. This is the conventional way of representing the genetic code because DNA is transcribed into RNA (which contains U in place of T) before it is translated into protein (see Sec. 13-2). In this "language," UCU, UCC, UCA, UCG, AGU, and AGC all code for the amino acid called serine. However, only UGG codes for the amino acid called trytophan. Most amino acids are specified by more than

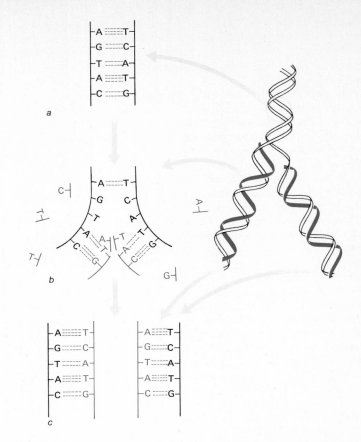

FIG. 9-24 The mode of replication of DNA. The two helical strands separate, and each can serve as a template against which new strands of DNA are formed. (*a*) Parental strands of DNA; (*b*) strands separate as hydrogen bonds break, and new strands are synthesized as nucleotides pair with separated strands; (*c*) two new strands of DNA, each consisting of one parental strand and one newly synthesized strand.

one triplet word. A few words are used for "punctuation" (for instance, to show where a gene begins and ends). Does DNA seem needlessly complex? After all, what other language could specify instructions for building all the living things on earth with only 64 words? Of course, DNA does this by specifying sequences of amino acids in proteins. (See Fig. 13-5.) There are 20 amino acids which can be put into millions of different arrangements, and the average protein is composed of several hundred of them. These proteins not only go into the structure of the organism, but some of them are also enzymes which make possible the production of all the chemicals the cells need to make an organism and keep it going with energy and raw materials from its environment.

9-10 Genes may change in a process called mutation

Genes are subject to occasional change. For instance, once in a great while, a cross between two short pea plants will produce a tall offspring. How is this possible when both parents must have genotype *tt*? The answer is that one of the alleles has changed from *t* to *T*,

TABLE 9-2 THE GENETIC CODE

RNA TRIPLET	AMINO ACID	RNA TRIPLET	AMINO ACID	RNA TRIPLET	AMINO ACID
UUU, UUC	Phenylalanine	CGU, CGC, CGA, CGG	Arginine	GCU, GCC, GCA, GCG	Alanine
UUA, UUG, CUA, CUU, CUG, CUC	Leucine	CAU, CAC	Histidine	GGU, GGC, GGA, GGG	Glycine
		CAA, CAG	Glutamine		
UCU, UCC, UCA, UCG, AGU, AGC	Serine	AUU, AUC, AUA	Isoleucine	GAU, GAC	Aspartic acid
		AUG	Methionine	GAA, GAG	Glutamic acid
UGU, UGC	Cysteine	ACU, ACC, ACA, ACG	Threonine	UAA, UAG, UGA	Termination signal
UGG	Tryptophan				
UAU, UAC	Tyrosine	AAU, AAC	Asparagine		
CCU, CCC, CCA, CCG	Proline	AAA, AAG	Lysine		
		GUU, GUC, GUA, GUG	Valine		

or as geneticists would say, it has mutated $t \to T$. **Mutations** are changes that occur randomly in the genes, that is, in the DNA sequence. When a mutation occurs, it *may* appear immediately in the phenotype of the next generation; that is, it may be dominant. More often, however, it is recessive. Because a mutation *changes* a protein that is already functioning in the complex system of the organism, it is apt to have a harmful effect. Often the modified proteins produced are inactive as enzymes or do not function properly as structural components of cells. Since most animals and plants are diploid, the normal allele of a pair, one of which has a mutation, may make enough of the needed protein to hide the fact that the organism is carrying a mutated recessive allele. When two organisms carrying the mutated allele mate, one-quarter of their offspring are likely to be homozygous for the defective allele.

Organisms ordinarily produce offspring with phenotypes that will work best in a particular environment. A mutation is a *random change* in the genotype. The chances that it will result in an improved phenotype for that environment are very slight. Consider the comparison of an organism to a television set. Suppose you have a TV set which is properly adjusted and is presenting a perfect picture. You then unplug the set, open the case, reach in with a pair of insulated pliers, and make a random change in the set without looking. What are the chances that you will improve the working of the TV set and get a better picture? Considering that the simplest bacterium is an enormously more complex mechanism than a TV set, you can see why random mutations are unlikely to improve an organism already functioning well.

If you had a population of millions of TV sets, however, and the best of them reproduced themselves rapidly, random changes would not be such a bad thing. Many sets would be damaged or ruined, but the few that were improved would reproduce the most and their offspring would become the most prevalent kind of TV set in the population of sets.

Cause of mutations Mutations may be caused by an error in the process of duplicating the genetic information, say the pairing of an A with a C, or by some kind of radiation knocking one or more subunits from the DNA. If the structure of the DNA is changed, a different sequence of nucleotides will be present to specify a different sequence of amino acids and a changed protein will be produced. Mutations can easily be produced in the laboratory by the use of x-rays or by treating organisms with certain chemicals. Anything causing mutation is called a **mutagen**. (See Sec. 10-3.) The way mutagens function is poorly understood; however, some chemicals have structures very much like DNA nucleotides and can be incorporated into the DNA molecule, which then behaves abnormally.

A disturbing number of herbicides, fungicides, and food additives have structures which resemble DNA subunits. A bacterial test system has been developed recently which shows that many such compounds cause mutations in bacteria by fitting in between DNA bases in the chain and causing the code to be misread because the sequence is shifted over by one base. A number of chemicals have been tested which are known to cause cancer in animals, and it has been found that most of them cause mutations in bacteria. Many scientists believe that only a single molecule of a mutagen is necessary to cause a mutation, and there is a possibility that at least some cancers are caused by mutations in body cells. Therefore, there is no lower limit (minimum level) of known mutagens which is safe.

9-11 The human species may change its hereditary endowment deliberately or accidentally

Since mutations are random changes of an already balanced system, almost all will certainly be harmful. More than 1,500 known single-gene human genetic variations have been catalogued. By far the majority are **genetic diseases.** About 100 of these diseases have had a single enzyme deficiency identified. Many other genetic diseases are known, some of which are polygenic in inheritance and some of which are caused by extra or missing chromosomes or chromo-

TABLE 9-3 SOME GENETIC DISEASES IN HUMAN BEINGS

DISEASE	MODE OF INHERITANCE	SYMPTOMS
Huntington's chorea	Dominant	Symptoms begin between ages 30 and 50; personality changes, nervous twitches, progressive deterioration, and death within 15 years
Cystic fibrosis	Recessive	Appears in infancy or later; mucus glands produce sticky mucus which clogs lungs and stomach
Muscular dystrophy	Several kinds; some on X	Appears in first 5 years of life; muscle tissue replaced by globules of fat; death by age 25
Hemophilia	On X chromosome	Blood lacks a vital clotting agent
Diabetes	Not known	Pancreas fails to produce sufficient insulin to utilize sugar properly
Phenylketonuria (PKU)	Recessive	Inability to metabolize the amino acid phenylalanine leads to mental retardation if not treated
Sickle cell anemia	Incomplete dominant	Blood lacks ability to carry oxygen; development impaired; many secondary symptoms leading to death before age 40
Down's syndrome (mongolism)	Extra chromosome no. 21	Physical and mental retardation; may survive to adulthood or old age
Cri du chat syndrome	Missing part of chromosome no. 5	Abnormal head and face; mewing cry of newborn; retarded mental and physical development
Breast cancer (female)	Probably polygenic	Cancer in breast at maturity; surgical removal of breast (mastectomy) may save life
Tay-Sachs disease	Recessive	Enzyme deficiency allows substances to accumulate in neurons, causing progressive brain deterioration and death by age 5

some parts. Table 9-3 lists a few human genetic diseases, including some in which the mode of inheritance is not known for certain.

It seems logical, therefore, that we should consider *any* environmental change which raises mutation rates as harmful. This is one reason why biologists are so concerned over the proliferation of nuclear power plants at a time when we do not have the technology for keeping mutagenic radiation out of the environment. Biologists are also greatly concerned about the proliferation of chemical mutagens produced by industrial society. The problems of human-caused mutagenesis will be discussed again in more detail in Chap. 17.

Naturally occurring mutation and recombination provide all the variability necessary for human evolution (see Sec. 10-3). Therefore, increasing the mutation and recombination rates would cause an increase in suffering and tragedy. More babies would be born defective, and more genetic diseases would appear in the population.

Approaches to the problem Presently the number of people born with genetic diseases is sufficiently large that attempts are being made to alleviate the situation. There are two ways usually proposed for doing this. One is **eugenics,** that is, designing the biological composition of future generations by goal-oriented processes of selective breeding. Unfortunately, we do not now know enough to plan a program which would maintain the diversity needed for changing environments even if such a design were acceptable to society—which it would not be. The second is **euphenics,** that is, treating the effects of a hereditary disease on an individual basis for the benefit of the afflicted person. This has the disadvantage that the person may be cured of his or her affliction but may still carry and pass on to offspring the allele that caused it. Some variations of both eugenics and euphenics are presently being practiced, and there is the possibility for the future of actually changing the genes of an individual.

In some cases, individuals heterozygous for defective alleles can be detected; they can then decide whether or not they want to take the risk of having a defective child. There are centers where genetic counseling is done to help people make such decisions. Pedigrees may be used, or the outcome of previous pregnancies, as well as tests which can detect when an individual is heterozygous for certain defective alleles. People who are heterozygous for Tay-Sachs disease or sickle-cell anemia can be detected by relatively simple tests. Since two heterozygotes who marry have one chance in four that their offspring will be afflicted, many people in the high-risk groups for these diseases are being tested.

In some cases a further step can be taken to prevent the birth of a child with genetic disease. If there is reason to suspect that a pregnant woman is carrying a defective embryo, a procedure called amniocentesis (see Fig. 17-8) can be done. A needle is inserted through the abdominal wall of the pregnant woman into the amniotic sac of the embryo and some amniotic fluid is withdrawn for testing. This fluid contains cells and enzymes from the embryo which can reveal the presence of certain genetic abnormalities. For example, Tay-Sachs disease and Down's syndrome can be detected. The parents can then decide whether the woman should have an abortion or bear the defective child. This method is also able to determine the sex of the embryo; female carriers of X-linked diseases could avoid having afflicted offspring by having all male embryos aborted.

In a few cases it is possible to prevent defective births resulting from the incompatibility of blood types. There are many human blood groups other than the A-B-O group which we studied earlier in the chapter; the best known seems to be Rh. When a woman who is type Rh^- has a child by a man who is Rh^+ the infant may be Rh^+. The first Rh^+ infant is normal, but, if there is a placental defect (probably also genetically controlled), the mother produces antibodies which may react with the Rh^+ antigens on the red blood cells of a second fetus, impairing their ability to carry oxygen. If this condition is not treated, the second infant will die. One treatment used in the past was a complete exchange transfusion of the infant's blood; a newer and better method is to prevent the buildup of antibodies in the mother. Specific anti-Rh gamma globulins (protein antibodies) are injected into the mother's bloodstream after the birth of an Rh$^+$ child; these destroy the Rh$^+$ cells of the child that have leaked into the mother's circulation and keep her from building up antibodies which would harm infants in her subsequent pregnancies.

There are a few genetic disorders that can be treated after birth. Phenylketonuria (PKU) can be detected in an infant, and the infant can be placed on a diet which will prevent it from growing up mentally retarded. Unfortunately, this test, which is mandatory in some states, often gives false positives which cause anxiety before they are ruled out in later tests; also, false positives are damaged by the special diet. Diabetes can be treated with diet and insulin; hemophiliacs are treated with transfusions and injections of a clotting factor taken from other people's blood. More such treatments are being developed each year, but they are often expensive, and they must be continued for a long period of time—often the lifetime of the patient—since they are treatments, not cures.

Changing our hereditary endowment We have seen that the genetic information resides in the nucleus in eukaryotes. There it can be duplicated and distributed to daughter cells by mitosis. In each new daughter cell the genetic material provides the master blueprints which cellular mechanisms "read" and use to produce all the substances necessary for the cell to grow and metabolize. Whether or not a particular unit of genetic information is utilized at a given time in the life of a cell depends on the environment of the cell at that time. Biochemists are just beginning to understand the mechanisms which turn genes on and off in prokaryotic organisms. Some of these mechanisms are discussed in Chap. 13. Biochemists are beginning to unravel the enormously intricate developmental systems of birds and mammals. When they do reach some understanding of these systems, the opportunity for "genetic engineering" in mammals may present itself. It may be possible for human beings to intervene in the gene-environment system in order, for instance, to prevent a child from growing up with an inherited deformity. Genetic engineering obviously could be of great benefit to the human species, but it could also be used to the detriment of humanity. In Chap. 17, some of the possible social consequences of genetic engineering are discussed.

QUESTIONS FOR REVIEW

1 What is the difference between linkage and sex-linkage?
2 What do these terms mean: homozygous, dominant, recessive, allele, gene, pedigree, heterozygous, genetic code, PKU, "nucleotide triplet"?
3 A certain plant has flowers that are always either red or white and stems that are always either smooth or fuzzy. The plant's color is controlled by two

alleles of a single gene, and the allele for red is dominant. The plant's stem texture is also controlled by two alleles of a single gene, and the allele for fuzziness is recessive. Both color and stem texture are independently assorting traits. If you cross a truebreeding red, smooth strain of the plant with a truebreeding white, fuzzy strain, what is the flower color–stem texture phenotype of each plant of the resulting (F_1) generation? What is the genotype of each? If you next cross several plants of the F_1 generation, how many different color-texture genotypes (allele combinations) do you almost certainly find in the F_2 generation? How many different phenotypes (flower color–stem texture combinations)? In what ratios are the phenotypes of the F_2 generation produced?

4 Cover the right-hand column in Table 9-1 without looking at it first. Then, keeping it covered while you study the uncovered columns, can you determine which blood type(s) makes acceptable transfusions for people with type O blood? People with type A? With type B? With type AB?

5 Why is it not useful, or even meaningful, to speak of nature *versus* nurture when considering the effects of genotype and environment on phenotype?

6 Why couldn't the "words" of the genetic code consist of "doublet" combinations of the base units A, C, G, and T—for instance, AG or GT or CG—instead of the "triplet" combinations they actually form? Why is a "quadruplet" code not used instead of a "triplet" code?

7 If mutations almost invariably cause defects instead of improvements in an organism's biological machinery, why are mutations essential to the long-term survival of species, since species do of course consist of individual organisms? (Hint: What besides the mutating organisms is always changing over the long run?)

READINGS

Beadle, G., and M. Beadle: *The Language of Life—An Introduction to the Study of Genetics,* Doubleday, Garden City, N.Y., 1966.

Crick, F. H. C.: "The Genetic Code," *Scientific American,* **207**(4):66–74, offprint 123, 1962.

———: "The Genetic Code III," *Scientific American,* **215**(4):55–62, offprint 1052, 1966.

———: "The Structure of the Hereditary Material," *Scientific American,* **191**(4):54–60, offprint 5, 1954.

Feldman, M. W.: "Basic Principles of Genetics," *Biocore,* Unit VII, McGraw-Hill, New York, 1974.

Friedman, T., and R. Roblin: "Gene Therapy for Human Genetic Disease," *Science,* **175**:949–955, 1972.

Helinski, D. R.: "Molecular Genetics," *Biocore,* Unit VIII, McGraw-Hill, New York, 1974.

Lerner, I. M.: *Heredity, Evolution, and Society,* Freeman, San Francisco, 1968.

McKusick, V. A.: *Human Genetics,* 2d ed., Prentice Hall, Englewood Cliffs, N.J., 1969.

Nagle, J. J.: *Heredity and Human Affairs,* Mosby, St. Louis, 1974.

Nirenberg, M. W.: "The Genetic Code II," *Scientific American,* **208**(3):80–94, offprint 153, 1963.

10

THE ORIGIN OF DIVERSITY

SOME LEARNING OBJECTIVES

After you have studied this chapter, you should be able to

1. Summarize the biological theory of the origin of life.
2. Demonstrate your familiarity with some of the basic factors involved in the diversification of life by
   a. Stating three reasons why environmental diversity leads to diversity of living things.
   b. Explaining why environmental change also leads to biological diversity.
   c. Giving the text's definition of organic evolution.
3. Explain why recessive alleles of individuals are the key to the ability of species to change (evolve).
4. Demonstrate your knowledge of variability in populations by
   a. Stating what variability in a population means in genetic terms.
   b. Naming and describing two major causes of variability in populations.
   c. Explaining why variability is a pre-

requisite for evolutionary change.
   d. Explaining why a severe environmental change, whether natural (such as a climatic change to much colder winters) or artificial (such as exposure to a pesticide), is less likely to exterminate a species if the species' population is large—even if all the organisms of the population are exposed to the severe environmental change.
5. Demonstrate your knowledge of the process of natural selection by
   a. Stating what, in genetic terms, happens in the course of natural selection.
   b. Explaining how natural selection is the major cause of the evolution of a population.
6. Name three other processes which affect evolution and give an example of each.
7. Outline a general approach to the sensible control of herbivorous insects which includes the use of insecticides.
8. Explain how the interaction of geographic variation and natural selection results in speciation—the "splitting" of one population into two or more populations.
9. State two main reasons why the human species, *Homo sapiens*, shows no signs of "splitting" into two or more species despite the fact that it is found in widely different geographic areas and is, like all species, subject to selection pressures.

As was suggested in Chap. 6, there *may* be as many as 10 million species of plants, animals, and other living things. It is clear that the biosphere is very diverse, that it contains many different species of organisms; biologists are curious about the source of this diversity. They are interested in why there is not just one kind of organism living everywhere on earth and whether there are biological mechanisms responsible for the production of diversity.

THE ORIGIN OF DIVERSITY

Biologists do not agree on how and where life originated. Many biologists believe that life originated here on earth from simpler chemical systems. A few speculate on the possibility that life came to our planet from elsewhere in our solar system or beyond. Since there seems to be no clear-cut evidence that life occurs anywhere in the solar system except on earth (Chap. 4), this theory seems unlikely. Certainly there is no evidence that life is arriving on our planet now in meteorites or any other way. There are some people who prefer to believe that life was created by a supernatural force or being; this idea is beyond the scope of science since it cannot be tested.

10-1 Most biologists agree that life originated from nonliving chemical systems

Two or three hundred years ago, many people believed that living things originated spontaneously from nonliving things such as dust, dirty underwear, piles of garbage, or carcasses. Since piles of garbage soon come to be inhabited by mice and since carcasses are soon full of maggots, it was rather natural that even such complex animals as worms, bees, and mice were thought to arise by spontaneous generation. Experiments, however, showed that no such thing happened, as all of us today would intuitively expect. There is no evidence that life in any form arises from anything other than preexisting life today.

Most biologists think that when life first appeared on earth more than 3.2 billion years ago, it must have originated from nonliving chemical systems. The prerequisites of life are (1) molecules made up principally of carbon, hydrogen, oxygen, and nitrogen atoms; (2) some means of obtaining and using energy; and (3) a boundary from other systems. It has been hypothesized, and also confirmed experimentally, that molecules such as those in living systems can be formed from simple inorganic molecules and energy. Certain kinds of molecules similar to these tend to separate out into little blobs in solution. Some sort of chemical blob very much like a cell in appearance may have been an early stage in the origin of life.

In laboratory experiments, amino acids and other organic compounds were formed when water vapor, methane, ammonia, and hydrogen were put together in a tube and exposed to electrical discharges (Fig. 10-1) or ultraviolet light. The mixture of substances used was similar to what scientists think the primitive atmosphere was like; in that atmosphere, there should have been plenty of energy available in the form of lightning and ultraviolet light from the sun. The primitive atmosphere presumably did not contain gaseous oxygen in any quantity since the oxygen in our atmosphere was produced by green plants and they, of course, had not yet evolved.

Experiments using mixtures of organic substances have produced blobs superficially resembling cells. The nature of certain of the molecules themselves causes them to become arranged so as to resemble cellular membranes. Thus the three prerequisites for life can be accounted for: organic molecules, energy, and a boundary separating living systems from their environment. There have now been enough experiments along these lines to convince biologists that, given enough time, life could have begun this way. Analyses of rocks dating prior to the first microfossils indicate that carbon compounds similar to those found in organisms were present. Therefore, there may have been a long period of chemical evolution before primitive cells appeared.

The need for an information system and an energy system Most biologists think that life on earth originated in the form of simple cell-like structures either in the ocean or in warm mineral springs. These primitive living systems were set apart from the environment by a boundary something like a cell membrane and thus could evolve energy-trapping and energy-using mechanisms. They could capture food in the form of energy-rich molecules from the environment and grow. Soon their volume would become so large, however, they could not have enough surface area to take in food (Fig. 10-2); therefore they could only divide or perish. These primitive systems could also have developed chemical regulators which would ensure that when they divided (reproduced), their "daughters" would be functional systems. Once the energy-utilization system got going, it had to become coupled with an information-transfer system which could duplicate itself (self-replicate). However, an information system such as DNA cannot work without an energy-utilization system to make more molecules, including DNA subunits. If the events postulated did occur, then these primitive living systems would be expected to evolve the systems found today in cells, one for replication and one for metabolism, each totally dependent on the other.

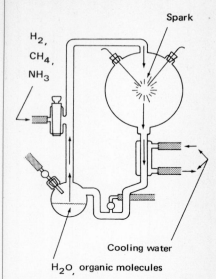

FIG. 10-1 Diagram of the apparatus used by Stanley Miller to form amino acids and other organic compounds from water vapor (H_2O), methane (CH_4), ammonia (NH_3), and hydrogen (H_2).

10-2 Life has developed many diverse forms

Diversity of environment If life could exist as such a simple cell-like chemical system, why is there today such a variety of living things? It is possible to imagine one form of life spread like a living blanket over the surface of the earth, but there are many reasons why this did not occur.

First, the range of environments on the planet is very great. Such a living blanket would have to be very adaptable, indeed. It would

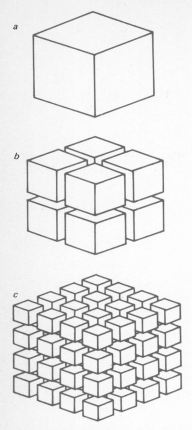

FIG. 10-2 The effect on surface area of increasing subdivision of a cube of hypothetical living substance. (*a*) A 1-meter cube has a surface area of 6 square meters and a volume of 1 cubic meter; the ratio of surface to volume is $6/1 = 6$; (*b*) if this cube is divided into 8 cubes, each 1/2 meter on an edge, a single cube has a surface area of 1.5 and a volume of 0.125, giving a ratio of $1.5/0.125 = 12$; the total volume remains the same, but the total surface area is 12 square meters; (*c*) if the original cube is divided into 64 cubes, 1/4 meter on an edge, a single cube has a surface area of 0.375 square meter and a volume of 0.015625 and a ratio of $0.375/0.015625 = 24$; once again the total volume is 1 cubic meter, but the total surface area has increased to 24 square meters.

exist at the same time in fresh and salt water, in deserts, and in tropic, temperate, and polar regions. Obviously one massive organism would find it difficult or impossible to live simultaneously in all these very different habitats. But there are even more complicated problems than living simultaneously in many different environments. Think of the difficulties of controlling such a monstrous living system. The activities of such a living blanket would have to be regulated over thousands of miles. Critical materials such as phosphorus would have to be moved great distances across the earth, bringing about enormous problems of transportation. How would this monster tell its photosynthesizing middle that its arctic edges needed more food to keep alive during the sunless arctic winter?

Finally, there is the problem of exchange with the environment. Exchange between living things and their environment takes place across surfaces. The surface area of a cell must be sufficiently large to supply the entire volume of the cell. This places a restriction on cell size. As the volume of a cell increases, its surface does not increase proportionately. Figure 10-2 shows the relationship between surface and volume. The small cubes have proportionately more surface area in relation to their volume. Eventually a limit is reached when the surface area is not large enough to supply the entire volume. Our hypothetical "world creature" would have the same problem. A planet-sized organism would have to be *very* thin!

There are many factors which led to the packaging of life into discrete packets—the plants, animals, and other organisms we see today. The surface of the earth, as has already been noted, is very heterogeneous. In order to exploit all the possible habitats and ways of life, it is probably most efficient to have a wide variety of organisms: unicellular and multicellular, photosynthetic and nonphotosynthetic, herbivores and carnivores. Energy and matter are most efficiently used by a wide variety of organisms doing different things in different ways, in slightly different habitats. Remember that ecosystems evolve toward the most efficient use of materials and energy in a process called succession.

Environmental change The environment is always undergoing change. No place on earth will be the same tomorrow as it is today. In some places change is very rapid; in others it is very slow; nonetheless, change is the rule. Organisms themselves create changed environments during ecological succession. The weather, the wearing away of the rocks, the flow patterns of rivers are all agents of change. If organisms did not continually change, they could not survive. These changes must not be confined to just an individual; there must be changes that are capable of being passed on from generation to generation. Such changes can only be genetic and must result from mutation and recombination. Diploid species have the ability to change because individuals can carry recessive alleles. Individuals who may not do well in today's environment may succeed in tomorrow's and may produce descendants who will do extremely well in the changed environment a few generations later. If the environment changes too rapidly for the rate of genetic change, individuals of the species must move or die. Therefore we would expect that the genetic rate of change would correspond to the rate of change in the environment. Inherited change in populations through time is called **organic evolution** or often just evolution. Evolution in a changing world of environmental heterogeneity has resulted in the diversity of life we see around us.

VARIABILITY IN POPULATIONS

The ability of a population to change depends upon its variability. In order to evolve, a population must be variable; that is, it must contain individuals of different kinds. This variability must involve changes in genotype. Individuals with different genotypes in the population provide the basis for evolutionary change. Organic evolution may be thought of as change in the genetic information of a population during the course of time.

10-3 There are natural sources of variability in populations

Sexual reproduction Let's review briefly. As we saw in Chap. 7, sexual reproduction requires the fusion of two gametes—usually a motile gamete fertilizes a nonmotile gamete. Gametes, you will recall, are haploid. Haploid gametes are produced from diploid individuals by the process of meiosis. In haploid organisms, gametes are produced by mitosis. In diploid organisms the set of chromosomes a gamete receives is never identical in genetic material with either of the parent sets. Diploid organisms have a set of chromosomes from their male parent and another set from their female parent. During meiosis when crossing over occurs, each chromosome pair behaves independently of every other in segregation. The odds *against* getting two gametes with exactly the same genetic endowment are impossibly large.

Recombination occurring in meiosis is a major source of genetic variability in populations. It assures that no two gametes will have exactly the same genotype. Furthermore, gametes combine pretty much at random in most organisms. Thus it is very unlikely that the fusion of two gametes will result in a diploid genotype which is like that of either of the parents. Existing differences between the genotypes of the two parents, their parents, etc., are recombined in meiosis. The basic source of newly arising differences among genotypes is, of course, mutation (Sec. 9-10).

Mutation Genetics is the science which studies how genetic information is expressed and how it is transmitted from generation to generation. If you think for a moment, you will realize that if individuals of a species were all alike *there would be no patterns of inheritance to study.* Geneticists at first worked with wild or cultivated species in which there were clear-cut differences among individuals that could be studied from generation to generation—Mendel's short and tall peas being the classic example. When organisms such as the fruit fly (*Drosophila*) were studied in the laboratory, it became obvious that characteristics often changed for no apparent reason. For example, in a population of the fruit fly which normally had red eyes, an individual with white eyes would appear. Suddenly, one seed from a plant of a population having only red flowers would mature into a plant with white flowers. Crossing experiments showed that these changes were not merely a change in the phenotype. They involved a change in the genotype as well.

Spontaneous changes in genes are called mutations; alleles are alternate states of a gene: therefore, a mutation is a change from one allele to another. For example, from time to time, the *T* allele for tallness in garden peas mutates to the recessive allele *t*. It has been shown that each gene has its own rate of mutation. Thus, for instance, the mutation from a normal allele in human beings to the one causing hemophilia occurs in about 30 of each 1 million gametes; that which produces the allele for retinoblastoma (a dominant gene which produces an eye tumor) occurs in 18 of each 1 million gametes.

Mutations, as we have seen, occur at random. It is not possible to predict *which* gene will mutate or *when* a gene will mutate—only what the *probability* of a mutation in a given population of genes will be. This is known as the **mutation rate** and is often expressed as the number of mutations that can be expected in 1 million gametes carrying this gene. Mutations are generally recessive to the dominant allele of a gene. Therefore the occurrence of a mutation can be known when *recombination* brings two recessive alleles together in the same diploid individual. Suppose you had a strain of guinea pigs which bred true for black coat color. You would assume that they all had the genotype *BB*. If a mutation occurred in one individual to change a *B* to *b*, you could not tell this by looking at its phenotype if *B* were completely dominant to *b*. *BB* and *Bb* individuals both have black fur. How could you identify an individual with the mutation? One way would be to cross it with a pure-breeding white individual (*bb*). This would be the easiest way to learn of the presence of the recessive allele *b*.

10-4 Variation may be produced artificially in populations

Artificially induced mutations Early in the study of genetics, it was found that there were ways of *increasing* mutation rate. Treatment of an individual with various chemicals or with certain kinds of radiation caused mutations to appear in its gametes. Such mutations could be identified by making the appropriate crosses. They proved to be no different from spontaneous mutations: they took place randomly, they were usually recessive, and they were usually harmful.
Radiations Radiations that act as mutagens are ultraviolet light, x-rays, and various kinds of radiations produced by radioactive substances such as uranium and radium. In the laboratory, scientists may irradiate organisms such as fruit flies or wheat with x-rays or ultra-

violet light. The offspring are then examined for mutations. Many of the irradiated organisms will die; those that survive may produce offspring carrying mutations. Ultraviolet light also causes skin cancer, presumably because it has caused mutations in somatic cells. (Farmers, sailors, and others who are much exposed to the sun sometimes have a great many such cancers; fortunately, if discovered early and removed, most skin cancers are not a serious matter.)

In people, radiation more penetrating than ultraviolet light may also induce cancers. Radiation from radioactive materials or x-radiation can kill a person, as can ingestion or inhalation of radioactive materials. People who are exposed to radiation for long periods of time or to intense radiation may die relatively quickly. Lighter doses of x-rays or radioactive materials may cause mutations in somatic cells which may lead to cancer; in the ovaries and testes mutations may occur in the sperm or eggs. These mutations will, as discussed above, usually be recessive and harmful, but they will not be expressed in the phenotype until the same two recessive mutations of a gene are combined in the same individual. Thus a man or woman who has experienced irradiation of the testes or ovaries may produce sperm or eggs containing hidden recessive mutations. The results may not show up until his or her distant descendants marry another person carrying the recessive.

NATURAL SELECTION

The major cause of evolution, **natural selection,** occurs when the genetic variability provided by mutation and recombination is reduced as some genotypes become scarcer and others more common. The genetic composition of populations changes from generation to generation because some genotypes produce more offspring than others. This differential reproduction of genotypes is natural selection. (In Sec. 9-10, we described natural selection of TV sets.) Differential reproduction of genotypes is the result of the action of **selective agents** such as other organisms, chemicals in the environment, and factors of soil and climate. Evolution is the genetic change in populations through time resulting from mutation, natural selection, and other evolutionary forces. Human beings themselves have frequently acted as an important selective agent in determining which genotypes will reproduce. This is called **artificial selection.**

10-5 Selection can be demonstrated artificially

Selection in the laboratory One way to demonstrate evolution in the laboratory is to expose houseflies to DDT and allow only the survivors to be parents of the next generation. In each generation a test is run to determine how sensitive the flies of that generation are to DDT.

In one study (Fig. 10-3) houseflies which were suspected of having some degree of DDT resistance were collected from a barn. These

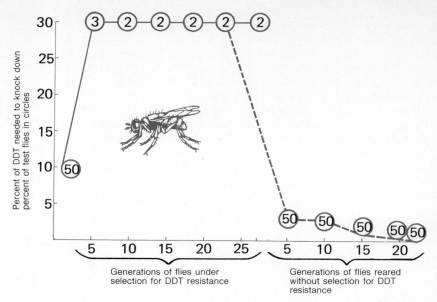

FIG. 10-3 The results of an experiment to test resistance to DDT. The solid line shows the development of DDT resistance (measured in test cages) by a strain of houseflies exposed to heavy doses of DDT in their rearing cages. The dashed line shows the loss of DDT resistance (measured in test cages) by descendants of the resistant strain when these were reared without DDT. The numbers in the circles indicate the percent of flies knocked down in 12 hours by the indicated doses of DDT in the test cage. (*Data from D. Pimentel, H. H. Schwardt, and J. E. Dewey,* Journal of Economic Entomology, **46**(2):295–298, 1953.)

flies were inbred for three generations in the laboratory in a medium not containing DDT. Ten groups of the third generation of 100 flies each were exposed to each of five concentrations of DDT in special test cages for 12 hours. It took 10 percent DDT to knock down 50 percent of these flies in the 12 hours. Flies from each of five successive generations were allowed to remain in DDT-treated cages until 98 percent were knocked down. The remaining 2 percent were allowed to be the parents of each next generation, and then the next 20 generations were reared directly in heavily DDT-treated cages. By the fifth generation, resistance has increased so that, in the test cages, it took 30 percent DDT to knock down only 3 percent of the flies instead of the 10 percent which knocked down 50 percent of the flies tested in the original 12-hour test period. By the tenth generation only 2 percent of the males of this strain of housefly were knocked down by 30 percent DDT. That is, they were essentially *resistant* to 30 percent DDT. At the twentieth generation, 5,000 flies of this resistant strain were removed from the cages containing DDT and reared without exposure to DDT. Within five generations they had lost most of their DDT resistance and were more susceptible, when tested in the test cages, than their ancestors from the barn had been. In 22 generations of relaxation of selection they had lost all resistance to DDT. In this experiment, there has been a genetic change in the population as a result of the differential survival and reproduction of flies with genotypes that make them relatively resistant to DDT.

Insecticide resistance The original batch of flies with which the line was started must have contained genetic variability with respect to DDT resistance. Otherwise a resistant strain could not have evolved. Some of the genotypes were, *by chance*, slightly more resistant to DDT than others. When the flies were exposed to DDT, the individuals with those genotypes were more likely to survive than the others. They lived to breed, and their offspring inherited their resistance. Slightly resistant flies mating with other slightly resistant flies produced some flies more resistant than either parent. These, of course, are most likely to survive to breed in the next generation.

Thus, each generation of flies treated with DDT becomes genetically more and more resistant. But, without exposure to DDT in the rearing cages, resistant genotypes would have no advantage over others, and you would not expect DDT resistance of the population to increase. Notice in Fig. 10-3 that the resistant flies lost their resistance when selection was "relaxed"—that is, when the flies were no longer exposed to DDT in their rearing cages. Such a return to control level is often found when selection in the experiment is relaxed. This shows that, in a sense, "nature knows best." What is happening, of course, is that natural selection is occurring, with nonresistant genotypes outreproducing resistant ones. This shows that the resistant recombinants or new mutants were actually at a selective disadvantage in the environment without pesticides. In other words, mutations conferring resistance were harmful until the environment was changed to one containing DDT.

Selection in the field The development of insecticide resistance under field conditions follows this general pattern. Most often people are attempting to protect their crops from herbivorous pests that have large populations (that, after all, is why they are pests). Large populations of herbivorous insects are much more difficult to kill off with pesticides than the smaller populations of predators and parasites that attack them. First of all, having large populations gives the pests a *greater store of genetic variability* and thus makes it easier for them to develop resistance. Secondly, most plants manufacture "insecticides" of their own with which to poison herbivorous insects, for example, nicotine (in tobacco) and many spices (see Fig. 14-17). The plant-eating insects have, therefore, had many millions of generations in which to evolve methods of dealing with poisons. Natural selection often results in insects that can detoxify (make harmless) the poisons people manufacture or otherwise avoid them.

As you can see, the insecticide industry has a good thing going. Pests quickly become resistant to their products, while the predators

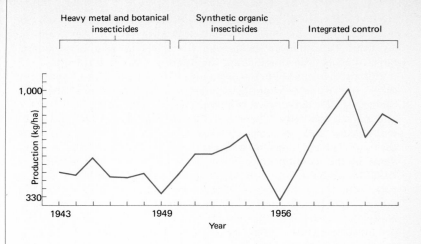

FIG. 10-4 Effects of spraying cotton with insecticides in the Cañete Valley of Peru. From 1943 to 1949, yields of cotton were stable but low, and only 7 species of insect pests were found; heavy metals (e.g. lead arsenate) and botanicals (e.g. pyrethrins) were being used as insecticides. In 1950, synthetic insecticides began to be used; the yields went up at first, but dropped to a new low by 1956; 6 new species of insect pests were found, making a total of 13 pest species, and by 1956 many were resistant to insecticides; 50 percent of the crop was destroyed, and insect-eating reptiles and birds were gone. In 1957, integrated control (no synthetic insecticides, hand picking of pests, biological control, etc.) and better growing practices (cleaning up fields, better irrigation, crop rotation, etc.) were begun; the number of pest-insect species dropped to 8 and crop yield increased dramatically. (*Data from T. Boza Barducci in T. Farvar and J. Milton, eds.:* The Careless Technology, *Natural History Press, New York, 1969.*)

and parasites that normally attack the pests are killed off. If one stops spraying, the pest populations have enormous *outbreaks* because their natural enemies are absent (Fig. 10-4). The pesticide industry has a "solution" to these problems—bigger doses of pesticides and more powerful pesticides. In short, the industry manages to sell more and more of its products *because they do not work.* If they worked, of course, less and less would be sold each year because the insecticides would be wiping out the pests.

Sensible control of herbivorous insects involves a mixture of methods, including, when necessary, the use of pesticides. This is called *integrated control.* Such control may involve improving conditions for predators and parasites, reducing the size of monocultures, and using methods of cultivation which discourage pests. When pesticides are used, they should be used only for short periods to control serious outbreaks. Whenever possible, different pesticides should be used for successive treatments. In that way, the chances for serious resistance to develop are minimized. Whatever resistance is developed in one or two generations usually will be lost when pesticide use is discontinued (that is, when selection is relaxed). Resistant genotypes often are less likely to survive and reproduce in an environment free of pesticides than are nonresistant genotypes. Furthermore, if two different pesticides are used in successive treatments, then there is always a chance that the mechanisms evolved to produce resistance to the first will not work on the second.

10-6 Microorganisms evolve resistance to antibiotics

Resistance of bacteria to antibiotics Even disregarding the severe ecological consequences of insecticide pollution, it is clear that frequent spraying with the same insecticide is hardly the ideal way to control a pest. Thus it does not seem likely that the widespread and continuous use of the same antibiotic is the ideal way to control bacteria causing a serious disease. Bacteria may have *more than one generation every hour*. It is hardly surprising that they can become resistant to antibiotics quite rapidly. The bacteria which cause a great many diseases have, for instance, developed strains that are resistant to penicillin.

Misuse of antibiotics. Needless to say, it was hardly wise of physicians to try to use penicillin to cure even minor ailments, as they did when it was first discovered. One of the most serious problems of modern medicine is the evolution of antibiotic resistance in microorganisms which attack people. Our problem is the microorganisms' solution. Natural selection permits them to live in new environments, which include antibiotics. One particularly stupid thing we do to aid our microbial enemies is to add antibiotics to feed for cattle, chickens, pigs, and sheep. It usually takes special effort to buy feed *without* antibiotics. Since we share many bacteria with our domestic animals, this practice represents a selective agent which could lead indirectly to some of our most dangerous bacteria becoming resistant to antibiotics. In fact, farm workers who work around antibiotic-fed cattle have been tested and show more resistant strains of bacteria than comparable controls. Even worse, in bacteria, resistance to an antibiotic can be transferred from one strain to another. It has been suggested that the transfer of resistance to chloramphenicol from bacteria of cattle to those of people may make that antibiotic nearly useless in the future. Although it is too toxic to people to use for mild diseases, chloramphenicol is the most useful drug in the treatment of typhoid fever.

10-7 Other examples of evolution have been observed

Industrial melanism in moths In many industrial areas of the world, there are moth species that were light-colored or speckled before 1850, but are now **melanic** (very dark). The most famous case is that of the peppered moth in England. Melanic forms occurred as occasional mutants before 1850; today in areas of heavy industrial pollution, it is the peppered form that is the rare mutant. A series of studies has shown that birds are more easily able to see peppered moths than melanic moths on a soot-covered tree trunk (Fig. 10-5a). On a clean, lichen-covered tree trunk, the peppered moths are nearly invisible and the melanic forms are conspicuous (Fig. 10-5b).

In industrial areas, birds are more apt to see and eat the peppered form; in nonpolluted areas, they tend to see and eat the melanic form. This selective advantage of the melanic form has made it the commonest form in polluted areas, but the selective advantage of the peppered moth has prevented this elsewhere. In the new "smokefree" zones of England, the peppered moths are once again becoming abundant and the melanics are disappearing because they are now easily visible on the lichen-covered tree trunks.

Natural selection may not involve environmental effects caused by people. For instance, selection has been shown to maintain certain

FIG. 10-5 Light and dark forms of the peppered moth (*Biston*): (*a*) two moths on lichen-covered bark; (*b*) two moths on pollution-darkened bark of individual areas.

banding patterns in land snails and water snakes (Fig. 10-6). Biologists think that virtually all patterns observed in nature are the result of selection, but only under special circumstances can the action of selective agents be directly observed. Selection in nature may occur over a long time, and it is usually too difficult to observe natural populations generation after generation. Also, very slight selective advantages may cause spectacular changes over thousands or millions of years—but we do not live long enough to observe them. We do not even have records long enough, since recorded history is only about 6,000 years. Only where selective advantages are very great, or where the selective agents (the birds in the case of industrial melanism) can be easily observed in action, do we have a good chance of studying selection under natural conditions.

10-8 People are subject to natural selection

The sickle-cell gene Some human genotypes are more likely to survive and reproduce than others. That means that human populations

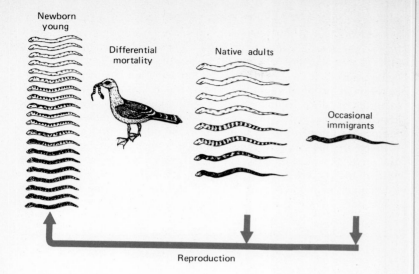

FIG. 10-6 Differential mortality in water snakes of the genus *Natrix*. On islands in Lake Erie, water snakes generally lack the banded pattern of mainland snakes or have fewer bands. It is thought that visual predators (sea gulls) consume more banded than unbanded young, and therefore the proportion of unbanded adults increases on the islands. Occasional banded immigrants keep the genes for banding in the populations.

are changed by natural selection. For instance, in some populations in Africa and the Mediterranean area, a relatively high proportion of individuals have a genotype making them resistant to malaria as well as causing a condition called **sickle-cell trait.** The genetics of the situation have been worked out, and the role of malaria as a selective agent has been demonstrated.

Every human being has a pair of alleles which determine the kind of **hemoglobin** that develops in the red blood cells. The red pigment hemoglobin functions in the transport of oxygen from the lungs to the cells of the body. In most individuals, only hemoglobin A, the normal adult hemoglobin, is found. The genotypes of such individuals are traditionally symbolized as $Hb^A Hb^A$ which we will simplify to AA. Individuals with sickle-cell trait have the genotype $Hb^A Hb^S$ (AS). The allele Hb^S (S) leads to the formation of hemoglobin S, which is identical to hemoglobin A except for one of its amino acid subunits. Thus persons with sickle-cell trait have a *mixture* of two different kinds of hemoglobin, A and S. If their blood is subjected to low oxygen concentration, the red blood cells become sickle-shaped (Fig. 10-7). Under most conditions, persons with sickle-cell trait live normal lives and may never know they carry the S allele. Under certain conditions—for example, exercising at high altitudes, they may suffer painful cramps or more serious problems, such as blocking of blood vessels.

If there are AS individuals in a population, it is inevitable that some SS genotypes will be produced. Individuals with this genotype have *only* hemoglobin S and suffer from a severe anemia called **sickle-cell anemia.** Ordinarily such persons die before they reach reproductive age. The question arises, then, why do AS genotypes occur in such high numbers in some populations when SS individuals do not reproduce? After all, do not the AS individuals "waste" some

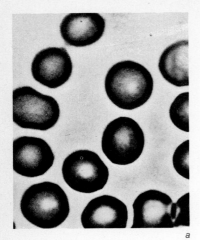

FIG. 10-7 (a) Normal human red blood cells; (b) red blood cells from a person with sickle-cell disease showing the change in shape when oxygen concentration is reduced. [(a) *Eric V. Gravé*; (b) *Jeroboam*.]

gametes which go into SS offspring, while AA individuals do not? Doesn't that mean that AA genotypes are outreproducing AS genotypes as well as SS genotypes (which normally do not reproduce at all)? The answer lies in a consideration of gene frequencies.

Gene frequencies in populations So far selection has been discussed in terms of proportions of genotypes. But it can also be considered in terms of allele frequencies in populations. By convention, these are called *gene frequencies.* The gene frequency of S alleles is simply the proportion of all the alleles in the population that are S. Thus, if there are 100 AA genotypes and 100 SS genotypes, the gene frequency of S would be 0.50. The gene frequency of A would also be 0.50. If there were 50 AA genotypes, 100 AS and 50 SS, the gene frequencies of both A and S would still be 0.50. Note how this is calculated:

GENOTYPE	NUMBER OF GENOTYPES	NUMBER OF GENES			FREQUENCIES OF GENES	
		A	S	TOTAL	A	S
AA	50	100	0	100		
AS	100	100	100	200		
SS	50	0	100	100		
	200	200	200	400	200/400 = 0.50	200/400 = 0.50

Thus populations can have *different* frequencies of genotypes at a locus and the *same* gene frequencies.

Malaria Geographic variation in the frequency of the S allele parallels the distribution of the protozoan parasite that causes one type of malaria. Studies have shown that AS individuals are *more resistant* to this type of malaria than are AA individuals. This is because the parasite is not able to survive as well in the red blood cells of AS individuals. Therefore, in the presence of the malarial parasite, the heterozygotic AS individuals are at a reproductive advantage compared with either homozygous AA or SS. The malarial parasite thus acts as a selective agent keeping S alleles in the population, despite the periodic occurrence of SS persons. Indeed, whenever heterozygotes outreproduce homozygotes, both alleles are retained in the population. If AA outreproduced both the heterozygotes and SS individuals, the frequency of the S allele would become very low until its presence in the population would be only the result of mutation changing A alleles to S.

The S allele is found in some black people, and in other peoples who come from malarial areas as well. Steps are being taken to treat the effects of sickle-cell trait and to identify individuals that carry one S allele. In parts of the world where malaria does not occur, and in places where the disease and the insects that transmit it (mosquitos) are being controlled, the S allele should eventually reach very low frequency as selection favoring the heterozygotes is relaxed.

10-9 People use evolution to their own ends

Breeding desirable organisms The malarial parasite has caused changes in the genetic information in populations without purposeful intervention by people; this is an example of natural selection. But people often wish to change populations in a direction useful to them. They do this by **artificial selection;** they choose which genotypes they will use as the parents for the next generation: the cow that gives the most milk, the fattest pig, the chicken that lays the most eggs, the prettiest rose, the corn with the biggest ears. By selecting, generation after generation, people have produced high-yield crops which are the cornerstone of Western agriculture. They have produced a wide variety of breeds of dogs (Fig. 10-8); they have produced fast race horses; they have produced chickens that are veritable egglaying machines.

People do this both by selection and by providing a proper environment. If people were to disappear tomorrow, so would most breeds of dogs, high-yield crops, fat pigs, and so on. Under natural conditions, all of these would be at a selective *disadvantage*. The high-yield grains would be subject to heavy insect attack and would be unable to thrive without fertilizer. They would quickly disappear. Fat pigs would be at a great disadvantage compared with thin ones—fat pigs cannot run away from predators or fight them off as well. As in experimental situations, when artificial selection is relaxed, natural selection will tend to restore the genotypes to something like their previous state.

Differential reproduction of genotypes In summary, whenever individuals with one genotype tend to leave more offspring than those with another genotype, we say that *selection* is occurring. If people are choosing the favored genotypes, then it is *artificial selection;* otherwise it is *natural selection*. Selection is simply the *differential reproduction of genotypes*. Death of individuals is not necessary for selection to occur, although obviously, if individuals with one genotype tend to die before they can reproduce, that genotype will be at a selective disadvantage. But if individuals of one genotype have, on the average, one offspring and those of another average three, the first will be at a selective disadvantage even if *all* individuals in the population live to reproduce. Considering the definition of natural selection, it would not make much difference if individuals with a

FIG. 10-8 Some of the many breeds of dogs which are the result of artificial selection by dog breeders. Dogs may have been first domesticated from a small wolf at least 12,000 years ago in the Near East; however, some people believe that many other species of the genus *Canis* may have been domesticated in other areas. Over the thousands of years of selective breeding, people have encouraged tremendous diversity. In 1962, there were some 800 true-breeding types throughout the world. All these types are able to interbreed, although certain combinations present obvious difficulties in mating and birth. (Shoulder heights are given for the following examples.) (*a*) A borzoi (31 inches); (*b*) a bull mastiff (27 inches); (*c*) a dachshund (12 inches); (*d*) a lhasa-apso (10 inches); (*e*) a chihuahua (6 inches).

certain genotype were sterilized rather than killed. Being unable to reproduce is the equivalent of being genetically dead, unless an organism is in a position to help related individuals (those with similar genotypes) reproduce.

10-10 Factors other than selection affect evolution

In addition to mutation and natural selection, there are other factors affecting evolutionary change. In a population, individuals may not mate at random. Individuals with similar phenotypes may prefer to mate with each other; or, individuals with different phenotypes may prefer to mate with each other. If either of these things is happening in a population, *differential reproduction* will obviously occur. If peppered moths had preferred to mate with other peppered moths instead of melanics, this would have affected the rate at which industrial melanism spread through populations.

The gene frequency will also be changed if a group of individuals with a particular gene frequency migrates into or out of the population. Finally, the size of a breeding population may be important also. In a very small population, changes in gene frequency leading to some alleles being lost or fixed (become 100 percent) will occur purely as a result of chance. The following analogy explains why one or the other allele at a locus is more likely to be lost in a small population than in a large one. A game of matching pennies played by millionaires would be very boring to watch, because they could go on playing indefinitely without either one losing a significant (for them) amount of money. However, if two men with only four pennies each begin to play, their game is unlikely to go on indefinitely. Obviously, the probability is high that one or the other will soon have all the pennies. Random processes, analogous to penny matching, occur in the formation of fertilized eggs in very small populations. The resulting random changes of gene frequency, often leading to loss of alleles, is called **genetic drift.**

Genetic drift has been held partly responsible for some small, isolated groups of people being quite different genetically from the larger populations from which they originally separated. For instance, there is a very small religious group in Pennsylvania called the Dunkers which has a different frequency for one blood-group gene. Dunkers have strict rules about marriage outside the group. Apparently drift has caused this change in gene frequency of one blood-group gene, but not others, tested.

The loss of alleles from small populations is called **decay of variability.** In Chap. 17, the problems of maintaining genetic variability in plants grown as food crops are discussed. Growing small samples of different strains of crops in experimental gardens is a less-than-satisfactory way of preserving samples of their genetic variability because genetic drift will lead to decay of that variability.

DIFFERENTIATION OF POPULATIONS

Natural selection and the other factors mentioned above are the mechanisms of evolution within a population. Do they also account for the great diversity of species? Why are there millions of insect species instead of just a few? If natural selection results in one population changing through time, how does one population become two or more populations? Figure 10-9 illustrates one theory of the *splitting* process of evolution.

10-11 Geographic variation often precedes the splitting process of evolution

Different populations of the same species of organism are normally genetically different from one another. For instance, the map in Fig. 10-10 shows the frequencies of alleles which control a human blood type. Notice that different populations have different frequencies of the gene. Figure 10-11 shows geographic variation in the proportion of melanic forms of the peppered moth. Figure 9-21 shows geographic variation in the height of a plant; plants of different heights have different genotypes.

Whenever geographically separated populations of sexually reproducing organisms are studied, they are found to be different genetically. **Geographic variation** is one of the most widespread biological phenomena. Sometimes variation forms easily understandable patterns. For instance, mammals and birds that live in cold regions tend to be larger than those of the same species that live in warm regions. A large mammal or bird has proportionately less surface area than a small mammal or bird. Large mammals and birds thus do not have as great a problem with heat loss in cold climates as do smaller ones.

Natural selection Different environments put different **selective pressures** on populations. Pollution on tree trunks leads to one pattern of selection in peppered moths, clean tree trunks to another. In the Arctic, large mammals are at a thermal advantage because they more readily retain their body heat than smaller ones. In the moist tropics, the situation is reversed. Large mammals have trouble getting rid of excess heat and are at a thermal disadvantage relative to smaller ones.

In people there is a great deal of geographic variation in characters such as skin color, nose length, hair type, body size, blood groups, and so on. These too are believed to be largely under selective control. Skin color seems to be related to the effects of the sun's radiation on the skin. Short limbs in peoples of the Arctic and sub-Arctic may be related to the possibility of frost bite on appendages. Big teeth should have a selective advantage in human groups that use teeth as tools. But the great mobility of people, their use of clothes and other artifacts, and the multiple environmental factors influencing each character have confused things to the point where the selective significance of much of our geographic variation is difficult to determine.

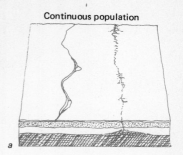
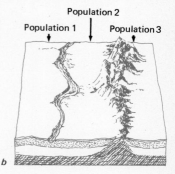
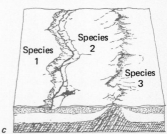

FIG. 10-9 Stages in the evolution of species. Isolation of populations leads to their divergence from the ancestral population as a result of natural selection. (*a*) A single species occurs in an area with a meandering river and low hills; (*b*) millions of years later, the river has eroded a deep canyon and a mountain range has developed, dividing the original population into three, which are genetically isolated from one another; (*c*) millions of years later, the mountains have eroded away, the canyon is broad and shallow, and the three populations can come together; selection has led to differences among the populations so that they look different and do not interbreed.

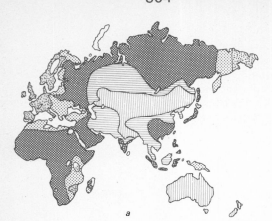
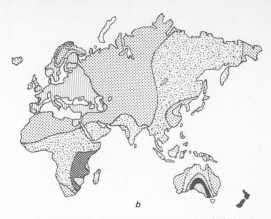

FIG. 10-10 Variation in blood-group genes in aboriginal human populations. (a) Gene for type B; (b) gene for type A. Degree of shading indicates area occupied by populations with approximately the same gene frequency. (*Adapted from A. E. Mourant,* The Distribution of Human Blood Groups, *Blackwell Scientific Publications, Ltd., F. A. Davis Company, Philadelphia, 1954.*)

A number of factors affect skin color; for example, excess solar radiation and the amount of exposure to the sun varies with climate. Although the lightest people tend to live nearer the poles and the darkest in tropical areas with abundant sunlight, there are many exceptions to these trends. Eskimos are not very light-skinned; and Bushmen, living under the blazing sun of African deserts, are quite light-skinned in comparison with the Zulus, for example.

Persons in cold areas cover their skin with clothing, but in humid tropical areas they do not. Vitamin D is synthesized in the skin only after exposure to the sun, which means that people in areas with abundant solar radiation might synthesize an excess of vitamin D (toxic in large amounts) and people in areas with less radiation might suffer a lack of vitamin D. Very dark-skinned people who move to areas with little sun sometimes suffer vitamin D deficiencies. A lack of sun-induced vitamin D, however, can be made up for in the diet. Eskimos consume oily foods from the sea which contain vitamin D and are therefore able to cover most of their skin with clothes without suffering vitamin D deficiency.

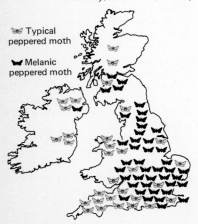

FIG. 10-11 Distribution of the peppered moth (*Biston*) showing the proportion of light and dark types. The dark forms are more numerous in industrialized areas. (*Data from H. B. D. Kettlewell,* Heredity, **12:**51–72, 1958.)

10-12 The splitting process of evolution results in the formation of new species

Populations of organisms living in different places usually are subject to different selection pressures and evolve in response to them. Populations that are geographically separated will become differentiated genetically. Then, when their structural and physiological differences have become sufficiently great, biologists consider them to be two different **species.** What was once considered to be a single species evolves into what are now considered two species. Different species have had separate evolutionary histories for long periods of time. Therefore, if the two new species ever come to live in the same place, they would not be expected to interbreed. This process of differentiation of populations is called **speciation**. Speciation is the inevitable result of natural selection in a world in which the environment varies from place to place, both physically and biologically.

Differences in the biological components of the environment may be very important in speciation. Many of the selective pressures to which a population of organisms is subjected originate with other organisms, not with the physical environment. A plant population being attacked by a herbivore population will be under selective pressure to develop defenses, for example, to evolve poisons or spines. The herbivores in turn will be under selective pressure to get through or around the defenses. They may evolve enzymes which catalyze the destruction of the poison, they may evolve tough mouths for chewing spiny leaves, or they may switch their attack to a different plant.

A system in which two populations are operating as selective agents on each other is an example of **coevolution**. Coevolutionary interactions may run entirely different courses in two different areas, hastening the process of speciation in both participants.

10-13 Present-day human beings all belong to a single species

There is no evidence that new species of the genus *Homo* are appearing today. Why, if human populations are subject to different selection pressure and if the human species shows geographic variation, has not *Homo sapiens* divided into two or more species? As far as we know, all individuals of human populations are perfectly capable of interbreeding with every other, and some have done so whenever the opportunity presented itself. Furthermore, although *Homo sapiens* shows a great deal of geographic variation, variation in one character tends to show little similarity to variation in other characters. Thus, for instance, skin color, and shape of nose vary quite independently of one another (Fig. 10-12). And there is little sign that *Homo sapiens* is about to undergo a process of speciation; no biological subunits of *Homo sapiens* have been detected that could be considered incipient species. It is possible that if travel among human populations had not proceeded so rapidly, speciation would have occurred.

Genetic and cultural evolution There are two main reasons which may explain why humanity is not dividing into two or more species: (1) cultural evolution and (2) mobility. One is that selection pressures in various geographic areas are not strong enough. Human beings, through culture, exercise more control over their environments than any other organism. In addition to genetic evolution, which is always going on, human beings experience another kind of evolution called

CHAPTER TEN

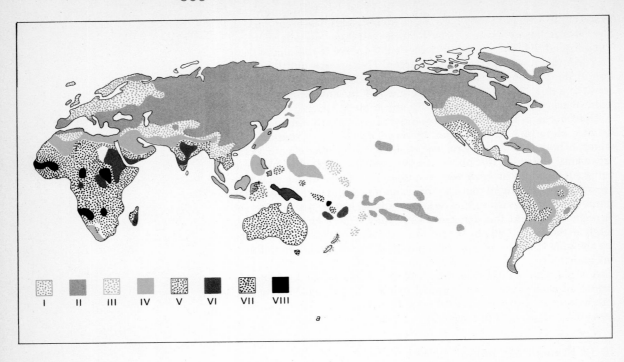

a

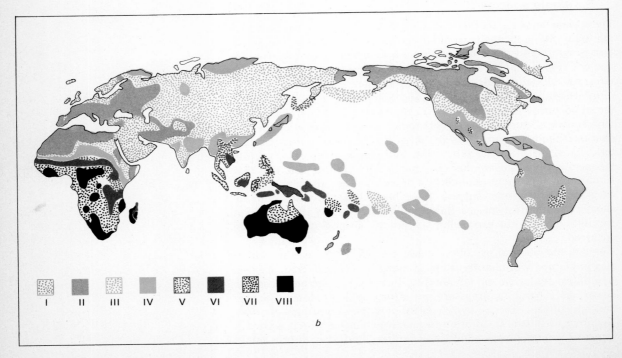

b

FIG. 10-12 Geographic variation in two characters in original populations of *Homo sapiens*: (*a*) skin color (I is lightest, VIII is darkest); (*b*) shape of nose (nasal index, with I the longest, narrowest noses and VIII the shortest, broadest). (*Adapted from R. Biasutti*, Razzee Popoli Della Terra, *2d ed., Unione Tipografico, Turin, 1953.*)

cultural evolution. Genetic evolution involves the transmission of a body of *genetic information* through time with change. Cultural evolution involves the transmission of a body of *nongenetic information*—culture—through time with change. In other words, invention of new techniques and new ideas and their transmission through time are part of cultural evolution. Because it is not restricted to transfer between parent and offspring, cultural evolution can take place much more rapidly than genetic evolution. Birds evolved from reptiles, developing wings over a period of millions of years. People "evolved" flight in less than 100 years through a series of inventions leading from the Wright brothers' plane to modern jet planes. In addition, cultural inventions, unlike genetic "inventions," can spread much more rapidly than individuals—indeed, today they can be transmitted virtually instantaneously around the world by radio and television.

Nonreproductive individuals do not contribute *directly* to the genetic information of future generations. A plant, a human being, or another animal that has no offspring does not have any direct *genetic* effect on future generations. This does not mean, however, that such individuals have no effect at all on future generations. People who do not have children have effects on cultural evolution in many different ways. Such popularly known historical figures as Jesus Christ, Joan of Arc, and Queen Elizabeth I obviously affected future generations. But so also did the scientists Isaac Newton, Francis Bacon, and Ruth Benedict; the writers Gertrude Stein and George Bernard Shaw; the entertainers Janice Joplin and Greta Garbo; the artists Vincent van Gogh and El Greco; and the musicians Ludwig van Beethoven, Wolfgang Mozart, and Peter Tchaikovsky. Important contributions to cultural evolution were made in many parts of the world by celibate religious orders (Christian, Buddhist, etc.) by preserving books, music, and art objects. Adolph Hitler fathered no children; however, his program of genocide which led to the slaughter of 6 million Jews, most of the Gypsies, and many Slavic peoples, has profoundly affected the genetic variability of many European countries for many generations. Pope Paul VI has fathered no children; his decree forbidding the use of birth control methods by Catholics has led to many births that otherwise would not have occurred, especially in the Latin American countries.

In the tropics, people shelter themselves from the heat, in the arctic from the cold. People have learned ways to prepare food which, if eaten unprepared, would be poisonous. With cunning, skill at making weapons, and social organization, people have removed most of the threat of large predators. The worst enemy of most human populations is the same—other people. In spite of superficial diversity, most human populations live in relatively similar environments. They are subject to similar selection pressures—selection pressures *not to change*, but to maintain that same very successful basic human genotype.

People in industrial areas may be more subject to the selection pressures of air pollution and the fast pace of life than people in rural areas. Like most human inventions, pollution is quickly spreading to all parts of the globe, including rural areas; and the stress of air pollution is now changing selection pressures all over the world. Many people will suffer and die from such pollution stress, but hundreds of generations, and thus thousands of years, would be required for humanity to evolve resistance. This is because most people who die of air pollution die late in life of respiratory and heart disease, *after they have reproduced*. Obviously, this means that the new selection pressures are relatively weak and evolution will be slow. Most air pollution today comes from fossil fuels, directly or indirectly. Fossil fuel supplies will run out in a few hundred years. It is unlikely *Homo sapiens* will become strongly pollution-resistant.

Mobility of people The second reason, the great *mobility* of human beings, reinforces the first. Human individuals, as well as human inventions, are mobile, and both these movements subject people to new selective agents and provide a great deal of interbreeding among populations. Africans who were moved as slaves to North America, for example, are losing their genetic resistance to malaria now that North America is largely free of malaria. Some also interbred with Europeans who had moved to North America, creating a new series of gene combinations on which selective agents can operate.

If it were not for the continual genetic mixing of the human populations, the differences in selection pressures among them might have led to speciation. But with the mixing, these relatively weak selection pressures have produced only a somewhat bewildering pattern of variation of such superficial characters as skin color, hair color and shape, blood types, etc.

10-14 Races are arbitrary subdivisions of humanity

By selecting one or two characters, people have attempted to divide the human species into a series of races. Since, of all the five senses, people depend most heavily on *sight,* it is not surprising that the principal character selected has usually been skin color. This has led to discussions of "white," "black," "red," and "yellow" races; or "Caucasoid," "Negroid," "American Indian," and "Oriental" races, and so on. Because in America our attention has been called to differences in skin color since we were very young, these subdivisions seem natural to us. But they are no more natural than subdivisions based on height or blood type or head shape or any other character which varies geographically. Races are primarily *social* rather than biological units. For example, some people in the United States believe that anyone who can be shown to have one dark-skinned ancestor belongs to the "Negro race." In Brazil, a very poor person with a white skin is considered a "Negro" while a very rich person with black skin belongs to the "white race." Clearly this is a matter of social definition, not biology. *Homo sapiens* is a very variable species, and any two groups of people from different localities will differ in many characteristics. But this does not mean that people can be divided biologically into a series of races (Fig. 10-13).

Racist myths   Years ago most light-skinned people in Western societies thought that people with light skins were naturally superior to everyone else. That is not surprising. Human groups have almost always thought that members of their group were superior to members of other groups. The Eskimos' name for themselves is *Innuit,* meaning "*the* people." Many Germans at one time thought they were a "master race." The Jews were the "chosen people." The British thought colonial peoples to be "lesser breeds." The list goes on and on.

All this is utter rubbish. There is not the slightest sign that any group of people, however defined, has a monopoly on genetic quality, however defined.

Recently, old racist ideas have been revived in a discussion of whether or not blacks may have less innate (inborn, inherited) intelli-

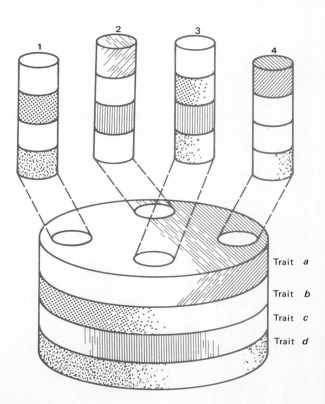

FIG. 10-13 A cylinder as a model of a population varying in four characteristics. Cores taken from the cylinder in different places show that the characters vary independently. The layers represent four different traits which vary geographically. Core 1 differs from core 4 in traits *a* and *b*, but is the same with respect to trait *c*. Similarly, cores 2 and 3 are similar for trait *c*, but not for the others. This shows that the population of the cylinder cannot be divided into races (cores).

gence than whites. As you know, blacks and whites are arbitrary groups so this is pretty much a nonquestion. It is true, however, that on standard intelligence tests, people in America classified as black tend to score lower than those classified as white. It is not known what all the things are that contribute to intelligence. However, it is generally assumed that intelligence tests are a measure of what we think of as "smartness," that is, the ability to do well in the academic subjects studied in school. Just because blacks score lower, on the average, than whites on this test, does that indicate that blacks are less well-endowed genetically than whites? The answer is *no*. Blacks from the northern part of the United States do better than whites from the South on the average.

Most scientists agree that "black-white" differences on intelligence tests can be explained easily on the basis of the environmental deprivation that blacks suffer in a white, racist society and on the cultural bias in tests written and given by white examiners. In the United States, blacks on the average have less adequate diets, less stable and stimulating home environments, and lower-quality education than whites. Furthermore, they live in a black subculture, and the intelligence tests most widely used were designed by and for the white subculture. Recently a black psychologist devised a black intelligence test based on the black subculture. Most white people would do poorly on it.

This does not mean, of course, that differences do not exist among different groups of people in genes affecting IQ, just that we cannot now measure such differences. It is entirely conceivable, for instance, that dark-skinned people in the United States are "genetically smarter" than light-skinned. Generations of environmental deprivation may have produced strong selection for intelligence, selection not present in the white population, many of whom have had relatively easier lives than blacks. If this is the case, once *environments* were equalized, we could expect average performance on IQ tests to be higher in blacks than in whites. Of course, it is also conceivable that in equal environments the average white person's performance would be higher.

In order to answer the "who is basically smarter" question with assurance, we would have to be able somehow to determine exactly the innate ability of individuals. If we could find out each person's "genetic IQ range" (a simplified abstraction, as you will recall from Chap. 9), we would only have to take the average of those IQs for a large number of dark- and light-skinned people to get our answer. But what would be the point? Once we knew a person's innate potential we would deal with him or her on *that basis*. If it was high, we would expect great things. If it was low, we would put a lot of effort into creating an environment in which a maximum of the possible genetic range was realized phenotypically. The question of how "genetic IQ range" is associated with skin color, head shape, height, or any other character would then be of little interest.

You might wish to think about one final point. IQ, as it is usually measured in our society, predicts *school performance* very well, but is rather poorly correlated with *success in life*. Is it possible that we put too much emphasis on IQ and too little on human characteristics less frequently measured—ambition, compassion, empathy, dependability, sense of humor, sensuality, and stability—just to mention a few?

QUESTIONS FOR REVIEW

1 Supernatural theories (explanations) of the origin of life cannot be ruled out as impossible. Why, then, does biology rule them out as theories worthy of investigation by the scientific method (see Chap. 17)?

2 What three *prerequisites of life* are noted in the text discussion? All three are available in one form or another today. Yet, as the text notes, it is believed that life is no longer originating from nonliving chemical systems; all life, biologists now think, issues from already existing organisms. Can you think of any reasons why life can no longer "get started on its own"? (A later chapter will note one reason.)

3 What do these terms mean: artificial selection, resistant genotype, speciation, genetic drift, selective agent, differential reproduction, selective advantage?

4 Sickle-cell anemia kills its victims before they can reproduce. Yet the allele which causes the disease is not being effectively selected against in some areas—its frequency in the populations, some of whose members die from the disease, is not decreasing from generation to generation. Why not?

5 A good little book by the British archaeologist V. Gordon Childe is titled *Man Makes Himself*. The title refers to one of two processes of evolutionary change discussed in this chapter. What are the two processes? With which do you think Childe's book is mainly concerned? Which of the two processes would you say has played the larger role in *making* "man" over the last 100,000 years? (The next chapter will provide some more information that you can bring to bear on this question.)

6 Why since so many people die of respiratory diseases, cancer, and heart disease, do these diseases *not* apply strong selection pressure against genotypes that produce phenotypes with greater than average susceptibility to these diseases?

7 What we call *race* clearly involves biological (genetic) factors—yet "Races are social not biological units." Why?

8 As the text notes, mammals and birds that live in cold regions tend to be larger than those that live in warm regions mostly because larger mammals and birds do not have as great a problem of heat loss as do smaller ones. That is, the larger animal's ratio of heat generated to heat radiated is more favorable, and so they can stay warmer. The same *ratio rule* holds for reptiles too: larger reptiles more readily retain their body heat than do the smaller ones. Yet reptiles of cold regions tend to be *smaller* than their tropical relatives: the largest lizards and snakes in Canada, for instance, are much smaller than even the smaller-than-average-sized alligators and anacondas in Brazil. Why do you suppose this is so?

READINGS

Bishop, J. A., and L. M. Cook: "Moths, Melanism and Clean Air," *Scientific American*, **232**(1):90–99, 1975.

Bodmer, W. F., and L. L. Cavalli-Sforza: "Intelligence and Race," *Scientific American*, **223**(4):19–29, offprint 1199, 1970.

Cavalli-Sforza, L. L.: "Genetic Drift in an Italian Population," *Scientific American,* **221**(2):30–37, offprint 1154, 1969.

Ehrlich, P. R., and R. W. Holm: "Evolution," *Biocore,* Unit XXII, McGraw-Hill, New York, 1974.

Ehrlich, P. R., R. W. Holm, and D. R. Parnell: *The Process of Evolution,* 2d ed., McGraw-Hill, New York, 1974.

Kamin, L. J.: *The Science and Politics of I.Q.,* Lawrence Erlbaum Associates, Publishers, 1974.

King, J. C.: *The Biology of Race,* Harcourt, Brace, Jovanovich, New York, 1971.

Moore, R., et al.: *Evolution,* Life Nature Library, Time-Life Books, New York, 1968.

Shepard, P. M.: *Natural Selection and Heredity,* Harper & Row, New York, 1960.

11

THE HISTORY OF LIFE

SOME LEARNING OBJECTIVES

After you have studied this chapter, you should be able to

1. Demonstrate some of what you have learned about fossils and the fossil record by
   a. Explaining what fossils are, where and how they are formed, and how they become available for study.
   b. Giving two reasons why the fossil record does not provide a very complete picture of life in the past.
   c. Stating what is meant by the relative dating and the absolute dating of fossils, naming one method by which each type of dating may be done, and saying why each is useful to the paleontologist.
2. List, in order, the major geological periods and name a form of life which originated during each.
3. Further demonstrate your knowledge of the history of some major life forms by
   a. Outlining the development of life in the early seas.
   b. Stating about how long ago plants invaded the land and listing the major land-plant types in order of their evolutionary appearance.
   c. Stating about how long ago animals moved onto the land and briefly explaining how this first move, as well as the subsequent evolution of terrestrial animal life, was dependent on selection pressures created by the prior diversification of land plants.
   d. Listing in order of their evolutionary appearance the major vertebrate animal groups discussed in this chapter, saying about how long ago the first forms of each group appeared, and naming the major adaptive biological *innovations* (changes) each group developed that permitted each in its turn to become the dominant terrestrial vertebrate group.
   e. Describing two types of coevolutionary interactions between plants and insects—one involving plant defense, the other involving plant fertilization.
4. Name three types of fossil fuel and describe from what, and how, each is believed to have been formed.
5. Demonstrate some of what you have learned about the evolution of our own species by
   a. Tracing the evolution of *Homo sapiens* in terms of the major habitat changes involved (beginning with our species' insectivore ancestors) as discussed in the chapter.
   b. Stating the factors now believed responsible for these changes of habitat.
   c. Naming, and arranging in proper evolutionary sequence, those of our fossil ancestors which are discussed in the text, and describing some major biological and cultural characteristics of each.

Biologists who study fossils are called **paleontologists;** paleontologists have been able to reconstruct the history of life on this planet. We saw in the last chapter that life has probably existed on this planet for a few billion years. Study of the vast sweep of geological history gives us a new perspective on our species and other present-day organisms. In comparison with most organisms, we are very recent and have been on the scene only a short time. In this chapter we explore backward in time and reconstruct a few of the multitudes of organisms which existed before us, some of which are ancestors of present-day species.

THE FIRST SIGNS OF LIFE

The first signs of life which biologists have found in ancient rocks are not fossils of organisms but certain chemicals characteristic of living things; these compounds are often called **chemical fossils.** They may have originated in the long period of chemical evolution which probably preceded the development of organisms, or they may be the remains of the earliest organisms which were probably small and soft-bodied and did not make good fossils.

11-1 Fossils are records of organisms of the past

Formation of fossils The changing conditions on the earth often allow fossils to be formed. Mountains are eroded away by wind and water and are washed into oceans and other bodies of water where they settle out and form beds of rock once more. Volcanoes throw out great quantities of dust and ash which also settle down. These materials often form layers of rock called **strata** (sing., **stratum**). In each of these situations, chances exist for the preservation of organisms as fossils. Remains of whole animals or groups of animals and entire plants or forests may be preserved (Fig. 11-1). Separate parts, such as teeth and bones, leaves, and even pollen grains, may become fossils (Fig. 11-2). Plant or animal tissues may be partially replaced by minerals in such a way that details of the original structure remain. They are then said to be **petrified.** Stems or shells buried in mud may be replaced by minerals so that the fossil is actually a **cast** formed in the mold of the surrounding mud.

Fossils are not necessarily the remains of animals or plants or their parts, but may be some other record of their lives. Animals buried in volcanic ash may be burned, but their shape may be visible. Even the footprints (Fig. 11-3) and droppings of dinosaurs and the tracks of worms have been preserved as fossils, and they may tell us a great deal about the organisms that made them. These remains of past organisms become available for study when, after hundreds of thousands or millions of years, geological processes push rock beds up from water or expose crumbling cliff sides. Fossils also may be found when deep wells or mines are drilled or dug or when excavations for buildings or roads expose strata that have been pushed near the surface by geological processes.

FIG. 11-1 (a) The remains of trees of a forest in Arizona in the Triassic about 225 million years ago; (b) this skeleton of a brontosaurus is being carefully uncovered by scientists at a site called Bone Quarry Cabin, Wyoming. Before being dug up in 1899, it had been buried in the earth for about 150 million years. (*American Museum of Natural History.*)

An incomplete record The fossil record does not give a very complete picture of life in the past. First, not all organisms make good fossils and not all parts of an organism become fossilized. Generally, only hard structures, such as shells and bones, fossilize; soft-bodied organisms, such as jellyfish and worms, may not be preserved at all. They are easily decomposed and have no hard parts so that they may not even leave a cast in the mud where they die. When an animal such as a dinosaur becomes a fossil, only the hard parts are preserved. The skeleton often is very well preserved. Even pieces of scaly skin

FIG. 11-2 This leaf fossil, about 12.5 centimeters long, is an imprint of a leaf from a willow which grew about 135 million years ago in the area of Denton, Texas. (*Ward's Natural Science Establishment, Inc.*)

FIG. 11-3 These fossilized dinosaur footprints show that a giant reptile walked across these former mud flats near Holyoke, Massachusetts, about 190 million years ago. (*Ward's Natural Science Establishment, Inc.*)

may become fossils. We almost never have any record of soft tissues, such as intestines or heart, however. Another problem is that when a mass of animal or plant fossils is found, it may be difficult to decide which parts go together: which leaves belong to which stems and which teeth belong to which skull.

Fortunately, our study of organisms existing today can help us in interpreting fossils from the past. By knowing how the bones of present-day reptiles are put together, biologists can more easily put together those of a fossil reptile, such as a dinosaur. By knowing how, in mammals of today, the teeth of herbivores differ from those of carnivores, we can make decisions about the way of life of a past mammal known only by its skull or teeth.

The older the rocks in which fossils have been preserved, the more likely they are to have been worn away by erosion. Therefore, the fossil record becomes progressively more difficult to interpret the further back it goes in time. The fossil record of animals and plants that lived only a few thousand years ago is often much more complete. We may be able to learn a great deal about their mode of life and even something of their population size and genetic variability. Finally, upland areas are less suitable for fossil preservation than lowland areas. Often, only badly worn parts of organisms of upland areas become fossilized after they have been transported to the lowland.

11-2 Fossils are dated in many ways

Stratification of rocks   It is not especially difficult to tell the **relative age** of the rocks containing fossils. The geological strata in which fossils are found are *laid down in order of age,* the most recent being the uppermost. Even if they are twisted and distorted by the heavings of the earth, their proper sequence can be worked out. The time scale of the geological record of the past is divided into a series of eras, periods, and epochs (Table 11-1), based upon the layering of rocks over the earth's surface. As we discuss the history of life, you may find it helpful to refer to this table.

Constructing a geological time scale for the entire planet is not without its difficulties. For example, how can it be determined that certain strata in the United States are the same age as strata in Europe? In part, this can be done by the use of fossils. There are some animals that are so widespread and so characteristic of a particular age that they can be used to show that rocks found in different places are of the same age. An easier way to cross-date strata, however, is to determine their *absolute* age.

Radioactive dating   In recent years, a number of different methods of determining the **absolute age** of rocks or of the fossils in them have been developed. These are based upon the rate of radioactive change of various chemical elements. Chemists have determined the time it takes for certain radioactive elements to decay into others. It is, therefore, possible to look at the relative proportions of the elements, e.g., potassium and argon, in a fossil or in rocks. From this

TABLE 11-1 GEOLOGICAL TIME TABLE

ERA	PERIOD	EPOCH	APPROXIMATE AGE IN MILLIONS OF YEARS
Cenozoic (recent life)	Quaternary	Holocene	0.01
		Pleistocene (most recent)	2.0–3.0
	Tertiary (from the third part of an eighteenth-century classification)	Pliocene (very recent)	7
		Miocene (moderately recent)	25
		Oligocene (slightly recent)	40
		Eocene (dawn of the recent)	60
		Paleocene (early dawn of the recent)	68–70
Mesozoic (intermediate life)	Cretaceous (from chalk)		135
	Jurassic (from Jura Mountains, France)		180
	Triassic (from threefold division in Germany)		225
Paleozoic (ancient life)	Permian (from Perm, a Russian province)		270
	Carboniferous (from abundance of coal)		
	Pennsylvanian		325
	Mississippian		350
	Devonian (from Devonshire, England)		400
	Silurian (from the Silures, an ancient British tribe)		440
	Ordovician (from the Ordovices, an ancient British tribe)		500
	Cambrian (from Cambria, a Latin form of the native Welsh name for Wales)		550–600
Precambrian	No worldwide classification established		3,500 or more

After Bernhard Kummel, *History of the Earth*, 2d ed., Freeman, San Francisco, 1970, p. 14.

value, the age of the fossil or rock can be accurately calculated. Table 11-1 shows the duration of the geological periods and the time in the past at which they began.

11-3 It is difficult to visualize the enormous time periods represented in the geological record

What does it mean to say that plants first grew on land about 420 million years ago, whereas dinosaurs appeared 225 million years ago?

Zero hours
Life begins

3:00 A.M.
Photosynthesis begins

4:00 P.M.
Probable eukaryotes

7:18 P.M.
First multicellular organisms

9:12 P.M.
Plants invade land

9:40 P.M.
Insects diversify

10:20 P.M.
First flowers

11:02 P.M.
Middle of Age of Reptiles

11:48 P.M.
Middle Age of Animals

11:59.5 P.M.
First human beings

FIG. 11-4 The history of life as if it had occurred during a 24-hour day.

A million years is such a long time that it is hard to imagine how much longer 420 million years is than 225 million years. Perhaps it will be helpful to your understanding of the fossil record to visualize the whole history of life as if it had occurred within a 24-hour day (Fig. 11-4). More than three-quarters of this 24-hour day would have passed before fossils even became abundant. Life was not found on land until approximately 9:12 P.M., and the ages of the reptiles, amphibians, etc., of course, occurred after that. Finally, near 11:00 P.M., the mammals appeared. The earliest appearance of human beings would have been at one-half second before midnight. Recorded history began so near midnight on this scale that it is impossible to show in the figure. Perhaps the time it takes to blink your eye—one-tenth of a second—would represent *all of recorded history.*

11-4 The oldest known fossils are very ancient

Molecules that could well be a remnant of ancient chlorophyll, together with other apparently organic molecules, have been found in very early strata. Molecular paleontologists have very complicated but convincing reasons to believe that these compounds were produced by living systems at the times the strata were laid down. One of the most convincing is that nonbiological processes form mixtures of different molecular arrangements of a compound; however, living systems produce only one or two types.

Fossils that look very much like bacteria and algae have been found in Africa in Precambrian rocks 3.2 billion years old (Fig. 11-5). This is less than 2 billion years after the earth is thought to have been formed. If these are in fact fossil cells of photosynthetic organisms, this means that photosynthesis was going on at least 3.2 billion years ago.

Other Precambrian organisms have been found in central Australia, Canada, Siberia, and England. *Kakabekia* (Fig. 11-5b) is especially interesting. It consists of a bulb, a short stalk, and an umbrellalike top, and it is found in rocks about 1,900 million years old in Ontario, Canada. In 1964, in the moat at Harlech Castle, Wales, an organism was found which structurally resembles *Kakabekia* closely. It requires no oxygen, but does require ammonia and will live in an atmosphere of ammonia, methane, and nitrogen. A prokaryote, it does not have obvious relationship to any known group of organisms.

LIFE IN THE EARLY SEAS

As we discussed in Chap. 10, it is generally thought that life originated in the sea. Indeed, on the basis of theories about the formation of our planet, the land was a very inhospitable place for life. About 2 billion years ago, there were a number of kinds of marine algae. Subsequently, there appears to have been a fairly rapid evolution of other algae, followed by animal life.

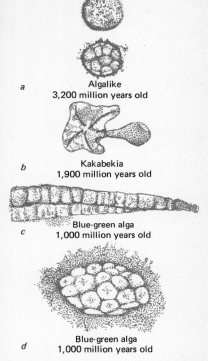

a Algalike
3,200 million years old

b Kakabekia
1,900 million years old

c Blue-green alga
1,000 million years old

d Blue-green alga
1,000 million years old

FIG. 11-5 Some Precambrian organisms.

11-5 Algae probably were the first photosynthetic organisms to appear in the fossil record

The first algae known in the fossil record were marine algae. These Precambrian (Table 11-1) forms of algae were similar to the algae we today call blue-green algae. Some of them, as old as 1 billion years, appear to have undergone very little structural change up to the present. Filaments of cells apparently with nuclei and organelles were also found. These probably were decomposed remains of blue-green algae and not eukaryotes.

Effects of ancient algae Before photosynthetic organisms began to produce oxygen, the atmosphere was made up mostly of methane, water vapor, hydrogen, and ammonia. With the appearance in the early seas of photosynthetic algae, the stage was set for the change of the earth's atmosphere from an oxygen-free state to a composition of almost 21 percent oxygen. The build-up of oxygen was slow and, until there was sufficient oxygen, air-breathing and land-living organisms could not survive. Animals must have oxygen for the slow-burning process by which they extract energy from their foods. Furthermore, it was not safe for organisms to come out on land or live in water that was too shallow. They would have received the "fatal sunburns" which we discussed in Chap. 4 since, until oxygen was formed in sufficient quantities, the protective ozone (O_3) layer could not form. This ozone layer now absorbs most of the harmful ultraviolet wavelengths of the sun's radiation. Although once the energy needed for the origin of life may have been supplied by massive amounts of ultraviolet rays, today, without the ozone barrier to this radiation, life would be impossible.

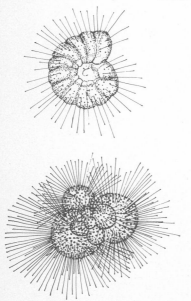

FIG. 11-6 Some representative foraminiferans.

11-6 Many kinds of invertebrates were found in the early seas

The first animals Since animals require oxygen produced by green plants in order to carry on respiration that powers their activities, it is clear why animals did not appear until very late Precambrian times. The first animals included jellyfish and worms, as well as many invertebrates quite different from those appearing in later Cambrian rocks.

In Cambrian times, some 550 million years ago, protozoa, sponges, mollusks, arthropods, and echinoderms existed. Among the protozoans were forms known as foraminiferans (Fig. 11-6), with a chalky shell which makes excellent fossil material. Today there are extensive beds of limestone and chalk made up of foraminiferans. Blackboard chalk may contain the shells of these ancient one-celled animals.

During the following periods of the Paleozoic, most of the phyla of invertebrates developed a wide variety of forms all over the world. The arthropods and mollusks had a particularly dramatic period of evolutionary diversification in the early seas around the world in Paleozoic and Mesozoic times, as we shall see.

Arthropods Arthropods known as **trilobites** (Fig. 11-7) appeared in the early Cambrian period and diversified greatly before they finally became extinct in the Permian. Trilobites were small, flattened animals (although some exceeded 30 centimeters in length) that presumably crept about on the ocean floor. Some may have been burrowers or swimmers. They were at one time perhaps the most abundant and diverse of all animals. They are such common fossils that they are often made into jewelry, such as belt buckles or tie clips. Trilobites give us some idea of what the probable ancestors of the arthropod line as a whole looked like.

There were other kinds of arthropods in the Paleozoic seas, including some related to present-day spiders and scorpions. Perhaps the most dramatic of all, however, were 3-meter (10-foot)-long creatures called **eurypterids** (Fig. 11-8). Eurypterids were common in Silurian and Devonian times. In addition to walking and swimming appendages, many had long, crablike pincers. The eurypterids are thought to have lived in brackish waters; they were very large, as invertebrates go. All the evidence suggests they were important predators in ecosystems some 400 million years ago; however, the eurypterids are now extinct. Also playing important roles in marine ecosystems were many kinds of mollusks.

Mollusks There are a number of different kinds of **mollusks**, including snails, clams, and squid, all of which were found in great abundance and variety in Paleozoic and Mesozoic waters. The squidlike forms are especially interesting, since a number of unusual types developed (Fig. 11-9), but most became extinct by the end of the Mesozoic. These mollusks resembled present-day squid, but had straight or curved shells. If they behaved as do modern squid, they must have been important carnivores in Paleozoic and Mesozoic times, playing the ecological roles that fishes later came to play. The only living representative of this once important group is the chambered nautilus (Fig. 11-10).

11-7 The first vertebrates were fishes

Vertebrates are not common in the fossil record until late Silurian and Devonian times, but they first appeared in the Ordovician, about

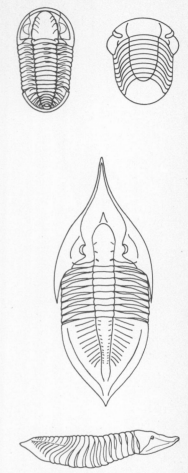

FIG. 11-7 Some representative trilobites.

FIG. 11-8 Eurypterids were huge carnivorous arthropods.

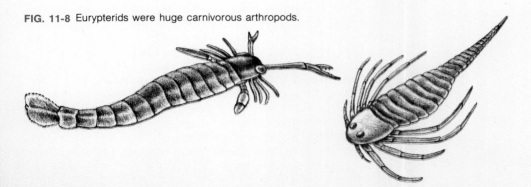

FIG. 11-9 A Paleozoic cephalopod mollusk.

FIG. 11-10 A model of a chambered nautilus, the only living representative of a group of shelled, squidlike mollusks that otherwise became extinct at the end of the Mesozoic. (*American Museum of Natural History.*)

500 million years ago. Fossils of Ordovician vertebrates are of strange fishes with armor plating (Fig. 11-11). Details of their anatomy can be determined from these fossils, and it is known that these fishes *did not have jaws* as do the vast majority of vertebrates today. Today only the lampreys and hagfish (Fig. 11-12) are jawless.

While the jawless fishes were developing, another group of primitive fishes *with jaws* began to evolve and diversify in the late Silurian and Devonian. Although jawless fishes were common and important in Paleozoic times, most became extinct at the end of that era. The early jawed fishes varied in size from a few centimeters to over 10 meters (Fig. 11-13). It is thought that an early member of this group gave rise to the two groups of more recent fishes: the **sharks** and their relatives, and the **bony fishes.** Sharks appeared in the Devonian and, of course, are still in existence today. They do not show the evolutionary diversity of the bony fishes, however, which have more species than all other vertebrates combined.

Bony fishes early separated into two main lines of evolution. The common fishes, such as goldfish, trout, tuna, and perch, belong to the group of **ray-finned fishes.** Their bodies are streamlined for efficient swimming, and they usually have conspicuous fins supported by slender, bony struts called rays.

From the point of view of evolution, however, the second evolutionary line may be more interesting. This group includes the **lobe-finned fishes** and the **lungfishes.** Both kinds are found in Devonian strata, and three genera of lungfishes are alive today (Fig. 11-14). For years it was thought that the lobe-finned fishes became extinct in the Cretaceous, but a living species of lobe-finned fish was discovered in the 1950s living in deep waters off southern Africa (Fig. 11-15).

Lobe-finned fishes have paired, heavy fins with a bone structure similar to that of the legs of modern four-legged vertebrates. The skull of these fishes also closely resembles that of amphibians. It is thought that the amphibians arose in Devonian from lobe-finned ancestors. It is possible that if the habitats of these fishes began to dry out in the late Devonian, they could, using their lobed fins, crawl from one pool to the next. Some living fishes, for example, the walking catfish (Sec. 5-15), do just this today, although their fins are smaller.

It is possible this ability of some fishes to crawl about on land may have set in motion the evolution of the early amphibia, which first appeared in the middle Devonian. These very early amphibians became extinct in the Triassic. The amphibia that live today—salamanders, toads, frogs—are derived from late Paleozoic amphibia. Later in this chapter, the history of the animals as they diversified on land will be continued. First, however, it is necessary to consider the evolution of the plants. Until plants could survive on land far from oceans or lakes, animals would have no source of food if they tried to move far from the water. Being an amphibian and having to return to the water to reproduce was a successful strategy. Plants soon dispersed, however, far from bodies of water, and new habitats opened.

PLANTS MOVE ONTO LAND

We have earlier described the process of ecological succession (Sec. 3-8). The land of the young planet earth was a very inhospitable place for life, which existed in abundance only in the seas. The gradual invasion of the land was the first instance of terrestrial ecological succession. As with succession on bare rock, the photosynthesizers came first, to be followed by decomposers, herbivores, and carnivores.

11-8 Vascular plants developed in marshy areas

It is thought that land plants developed from aquatic green algae, probably at the borders of the sea. The earliest land plants (Fig. 11-16) are found in the Silurian and Devonian (440 million to 400 million years ago). These *psilophytes* were small, rarely exceeding 1 meter in height. They had a waterproof layer on the outside (called a **cuticle**) to reduce water loss and vascular tissues, xylem and phloem, which conducted water and food. Leaves were usually lacking, and there were no roots. Underground, creeping stems served to anchor the plants and bore single-celled, absorbing structures called **rhizoids**.

Following the appearance of the earliest vascular plants, the plants rapidly diversified in the Devonian. Ferns and relatives of ferns—club mosses and horsetails—became dominant features of the landscape (Fig. 11-17). True ferns and primitive seed plants were also found. Colonization of the land by plants made possible the development of soil, which permitted larger and larger plants to grow. Terrestrial photosynthesis led to increasing amounts of oxygen in the atmosphere. More and more kinds of animals could evolve structures that would allow them to survive out of water. The new habitats opened up on the land by plants created an environmental change which led to a change in selective pressures. An animal that could remain on land seeking food was not competing with its aquatic relatives. It thus had a very good chance of surviving to leave offspring that inherited any genetic components of this ability. Thus over millions of years the first terrestrial ecosystems evolved. The animals interacted with the environment, the plants, and with each other. The fossil record suggests that these ecosystems approached those of today in structure and complexity.

FIG. 11-11 A jawless armored fish that presumably fed on the bottom of shallow, muddy seas.

FIG. 11-12 Lampreys attach to rocks when moving upstream. The lamprey has many teeth in a jawless, suckerlike mouth. It rasps a hole through the skin of a fish and feeds on its blood. The sea lamprey invaded the Great Lakes over 100 years ago when the Welland Canal was opened so that ships could bypass Niagara Falls and enter Lake Erie. Lampreys almost destroyed populations of two commercially important fishes in the western Great Lakes until the lampreys were destroyed by chemical treatment in the 1950s. (*Carolina Biological Supply Co.*)

FIG. 11-13 *Dinichthys* was a 9-meter-long (30-foot-long) carnivorous fish of the Devonian.

11-9 There were extensive forests in the Carboniferous

By the time of the Carboniferous period, some 350 million years ago, the club mosses and horsetails had become treelike and some were nearly 50 meters (150 feet) tall. There were also treelike ferns, as well as several kinds of primitive plants. As you can see in Fig. 11-18, most of these trees are not at all like our modern oaks and maples (which had not yet come into existence). All the genera found in the Carboniferous forests are extinct. Today only a few genera of small plants are left of the once important club mosses and horsetails.

FIG. 11-14 A lungfish from Gambia, West Africa. African lungfish have gills and lungs. They can survive periods of drought by secreting a leatherlike cocoon and burying themselves in the mud when their ponds dry up. (*American Museum of Natural History*.)

FIG. 11-15 *Latimeria* is a lobe-finned fish of a type thought to have become extinct in the Cretaceous. The first living one known to modern science was caught off the coast of South Africa. Fourteen years passed until the second was caught. Since 1952 over a dozen have been taken, all in waters near the coasts of South or East Africa. (*American Museum of Natural History.*)

Animals in the Carboniferous forests Living in the Carboniferous forest were amphibia and the primitive reptiles that evolved from them or their ancestors. The reptiles, as discussed in Chap. 6, developed the shelled egg, which freed them from an aquatic environment. They began an evolutionary diversification and became the dominant vertebrates for millions of years during the Mesozoic. Also present in these forests were early insects, including giant dragonflies with a wingspan of 30 centimeters (1 foot). The insects eventually came to be the dominant terrestrial arthropods during the Carboniferous. Cockroaches almost identical to those that plague us today were common in the Carboniferous. It is indeed probable that in terms of both *number of species* and *number of individuals* the insects were then and are still the *dominant life form*.

11-10 Fossil fuels were formed from incompletely decomposed organic matter

The climate of the Carboniferous is thought to have been warm and humid over much of the world. The trees of the Carboniferous forests grew in low-lying marshy areas. This sort of climate and this kind of habitat are perfect for the accumulation of partially decomposed plant materials. As the plant parts decomposed, they formed a brown, loosely compacted deposit of organic matter. If such material is covered by mud, decomposition is greatly reduced because of lack of oxygen. In a few places, such as Mud Lake, Florida, similar processes are occurring today; however, they are happening very slowly and in such a few places that very little fossil fuel is being formed today. Coal As it happened, during the Carboniferous, and in succeeding years, alternating layers of plant materials and inorganic sediments were laid down. The organic materials were buried deeper and deeper, and they were subjected to increasing pressure and increasing heat. This resulted in chemical changes in the plant materials. Water and volatile products such as oils, turpentines, and resins were given off. The eventual result was a dense, black rock which is combusti-

FIG. 11-16 The earliest land plants are psilophytes; four types are shown. (*After Z. Burian.*)

FIG. 11-17 Some ferns and fern relatives of Devonian times.

ble—coal. Two thirds of the coal deposits of the world were formed from plant materials of Carboniferous forests in both the Northern and Southern Hemispheres.

The degree of hardness of coal and the relative lack of water and volatile (readily vaporizing) substances depend upon the age of deposit and the amount of heat and pressure to which the plant remains were subjected. The most ancient coals and the Carboniferous coals are the hard coal called **anthracite.** Less valuable are more recent **bituminous coals** having a higher percentage of volatile substances (which cause greater air pollution), as well as water. The most recent coals (Cenozoic in age) are brown coals called **lignite** with a high percentage of water and volatile materials. The oldest coal deposits were found in Michigan in strata laid down in Precambrian times and are about 1.5 billion years old. Since vascular plants had not evolved in the Precambrian, this anthracite coal was formed from algae.

Oil and gas The origin of oil and natural gas is a matter of some controversy. Practically all geologists believe that petroleum originated in sedimentary rocks—compacted accumulations of mud, silt, and sand. The source material seems to have been marine ooze, a complex mixture of the remains of many organisms. Diatoms (single-celled algae surrounded by walls containing the element silicon) were probably the original synthesizers of the organic materials, but other organisms also may be a source of materials. It is thought that these materials were repeatedly worked over by the digestive tracts of many animals on the ocean bottom. They were finally buried under more ooze and then further altered by bacteria working in the absence of oxygen. The final transformation into oil and gas remains a mystery. Oil is often associated with saltwater deposits underground, suggesting its marine origin. Sometimes oil and coal are found in the same sedimentary sequences in the United States, but there seems to be no evidence that oil is a natural derivative of coal. No oil has been found in Pleistocene sediments (formed during the last million years or so), suggesting that no significant amount has been formed recently. Certainly it is not being formed at a rate anywhere near the present rate of *consumption*. The United States passed its peak of oil production in about 1971. Although the world has reserves that will last into the twenty-first century, the era of fossil fuel use by human beings is very brief in relation to our history as a species and to the length of time required for fossil fuel formation (Fig. 11-19). Our entire technological society has been made possible by the failure of the biosphere to recycle carbon stored in organic compounds. The energy in the bonds of these compounds became available to us when we developed the ability to mine coal and drill for oil.

11-11 Ferns and seed plants replaced earlier land plants

After their period of dominance in the Carboniferous, the club mosses and horsetails were found in greatly reduced numbers and variety,

FIG. 11-18 Artist's reconstruction of a Carboniferous forest: The tall trees are mainly extinct relatives of club mosses and horsetails; their remains make up much of the coal we use today. (*After Z. Burian.*)

and they were smaller. The true ferns and early seed plants appeared in the Mesozoic and became more abundant. Primitive ferns were small to medium-sized trees with very large, much divided leaves (Fig. 11-20). Impressions of these leaves are common fossils of the Mesozoic. One group of early seed plants also had large fernlike leaves closely resembling those of true ferns (Fig. 11-21). These plants, however, bore seeds, as well as pollen-producing structures, on their leaves, so we know that they were not ferns.

Plants living a terrestrial life face most of the same problems faced by terrestrial animals. As we have seen, gametes of plants and animals must meet in a fluid environment; for example, the amphibians must return to water for reproduction. Some vascular plants, however, such as club mosses, horsetails, and ferns, in order to reproduce, must wait for water to come to them and float their gametes together. Reptiles and their descendants, the birds and mammals, solved this

FIG. 11-19 Oil production for the United States and the world with projections into the future. Fossil fuels were not widely used until the second half of the nineteenth century. Because of the rapid consumption of these nonrenewable resources, the entire era of fossil fuel use will not last more than a few hundred years—a brief moment in time compared with even the time people have practiced agriculture.

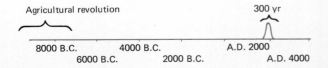

FIG. 11-20 A tree fern.

FIG. 11-21 A seed fern.

problem with internal fertilization. And, as we saw earlier, the seed plants evolved the pollen tube through which male gametes can move in a moist environment to the egg.

Many of the ferns became extinct, and many of the more specialized seed plants also became extinct. Some of the seed plants that were trees were ancestors of our present-day cone-bearing trees (conifers), such as pines, spruce, and redwoods. An unknown group of early seed plants was the ancestor of the first flowering plants.

11-12 Flowering plants are the most recent seed plants

It is thought that the flowering plants originated in upland areas where there was a diversity of habitats and the ancestral populations were divided up into a series of small, partially isolated subpopulations. Such a situation would be expected to lead to rapid evolutionary change. A rapid rate of evolution in upland habitats, where chances of fossilization are not good, may explain why the early record of the flowering plants is very poor.

Although they probably arose in Permo-Triassic times, flowering plants are not abundant in the fossil record until the Cretaceous period, 135 million years ago. When they do appear commonly, virtually all the basic types are found: grasses and daisies, oaks and willows, members of the rose family and the magnolia family. Judging from their modern counterparts, they were almost all (except for such plants as grasses, sedges, and some trees) pollinated by insects.

Coevolutionary interactions The evolution of the flowering plants was closely interrelated with the evolution of other organisms. In particular, the insects, birds, and mammals have long been involved in complex **coevolutionary interactions** with flowering plants. These interactions involve, in particular, three structures that make flowering plants successful. These three are (1) efficient leaves for photosynthesis, (2) pollen grains, and (3) seeds. There are interactions that occur between insects and leaves, between insects and pollen, and between insects and seeds, and there are similar relationships for birds and mammals. People also had coevolutionary interrelationships with the flowering plants as food crops and ornamental plants.

ANIMALS MOVE ONTO THE LAND

The Paleozoic seas were the home of the first plants and animals. In the course of time, group succeeded group and more and more specialized forms evolved. In the Silurian or Devonian, the plants first began to invade the land, changing the terrestrial environment and making it suitable for animals. The animals were not far behind.

11-13 The vertebrates evolved land-dwelling forms

Amphibians As we saw earlier, the lobe-finned fishes gave rise to the amphibians in the late Devonian, some 375 million years ago. The first amphibians probably spent most of their time in the water, for they were little more than four-legged fishes. In fact, there was probably little food for them on the land. Their teeth were simple, cone-shaped structures which were very similar to those of the lobe-finned fishes.

By Permian times much larger, more clearly terrestrial amphibians had appeared in the form of 2-meter-long (6-foot-long) creatures of rather massive proportions (Fig. 11-22). As far as we know, these amphibians, like their modern relatives—salamanders, toads, and frogs—did not lay shelled eggs. Presumably, they required water for breeding as do amphibians today. One group of amphibians was apparently ancestors of the reptiles of the early Carboniferous.

Reptiles In contrast to the amphibians, reptiles show many specializations for terrestrial life. Those with legs have legs that are longer and more efficient for walking than those of amphibians. All reptiles have more efficient lungs for breathing. Their hearts have a more efficient separation of oxygen-carrying blood from the lungs and that returning from the rest of the body. Reptiles have scaly skin which is resistant to drying, and finally, they have internal fertilization and shelled eggs. Most of these characteristics are not suitable for fossilization! It should not be surprising, therefore, that it is not an easy matter to determine whether fossil bones are those of a reptile or an amphibian.

FIG. 11-22 *Eryops,* an amphibian which lived in the Permian.

11-14 Reptiles became the dominant land vertebrates

In the Carboniferous, the earliest reptiles began an amazing evolutionary diversification. The present-day reptilian groups were established; a number of lines, which are now extinct, expanded greatly in the Mesozoic era; and the ancestors of the birds and mammals were produced. During the Mesozoic, the reptiles were the dominant land vertebrates, a position they held for *millions of years.* Indeed, the Mesozoic era is often referred to as the Age of Reptiles. In terms of the 24-hour clock, reptiles were dominant for an hour (3,600 seconds) while humans have been around for only one-half second!

An early unspecialized land-reptile group was ancestral to the dinosaurs, the flying reptiles, the crocodilians, and the birds. The early unspecialized reptiles at first appeared to have walked on four legs, but later forms had forelimbs much shorter than the hindlimbs. This indicates that they probably walked on their hindlimbs only, at least part of the time. Thus they foreshadowed many of the most famous of all prehistoric reptiles, the **dinosaurs.**

Dinosaurs Dinosaurs are sometimes called the "ruling reptiles" because of their spectacular development in the Mesozoic. Not all dinosaurs were giants, however; in addition to 25-meter (82-foot) monsters, there were dinosaurs the size of chickens.

Two main groups of dinosaurs are known. They differ in the way the bones of the hips are arranged. We may refer to these as *reptile-hipped* dinosaurs and *bird-hipped* dinosaurs (Fig. 11-23). The reptile-

FIG. 11-23 Diagram showing the difference between reptile-hipped (*a*) and bird-hipped dinosaurs (*b*).

hipped dinosaurs include both herbivores and carnivores. The early carnivorous reptile-hipped dinosaurs were small in size, but there was a trend toward increased size that resulted in the giant *Tyrannosaurus* (Fig. 11-24), which lived in the Cretaceous. This enormous beast was 15 meters (50 feet) long and stood 6 meters (20 feet) high on its huge hind legs. The forelimbs were tiny by comparison. The head was amply supplied with biting teeth. Also in the Cretaceous, as well as the Jurassic, were smaller carnivores similar in overall pattern that also walked on their hind legs. An adult *Tyrannosaurus* must have weighed as much as 10,000 kilograms (10 tons). By comparison, a large elephant weighs perhaps 5,000 kilograms. Impressive as

FIG. 11-24 An artist's reconstruction of *Tyrannosaurus*.

FIG. 11-25 An artist's reconstruction of *Brontosaurus*.

this seems, the herbivorous reptile-hipped dinosaurs were even larger.

Perhaps the most familiar of the reptile-hipped herbivores was *Brontosaurus* (Fig. 11-25). For one specimen, for which there is a complete skeleton, it is estimated that the live weight was 50,000 kilograms (55 tons). The skeleton measures nearly 25 meters (82 feet) in length, with a height of over 12 meters (40 feet). *Brontosaurus* walked on all four legs and probably spent most of its time in marshes and swamps. In contrast to those of *Tyrannosaurus*, the teeth of this herbivore were flattened, grinding teeth for dealing with the aquatic and subaquatic plants on which it fed.

The bird-hipped dinosaurs also included several other well-known forms. All were herbivorous; some walked on their hind legs, while others were four-legged. *Stegosaurus* (Fig. 11-26) lived in the Jurassic.

FIG. 11-26 An artist's reconstruction of *Stegosaurus*.

FIG. 11-27 An artist's reconstruction of *Triceratops*.

It had a double row of protective, bony plates from one end to the other, and the tail had four long, spikelike projections. The animal was 6 meters (20 feet) long, and the spikes, swung by a powerful tail, were undoubtedly protection against carnivorous dinosaurs, as were the protruding plates.

Also well protected were the bird-hipped horned dinosaurs, of which *Triceratops* is probably the most famous (Fig. 11-27). It had three large horns on a massive head and a large bony shield extending backward from the head over the neck. Even a hungry *Tyrannosaurus* must have had second thoughts when facing a defensive *Triceratops*. An earlier, smaller relative, *Protoceratops*, of this 9-meter (30-foot) herbivore reached a length of only 2 meters (7 feet). On an expedition to Mongolia, eggs of this ancestral form were found, along with skeletons of the dinosaurs (Fig. 11-28). Some of the eggs had cracked and the tiny bones of the unhatched offspring could be seen within. *Triceratops* and its relatives lived for a short time during the late Cretaceous.

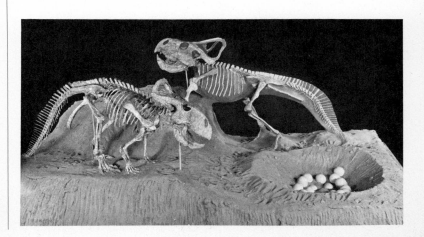

FIG. 11-28 Reconstruction of *Protoceratops* with eggs. (*American Museum of Natural History.*)

FIG. 11-29 An artist's reconstruction of *Pteranodon*.

Reptiles in the water and the air In addition to the dinosaurs, there were other groups of reptiles important in the Mesozoic but now extinct. First, there were several lines which reentered the water. Some were streamlined and fishlike in appearance; others had long necks. It is interesting to note that females of one of the aquatic forms, *Ichthyosaurus*, did not lay eggs, but, like some snakes today, retained the eggs within their body. The young were born alive when they reached the proper stage of development, as one specimen seemingly shows.

In Jurassic times the flying reptiles appeared (Fig. 11-29). These had hollow, lightweight skeletons and were called **pterosaurs.** Early forms were small; but one type, common in the Cretaceous, had a wingspread of 9 meters (30 feet). Wings in the flying reptiles were membranous and supported largely by the greatly elongated fourth finger of each forelimb. Recently, fossils of the largest known flying animal were found in Texas. From study of three partial skeletons, this pterosaur is thought to have had a wingspread of 15.5 meters (over 50 feet)! This is considerably greater than that of many light airplanes. It is not known exactly what the way of life of this creature was, but there are suggestions it was a carrion-feeder with toothless jaws a meter long. Thus it would be analogous to a vulture or condor in behavior. The condor is a bird with a wingspread of 2.9 meters (9.5 feet). The greatest wingspread in birds is found in some albatrosses, which may exceed 3.5 meters (11 feet). The flying reptiles, like the dinosaurs, became extinct at the end of the Mesozoic.

11-15 Reptiles gave rise to birds and mammals

Birds The first fossil bird was found in Jurassic rocks. This creature, named *Archaeopteryx* (Fig. 11-30), was distinctly reptilian in its anatomy. For example, it had a hip structure close to that of the bird-hipped dinosaurs. Fortunately, impressions of feathers were found with the skeleton so that it was clear that a bird, and not a reptile, had been found. *Archaeopteryx* had a long, bony, feathered tail and

FIG. 11-30 An artist's reconstruction of *Archaeopteryx*.

a tapered beak with teeth. Judging from its anatomy, this bird probably could not fly as well as modern birds, which did not appear until the Cretaceous.

Mammals At the time the unspecialized reptiles were giving rise to the ruling reptiles—the dinosaurs—they also gave rise to several other evolutionary lines. Among these was the line leading to the mammals. The mammals became the dominant terrestrial vertebrates during the Cenozoic era, which is often called the Age of Mammals. When they first appeared in the Late Triassic, however, the mammals were small, insignificant, and are rarely found as fossils. Only after the dinosaurs became extinct at the close of the Cretaceous did the mammals begin their great evolutionary elaboration.

There is no general agreement as to why the dinosaurs became extinct. Many biologists believe climatic change may have been responsible. No theory, however, seems adequate to explain the relatively sudden disappearance of these massive reptiles. Some think it could have been disease or competition from the brighter, more agile mammals that were able to maintain a high body temperature and exploit a nocturnal life that was probably too cold for reptiles. Some biologists believe that dinosaurs were warm-blooded, however.

Mammals show many advances over reptiles. The legs, instead of extending sidewise from the body, as in most reptiles and amphibians, are rotated so that they extend downward. This makes for more efficient movement and also helps control of body temperature since the body is held up off the ground. Internal means of maintaining a warm body temperature evolved, along with hair, which serves to insulate the body better than scales. Regulation of a constant body temperature makes it possible for mammals to be active at night when it is cool or cold, as well as to occupy habitats at higher elevations and latitudes than is possible for reptiles.

There are many important anatomical differences between reptiles and mammals; for example, in mammals, the heart is four-chambered and the lung circulation and that of the rest of the body are completely separate. As we have seen earlier, except for two species, mammals do not lay eggs. The young are kept in the uterus, nourished by a placental attachment to the mother until birth. After birth they are nourished by milk from mammary glands of the mother. Finally, the size of the brain is relatively larger in mammals and the "thinking portion," the forebrain, is much more highly developed.

THE TRIUMPH OF THE MAMMALS, BIRDS, ARTHROPODS, AND FLOWERING PLANTS

As we have seen, no one knows why the ruling reptiles became extinct at the close of the Cretaceous. The dinosaurs ruled for some 135 million years, and with their leaving, many habitats were open to the mammals, who spent some 40 million years diversifying to fill them. At the same time, the flowering plants virtually replaced the older seed plants. The insects and flowering plants became involved in a series of striking coevolutionary relationships, as did the mammals and the flowering plants.

11-16 The mammals became the dominant land vertebrates

Differentiated teeth One of the features of mammals that distinguishes them from reptiles is the differentiation and specialization of their teeth. Reptiles generally have all their teeth alike: conical, pointed, biting teeth in the flesh eaters and flattened, grinding teeth in the herbivores (see Fig. 6-20). Most mammals have teeth specialized for several different functions, as a glance at your own teeth in a mirror will show (see Fig. 6-20). Toward the front of the mouth are vertically flattened or pointed biting teeth, the **incisors** and **canines.** In the rear are the **molars,** which are flattened horizontally so as to have a combination of grinding and shearing effects. The molars of many carnivores, such as dogs and cats, work like the blades of a scissors, slicing up food into smaller pieces that can be swallowed (Fig. 11-31). In a real sense, the history of evolution of the mammals is the story of the evolution of teeth—especially since teeth are the commonest mammalian fossils and are a good indication of what kind of food the creature ate.

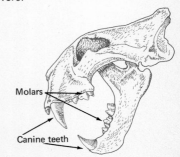

FIG. 11-31 Skull of a cat showing the differentiated teeth found in this carnivore.

FIG. 11-32 Artist's reconstruction of giant mammals. (*a*) A wooly mammoth (*Mammuthus*) was over 4 meters (14 feet) tall and was a contemporary of human beings who included the mammoth, along with other extinct mammals, in their cave paintings; (*b*) *Megatherium* was a 6-meter-tall (20-foot-tall) ground sloth; (*c*) the wooly rhinoceros (*Opsiceros*) was a contemporary of *Mammuthus* and was 2 meters (7 feet) tall.

Gigantic species There are many different groups of mammals and there is an interesting paleontologic record for nearly all these groups. Here it is important to note that, in many of these groups, there was a tendency toward **gigantism**—that is, to produce animals of very large size (Fig. 11-32). Some of these large, slow-moving forms were fed upon by highly specialized carnivores, e.g., the saber-toothed cats (Fig. 11-33). As we have seen, their most ruthless predator, however, appears to have been early *Homo sapiens*. When people became common in North America, Europe, and Africa, most of the giant forms were killed off. With the changing climate of the Cenozoic and the glaciations which occurred toward its end, the familiar mammals which we know today developed (see Chap. 5).

11-17 The evolution of mammals, birds, flowering plants, and insects is closely interrelated

One of the important evolutionary changes in mammals concerned the herbivores. The early herbivores presumably lived in the forest or savanna biomes. They were **browsers,** feeding upon leaves, buds, and small branches—especially of shrubs and trees. Their teeth were a mixture of biting and grinding teeth. The relatively recent grassland biome was difficult for mammals to invade because the creeping underground stems of grasses are difficult to get at to eat, and their leaves and stems contain silica, a glasslike material which acts like sandpaper on the teeth of herbivores. Nevertheless, in response to this rich new source of food, many mammals changed from browsers to **grazers,** feeding upon low-growing plants. Their teeth became specialized for eating grasses without being ground down to the gums. Indeed, many of them took advantage of grasses as ideal tooth-sharpeners. Browsers and grazers often have teeth that grow continuously or are replaced when they wear down (Fig. 11-34).

Seeds and seed eaters As discussed in Chap. 6, flowering plants produce seeds consisting of an embryo plant surrounded by food reserves for its growth upon germination. Both the mammals and insects were quick (in an evolutionary sense) to identify these little pellets of energy as excellent sources of food for themselves. Seeds have the advantage, as food for herbivores, of being small, easily transported, and concentrated high-energy foods. In addition, they keep well when stored. Of course, seed-eating birds also evolved to utilize this evolutionarily relatively new commodity.

Flowering plants evolved a number of techniques for countering

FIG. 11-33 A saber-toothed cat: many catlike forms with long, saberlike teeth preyed upon the giant herbivores. There was even a saber-toothed marsupial whose teeth were partially enclosed by fleshy sheaths growing from its lower lips.

the attacks of seed eaters. Seeds with very tough seed coats were evolved, as well as seeds with poisons stored along with the food. Many mammals, birds, and insects developed means of getting around these defenses. Some flowering plants evolved nutritious fruits of a color attractive to mammals and birds. They produced seeds which were resistant to digestion by vertebrates. In the course of eating the fruits, the vertebrates dispersed the seeds of the plants, which passed through their digestive tracts. Evolution of fruits made possible the development of a new way of life for many herbivores.

Flowering plants and pollinating insects A spectacular coevolutionary interaction, or set of interactions, went on between the insects and the flowering plants. Most of the flowering plants are insect-pollinated, attracting insects by their color and fragrance, and often providing a carbohydrate and amino acid–containing nectar as insect food. The story of the evolution of the flower is generally agreed upon. Early in the evolution there were large, open flowers pollinated by beetles. More advanced flowers have the nectar enclosed and protected. They are pollinated by bees, flies, butterflies, and moths.

The larvae of many beetles and flies and most butterflies and moths are herbivores feeding on flowering plants. In many groups of flowering plants, mechanical (hairs and spines) and chemical (poisons) defenses have evolved. And certain groups of insects have evolved the means to get around these defenses and to specialize in feeding on those plants protected from most other herbivores.

HUMAN EVOLUTION

In forests made up of flowering plants, mammals which would eventually give rise to *Homo sapiens* developed. Indeed, living in the treetops and feeding on buds, seeds, and fruits was an important step in our evolution.

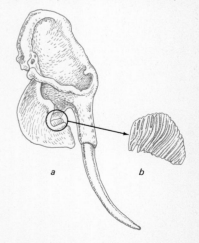

FIG. 11-34 Teeth of a browsing animal such as this elephant (*a*) are usually specialized for grinding grasses and other plants and for resisting being worn down; elephant teeth (*b*), seen in top view, do become worn down and are replaced by new teeth as necessary.

FIG. 11-35 These four primates, often called the great apes, are the nearest living relatives of *Homo sapiens*: (*a*) chimpanzee, (*b*) gorilla, (*c*) orangutan, (*d*) gibbon. [(*a*) *American Museum of Natural History;* (*b*) and (*c*) *Barry Edmonds, Image;* (*d*) *Leonard Rue Enterprises.*]

Our nearest living relatives are the great apes, the orangutan, gibbon, chimpanzee, and gorilla (Fig. 11-35). Human beings, chimpanzees, and gorillas are primarily land-dwelling animals. Orangutans spend much time in the trees, and the gibbons spend essentially all their life above ground level. But all these organisms are descended from ancestors who lived in trees.

11-18 Many of our most important characteristics can be traced to our tree-dwelling ancestors

It appears that about 25 million years ago our early ancestors, small insect-eating mammals, moved from the ground into the trees. A whole series of changes evolved which specialized our ancestors for living in trees. Their sense of smell would have been less important than their sight. Odors are quickly dispersed in the breezy treetops, and it is difficult to smell the position of a branch you are going to leap to! Vision, on the other hand, would have become more important than it was for animals living on the dark, odor-rich forest floor. Forelimbs modified for grasping could be used both to hold on to branches and to manipulate objects. Apparently, forelimbs could now replace the snout as a manipulator. This, and the reduction in need for a keen sense of smell, permitted a reduction in snout length (Fig. 11-36). In turn, shortening of the snout made possible an evolutionary

FIG. 11-36 A tarsier (*a*) and a lemur (*b*). The lemur has a longer snout since the lemur line of descent emphasized acute sense of smell. The tarsier line emphasized binocular vision. (*A. W. Ambler, National Audubon Society.*)

a b

change in the position of the eyes. Successive generations had eyes closer and closer to the front of the head.

Eyes on the front of the head provided **binocular vision,** which greatly improves judgment of distance. (Try to bring two pencil points together a foot or so in front of your nose with one eye closed.) Good distance judgment is very convenient, to say the least, for leaping from limb to limb. Tree dwelling also brought about increased development of the centers of the brain where coordination occurs. If you judge the distance correctly and then leap too far, you are out of luck. Of course, all these changes took place as a result of natural selection over many millions of years. Individuals that happened to have genotypes with the eyes a little more toward the front, or that had a slightly higher innate ability to coordinate leaps, tended to leave more offspring than their less fortunate companions.

Year-round sexuality and single births Perhaps our greatest legacy from the time our ancestors spent in the trees is in our reproductive behavior. Infant care is a difficult problem in treetops, and the production of offspring one at a time perhaps was an advantage for our ancestors. In fact, one is tempted to interpret the old nursery rhyme about "rock-a-bye baby in a treetop" in the light of our ancestry—especially the part about "when the bough breaks, the baby will fall." Caring for an infant is difficult enough on the ground, let alone in a treetop. There were advantages, however, to living in the tropical forest. Food was available the year round since one or another tree species was always in bloom or producing fruit. Therefore, there seems to have been no selective advantage for infants to be produced at any particular season of the year. In fact, it seems likely that the first steps toward year-round sexuality in human beings developed during this period.

11-19 The descent from the trees was another important step in human evolution

Civilization required dense populations and raw materials. It seems unlikely that a complex civilization could ever have developed in the trees. The food supply in the treetops is too sparse to support a dense population at one place indefinitely. Furthermore, the raw materials available for a civilization in the trees are too limited.

We do not know why our ancestors descended from the trees; perhaps a climatic change reduced the extent of the forests. Our newly terrestrial ancestors may have lived much as modern baboons do. Baboons occur in troops or bands dominated by powerful males (Fig. 11-37) and feed on roots and berries, and occasionally hunt dead or new-born animals. They spend the night in trees for safety. We may presume that our ancestors began to hunt more actively than do baboons today. Individuals who could run swiftly on their hind legs were at an advantage, especially since their grasping hands could hold sticks or stones to use as weapons.

THE HISTORY OF LIFE 341

FIG. 11-37 A troop of baboons in Nairobi, Africa. Males are seen along the edges of the road, while females and juveniles are in the center of the group. (*Irven DeVore, Anthro-Photo.*)

11-20 The first human fossils we know much about were of upright people with small brains

Australopithecus Except for some jawbones about 15 million years old, we have no significant fossil record of our early evolution. The first of our ancestors of which we have abundant fossil material is an individual called *Australopithecus africanus* who lived in Africa 2 million to 5 million years ago. *Australopithecus* walked fully upright (Fig. 11-38). The brain size of *Australopithecus*, however, was only less than half that of modern people—about the size of that of a gorilla. *Australopithecus* was small, only a little more than a meter (4 feet) tall. Therefore the *ratio* of brain size to body size was considerably higher than that of a gorilla.

Australopithecus had another feature in common with modern people: the teeth were virtually the same. Look at Fig. 11-39, which shows the jaws of a gorilla, *Australopithecus*, and *Homo sapiens*. Notice the large, flat surfaces of the gorilla's teeth, suitable for grinding its coarse vegetable food. *Australopithecus*, however, had teeth which are suitable for eating meat as well as plants; and large numbers of animal bones are often found in strata with the remains of *Australopithecus*. The teeth of fossils of a slightly larger relative of *Australopithecus* suggest that it was apparently a vegetarian.

Homo erectus *Homo erectus* is intermediate in most characteristics between *Australopithecus* and *Homo sapiens*. *Homo erectus* is the scientific name given to a group of individuals who first appeared about 1 million years ago; sometimes they are referred to by the place they were found, for example, Java man and Peking man. Fossils of *Homo erectus* were first found in Asia, but additional fossils have been found in Africa. *Homo erectus* stood fully upright, was about a meter and a half (5 feet) tall, and had a brain capacity larger than that of *Australopithecus* but smaller than that of *Homo sapiens*. *Homo erectus* used tools and fire, ate meat, and had a culture. That is, we can infer from their tools, and remains of their huts and

FIG. 11-38 An artist's reconstruction of *Australopithecus*.

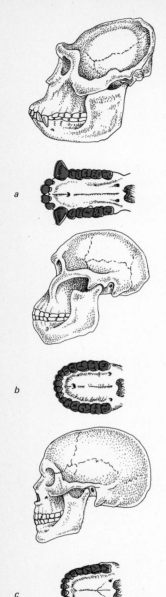

FIG. 11-39 Skulls of (*a*) a gorilla (*b*) *Australopithecus* and (*c*) *Homo sapiens*.

clothing that they had a large body of nongenetic information which was transferred among individuals in their populations. On the basis of the structure of their skulls and the presence of culture, it is thought that *Homo erectus* had a well-developed language, for language is the means of the transmission of culture. There is evidence that occasionally their culture included cannibalism.

11-21 Homo sapiens first appears in the fossil record about 100,000 years ago

Neanderthals The earliest modern people whose remains have been found in sizable numbers were somewhat shorter, more muscular, and had coarser features on the average than people living today (Fig. 11-40). They were, however, well within the range of variation found today. These people are known as Neanderthals, after the German for "Neander Valley," where their first fossils were found. Remains of Neanderthal people found in 1856 were the first human fossils to be recognized as such. Since then, abundant Neanderthal remains have been discovered all over Eurasia (and possibly Africa). Neanderthals were "human" in every respect. They made fine tools, wore clothes, were talented hunters, and buried their dead in special graveyards accompanied by provisions and flowers, suggesting that they had religious ideas, such as, perhaps, the concept of an afterlife.

Cro-Magnon people About 35,000 years ago, the Neanderthals disappeared, as a group, in a relatively short period of time. They were replaced in the fossil record by populations of people essentially identical to modern people. The reason for the disappearance of the Neanderthals is unknown. Perhaps they were exterminated by the new people, perhaps they interbred with them, or, more likely, perhaps both occurred.

One scientist has proposed that the disappearance of the Neanderthals may have resulted indirectly from their practice of cannibalism. Neanderthal burial sites often reveal skulls that had been opened to remove the contents. Modern tribes which consume the brains of their dead as part of their burial rituals may have a virus disease called *kuru* which is believed to be transmitted in this fashion. The kuru virus leads to a complete and fatal degeneration of the central nervous system; however, it may take many years after infection before the disease becomes apparent. Prevalence of a similar disease along with increasing cannibalistic rituals could have led to the end of the Neanderthals, as it is presently leading to the extinction of the Fore peoples in New Guinea.

The peoples that replaced the Neanderthals are usually called **Cro-Magnons,** named after a cave in France where their remains were first discovered. We have more than skeletons and tools to tell us about Cro-Magnon people. There are remains of body ornaments and, above all, magnificent paintings which they left on cave walls (Fig.

11-41). The grace and style of these paintings give us every reason to believe that human artistic talents were as fully developed tens of thousands of years ago as they are today.

These paintings were mostly done deep in caves, far from where the people lived their everyday lives. With rare exceptions, they did not portray people but rather the animals that were hunted for food. For these reasons, it is thought they may not have been made for decorative effect. Rather they may have been part of a religious ritual designed to assure an abundance of game or good luck in the hunt. This idea is supported by the existence of animal painting and sculptures which show the marks of spear thrusts. However, some anthropologists feel, because of the spear marks, that they were practice targets for young apprentice hunters.

11-22 Brain and culture developed together

At one time it was thought that human beings evolved a large brain and then "invented" culture. Now we know that this was not the case. Relatively small-brained people had culture. It seems clear that the enlargement of the brain and the enrichment of culture went on in parallel. People with the largest brains could take the greatest advantage of cultural information and were thus at a selective advantage. Over thousands of generations of selection, their brain size gradually increased. Larger-brained people could develop more complex cultures. This, in turn, gave them further selective advantage. Thus culture promoted larger brains and larger brains promoted culture.

The large brain of *Homo sapiens* grows to 80 percent of its adult size in the first three years after birth. If the brain grew to its full child size before birth, women's hips would have to be enormously wide in order to permit children to be born. During the period of brain growth after birth, the human infant is extremely helpless. It requires almost constant attention from the mother if it is to mature successfully. The presence of the male is an enormous help in providing food, shelter, and defense for the mother and child. This gives a strong selective advantage to the intense year-round sexuality which distinguishes people from virtually all other mammals and which plays an important role in all human cultures.

It might be tempting to conclude, on considering the vast panorama of evolutionary history, that a large-brained, cultured ape was the end or goal of evolution. It would be equally logical for a housefly, a member of a more abundant and equally widespread species, to conclude that *it* was the goal of evolution. Certainly a cockroach, whose species has been around a hundred times as long as ours, could

FIG. 11-40 An artist's representation of a Neanderthal.

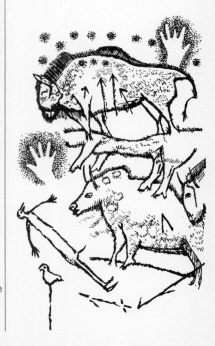

FIG. 11-41 Artist's sketches of paintings done by Cro-Magnon peoples on the walls of different caves in France. The bison painting at the top was found on the wall of a cave at Lascaux, as was the male figure and bird on the lower left. The lower bison is painted on a wall of Niaux cave. The hand prints were found at Pech Merle cave.

see itself as a great evolutionary success. Just as our intelligence has led some people to conclude that we are the goal of evolution, it has led other people to conclude that people, flies, and cockroaches are just different ways DNA has evolved for making more DNA. These are philosophical, not scientific, matters. There is no scientific way to frame a hypothesis about the goal of evolution or to test the hypotheses proposed above (see Chap. 17). How we choose to view ourselves in the scheme of evolution is a social, not a scientific, issue. In this chapter we have seen that for the most part life evolves *away* from the forms of the past, but we do not know whether it will evolve *toward* any particular form in the future.

QUESTIONS FOR REVIEW

1 Why do biologists believe that photosynthesis was taking place over 3 billion (3,000 million) years ago?

2 What important change in the earth's atmosphere resulted from activity of photosynthetic organisms, and why did the invasion of the land by both plants and animals depend on the prior occurrence of this atmospheric change?

3 What do these terms mean: paleontologist, Paleozoic, Mezozoic, Cenozoic, fossil, anthracite, coevolution, chemical fossil, lignite, petrified, bituminous coal?

4 Our ancestors' life in the trees favored—that is, selected for—year-round sexuality and single births. One of these two characteristics may be thought of as the result of an *increase* in a particular selection pressure; the other may be thought of as resulting from a *decrease* in a particular selection pressure. What were the two changing selection pressures, and which of the two just-mentioned ancestral characteristics did each help bring about?

5 Why did the development of civilization necessarily follow, not precede, our ancestors' descent from the trees?

6 Organic evolution is a major field of biological study. Why, then, is the question of what is or may be evolution's goals of no concern in biology?

7 Can you rearrange the following in order of their evolutionary appearance: reptiles, lobe-finned fishes, insectivores, culture, trilobites, mammals, Neanderthal people, Cro-Magnon people?

READINGS

Constable, G., et al.: *The Neanderthals,* Time-Life Books, Time Inc., New York, 1973.

The First Men, Time-Life Books, Time Inc., New York, 1973.

Hedgpeth, J. W.: "Animal Diversity: Organisms," *Biocore,* Unit XIV, McGraw-Hill, New York, 1974.

Howells, W. W.: "*Homo erectus,*" Scientific American **215**(5):46–53, offprint 630, 1966.

Laughlin, W., and R. H. Osborne (eds.): *Human Variation and Origins, Readings from Scientific American,* Freeman, San Francisco, 1967.

Pilbeam, D.: *The Evolution of Man,* Funk & Wagnalls, New York, 1970.

Rickson, F. R.: "Plant Diversity: Organisms," *Biocore,* Unit XV, McGraw-Hill, New York, 1974.

Spinar, Z. V.: *Life before Man,* McGraw-Hill, New York, 1972.

12

OBTAINING ENERGY

SOME LEARNING OBJECTIVES

After you have studied this chapter, you should be able to

1 Demonstrate your knowledge of certain energy-related cell structures and chemical compounds by
   a Naming the sites where energy is captured or released in usable form in prokaryotic cells, in plants, and in animals.
   b Naming the four major groups of organic compounds in cells and listing the main chemical elements involved in the makeup of each.
   c Stating which of the four major groups of organic compounds are primarily employed in the storage of food by (1) plants and (2) animals.
   d Naming and describing each of the three processes by which materials enter and leave cells.

2 Show your understanding of the photosynthetic process by
   a Writing the general equation for photosynthesis.
   b Describing what happens in the light reaction stage, listing inputs and outputs.
   c Describing what happens in the dark reaction stage and listing its inputs and outputs.

3 Demonstrate your understanding of the basic processes of cellular respiration by
   a Listing the inputs and outputs of cellular respiration.
   b Stating the role of the *energy*

currency (ATP) in cellular respiration and describing how ATP is formed in the electron transport chain.
   c Naming two types of cellular respiration (one using oxygen, one not using oxygen), stating which type is more efficient, explaining how glucose is converted into useful energy in each type, and naming the "waste products" of each type.
   d Naming the three types of fuel (organic compounds) employed in cellular respiration, and stating how they compare in terms of energy yielded.

4 Name the three basic modes of nutrition (ways organisms obtain their energy) and state which general groups of organisms employ which mode.

5 Show your knowledge of human digestive structures and processes by
   a Naming four specialized regions of the human digestive tract.
   b Stating which kinds of foods are digested in each region, and describing the processes which take place in each region.

6 Demonstrate your understanding of the principal functions and processes of the human blood vascular system by
   a Naming three principal components of the system.
   b Naming the special oxygen-carrying molecule in red blood cells and indicating its role in oxygen transport.
   c Tracing the route of blood through the system and describing the role it plays in transporting two main substances (name them) *to* the body cells and carrying one important substance (name it) *from* the body cells.

7 Show your knowledge of the human lymphatic system by
   a Stating where the lymphatic fluid is found in the body, how it is circulated, and the route it follows back into the blood vascular system.
   b Listing three main functions of the lymphatic system.

8 Demonstrate your understanding of the basic principles of human nutrition by
   a Naming three major chemical components of the human body, saying which one is the body's main structural material and thus

essential for growth and repair, and saying which two are mainly sources of energy for growth, repair, and other metabolic activity.
b Stating the main nutritional contributions of carbohydrates and fats, listing some common food sources of each, and describing the serious consequences of improper amounts of each in the diet.
c Naming the vitamins discussed in the text. Indicate which are fat-soluble and which are water-soluble, and state one important practical implication of this difference. List also the main dietary sources of each vitamin and describe the possible results of ingesting too little (deficiency) or too much (poisoning) of certain vitamins as discussed in the text.
d Stating why the body needs each of these essential minerals—calcium, phosphorus, iron, sulfur, salt (sodium chloride), potassium, iodine—and naming a food rich in each. Also name four of the essential *trace elements*.
e Naming the four basic food groups and stating which essential nutritional materials are obtained from each.

We have now discussed human ecosystems—cities and farms—and considered their relationship to other ecosystems and the major biomes of the earth. To put our species in perspective, we looked at the place of the earth in the solar system and the origin and evolution of life on earth. We share this planet with millions of different kinds of organisms. Because of their evolutionary interrelationships, these organisms have in common many fundamental features of organization and functioning. We will now consider some of these shared properties of organisms and will begin with a consideration of energy, as we did in Chap. 1.

Energy was discussed in the earliest chapters of this book because of its basic importance. All living things must have energy in order to carry out their life processes. Energy is necessary for growth; it is required for movement and to maintain the organism. And, without energy, reproduction could not take place. The problem of getting and using energy is therefore a critical problem of life.

ENERGY AND CELL STRUCTURE

The chemical reactions of living things that produce or use energy all take place in cells. Therefore, before we discuss these reactions, it will be helpful to understand the structure of cells in greater detail.

12-1 All cells have complex internal structures

Prokaryotic cells In Chap. 6, we studied the structure of cells. Figure 6-2 shows the structure of a bacterium, a typical prokaryotic cell. This type of cell lacks a nucleus and its genetic material is not separated from the rest of the cell. The energy reactions of the cell are carried out by membrane systems that are not organized into organelles.

Eukaryotic cells Figures 6-1 and 6-3 show the structures of typical animal and plant cells, respectively. These are eukaryotic cells, and they are more complex in structure (Fig. 12-1) than prokaryotic cells (see Fig. 6-2). They have a nucleus which is surrounded by a nuclear envelope made up of two membranes. The chromosomes are in the nucleus, and materials can enter and leave the nucleus through many

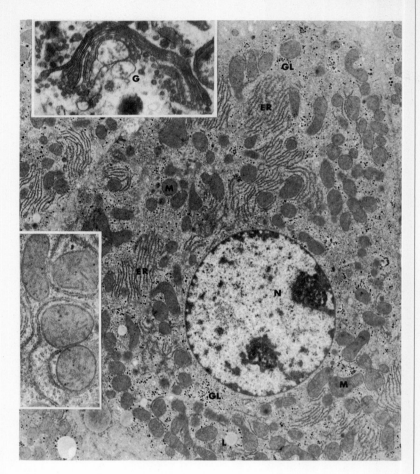

FIG. 12-1 A liver cell that stores glycogen demonstrates many of the specialized organelles found in animal cells, including a nucleus (N), mitochondria (M), and endoplasmic reticulum (ER). L indicates an enzyme-containing organelle (lysosome), GL regions of glycogen storage. (Magnification ×11,500.) One insert shows an active Golgi complex (G) as seen in a cell which will develop into a sperm. (Magnification ×46,000.) The other insert shows mitochondria and endoplasmic reticulum. (Magnification ×33,000). (*Courtesy of J. Belton and E. Lyke.*)

small pores in the nuclear envelope. The portion of the cell outside the nucleus is called the cytoplasm, and many different kinds of organelles are found there. Some are involved in synthesis, putting compounds together, and are discussed in Sec. 13-3; others are involved in energy reactions. Mitochondria and chloroplasts are sites of energy reactions.

The cytoplasm of both plant and animals cells contains many short, cylindrical or rod-shaped organelles called **mitochondria** (singular, **mitochondrion**). Mitochondria are the powerhouses of cells in which energy is made available to carry on the functioning of the cell. Electron microscope studies have shown that each mitochondrion is made up of an inner membrane and an outer membrane. As you can see in Fig. 12-2, the inner membrane of a mitochondrion is folded into a series of projections which extend into the cavity of the organelle. Imagine trying to get a large paper bag inside a shoe

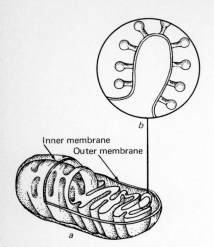

FIG. 12-2 (*a*) Diagram of a mitochondrion showing the inner and outer membranes. (*b*) Enlargement of inner membrane showing particles that may be important in energy transfer.

box. One way of doing this would be to fold it just as the inner membrane of a mitochondrion is folded.

The cells of plants which carry on photosynthesis contain organelles called **chloroplasts** (Fig. 12-3). Chloroplasts consist of a great many flattened, closed sacs of membrane arranged in a very complex manner. These are **grana**. Chlorophylls associated with the membranes absorb light energy. The structure of chloroplasts aids transformation of light energy into chemical energy (photosynthesis). It is in chloroplasts, then, that the basic chemical reactions necessary for practically all life on the planet take place. Many plant diseases and synthetic chemicals adversely affect the ability of chloroplasts to photosynthesize. Lead, DDT, sulfur dioxide, nitrous oxides, and ozone (O_3) have all been implicated in reducing photosynthesis in phytoplankton, with the potential, as we noted in Sec. 3-10, of changing the relative abundance of species.

12-2 The chemical structure of cells is important in their energy reactions

The four major groups of organic compounds in cells At an even more microscopic level than organelles, cells have a great deal of internal structure; in fact, their organelles themselves have complex arrangements of organic compounds. There are four main kinds of such compounds: nucleic acids, proteins, carbohydrates, and lipids. Nucleic acids were discussed in Chap. 9. **Proteins** are very large molecules. They are made up primarily of the elements carbon, hydrogen, oxygen, nitrogen, and sulfur. Hair is largely protein, as are most "meats"—the muscles of animals that you eat. **Enzymes,** the

FIG. 12-3 A thin section of a chloroplast as seen with the electron microscope. The stacks of membranes in regular arrays are grana. (Magnification ×35,000.) (*Courtesy of Dr. David Bishop, C.S.I.R.O. Division of Food Research.*)

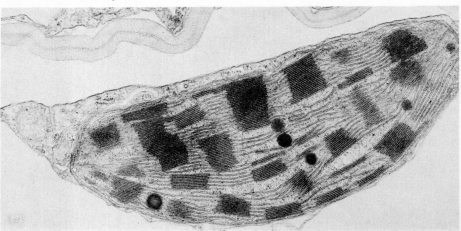

organic catalysts (Chap. 13) that are necessary for the chemical reactions of living systems, are proteins. Proteins are made up of a long chain of subunits called **amino acid residues** (Fig. 12-4). When amino acids become joined, water molecules are released and the remainder of each amino acid is called a residue. The cell membrane and all the internal membranes of a cell and its organelles contain protein.

A second important component of membranes are **lipids.** Lipids are a class of organic molecules which do not readily dissolve in water. They are compounds composed mostly of carbon, hydrogen, and oxygen in which the ratio of hydrogen to oxygen is much greater than 2 to 1. Fats and oils are examples of lipids, as is butter. Lipid molecules have a structure which permits them to associate with protein molecules in a particular way, and lipids and proteins make up the membranes of cells.

Carbohydrates are organic molecules containing carbon, hydrogen, and oxygen. The hydrogen and oxygen are in the ratio of two hydrogen atoms for every oxygen atom, just as they are in water. In fact, the term *carbohydrate* can be loosely translated "watered carbon." Sugars are carbohydrates; a common sugar is the one called glucose. Common table sugar, called sucrose, is made up of one glucose subunit and another subunit of the sugar fructose (Fig. 12-5). When carbohydrates are taken apart, their sugar subunits are called residues just as are the amino acid subunits of proteins. In Chap. 6, you learned that plant cells have a cell wall surrounding their cell membrane. This cell wall is made up in part of a carbohydrate called **cellulose.** Cellulose molecules are very large, being made up of a great many glucose residues attached together in long chains. Such long chains of molecules are called **polymers.** These long-chain polymers are woven together to form the tough cellulose cell walls of plants (Fig. 12-6).

Energy for metabolism **Metabolism** is the sum total of all the chemical processes used by organisms to maintain life and growth. Glucose, one of the carbohydrate molecules manufactured in photo-

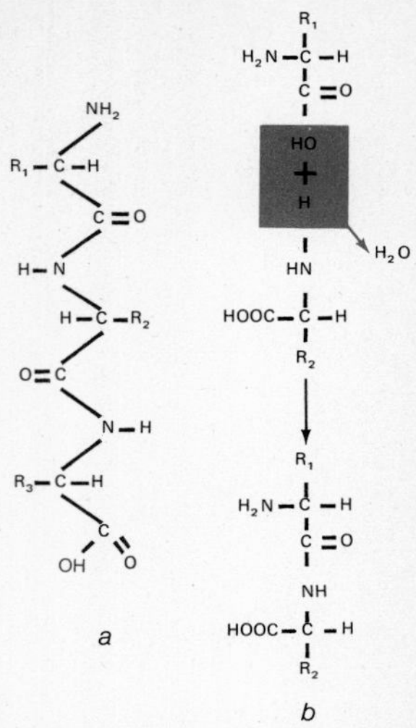

FIG. 12-4 Proteins are made up of long chains of amino acids joined together by peptide bonds. (*a*) A tripeptide. (*b*) Formation of a peptide bond with the loss of a molecule of water. R can stand for any of some 20 different chemical groups which give amino acids their individual properties.

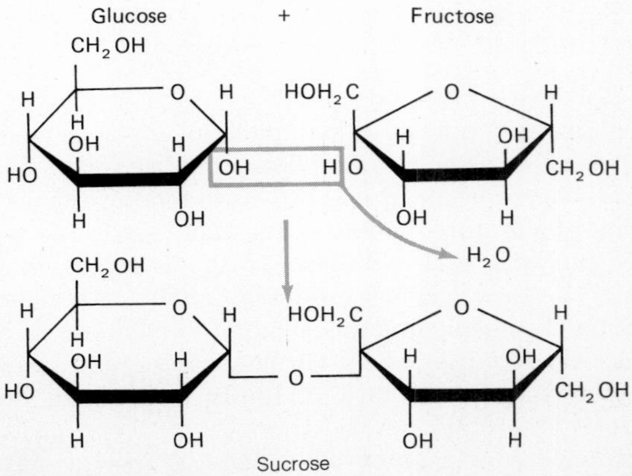

FIG. 12-5 Sucrose, common table sugar, is made up of a glucose subunit and a fructose subunit.

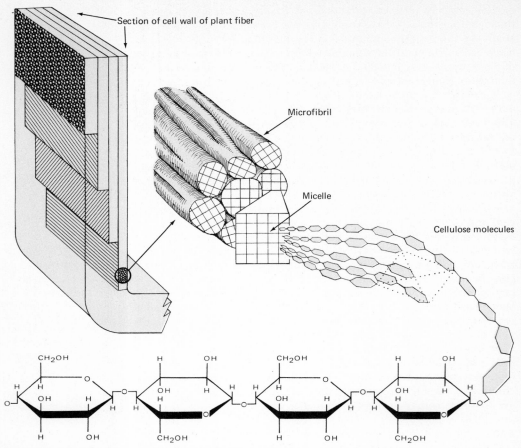

Glucose units joined by oxygens to make long-chain cellulose molecules

FIG. 12-6 The cellulose cell wall of plants is made up of cellulose subunits joined in long chains. The chains are joined into *micelles* and the micelles into *microfibrils*. The microfibrils are commonly arranged spirally. Different layers of the wall may have spirals running in different directions.

synthesis, plays an important role in metabolism as an energy source. Most organisms, however, store food as molecules other than glucose. Many plants—for example, potatoes—store food as the carbohydrate **starch.** Other plants store food as lipids; many seeds and nuts contain stored fats and oils. Like cellulose, starch is made up of glucose residues; starch, however, has branching chains. Animals store carbohydrates as **glycogen,** sometimes called animal starch because its structure is very like that of starch molecules found in plants. Animals also store food in the form of lipids. Both plants and animals can store food as protein.

Stored food contains potential energy. When an organism eats another organism, however, it must be able to digest the molecules present or it cannot utilize the stored energy. Most animals can digest starch, glycogen, lipid, and protein; however, only some animals can digest cellulose, and they must have help from unicellular organisms that live in their digestive tracts. In fact, most herbivores have microorganisms that produce cellulase, an enzyme which helps break down cellulose. Each generation of herbivores must be "infected" anew with these beneficial microorganisms.

The chemical structure of cell membranes In order for cells to carry out metabolism, a great many things must enter and leave across cell membranes. Figure 5-6 shows how molecules move around in fluids and gases until they equalize their concentrations. The process of diffusion occurs continuously in the fluids around and inside cells. Osmosis occurs across cell membranes since they allow some molecules to pass but not others. Very large molecules such as proteins cannot pass through cell membranes readily, while small molecules such as water and minerals get through easily. For example, water and minerals in the soil enter a green plant through root hairs, and carbon dioxide enters the air spaces of leaves through stomata; there the carbon dioxide can diffuse into the cells. Osmosis and diffusion occur without any use of energy on the part of the cell. Another process, **active transport,** requires energy. Using active transport, a cell can move substances across its membrane in a direction opposite to that they would take by diffusion or osmosis. This process can be used to protect organisms in estuaries which are alternately exposed to fresh and saltwater and thus in danger of either bursting or shriveling up due to osmotic forces. It can also be used inside animals and plants whenever cells of different tissues maintain different concentrations of materials.

PHOTOSYNTHESIS

The basic process by which nearly all organisms are supplied with energy is the photosynthetic process of green plants, as you will recall from Chap. 1. In the process of photosynthesis, raw materials containing low energy, water, and carbon dioxide are transformed with the help of the sun's light into oxygen and energy-rich sugar. A simplified chemical equation of photosynthesis is:

$$\underset{\text{(low energy)}}{6H_2O} + \underset{\text{(low energy)}}{6CO_2} \xrightarrow[\text{green plants}]{\text{light energy}} 6O_2 + \underset{\text{(energy rich)}}{C_6H_{12}O_6}$$

12-3 Chlorophyll pigments are necessary for photosynthesis

Light energy There are several different kinds of chlorophylls in different organisms which reflect light of slightly different wavelengths. They all absorb light in the reds and blues and reflect green light back to our eyes; thus they appear green. When a chlorophyll molecule absorbs light energy, its electrons move to a higher energy state and we refer to it as an *excited* chlorophyll molecule. Its excited state is temporary, and it quickly gives off its excited electrons to other molecules which thereby become energy-rich. Water is split into oxygen, hydrogen (as protons), and electrons. Electrons and hydrogens move through the various reactions together, so that in the figures either H or e^- indicates a change of electrical potential.

Transfer of energy A complex series of steps is involved, at each of which energy is transferred. The energy output of the system is a flow of molecules of the chemical **adenosine triphosphate,** known as **ATP.** ATP is sometimes called the primary energy "currency" of cells because it is used for all kinds of biological work. To put it simply, part of the photosynthetic system converts the energy of light into the energy of the chemical bonds which hold ATP molecules together (Fig. 12-7). In the manufacture of ATP in a cell, an additional phosphorus (P_i) atom is joined to low-energy **adenosine diphosphate,** or **ADP,** to produce high-energy ATP molecules: $ADP + P_i +$ energy $\to$ ATP. Thus light energy is used to make a high-energy chemical bond between ADP and the additional phosphate.

FIG. 12-7 The first stage of photosynthesis uses energy from the sun ("sun wheels" drive "chlorophyll pumps") to make ATP or carrier molecules with hydrogen, and these are used in subsequent stages. ATP can be made by a cyclic pathway (*a*) or a noncyclic pathway (*b*). Carrier molecules with hydrogen are made only in the noncyclic pathway, and water is split to provide the hydrogen needed.

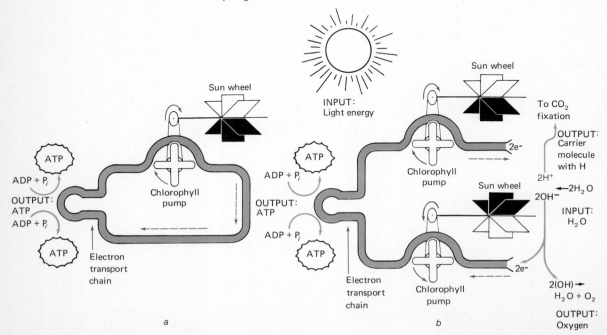

A second, related series of reactions also takes place as excited chlorophyll molecules return to their unexcited state. Water molecules are split into their components, hydrogen and oxygen. Some of the hydrogen atoms are picked up by special **carrier molecules.** (Carrier molecules are able to combine reversibly with hydrogen; there are several different kinds of carrier molecules.) The hydrogen atoms carried off are used in the building of complex carbon compounds, and some are recombined into water. The leftover oxygen is given off as a by-product of photosynthesis.

12-4 Photosynthesis occurs in two stages

Light reactions Together, making ATP and splitting water make up the **light reactions** of photosynthesis. In these reactions, the inputs are water and light energy. The outputs are oxygen, hydrogen attached to carrier molecules, and energy. This energy is chemical energy in the bonds of ATP *and the bonds of the carrier molecule–hydrogen combination.* The ADP + P_i + energy → ATP reaction is reversible: ATP → ADP + P_i + energy. You may think of ATP as a charged battery, ADP as a discharged battery, and P_i as the charge.

Dark reactions The light reactions of photosynthesis are connected to reactions which do not require the presence of light (Fig. 12-8). These **dark reactions** are the ones that fix carbon from the atmosphere and produce carbohydrate, which is a major product of photosynthesis. The dark reactions have inputs from the environment and inputs from the light reactions. Carbon dioxide is the input from the environment. The inputs from the light reactions are energy from ATP and hydrogen from the carrier molecule–hydrogen combinations. The eventual output is the sugar glucose, $C_6H_{12}O_6$. The transformation of the carbon in CO_2 to the carbon in glucose involves an extremely complex, cyclical series of separate reactions, each controlled by an appropriate enzyme. Energy for ATP plus high-energy hydrogen carriers are required to drive this cycle. The dark reactions cannot occur without a preceding period of light which provides some of the substances that react during the dark reaction. ATP and hydrogen atoms are produced in the light but can be used in the dark. The dark reactions are often referred to as **carbon fixation.** Cells of animals, protistans, and monerans can also carry out carbon fixation.

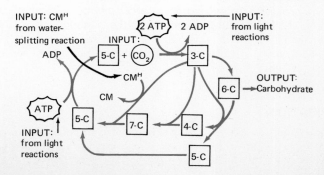

FIG. 12-8 Carbon fixation: ATP and carrier molecules with hydrogen are used in the production of carbohydrate from CO_2 (only the carbons are shown for simplicity).

12-5 Photosynthesis provides almost all the energy for all organisms

Some plants have photosynthetic pathways different from those discussed above. When the complex, cyclical series of separate reactions was discovered, it was named the Calvin cycle after its discoverer. It is sometimes also called the **C_3 pathway** because the first compound produced in the cycle has three carbon atoms. Recently, a new cyclical pathway was discovered in some plants. In this pathway, the carbon dioxide from the air reacts with a three-carbon compound to form a four-carbon compound. Plants with this pathway are called C_4 *plants*. They have leaves which are structured differently from those of C_3 plants, and at last report, the C_4 pathway has been found in nearly 100 genera in at least 10 different families of plants. The discovery of this new pathway is particularly exciting because C_4 plants are more efficient photosynthesizers than C_3 plants under certain conditions, particularly when the light is intense and the temperature is high. Many C_4 plants do best in arid regions, and more use of them could help provide food in deserts and in times of drought. Interestingly enough, some of our crop plants are C_4 plants, for example, corn, sugar cane, and sorghum. Other C_4 plants such as amaranth, which were once used as food plants, may be domesticated again because of their high-efficiency photosynthesis and more efficient use of water. Unfortunately, C_4 plants are not more efficient than C_3 plants in the usual temperate growing regions where the temperature is mild, the light intensity medium, and the water supply abundant.

Within the confines of chloroplasts, then, plants solve what may be considered *the* most basic problem of life: *obtaining usable energy*. They take the energy of sunlight, which is not directly useful for supplying energy for living processes, and convert it to ATP and sugars, which are useful (Fig. 12-9). The primary site of the transformation of light to chemical energy is the chloroplasts, and a long and complex series of enzyme-controlled steps is required before the chemical energy is stored in the bonds of glucose molecules. These involve the generation of energy-rich ATP molecules and the splitting of water to obtain hydrogen (with the release of oxygen from the water) in the light reactions and the combining of CO_2 and hydrogen to produce glucose in the dark reactions.

In the process of photosynthesis, green plants take simple compounds and use them to manufacture the complex organic compounds, which are the basis of all life on earth. Water and CO_2 are combined to make glucose, and on this foundation is built the entire structure of life.

RESPIRATION

Once the energy of the sun is stored in glucose and other compounds, how do plants and other organisms make it available again to do

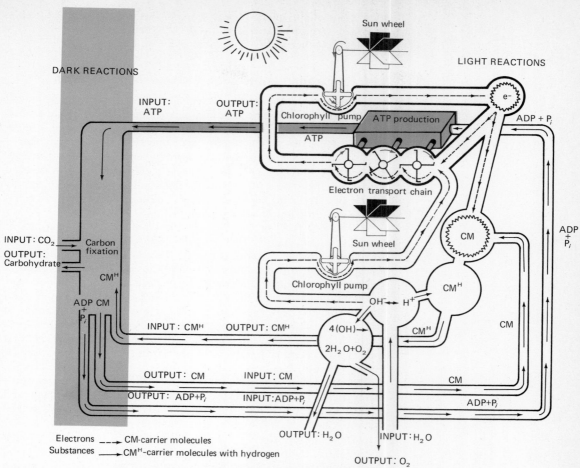

FIG. 12-9 Summary of the reactions of photosynthesis, showing the relationship between the light and dark reactions. Solar energy powers the "sun wheels," which in turn drive the "chlorophyll pumps" that move electrons and hydrogens through the system. ATP is made from ADP and P_i; carrier molecules (CM) pick up hydrogens (becoming CM^H). These outputs of the light reactions become some of the inputs of the dark reaction. The dark reaction uses CO_2 plus energy from ATP and CM^H to fix carbon as carbohydrate. The ADP, P_i, and CM molecules then become inputs for the light reactions. The water-splitting reaction produces oxygen as an output.

the work of life? One way people do it is by burning fossil fuels. However, this way does not help *cells* do the internal work necessary to keep themselves alive and functioning. The process by which cells get the use of some of the energy stored in compounds is called **respiration**. Some biologists call it **cellular respiration** to distinguish it from breathing, which is also called respiration by many people. Breathing is necessary for cellular respiration in the vertebrates since it brings oxygen into an animal and takes away carbon dioxide. However, respiration consists of a great deal more than this; it is a series of reactions that go on in cells of *all* types.

12-6 Respiration is the controlled oxidation of foods

If you were to burn a cube of sugar in the air, the energy produced would do no useful work. It would be lost, heating and lifting a small column of air over the flame. But in *controlled* oxidation of a molecule of glucose in a cell, slightly over one-half of the energy stored in its chemical bonds can be transferred eventually into the energy-currency of ATP. The chemical energy in ATP can, in turn, be used to run life processes (metabolism). Metabolism is run with energy stored in ATP using molecular inputs obtained through cell membranes by diffusion or active transport.

Production of ATP In greatly oversimplified terms, the outputs of respiration are roughly the inputs of photosynthesis and vice versa. In photosynthesis, water and carbon dioxide are "combined," with energy added, to produce oxygen and glucose. In complete respiration, oxygen and such compounds as glucose react to *release* energy, water, and carbon dioxide. The glucose molecule is taken apart, and the carbon and hydrogen atoms from it are carried, in a series of enzyme-controlled reactions, to be eventually combined with oxygen to form CO_2 and H_2O. It is important to remember, however, that the detailed chemical pathways of photosynthesis and respiration are very different.

The process of combining hydrogen with oxygen produces some of the yield of energy from respiration. If a mixture of hydrogen and oxygen is ignited, it will explode. The hydrogen will be oxidized instantly, with the release of a great deal of energy. What is left over is hydrogen oxide, namely water. In cellular respiration, the hydrogen is brought to the oxygen in a series of steps, with some energy being released at each step. Most of this energy is captured in the reaction $ADP + P_i + energy \rightarrow ATP$. The energy release of respiration is measured by the production of ATP.

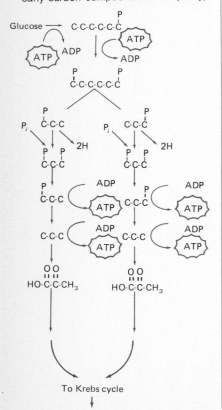

FIG. 12-10 Diagram of the reactions of glycolysis (hydrogens are omitted from early carbon compounds for simplicity).

The process of respiration actually begins (Fig. 12-10) with two ATP molecules reacting with each glucose molecule, supplying it with enough energy, by adding P, to take part in later reactions. The six-carbon glucose molecule then breaks down into two molecules, each with three carbon atoms and one P. *Each* of these glucose fragments then undergoes a series of chemical changes, in the course of which the P's are used to produce two ATP molecules. Two hydrogens go to carrier molecules and may be used to produce ATP in later reactions. Up to this point there has been a net gain of *two* ATP molecules. Four have been produced, two by the transformations of each of the three-carbon molecules. But two ATP molecules had to be used to start the whole process. The breaking down of a six-carbon glucose molecule into three-carbon molecules is called **glycolysis** ("glucose-splitting").

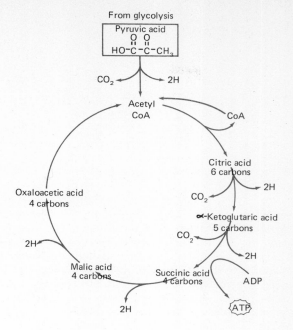

FIG. 12-11 The Krebs cycle, including the prior conversion of pyruvic acid to acetyl CoA.

The Krebs cycle The three-carbon molecule, called pyruvic acid, then gives off a CO_2 molecule and becomes a two-carbon compound. This two-carbon molecule combines with a carrier molecule called coenzyme A (CoA). The combination, called acetyl CoA, then enters a circular series of chemical reactions known as the **Krebs cycle** (Fig. 12-11). First, the acetyl CoA reacts with a four-carbon compound called oxaloacetic acid. The CoA separates and the two-carbon compound that has been carried into the Krebs cycle combines with the oxaloacetic acid to produce a six-carbon compound (citric acid). Next, a molecule of carbon dioxide is split off the six-carbon compound to produce a five-carbon compound (α-ketoglutaric acid). Then another carbon dioxide splits off, producing the four-carbon oxaloacetic acid again. Thus, in a series of steps the original two-carbon compound has been broken down and the oxaloacetic acid is ready to combine with the next two-carbon compound to enter the cycle.

We have seen the paths taken by carbon on its way from glucose to CO_2, but what about the hydrogens? These are pulled out at several points. In glycolysis, two are taken away, by carrier molecules, from each of the three-carbon glucose fragments just before they undergo the reactions which yield ATP. Two more are taken away when the three-carbon molecule is changed to a two-carbon molecule. And eight more are removed in the course of a single turn of the Krebs cycle. All these are doubled, producing 24 hydrogens.

The electron transport chain All these 24 hydrogens are passed, by way of two different kinds of carrier molecules, to a series of hy-

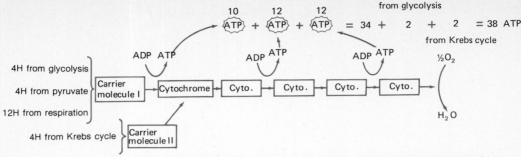

FIG. 12-12 Diagram of the electron transport chain. Each cytochrome is a hydrogen-carrier enzyme.

drogen–carrier enzyme molecules making up the **electron transport chain** (Fig. 12-12). Ultimately they are combined with oxygen, producing water. At three different steps, energy is released and stored in ATP. The passage of every 24 hydrogens along the chain of enzyme molecules results in the production of 34 ATP molecules. Two ATPs are produced in glycolysis and two more in the Krebs cycle. These, plus the 34 mentioned above, make a grand total of *38 ATP molecules* for each glucose molecule "burned" (Fig. 12-13).

12-7 Not all respiration requires oxygen

Fermentation does not require oxygen and is therefore sometimes called **anaerobic respiration.** The course of fermentation is the same as glycolysis until the production of the three-carbon molecule. That molecule then may be converted into various kinds of substances, such as ethyl alcohol, acetic and lactic acids, and methane.

Products of fermentation   The fermentation process which produces ethyl alcohol is called **alcoholic fermentation;** for example, yeast cells ferment grape juice to produce wine (Fig. 12-14).

Fermentation does not necessarily produce ethyl alcohol as an end product. In some bacteria, for instance, the end product of the breakdown of glucose is CO_2 and a different two-carbon molecule, **acetic acid.** When these bacteria get into grape juice, the result of fermentation is not wine but *vinegar* (from the French *vin aigre,* "sour wine!"). When you exercise strenuously and your muscles do not get enough oxygen, the cells switch from aerobic respiration to fermentation. Then the hydrogens are returned to the three-carbon molecule and no CO_2 is given off. The end product is the two-carbon **lactic acid** (Fig. 12-14). Lactic acid causes the feelings of pain and stiffness in muscles after exercise. Exercising these sore muscles speeds their recovery because it increases the blood flow which carries the lactic acid to the liver. The liver is the site of lactic acid breakdown. Other animals and some microorganisms also produce lactic acid as the end product of fermentation.

Anaerobic respiration by bacteria while decomposing sewage and garbage can produce methane. In some parts of the world, this methane is used for fuel; in others, it is allowed to form pockets of gas in old garbage dumps, where it may explode. Many cities are now considering using the methane from their garbage dumps for fuel; however, in most cases, the dumps are so constructed that it is difficult to pipe off the methane. Methane from this source is also contaminated with other gases, some highly toxic, which must be removed before it can be used for domestic purposes such as heating and cooking. Methane is also the major component of natural gas.

Some microorganisms employ only anaerobic respiration, even in the presence of abundant oxygen. Others could live their entire lives employing fermentation, but will switch to aerobic respiration when oxygen becomes available. Plant and animal cells can carry out fermentation under certain conditions, but they require oxygen to live. If, for instance, brain cells are deprived of oxygen for even a few minutes they will begin to die.

Health problems Many anaerobic microorganisms are problems for human health. Botulism bacteria can form their poison only in the absence of oxygen (Sec. 6-13). The tetanus bacterium is another organism that requires anaerobic conditions. It is especially prone to grow in deep puncture wounds where oxygen cannot penetrate in sufficient quantities. If the tetanus organism is allowed to grow unchecked in a wound, it produces a poison which causes convulsions and death. The muscles go into such spasms that a person's muscles may break his or her own bones. Often the jaw muscles become rigidly contracted, accounting for the common name of this disease—lockjaw. Today it is possible and desirable to be immunized for tetanus and to keep the immunization up to date. A deep puncture wound, however, should be opened surgically.

Another group of anaerobic organisms cause gas gangrene. Infections of wounds by these organisms are rare in peacetime but common in war wounds. Toxins from this organism may cause death; however, prompt treatment with antitoxin and surgery can prevent gas gangrene. Unfortunately, under war conditions, prompt treatment is often not available. Restriction of the flow of blood to legs and arms increases the possibility that gangrene may develop. Improper use of tourniquets, now usually discouraged in first-aid programs, is especially dangerous. Frostbitten fingers and toes are also likely candidates for gangrene.

12-8 The efficiency of respiration can be calculated

One way of measuring the amount of energy in a substance is to calculate the amount of heat it would give off if burned. The standard unit of measurement is the **calorie**. A calorie is the amount of heat it would take to raise the temperature of 1 gram (0.035 ounce) of water 1°C (1.8°F) at 1 atmospheric pressure. This is known as the *small calorie* (cal). Often it is more convenient to use a larger unit, which is the amount of heat required to raise 1 kilogram (2.2 pounds) of water 1°C (1.8°F). This is the *large calorie* (kcal). Caloric contents of foods are calculated in large calories. In the study of chemical

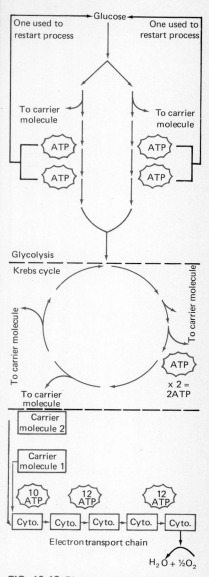

FIG. 12-13 Diagram summarizing the reactions of respiration and the formation of 38 ATPs.

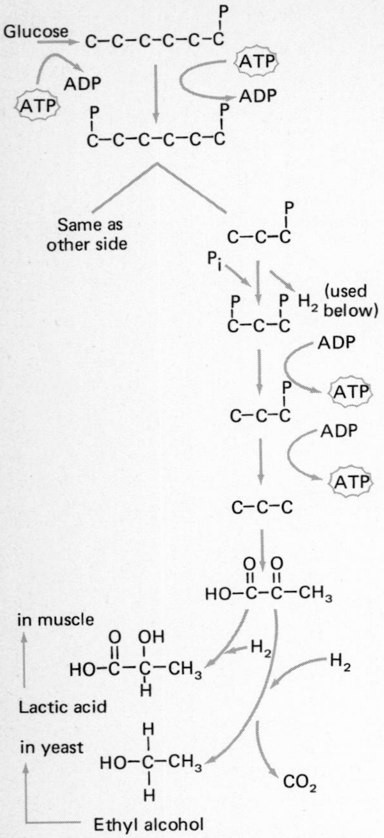

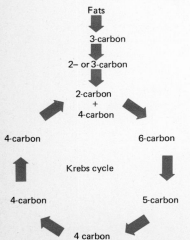

FIG. 12-14 Diagram of alcoholic fermentation.

FIG. 12-15 Diagram of lipid metabolism.

reactions in cells, however, small calories are more useful. To calculate the efficiency of respiration, one uses a standard amount of glucose called a **mole**. A mole is the weight of a substance in grams equal to the weight of its molecule relative to an atom of hydrogen. One mole of glucose weighs 180 grams and contains 686,000 calories.

Oxygen is necessary for efficient respiration. In cellular respiration using oxygen, 266,000 calories are obtained as ATP for each mole of glucose; 266,000 out of 686,000 is about 40 percent. This efficiency is about the same as that of the most efficient machines made by human beings. Many of our machines are less efficient, however; an automobile engine, for example, extracts only 15 percent of the energy potentially available in its gasoline.

The presence of oxygen is critical to obtaining this high yield of energy. Without oxygen there would be a net gain of only two ATP molecules for each molecule of glucose processed, or only 14,000 calories instead of 266,000 calories per mole of glucose.

12-9 Fats and proteins can also be used in respiration

The process of respiration involves much more than simply the "burning" of glucose. Fats may also be burned in the cellular furnace. Fats are broken down in a series of steps into three-carbon and two-carbon molecules which are then fed into the *metabolic pathway* of respiration as indicated in Fig. 12-15. The complete oxidation of a gram of fat yields more than twice as much energy as the complete oxidation of a gram of glucose. This is the reason why animals store energy in the form of fat. Some migratory land birds can store enough fat for a 1,600-kilometer (1,000-mile) nonstop flight over the sea. Many plants store fats in fruits and seeds.

Proteins are broken down into their component amino acids before they are fed into the metabolic furnace. The amino acids are then converted into a variety of compounds which enter the pathways of respiration at the appropriate places. Oxidizing a gram of protein yields about the same amount of energy as oxidizing a gram of glucose. Proteins are usually more important as building blocks of cells than as fuel. They, like all body components, are being replaced continuously. They play a major role in the energy-providing proeesses of respiration only when a person is on a high-protein diet or is starving.

OBTAINING ENERGY FROM OTHER ORGANISMS

In Table 6-2 organisms were classified by three basic modes of nutrition: photosynthesis, absorption, and ingestion. Organisms that can photosynthesize do not have the problem of obtaining energy from other organisms; they produce carbohydrate in many of their cells and thus supply the crucial raw material for respiration right on the spot. Vascular plants have specialized tissues for transporting sucrose to other plant cells. Organisms that cannot photosynthesize must obtain their energy by special chemical means or from other organisms. There are two main ways in which they can do this: absorption and ingestion.

12-10 Organisms have many ways of carrying out absorption and ingestion

Absorption Fungi, some bacteria, and most parasitic organisms obtain their food by absorbing it directly across their cell membranes. Sometimes these organisms live in a solution which contains suitable nutrients and they have only to absorb them. Other times, the medium in which they live is not suitable for immediate absorption and they must secrete enzymes into their environment to "digest" part of it so it can be absorbed. Parasitic organisms often absorb so much of their host or their host's food that they damage it severely. For example, the fish tapeworm, which may live in the digestive tract of people who eat raw fish, can absorb so much vitamin B_{12} that its host suffers a deficiency of this vitamin.

Ingestion Most animals, some protistans, and some bacteria ingest their food. Ingestion usually involves some sort of search behavior to find suitable food and specialized digestive organs or organelles. In protozoa, food is often taken into a temporary organelle called a **vacuole,** where it is digested. The vacuole which then contains the undigestible parts of the food moves through the cytoplasm to the cell membrane, discharges its waste to the outside, and disappears. In some of the simpler invertebrates, the digestive system is a blind tube, and food and wastes are taken in and expelled through the same opening. This type of digestive system is usual in the phylum Coelenterata. In the two-shelled members of the phylum Mollusca and in certain other organisms, food is brought in by water currents and captured by mucus threads. Organisms which use this mode of ingestion are called **filter feeders.** In some other invertebrates, for example, starfish (phylum Echinodermata), the organism ejects part of its digestive tract to temporarily surround its prey while it is digested and absorbed. In the vertebrates, and in many invertebrates, the digestive tract is a hollow tube extending from mouth to anus with specialized regions along it (Fig. 12-16). Food passes into the mouth and, after being processed along the tract, the indigestible portions pass out the anus. This type of digestive system and all others that involve ingestion require that absorption must also occur so that the food can be used by the cells of the organism. The human digestive tract provides a good example of the specialized regions used to break ingested foods down into particles suitable for absorption.

12-11 The human digestive tract has specialized regions

The specialized regions of the human digestive tract are diagrammed in Fig. 12-17; these specialized regions take in food, reduce it to a form suitable to enter the cells, and dispose of the leftover "wastes." Muscular actions move the material along as a series of digestive enzymes (Table 12-1) are secreted into different parts of the digestive tract.

The mouth Digestion begins in the mouth, where food is ground up and mixed with **saliva,** which is manufactured in the **salivary glands.**

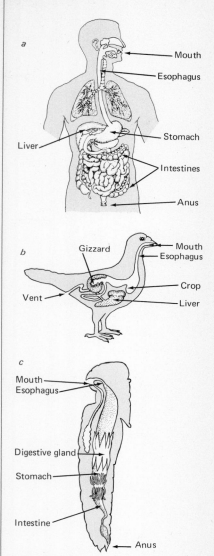

FIG. 12-16 Comparison of the digestive tracts of (a) mammal (human), (b) bird (pigeon), and, (c) insect (grasshopper). In the pigeon, the crop stores food and secretes a nutritious substance called pigeon milk which is fed to baby pigeons; the muscular gizzard grinds up seeds and grains with the aid of hard particles, such as gravel, which the pigeon swallows; the vent is a common opening of the digestive and urinary systems.

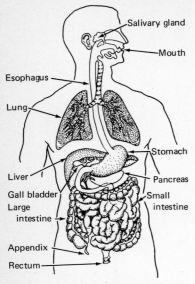

FIG. 12-17 Diagram showing the organs of the human digestive system.

TABLE 12-1 HUMAN DIGESTIVE ENZYMES

ENZYME	SOURCE	SUBSTANCE BROKEN DOWN	PRODUCTS OF BREAKDOWN
Ptyalin	Salivary glands	Starch	Sugars
Pepsin	Stomach	Protein	Polypeptides
Gastric lipase	Stomach	Fat	Glycerides* and fatty acids
Enterokinase	Duodenum	Trypsinogen	Trypsin
Trypsin	Pancreas	Polypeptides	Smaller polypeptides
Chymotrypsin	Pancreas	Polypeptides	Smaller polypeptides
Nucleases	Pancreas	Nucleic acids	Nucleotides
Carboxypeptidases	Pancreas	Polypeptides	Smaller polypeptides
Pancreatic lipase	Pancreas	Fat	Glycerides and fatty acids
Pancreatic amylase	Pancreas	Starch	Maltose sugars
Aminopeptidases	Intestine	Polypeptides	Smaller polypeptides
Dipeptidase	Intestine	Dipeptide	Amino acids
Maltase	Intestine	Maltose†	Glucose, galactose, fructose
Lactase	Intestine	Lactose†	
Sucrase	Intestine	Sucrose†	
Nucleotidase	Intestine	Nucleotides	Nucleosides‡ and phosphoric acid
Nucleosidase	Intestine	Nucleosides	DNA bases, RNA bases, sugars
Intestinal lipase	Intestine	Fats	Glycerides and fatty acids

* Glycerides are fats or breakdown products of fats; most fats are ingested as triglycerides and are broken down to di- and monoglycerides as fatty acids are removed from their molecules.

† Maltose, lactose, and sucrose are kinds of sugars.

‡ Nucleosides are compounds remaining when phosphate groups are removed from nucleotides.

After S. W. Jacob and C. A. Fracone, *Structure and Function in Man*, 2d ed., Saunders, Philadelphia, 1970.

Saliva serves for lubrication and also starts the digestion (the breaking up) of large *carbohydrates*. Before such long-chain carbohydrates as starch can be used by cells, they must be broken down to glucose. This process begins in the mouth, where starch and water react in the presence of the enzyme ptyalin in the saliva to produce a complex sugar. The conversion of that sugar to glucose occurs in another enzyme-controlled reaction in the small intestine.

The stomach The **esophagus** (Fig. 12-17) serves as a pipe leading to the **stomach,** where the first part of *protein* digestion occurs. The enzyme pepsin produced by stomach cells controls a reaction which chops up the long chains of amino acid residues of proteins into shorter chains. Thus one protein catalyzes the destruction of others. The process of protein digestion is completed in the small intestine. The enzymes trypsin and chymotrypsin, coming from the **pancreas,** catalyze reactions which break apart the chemical bonds holding together the amino acid residues of the protein fragments. In these reactions, which use water, short chains of amino acid residues are broken into single amino acids.

The small intestine The stomach empties into the long, tubular small intestine (see Fig. 12-17), in which *fat* digestion begins and digestion of other molecules continues. Fats do not mix well with water. Large fat droplets must be broken down into tiny droplets before the reactions controlled by fat-digesting enzymes can occur. The breakup of large fat droplets is aided by a substance called **bile.** This greenish-yellow liquid is produced by the **liver** and is stored in the **gall bladder** (see Fig. 12-17). Small droplets of fat are then broken down into two- and three-carbon molecules which can be absorbed by the digestive tract and enter cellular respiration.

The large intestine Human beings and many other animals mix a lot of water with the food they eat. There is water in various digestive "juices," which carry enzymes and lubricants. What happens to all that water? The final major unit of the digestive tract is the **large intestine** (see Fig. 12-17). Its major function is the reabsorption of water from the intestinal contents, thus preventing excessive water loss. After water is reabsorbed, the remaining materials, mixed with the bacteria which live and reproduce in the large intestine, are expelled as **feces** from the **anus.**

If the intestine is irritated by anything—for example, parasitic bacteria—its muscles may move the contents along very rapidly. There is not enough time for proper water absorption. Then the feces are watery and the person is said to have **diarrhea.** This is a very serious condition in young infants and elderly persons. Diarrhea is a major source of infant death in poor parts of the world. Death is caused by dehydration and is basically the same as dying of thirst.

If not enough roughage (material such as cellulose, found in leafy vegetables, bran, etc., which is indigestible) is consumed, the intestine may not be stimulated to move the contents along fast enough. This may result in too much water being absorbed. The feces become very hard and are expelled at longer than normal intervals. This condition is known as **constipation.** There is evidence that a deficiency of fibrous plant material in the diet may lead to cancer of the large intestine.

12-12 Circulatory systems transport molecules to and from the cells of multicellular animals

Once small molecules have been produced from the food that was eaten, we and other animals face the problems of getting them to all of our cells, getting oxygen there in order to metabolize them, and getting waste products carried away. These problems are solved for many animals by a **circulatory system.** In human beings, the

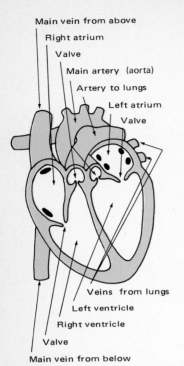

FIG. 12-18 The human heart as seen in longitudinal section. There are four chambers in a mammalian heart: the right atrium and right ventricle are separated by valves as are the left atrium and the left ventricle. Valves are also found at openings to the arteries. Valves control the flow of blood and are thus necessary for proper functioning.

principal components of the circulatory system are a powerful muscular pump, the **heart**; a series of **blood vessels** to and from the heart; and a special liquid tissue, **blood,** which is driven by the heart through these vessels (Fig. 12-18). The functioning of the heart is shown in Fig. 12-19.

There are three kinds of blood vessels in the circulatory system. **Arteries** carry blood away from the heart and **veins** return it to the heart. The main artery from the heart is the **aorta.** Connecting the two systems are very fine tubes, the **capillaries,** which come in extremely close contact with the cells (Fig. 12-20). Blood passes from the heart through the arteries to the capillaries and into the veins. (An exception to this is pulmonary circulation, discussed later in this chapter.) It then returns to the heart following a closed course consisting of many paths.

Food molecules The circulatory and digestive systems come close to one another in the walls of the small intestine, which has an enormous surface area (Fig. 12-21). These walls are folded and have many small, fingerlike projections, called villi, into the intestine. Cells of the villi have microvilli. These give the intestinal wall its very large surface area. Inside each villus is a system of capillaries. Many of the products of digestion pass through the cells of the intestinal wall, enter the capillaries, and thus enter the bloodstream. They are then pumped with the blood to all the tissues of the body. As the blood flows through the capillaries of those tissues, the food molecules pass through the capillary walls and through the cell membranes into the cells that require them. Sometimes large protein molecules get through the capillary walls and into the tissue fluid which bathes every cell. The lymphatic system (Sec. 12-13) eventually returns these to the blood.

Oxygen molecules To be able to get the maximum energy out of glucose and fats, cells must have a ready supply of oxygen. The circulatory system cooperates with the respiratory system to bring oxygen to cells. Figure 12-22 shows the special loops of the circulatory system that go to the lungs, the pulmonary circulation. The lungs are air sacs with an extremely large surface area. In fact, if you are an average-sized person, your lungs have 70 square meters (83.7 square yards) of surface, which they can expose to the air when you breathe deeply. This is about 40 times the surface area of the outside of your entire body and is approximately one-third the area of a tennis court. A large area is necessary because a great deal of oxygen must be taken into the bloodstream and a proportionate amount of CO_2 released. If you obtained oxygen through your skin, the exchange surface would have to be wet and extensively folded, or you would dehydrate rapidly. The large surface area of your lungs is inside your body and much less water is lost than, for example, in some amphibians which do get oxygen through their skin. Not only do the tiny sacs (called *alveoli*) in your lungs have an enormous surface area, they are also richly supplied with capillaries (Fig. 12-23). When you breathe and fill the alveoli with air, the blood in those capillaries

FIG. 12-19 Diagram to show the contraction of the heart. (a) Blood enters the relaxed right and left atrium from the main veins from the body (right atrium) and from the lungs (left atrium); (b) valves between atrium and ventricle on either side open, and blood enters the ventricles as the right and left atrium contract; (c) those valves close and the ventricles contract pumping blood into the arteries to the body (left ventricle) and the lungs (right ventricle).

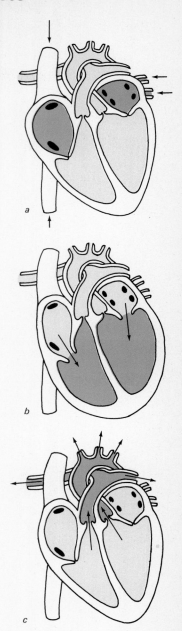

picks up oxygen from the air. However, the oxygen is not just dissolved in your blood; it is carried by a special pigment-protein hemoglobin, which was discussed in connection with sickle-cell anemia in Sec. 10-8.

Hemoglobin Hemoglobin is the pigment that gives blood its characteristic red color. Iron is an important part of the hemoglobin molecule. Hemoglobin is very neatly packaged in disk-shaped red corpuscles, which are cells that have lost their nucleus. These red blood cells are very rapidly destroyed and rapidly replaced. Each one lives only about 120 days; each second your spleen destroys about 3 million of them, yet the number circulating in your body remains fairly constant, enough so that their total surface area is about *1,500 times* the surface area of your body. Perhaps if you have donated blood, you wondered about how long it would take for your body to replace the red blood cells you lost. Since in a half-liter (pint) of blood there would be *only* about 2.5 billion red blood cells, at 3 million per second your body could replace them in about a day; however, it takes longer since your body also goes on replacing worn-out ones.

Each hemoglobin molecule in a red blood cell is able to combine with four oxygen molecules. When there is a lot of oxygen present, hemoglobin captures oxygen molecules. When there is little oxygen present, it releases them. Thus, hemoglobin is ideal for transporting oxygen from the oxygen-rich environment of the lungs to the oxygen-poor environment of the other tissues where oxygen is used.

Carbon dioxide molecules As we have seen, one of the main things which the blood transports is oxygen. Another is carbon dioxide. When oxygen-carrying blood reaches the cells of the body, it releases the oxygen and picks up carbon dioxide. Carbon dioxide is carried to the lungs, from which it is exhaled in several forms. Some of it forms a loose complex with hemoglobin, but most carbon dioxide is dissolved in the blood. Carbon monoxide, however, combines irreversibly with hemoglobin. Dissolved carbon dioxide is an important buffer system which maintains a specific acid-alkaline balance in the blood. There is a very narrow margin of safety in the acid-alkaline balance of the blood, and carbon dioxide and its compounds are important regulators of this balance.

It is often said that veins are blue and arteries are red. Interestingly enough, when hemoglobin is combined with oxygen, it has a slightly different shape than hemoglobin without oxygen. This changes its light-absorption qualities and gives it a different color. Hemoglobin with oxygen is a richer crimson. Thus oxygen-rich *arterial* blood is brighter red than oxygen-poor *venous* blood. Veins often are closer

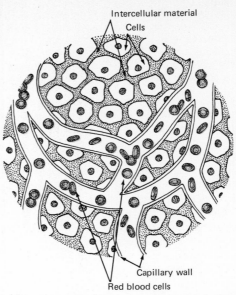

FIG. 12-20 Substances in the blood may move in and out of the wall of capillaries which are only slightly larger in diameter than red blood cells.

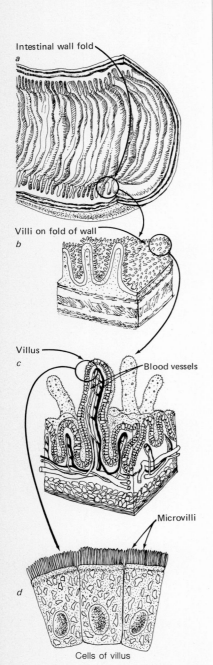

to the surface of your body than arteries. The blue-appearing vessels which may be seen on the hands or wrists of fair-skinned people are veins with their slightly more purplish blood containing less oxygen. Seen through several layers of tissue, such blood appears blue. As mentioned above, there is an exception to the statement that arteries carry oxygen-rich blood and veins carry oxygen-poor blood. This is the pulmonary (lung) circulation. Figure 12-22 shows that the arteries that carry blood from the heart to the lungs carry oxygen-poor blood. On the other hand, the veins that take the blood back to the heart (from which it is pumped out to the rest of the body) carry oxygen-rich blood.

The circulatory systems of human beings and other vertebrates must be able to pump the blood rapidly in order to move oxygen quickly to all tissues of the body. This rapid pumping maintains the blood at a high pressure in the blood vessels. If you increase the amount of work you are doing, for example, by running, you will need more energy. In order to obtain that energy, you will need more oxygen. Therefore, your rate of heartbeat will increase, pumping the oxygen-carrying blood faster. If blood is not provided rapidly enough to your muscles, some of your muscle cells will have to convert to

FIG. 12-21 The surface area of the intestine is greatly increased in three ways: folds (*a*), villi (*b* and *c*), and microvilli (*d*).

fermentation. As we saw, this will lead to an accumulation of lactic acid, which is one of the causes of sore muscles.

12-13 Vertebrates have another circulatory system, the lymphatic system

The circulatory system we have just described is sometimes called the **blood vascular system** to differentiate it from the other circulatory system of vertebrates, the **lymphatic system.** The lymphatic system is found only in the more advanced vertebrates having a completely closed, high-pressure circulatory system. This excludes the lampreys and hagfish, which have semiopen circulatory systems with blood sinuses similar to those of the invertebrates discussed in the next section. Although high-pressure blood vascular systems are completely closed, they tend to leak molecules. Among other functions, lymphatic systems recover fluids and proteins lost into tissues and get them back into the main circulation.

The lymphatic system of humans (Fig. 12-24) has no heart to pump fluid through it. Lymphatic fluid is circulated mostly by the squeezing of body muscles as they contract. Fluid which has come from the blood capillaries surrounds all cells of the body. This tissue fluid is picked up by the lymphatic system and returned to certain veins of the blood vascular system by lymphatic ducts in the chest area, and thus fluids and proteins are returned to the blood. This function of the lymphatic system is important in keeping tissues from becoming waterlogged. If the lymphatic system is blocked or cannot keep up with the fluid loss into tissues, **edema,** or swelling, may occur. This often shows up as puffy, swollen ankles.

The lymphatic system has branches in the walls of the digestive tract that absorb digested fats, and the lymphatic system also helps protect the body from infections. Part of the lymphatic system, the **lymph nodes,** are full of **lymphocytes** (a kind of white blood cell), which rush to the site of injury and engulf bacteria and other foreign matter. They also engulf body cells killed by bacteria. The cells and debris resulting from these battles are carried to the lymph nodes, where poisons are filtered out and prevented from entering the general circulation. This function of the lymphatic system leads to the sore and swollen lymph nodes which you may have experienced in your neck and armpits when you have had an infection. Unfortunately, cancer cells sometimes spread throughout the body through the lymphatic system.

The **spleen** is often considered part of the lymphatic system. It contains large numbers of lymphocytes called **macrophages,** which devour the 3 million aged and dying red blood cells each individual loses every second. Sometimes it is called the graveyard of the red blood cells; yet, one of the other functions of the spleen is to store a reserve supply of blood for emergencies. The **thymus gland** and the **tonsils** are also considered parts of the lymphatic system. They, too, are involved in protecting the body from infections.

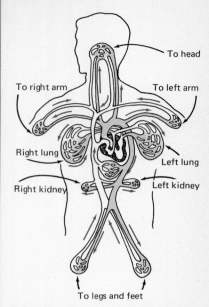

FIG. 12-22 Diagram showing the major features of the human circulatory system, including the pulmonary (lung) circulation.

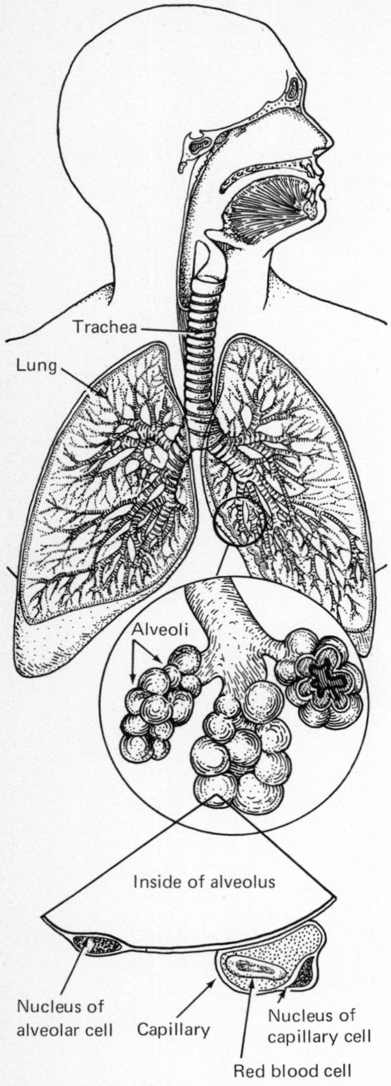

12-14 Invertebrates do not have a high-pressure circulatory system

Almost all vertebrates—and a few invertebrates—have a high-speed circulatory system that operates under the pressure created by a powerful pump. If an artery of a living vertebrate is cut open, blood will spurt out. Unless the leak is quickly stopped, the heart will pump out so much blood that not enough is left to transport oxygen and the organism will die. Insects and other invertebrates lack such a high-speed, high-pressure, closed circulatory system. A grasshopper's long, tubular heart (Fig. 12-25) is the only part of its circulatory system that is even partly closed. Inside its exoskeleton, an insect's blood supply simply sloshes around, bathing its intestinal tract and other tissues, stirred largely by the pulsations of its heart. The blood of insects carries food molecules from the digestive tract to the other tissues. Movement of food is not such an urgent matter as movement of oxygen, and the slow circulation of insect blood is quite sufficient.

One reason why this is possible can be seen in Fig. 12-26. Insects do not depend on their circulatory system to carry either oxygen to their cells or carbon dioxide from their cells. They solve this problem differently. They have a branching network of ventilation pipes called a **tracheal system.** The tracheal system carries air in tubes called *tracheae* from tiny holes (spiracles) in the exoskeleton to every cell in the body. This system works well for the insects. They are small and the oxygen does not have far to go. A system of small tubes would not work well for larger animals.

HUMAN NUTRITION

The major chemical components of the human body, like those of other organisms, are proteins, fats, and carbohydrates. To manufacture these, people must acquire the building blocks of these substances in their diet or make the building blocks themselves. Many building blocks are produced by digestion of food, as we have just discussed. The kinds of food we eat determine what kinds of building blocks will be produced by digestion. The chemistry of the body determines which will be made in the cells.

12-15 Proteins are essential for growth of all tissues

Of the some twenty building blocks of proteins, nine amino acids cannot be manufactured by the human body. Therefore, these nine are essential building blocks in the diet of growing children and pregnant women. Without these, necessary proteins cannot be made and growth stops. Eight of these amino acids are essential in the diet of adults because their bodies are continuously replacing proteins such as that lost when red blood cells are destroyed. The other eleven or so amino acids found in proteins can be manufactured by the human body from simpler molecules. Proteins are not only essential

FIG. 12-23 Lungs are connected to the mouth and nose through the windpipe, or trachea. They are divided into tiny alveoli in which gas exchange with capillaries takes place.

for growth and development but are also necessary for tissue maintenance and repair, healing wounds, and recovering from disease. Most of the nonliquid parts of your organs and tissues are mainly protein, from the hair on your head to your toenails. The structural material of your skin, your brain, your muscles, and your blood is mostly protein. Protein is also a major component of your bones. Bones are alive. They have nerves and blood vessels and cells arranged in concentric rings called Haversian systems (Fig. 12-27). When bone cells repair a broken bone, they need protein as well as minerals. People who are sick or injured, as well as infants, children, and pregnant or nursing women, need higher proportions of protein in their diets than do other persons.

Complete and incomplete proteins Proteins are found in virtually all foods, but in enormously differing quantities and qualities. *Complete protein foods,* those whose proteins contain all the nine essential amino acids, are primarily food of animal origin: meat, fish, poultry, eggs, and dairy products. Nuts and soybeans also provide complete proteins, but of *lower quality.* This means that while all essential amino acids are present, they do not occur in ideal proportions for meeting nutritional needs. Therefore, more of them must be eaten to supply the needed amounts or they must be combined with amino acids from other foods. The highest-quality protein is found in mother's milk. Eggs and cow's milk are second and third, respectively. This does not mean that an ideal diet consists of milk and eggs! Milk is a very suitable food for babies and children. Indeed, it evolved as such. For adults, milk may be less than suitable or even harmful. And the cholesterol content of egg yolks may rule them out of the diet of many. The amount of cholesterol in the blood serum is closely correlated with the occurrence of heart disease and may relate to a diet high in cholesterol.

Considerable amounts of incomplete proteins are found in **legumes,** such as peas and beans, and smaller amounts are found in **grains,** such as wheat and rice. By combining such lower-quality protein foods with each other, a meal of very high-quality protein can be achieved. For example, the beans and the corn tortillas that make up the basic diet in much of Latin America actually provide a substantial amount of high-quality protein. There are books that give ways to combine proteins from various sources to produce a diet with sufficient amounts of all the essential amino acids.

Lack of proper protein Every day a healthy adult needs about 1 gram of protein per kilogram of body weight. Children and pregnant women need more. A diet that lacks sufficient high-quality protein results in low levels of energy, stamina, and resistance to disease. This *protein malnourishment,* if it occurs during infancy and childhood, results in dwarfing, delayed physical maturity, and even death (see Sec. 8-6). This may occur even if the lack of protein is temporary and a normal diet is eaten later.

Some studies have indicated that protein malnourishment in infancy and early childhood may also result in permanent damage to

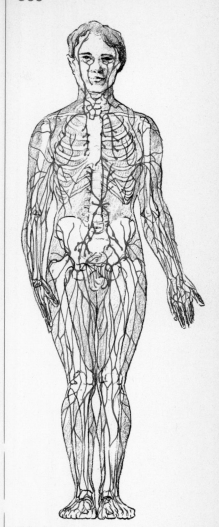

FIG. 12-24 A diagram of the human lymphatic system.

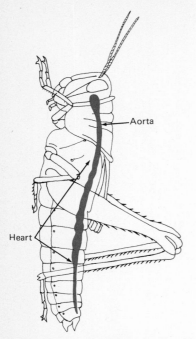

FIG. 12-25 The circulatory system of the grasshopper is not a high-pressure system as is that of humans.

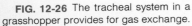

FIG. 12-26 The tracheal system in a grasshopper provides for gas exchange.

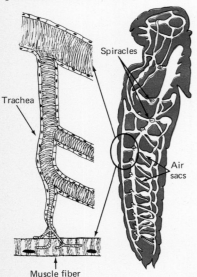

the brain. A child's body grows to 20 percent of its adult size in the first three years. During this time the brain grows to 80 percent of its adult size. This rapid growth is primarily a result of protein synthesis. More than 50 percent of the dry weight of brain tissue is protein. When protein is not available in the diet to supply the amino acids from which brain proteins are made, the brain stops growing, apparently irreversibly.

Other studies have shown a strong relationship between nutritional level and mental achievement. It is clear that if we wish young people to grow up as bright, useful citizens, it is important that they get adequate amounts of high-quality protein, especially early in life. Food programs for the poor that feature large amounts of such foods as lard and white flour but lack protein are inadequate.

Protein malnourishment may occur from simple ignorance, not just as a result of poverty. A great many people in America do not know how to choose an adequate diet, even if they have the money to afford it. The food industry has spent large sums developing and advertising good-tasting but *nonnutritious* food. **Obesity** (the state of being excessively fat) may occur as a result of protein malnourishment, and plumpness does *not* indicate that a person is well-fed. High carbohydrate diets, such as the poor must eat (and the better-off often choose to eat), may lead to fat people who are actually risking the dangers of protein malnourishment. A recent study showed that there are poor nutritional practices even in American hospitals; as many as one-third of the hospitalized patients in this study showed significant nutritional deficiencies as a result of their hospital stay. This is extremely serious because lack of adequate protein affects patients' immunity to infection.

12-16 Carbohydrates and fats are the chief energy sources in the diet

Since we must continually obtain energy to remain alive, energy extraction is a first priority in metabolic activity. Carbohydrates and fats are the chief source of energy. We take in our carbohydrates mainly as starches and sugars. Starches are found in grain and products made from grain, for example, spaghetti, which is made from wheat. Vegetables, especially root vegetables such as potatoes and yams, contain large quantities of starches. Sugars are found most commonly in fruits and, to a lesser extent, in vegetables, milk, and milk products. Refined (purified) sugars, which are used in cooking, baking, and preserving, come from sugar cane or sugar beets. Soft drinks (except for diet types) are made of refined sugar, water, and carbon dioxide, plus flavoring, coloring, and preservatives.

Lipids (fats, oils, and related compounds) are present in both plant and animal foods, as well as in eggs and dairy products. Fats are a source of energy; they can be used directly or stored as a reserve. They are also important structural parts of cell membranes and the sheaths that surround many nerves. Fats are made up of subunits called **fatty acids;** three fatty acids are attached to a glycerol molecule to make a fat. Fatty acids may be saturated or unsaturated. If a fatty acid is saturated, the maximum number of hydrogen atoms are attached to the carbon atoms. Fats which are solid at room temperature contain mostly saturated fatty acids, while those which are liquid contain unsaturated fatty acids. Margarine is produced from liquid fats by bubbling hydrogen through them. Most animal fats are saturated. The fats from plants, however, often contain many **polyunsaturated fatty acids.** Unsaturated fatty acids are essential to the human diet.

Lack of carbohydrates and fats When insufficient supplies of carbohydrates and fats are eaten, the body stores of carbohydrates, fats, and proteins are used up (in that order) and weight is lost. When the reserves are all gone, the individual has starved to death. Starvation from lack of energy is hard to separate, however, from starvation due to lack of protein, or due to enough protein without sufficient fats and carbohydrates. *All three are necessary for life.* If people get enough protein in their diet for normal growth and repair, but not enough fat and carbohydrate, some of the protein will be used for its energy value. There then will be a protein shortage for growth and repair. People vary greatly in their dietary requirements, so that some may benefit from unusual, unbalanced diets. However, such diets should be attempted only under the care of a competent physician who understands human nutrition.

Excess of carbohydrates and fats As we have seen, too much carbohydrate and fat with too little protein may lead to obesity. Obesity also results from the consumption of too much food for the amount of activity, even if the diet is perfectly balanced. Not only are severely overweight people handicapped in daily living by having to carry around extra weight, they are far more likely to get many serious diseases than people of normal weight. They run a very high risk of heart attacks, hypertension (high blood pressure), or cerebrovascular accidents (strokes, which are ruptures of a blood vessel in the brain). Or they may develop diabetes, a disorder of sugar metabolism. Fat people often eat large amounts of animal fats. Consumption of high amounts of *saturated animal fats* is strongly suspected of increasing risks of heart attack and stroke. The typical American diet has an excess of saturated fats. It is possible now to obtain beef (labeled calf or baby beef) that contains less fat than beef from grain-fed cattle.

Too much of certain carbohydrates can lead to other problems. There is increasing evidence that refined sugars play a role in generating heart disease and diabetes. Refined sugars appear to be almost addictive to some people; these people must frequently eat sweets or they go into a state of weakness, nervousness, depression, etc. Some physicians are now experimenting with low-carbohydrate diets in treating certain kinds of mental illness. Refined sugars are also a major cause of tooth decay. Almost all canned, baked, or processed food in America has refined sugar in it so that many people are

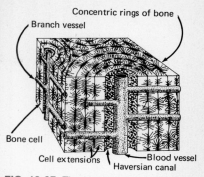

FIG. 12-27 The Haversian system is the structural unit of bone. It is composed of concentric rings of bone containing bone cells and a central canal with blood vessels which branch and supply the bone cells.

unaware of how much sugar they do eat. In summary, fats and carbohydrates are necessary for energy and a balanced diet; but taken in excess, they may be bad for you.

12-17 Vitamins are required in small amounts

In addition to protein, carbohydrates, and fats, all of which are required in relatively large quantities, a series of other nutrients are needed in small quantities. These are the **vitamins,** which fill a variety of needs in life processes. The lack of any of 13 or more essential vitamins leads to upsets of cellular metabolism which can, ultimately, end in death.

Vitamins fall into two general classes: **fat-soluble** (vitamins A, D, E, and K) and **water-soluble** (B-complex and C). Fat-soluble vitamins can be absorbed only in the presence of bile, which is in the digestive tract only when fats are present. Many vitamins, especially the water-soluble ones, may be destroyed by exposure to light, air, or heat. They are often lost from food through improper storage or cooking methods.

Vitamin A Substances called carotenes are present in some fruits and vegetables (for example, carrots). They are converted by cells in the intestinal wall into substances with vitamin A activity. Vitamin A is needed for normal vision, healthy skin and gums, and resistance to respiratory illness. Vitamin A can be stored in the body; however, the body has limited capacity to excrete an excess, and too much vitamin A can be harmful. Children who have received as much as 500,000 units per day over long periods have swellings of their long bones and impaired mobility of their joints. These children are often babies who have been cared for by their grandmothers. In grandmother's day, the only source of a vitamin A supplement was cod-liver oil. It had to be administered by the spoonful since it contained a relatively low concentration of vitamin A. Grandmothers today often cannot believe that one drop of a modern concentrated vitamin supplement is enough.

Adults receiving vitamin A in excess of 500,000 units per day show calcium being laid down in various structures other than bone. At even higher dosages, they get headaches, nosebleeds, nausea, etc. The Food and Drug Administration has recently restricted the amount of vitamin A present in a single vitamin pill to 10,000 units. Vitamin A poisoning is very unlikely in adults unless huge amounts of supplementary vitamins are taken. Deficiencies are much more common. Studies show that vitamin A–deficient rats produce abnormal fetuses. However, massive doses of vitamin A given to pregnant rats are teratogenic. Again, too much of a good thing may be bad.

Vitamins D, E, and K Vitamin D is sometimes known as the sunshine vitamin. When certain substances which we get in our foods are exposed to sunlight in our skins, they are changed into vitamin D. In the absence of sufficient sunshine, as in many of today's cities, vitamin D can be obtained only by drinking vitamin D–enriched milk

FIG. 12-28 A child with rickets. (*From "The Vitamin Manual," Published by the Upjohn Company; Courtesy of Rosa Lee Nemir, M.D.*)

or taking synthetic supplements or fish-liver oils, such as cod-liver oil. Vitamin D is involved in the metabolism of calcium and phosphorus and thus in the growth and maintenance of bone structure. A serious disturbance of bone structure known as **rickets** is found in children lacking vitamin D (Fig. 12-28). Too much vitamin D, however, can also cause serious health problems. The role that vitamin D synthesis may play in natural selection for human skin color is discussed in Chap. 10.

Vitamin E is found in vegetable oils, for example, wheat-germ oil. It seems to be involved in reproductive functions. It may have important functions in metabolism of fats. It is thought by many to have other significant roles, but these have not been critically demonstrated.

Vitamin K is essential to the clotting mechanism of the blood. This vitamin is ordinarily manufactured by bacteria that live in our intestines. It is also found in green leaves, fat, and egg yolks.

Vitamins B and C The B-vitamin complex consists of at least eight separate, essential vitamins. Riboflavin, B_2, for instance, is needed for the molecules of the electron transport chain. Even a partial list of the metabolic problems produced by deficiencies of B vitamins makes grim reading: central nervous system disorders, insanity, skin problems, diarrhea, swollen gums, indigestion, and dwarfism. One of the B vitamins, niacin, reportedly has been used with some success in the treatment of some mental disorders.

Vitamin C (ascorbic acid) is essential for healthy skin, gums, and blood vessels. It is also used in the treatment of mental disorders along with niacin. Vitamin C seems to be an important factor in resistance to stress and infections, especially colds. It is found widely in fruits and leafy green vegetables, the richest sources being citrus fruits. It cannot be stored for a long time in the body. Sailors and explorers who lacked fresh foods commonly used to show the symptoms of extreme vitamin C deficiency—*scurvy*. In the early days of the British navy, sailors on long voyages were given limes to prevent scurvy and thereby got the nickname *limeys*.

12-18 Minerals are necessary for normal growth

In addition to the three main classes of nutrients and the vitamins, people also require 17 essential minerals, some in very tiny quantities. Calcium is needed for bones and teeth, and plays many other vital roles. It is most easily obtained in milk, cottage cheese, and yellow cheeses. Phosphorus is also needed for bone building, for the functioning of nerves, for DNA and RNA, and to make ADP and ATP. It is found in dairy foods, meat, and eggs. Iron is an essential component of hemoglobin and the enzymes of the electron transport chain. Lack of iron leads to anemia from lack of hemoglobin. More iron is needed by women than by men because women have cyclic loss of iron due to the blood they lose during menstruation. Iron is found most abundantly in liver and other organ meats; it is also available from other meats, green vegetables, and egg yolks. Iron is excreted from the body with some difficulty so that, like vitamins A and D, too much iron may result in toxicity. Iron is being added to a number of foods, for example, bread, and some nutritionists are concerned that some diets may contain excess iron.

Sulfur is needed for protein building and is found in high-protein foods. Salt (sodium chloride) is found in seafoods, meat, some fruits and vegetables, and processed foods. Excess salt may lead to retention of water, affect heart function, and raise blood pressure. It is needed for nerve and muscle activity and many other metabolic processes, as is potassium. Potassium is widely available in meats, fruits, and vegetables. Iodine is essential for thyroid function. It is found in foods grown in iodine-rich soil and, most importantly, in iodized salt. Iodine shortage results in goiter, in which the thyroid

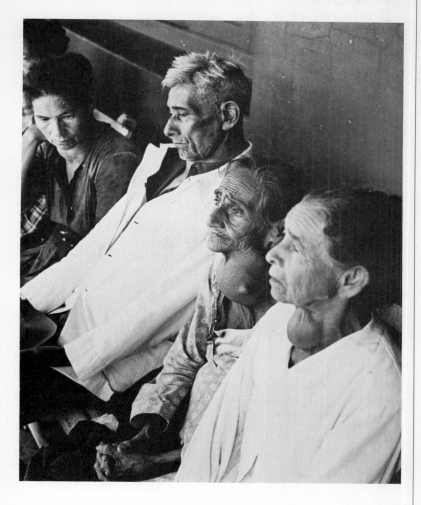

FIG. 12-29 Goiter patients awaiting examination in a village in the Asunción region of Paraguay, South America. Until recently a large part of the population of Paraguay suffered from goiter for which the remedy—known for over a century—is as simple as iodized salt. (*WHO photo by Paul Almasy.*)

gland of the neck swells (Fig. 12-29). Before iodized salt was introduced, goiter was very common in the central regions of the United States and other places where iodine is scarce and where seafood was not available. Zinc, magnesium, manganese, chromium, cobalt, copper, molybdenum, and selenium in very tiny amounts are also necessary for health, but large amounts may be toxic and many are becoming common pollutants of air and water. These elements are widely distributed in natural foods; however, they may be lacking in highly processed foods. There may be other elements which are necessary in trace amounts to all or to only some people. For example, some people with mental depression may respond dramatically to doses of lithium. Lithium, however, is highly toxic and must be used under close supervision by a physician.

12-19 It is possible to get an adequate diet by eating foods from four basic groups

There are conflicting claims about the adequacy and safety of the American diet. Because of the overprocessing of foods, vitamin supplements are often added to foods, and many people in America take vitamin supplements because they know or suspect that their diet is inadequate in vitamins. Mineral and protein supplements may also be taken. Often a physician will prescribe vitamins when a person is recovering from an illness. For years, representatives of the American food industry have told people that their diets were perfectly safe and adequate; however, the food industry continues to add vitamins and minerals to many of its products.

There is no doubt that the typical teenager's diet of soft drinks, fries, and hot dogs is a nutritional disaster. Many people feel that the government should not control the nutritional values of food and food supplements, but simply inspect to see that foods are not contaminated with poisons, disease organisms, or filth. Nevertheless, the Food and Drug Administration has recently limited the amount of vitamins A and D that can be sold in a single tablet. The government, however, allows the food industries to add preservatives, coloring agents, and many other synthetic chemicals to our foods. While some of these chemicals may keep food from spoiling and improve its appearance or taste, many scientists feel that some of them (for example, sodium nitrates and nitrites and food colorings) are harmful and should be banned.

For years food industry scientists have attempted to convince people that there is no difference between "organically" grown food and food from industrialized farms. "Organic" farmers use manure and compost, and perhaps rock phosphate, for fertilizer and do not use biocides. Some plants do accumulate excessive nitrogen compounds when heavily fertilized with synthetic nitrogen fertilizers. Both children and farm animals have become ill from eating such food. It seems likely that we should take the organic farmers seriously and that more research should be done on the nutritional value of our foods and the effects of various farming techniques on soils. The organic methods of composting and reutilizing organic material prevent pollution and recycle materials back to the soil where they are needed. Apparently these methods do reduce the damage done to soil by agriculture. The "energy crisis" may well lead to a return to organic farming. The price of synthetic nitrogen fertilizer has been increasing, and natural gas, which is used in fertilizer manufacture, will run out in this country well before 1990. No doubt food shortages and high prices will be the first indication of the necessary changeover from synthetic fertilizers. This will make it even more difficult for the poor to get an adequate diet.

The four basic food groups   In spite of impending shortages, high prices, and conflicting claims, it *will* be possible to get an adequate diet. Even though all the nutritional needs we have discussed may seem terribly complicated, they can be supplied by eating food from each of four groups daily. These are:

1 Milk and dairy products for protein, vitamins, calcium, and other minerals
2 Meat, fish, poultry, or eggs for protein, fats, and vitamins
3 Grains, nuts, and starchy vegetables for carbohydrates, fats, vitamins, minerals, and some protein
4 Fruits and vegetables, especially legumes, for carbohydrates, vitamins, minerals, and some protein

The foods in the first two groups are the luxury foods of the world. They are expensive in terms of both their price and the cost to the environment in producing them. The 10 million to 20 million people who die of starvation and malnutrition in the world every year die in large part from the lack of these foods, even though the proper combination from groups 3 and 4 is an adequate diet.

QUESTIONS FOR REVIEW

1 What may be considered "the most basic problem of life"? Why? Do all organisms solve it in the same way?
2 What is the text's definition of *metabolism*? *respiration*?
3 What is the practical implication of the discovery that many plants employ the so-called C_4 photosynthetic pathway?
4 How might anaerobic respiration, or fermentation, help ease the fuel shortage in some areas?
5 What has fermentation got to do with hard exercise and the way you feel afterward?
6 What are three grave dangers to human health which are caused by anaerobic microorganisms?
7 Why do insects not need a closed circulatory system? Why does their *solution* to the problem of transporting oxygen and CO_2, like their *solution* to the problem of skeletal design, ensure that huge, people-sized insects will continue to be a potential threat only in science-fiction stories?
8 Why is *organic farming* likely to be taken more and more seriously as time goes on?

READINGS

Björkman, O., and J. Berry: "High Efficiency Photosynthesis," *Scientific American,* **229**(4)80–93, 1973.
Calvin, M., and W. A. Pryor (eds.): *Organic Chemistry of Life,* readings from *Scientific American,* Freeman, San Francisco, 1973.
Ewald, E. B.: *Recipes for a Small Planet: The Art and Science of High Protein Vegetarian Cooking,* Ballantine, New York, 1973.
Gillie, R. B.: "Endemic Goiter," *Scientific American,* **224**(6):92–101, offprint 1223, 1971.
Hokin, L. E., and M. R. Hokin: "The Chemistry of Cell Membranes," *Scientific American,* **213**(4):78–86, offprint 1022, 1965.

Lappé, F. M.: *Diet for a Small Planet,* 2d ed., Ballantine, New York, 1975.
Mayerson, H. S.: "The Lymphatic System," *Scientific American,* **203**(6):80–90, offprint 158, 1963.
Miller, M. A., and L. C. Leavell: *Kinber-Gray-Stackpole's Anatomy and Physiology,* 16th ed., Macmillan, New York, 1972 (see especially chaps. 19–22).
Marchesi, V. T.: "Membrane Systems and the Organization of Cells," *Biocore,* unit V, McGraw-Hill, New York, 1974.
Reuben, D.: *The Save Your Life Diet,* Random House, 1975.
Stark, G. R.: "Enzymes: Function and Regulation," *Biocore,* unit IV, McGraw-Hill, New York, 1974.
Verrett, J., and J. Carper: *Eating May Be Hazardous to Your Health—The Case Against Food Additives,* Simon & Schuster, New York, 1974.
Williams, R. J.: *Nutrition against Disease—Environmental Prevention,* Bantam, New York, 1973.
Wood, J. E.: "The Venous System," *Scientific American,* **218**(6):92–101, offprint 1093, 1968.
Wood, W. B.: "Molecular Design in Living Systems," *Biocore,* unit II, McGraw-Hill, New York, 1974.
———: "The Molecular Basis of Metabolism," *Biocore,* unit III, McGraw-Hill, New York, 1974.

13

USING ENERGY

SOME LEARNING OBJECTIVES

After you have studied this chapter, you should be able to

1. Name four major ways organisms use energy.
2. Say what catalytic proteins in biological systems are called, state what their general function or role as a catalyst is, and list at least three kinds of reactions they catalyze.
3. List five steps of the *transcription-translation sequence* by which genetic information coded in DNA is decoded and used to direct the synthesis of proteins from amino-acid subunits.
4. Define transcription and translation as they apply to DNA, mRNA, rRNA, tRNA, and structural protein.
5. Name and describe two ways in which synthesis is controlled.
6. Demonstrate some of what you know about the maintenance processes of organisms by

   a. Explaining why maintenance is a problem and why it must be a continuous (nonstop) process.
   b. Listing some organism-environment exchanges which are employed by living things to maintain themselves.
   c. Saying why plants need CO_2 and O_2, and why animals need O_2.
   d. Describing the exchange of O_2 and CO_2 by (1) mammals, (2) fish, and (3) vascular plants.
7. State the text's definition of effectors, and give examples of one effector in plants and three effectors in animals.
8. Demonstrate your knowledge of the specialized effectors of *motion* in animals by
   a. Giving the general name of this effector system.
   b. Naming its energy source (the molecule that transports energy to the effector tissues' cells).
   c. Saying why one type of this effector tissue is called *voluntary* and referred to as *striated,* and why another type of this effector tissue is called *involuntary* and referred to as *smooth*.
   d. Describing some major tasks performed by each of the two types of tissue, and explaining why the label *involuntary* is not an entirely appropriate one.

CHAPTER THIRTEEN

In the last chapter, we studied some of the problems organisms have in obtaining energy. In this chapter, we explore some of the problems organisms have in using that energy to do the work necessary for their perpetuation. In order to perpetuate themselves, organisms must use energy to synthesize new molecules, cells, tissues, and organs. They must use energy for growth, maintenance, and reproduction. In animals, energy is necessary for the muscular activity involved in behavior, pumping of blood, and breathing, and for the proper functioning of their nervous systems. In people, as we saw in the last chapter, failure of one or more of the energy-supplying reactions of the brain such as those requiring the B vitamins can result in disorders of behavior and perceptions which may make it impossible for a person to function.

SYNTHESIS

Each gene in the nucleus of a cell controls the production of a polypeptide. Some proteins consist of a single polypeptide chain, while others are made up of several polypeptides. Proteins play major structural and metabolic roles in organisms. They are especially important as enzymes, for these biological catalysts control the vast majority of chemical reactions in metabolism, including those involved in synthesis. Indeed, enzymes catalyze the reactions involved in the synthesis of enzymes. Once the problem of making proteins is solved, so is a principal problem of synthesis.

13-1 Enzymes are proteins acting as catalysts in biological systems

Catalysts are substances which can change the rate of chemical reaction, but are not themselves changed in them. For instance, many chemical reactions which normally only take place rapidly at high temperatures will take place rapidly at body temperature if an appropriate catalyst is present. How does a catalyst work? Scientists generally believe that it may hold two molecules together long enough for them to react (Fig. 13-1).

Enzymes alone make possible the vast complex of chemical life activities which we call metabolism. They catalyze the reactions which break large food molecules down into components we can absorb; they catalyze the reactions by which these components are united into large molecules. They catalyze both the reactions which bind the energy of the sun in photosynthesis and the reactions by which that energy is freed to drive metabolic processes. Enzymes also catalyze chemical reactions which determine the sequence of amino acid residues in enzymes—that is, they catalyze their own assembly. Enzymes also control the reactions by which plant cells convert glucose molecules into the carbohydrates starch and cellulose.

13-2 The first step of protein synthesis is transcription of DNA to RNA

As we saw in Chap. 9, the genetic information is coded in the structure of DNA, in the sequence of the four types of subunits of DNA called nucleotides. That coded genetic information must somehow be decoded into the proper sequence of amino acids to make a polypeptide. Several stages are involved in the decoding process. The first stage consists of the transfer of a code sequence from DNA to a code sequence in a very similar molecule, **RNA (ribonucleic acid).** RNA, like DNA, also has four types of subunits. The backbone sections of the RNA subunits are slightly different from those of the DNA—they contain a single additional oxygen atom (Fig. 13-2). DNA is called *deoxy*ribonucleic acid because it lacks that oxygen atom.

Pairing of RNA and DNA bases The bases of three of the RNA nucleotides are identical with those of the DNA. These are adenine, cytosine, and guanine. Instead of the thymine nucleotide of DNA, RNA has a uracil nucleotide (a nitrogen-containing base very similar to the base thymine in the T subunit of DNA). The DNA bases are complementary and can pair together, A:T and G:C, by forming hydrogen bonds between their bases. The RNA bases are complementary to the DNA bases and can pair with them, A:U and G:C. Thus a section of DNA strand with the code sequence AAATGT can pair with another DNA strand with the sequence TTTACA *or* it can direct the synthesis of a new RNA strand with the sequence UUUACA (Fig. 13-3). The latter is called **transcription.**

When a protein is to be synthesized, the DNA double helix coding for it unwinds just as it does when DNA is to be duplicated. But instead of a new DNA strand being formed against the DNA template, RNA strands are formed. In this manner, portions of the genetic code are transcribed from the DNA to single strands of RNA. This process of transcription requires enzymes and energy from ATP.

Messenger RNA The newly formed strand of RNA then leaves the nucleus through pores in the nuclear envelope and moves into the cytoplasm to the place where protein synthesis will occur. Biologists think of this strand of RNA as a sort of chemical messenger that carries the genetic code from its storage center in the nucleus to the protein factory in the cytoplasm. For that reason they have named it **messenger RNA (mRNA).**

The protein factory in the cytoplasm must now carry out the job of assembling proteins, guided by the sequence of A, C, G, and U subunits in the messenger RNA molecules. The essential raw materials are, of course, amino acids. These have been ingested in the food or have been synthesized by the organism. Energy is also needed, and that is available from ATP molecules. Enzymes too are necessary to catalyze the various reactions, as are two other kinds of RNA.

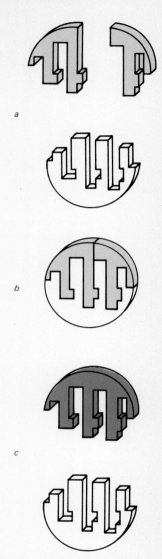

FIG. 13-1 Diagram of a possible mode of action of a catalyst. (*a*) Two substances (light color) are brought together on the surface of an enzyme, (*b*) react to form a new compound (dark color), and (*c*) the new compound leaves the enzyme, which has not been changed.

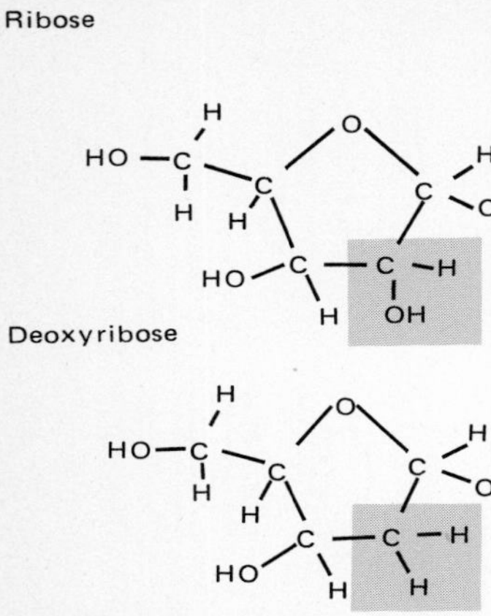

FIG. 13-2 The chemical structure of a ribose sugar and a deoxyribose sugar.

13-3 Translation is the second step in protein synthesis

Translation is the process by which the code sequence of the mRNA is converted into the sequence of amino acid residues in proteins. Translation is the second step in the conversion of genetic information into protein structure. As we have seen, the first is transcription, transferring of the coded message from DNA to messenger RNA.

Ribosomal RNA One of the most conspicuous structures in electron micrographs (pictures taken through an electron microscope) of cells is a system of tubes and sacs that run throughout the cytoplasm (Fig. 13-4) called the **endoplasmic reticulum.** On the walls of some of these tubes are small particles called **ribosomes.** Ribosomes are a central part of the machinery which a cell uses to put together proteins. They are composed of approximately half protein and half RNA. The RNA in ribosomes is called **ribosomal RNA (rRNA).** Ribosomal and messenger RNA are transcribed from different code sequences in the DNA.

Ribosomes are the physical sites for protein synthesis; they form a complex with messenger RNA strands and amino acids. One can picture ribosomes as moving along the messenger RNA strand and "reading" the code. Amino acids are brought to the ribosome and, at each triplet of nucleotides along the messenger RNA strand, the appropriate amino acid is added to a growing protein molecule.

Transfer RNA But how does the amino acid "recognize" the code triplet? How, for instance, does the amino acid called leucine get added onto the chain when the next code unit is UUA? The answer involves the third element in the system—a group of RNA molecules called **transfer RNAs (tRNAs).** A transfer RNA molecule is made up of about 80 nucleotides. The RNA strand folds back on itself because some of the subunits pair with one another and it takes on a cloverleaf shape (Fig. 13-5). One end of an appropriate tRNA molecule is able to recognize a leucine molecule, with the help of a special enzyme, and becomes attached to it using energy from ATP. The tRNA carrying the leucine then goes to the ribosome. At the opposite end of a tRNA molecule for leucine is the triplet AAU. This is complementary to the UUA on the mRNA. Thus it is the tRNA, carrying the appropriate amino acid, that recognizes the triplet on the messen-

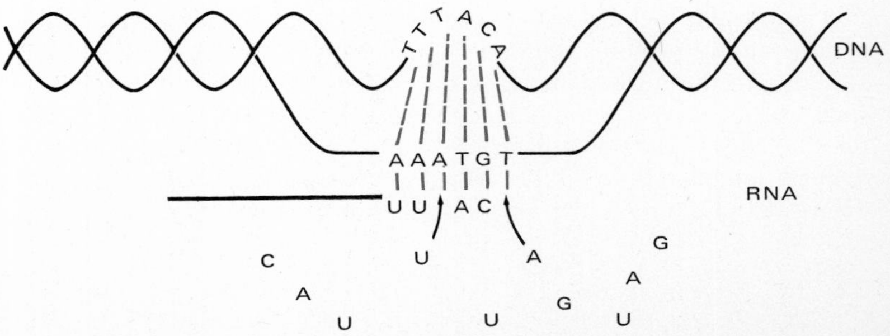

FIG. 13-3 Transcription of a strand of RNA from a strand of DNA.

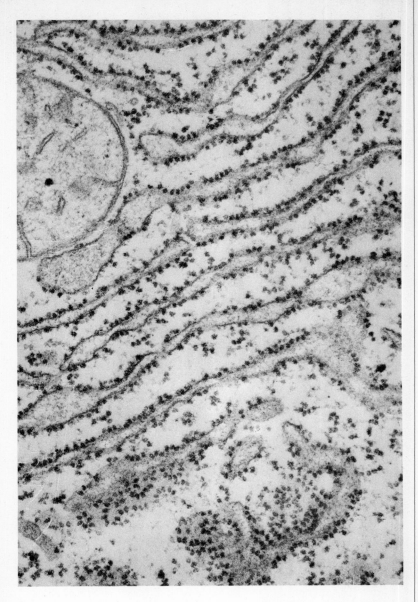

FIG. 13-4 Electron micrograph showing polyribosomes on the surface of endoplasmic reticulum. (Magnification about ×40,000.) (*Courtesy of G. E. Palade.*)

ger and positions the amino acid. Using the energy originally obtained from ATP and under the control of an enzyme, the leucine reacts with the amino acid previously added to the growing protein strand. A molecule of water is released and a strong chemical bond is formed between them (see Fig. 12-4). Thus the polypeptide chain grows, step by step, as each amino acid is added in the proper place.

Usually several ribosomes move along a messenger strand simulta-

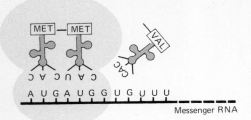 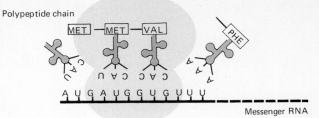

Direction of ribosomal movement ⟶

FIG. 13-5 Formation of a polypeptide chain as a ribosome moves along a strand of messenger RNA. A special amino acid associates with the first AUG which marks the beginning of the polypeptide.

neously. The coded message on that strand is being translated into protein structure by several ribosomes at one time. Each ribosome translates the entire mRNA strand.

13-4 Protein synthesis can be summarized briefly as a series of enzyme-controlled steps

Let's briefly review the transcription-translation sequence now. In a series of steps, controlled by enzymes and using energy from ATP, the following happen:

1 The coded information of the DNA is transferred to messenger RNA in transcription.
2 The messenger RNA moves from the nucleus to the cytoplasm.
3 The messenger RNA associates with ribosomes, where translation occurs.
4 In translation, amino acids are brought to the ribosome-messenger combination by specific transfer RNA molecules. These have the ability to pick up the correct amino acid and to recognize its coded site on the messenger.
5 Amino acids are brought into position and attached one at a time onto a growing protein strand.

Proteins may contain as few as 50 and as many as 60,000 or more amino acid residues. They do not normally function as simple strands. The sequence of amino acids with different chemical characteristics leads to coiling and complex folding of the molecule. Sometimes polypeptides join with other polypeptide molecules. The shape of the protein molecule, especially its pattern of folding, is essential to its functioning (Fig. 13-6). Heating above a certain temperature causes proteins to unfold and lose their ability to function; this is called *denaturation*. For example, heating causes enzymes to lose their ability to catalyze reactions. One of the main reasons that organisms are easily killed by high temperatures is the denaturation of heat-

FIG. 13-6 Proteins may have primary, secondary, tertiary, and quaternary structure. Primary structure (*a*) is the amino-acid sequence; secondary structure (*b*) is the spatial relationship of the amino acids; tertiary structure (*c*) is the way the chain of amino acids is folded; and quaternary structure (*d*) results when two or more polypeptides are associated. Hemoglobin, shown in (*d*), is composed of four polypeptides (disks are iron groups).

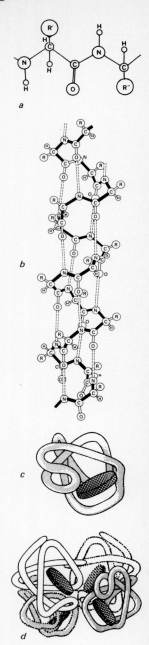

sensitive enzymes. When you boil water to kill bacteria or protozoa in it, you are doing so by denaturing their proteins.

13-5 Synthesis may be controlled in various ways

Negative feedback loops   In some systems that have been studied, accumulation of a certain amount of end product will shut down the assembly line leading to that product. One way this is thought to happen is by the product combining with a **repressor molecule** and thereby activating it. The repressor molecule is then able to combine with the DNA in a way which represses further production of one or more enzymes crucial to making the product.

This kind of control is an example of **feedback loop.** The amount of product produced under the control of a gene has an influence on that gene itself. Information on the amount of product in the system is *fed back* to the gene. In this case, we have **negative feedback** because the product of a sequence of reactions tends to turn off the sequence and to stop its own production.

Negative feedback loops are one way of regulating the production of substances to meet demand. If a product is used as fast as it is produced, the assembly lines will continue to run full blast. If a product starts to accumulate, enough may be around to combine with repressors and activate them. These activated repressors will start to shut the assembly lines down.

Scientists have recently been uncovering more and more details of the elegant, molecular controls of the synthetic activities of cells. For instance, it is now suspected that the different enzymes functioning in many synthetic pathways are controlled by genes which are adjacent stretches of subunits of the DNA. An entire collection of genes may be transcribed under the control of a special gene whose function is to turn production on or off as a unit. Unraveling the mechanisms that regulate gene activity is one of the most exciting challenges in biology today.

Induction   *Escherichia coli* lives in the intestinal tract of many vertebrates. It is a kind of bacterium which is often grown in cultures in the laboratory. All strains of *E. coli* can grow on a growth medium (substrate) containing glucose. In the presence of lactose, a disaccharide made up of one glucose plus one galactose, some *E. coli* begin to make the enzyme β-galactosidase. β-galactosidase catalyzes the splitting of lactose into glucose and galactose. This change in enzyme production, called **induction,** begins slowly as the first lactose molecules can enter the bacteria only by diffusion; shortly, however, there is rapid entry of lactose molecules into *E. coli* cells as another enzyme is induced which helps them cross the cell membrane.

Many bacteria and other cells have this ability to adapt their enzyme systems to their substrates. Certain bacteria, important in the nitrogen cycle, contain enzymes which can change nitrate to nitrogen gas or ammonia only when nitrate is present in their environment.

Certain cells of multicellular organisms also have the property of induction. The human liver, for example, is responsible for detoxifying and putting in suitable state for removal from the body many molecules that could be harmful. This is done by enzymes synthesized in *smooth endoplasmic reticulum* (endoplasmic reticulum without ribosomes). The smooth endoplasmic reticulum of liver cells contains a series of nonspecific enzyme systems which are capable of carrying out detoxification. Alcohol is a substance which is harmful when consumed in large amounts, and these liver enzyme systems are induced by excessive consumption of alcohol. An alcoholic thus can often consume greater quantities of alcohol before appearing drunk than other people because the alcohol is being broken down rapidly. The alcoholic, when sober, also requires greater doses of tranquilizers and certain other drugs than a nondrinker, if these are to be effective. Since the induced enzymes are nonspecific for alcohol, they also break down the tranquilizer or other drugs faster.

MAINTENANCE

Closely related to the problem of synthesis is the problem of maintenance. Organisms are open systems; both matter and energy are continually passing through them. They are something like the ripple that is always found behind a rock in a stream, the little whirlpool in a bathtub drain, or a flame of a candle. Although a wave, a whirlpool, or a flame may last for a period of time, the molecules of water or gases of which they are composed are continuously changing.

13-6 Organisms continually exchange matter and energy with their environment

With the exception of certain tissues like outer layers of the teeth, the molecules of your body are constantly being replaced by new ones acquired with the food you eat. The old ones are constantly being lost as you exhale and as you pass bodily wastes. Similarly, energy is constantly entering your body with food, doing work, and then being degraded to heat. Energy in the form of useless heat is constantly leaving your body. You can feel evidence of its departure when you blow your warm breath on your cold hands in winter or when you perspire. Sweating is a device for speeding the departure of excess heat when you are in a warm environment. When water evaporates, it takes heat with it.

All organisms have devices for keeping this one-way flow of materials and energy going. Animals must take in various molecules to supply them with energy and materials for synthesis. All organisms must take in water, since the chemistry of life can go on only in a watery medium. Water is also continually being lost with wastes and through evaporation. So all organisms drink or take in water through their cell membranes, skins, gills, roots, or leaves or (in some desert animals) make water chemically from their food.

The exchange of gases Beside food and solid and liquid wastes, all animals exchange gases with their environment. Inhalation involves the diaphragm and muscles of the chest and abdomen, which create a partial vacuum in the chest, drawing air into the lungs (Fig. 13-7). Oxygen enters the bloodstream and carbon dioxide leaves through the lung surface. The muscles which produced the inhalation relax, and exhalation takes place. This pushes out air which is now poorer in O_2 and richer in CO_2 than the air inhaled. Other mammals, reptiles, and birds breathe in a similar manner, although reptiles and birds lack the muscular diaphram of mammals.

Fishes get their oxygen by extracting dissolved O_2 from the water in which they live with their gills. The gills bring the blood into very close contact with the water (Fig. 13-8). The blood entering the gill has a much lower concentration of O_2 molecules in it than has the surrounding water. Oxygen diffuses in through the one- or two-celled layer separating the water from the blood and enters the bloodstream. moving from an area of high concentration to one of low concentration. Amphibians, such as frogs, get some of their O_2 through their simple lungs and some by diffusion through their moist skins.

Carbon dioxide is in higher concentration in the blood than it is in the water. Diffusion carries it out of the blood in the gills and into the water. Although oxygen makes up about 21 percent of the atmosphere, often less than 1 percent is found in water. Therefore, getting enough oxygen is a considerable problem for aquatic animals. Fishes must keep a rapid flow of water through their gills (see Fig. 13-8). If they did not, the water around the gills would rapidly lose its oxygen and the fish would suffocate. Some fishes, such as tuna and certain sharks, cannot pass water over their gills rapidly enough by pumping, as can many fishes. They must keep the flow going by swimming continually, and if they stop swimming (for example, by being held in a net), they will die.

Plants exchange gases with the environment without an elaborate "respiratory" apparatus such as lungs or gills. Plants need CO_2 in order to carry out photosynthesis. What they need diffuses directly into photosynthesizing cells from air spaces in their leaves, which are connected to the atmosphere by stomata. The oxygen produced by plant cells is in excess of what the plant cells use for respiration. In fact, all the oxygen in the atmosphere was placed there by photosynthesizing organisms. Monerans and protistans also exchange gases with their environment. Since they are often single-celled, oxygen and carbon dioxide can diffuse directly into their cells. Nitrogen from the environment diffuses directly into and out of those monerans which combine it with oxygen to make compounds that are used in growth and maintenance of themselves and other organisms.

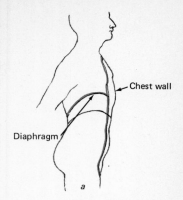

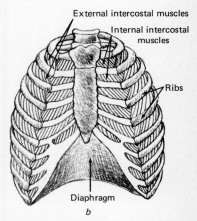

FIG. 13-7 Diagram to show inflation of lungs by expansion of chest cavity through muscular contractions which increase the volume of the rib cage and lower the diaphragm (a). Color shows body at exhalation, black at inhalation. The ribs (b) are moved up and down by the internal and external intercostal muscles.

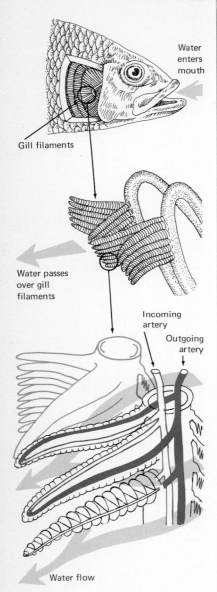

Water enters mouth

Gill filaments

Water passes over gill filaments

Incoming artery

Outgoing artery

Water flow

FIG. 13-8 Diagram to show current flow in the gills of a fish. Water enters a fish's mouth and passes over numerous gill filaments before leaving via the gills. Each gill filament has an incoming artery (bringing oxygen-poor blood) and an outgoing artery (carrying oxygen-rich blood). Gill filaments are arranged in such a way that the flow of water over them is in the opposite direction to the flow of blood in the blood vessels. This *counter-current* arrangement facilitates diffusion of oxygen in and carbon dioxide out of the blood.

13-7 Organisms continuously have to use energy to maintain themselves

Even when an organism seems to be doing nothing at all, it is still using energy. Seeds and spores have reduced their energy use to the absolute minimum, but they still must use some energy to keep alive while waiting to germinate. Why don't people just "turn off" completely at night instead of going to sleep? Sleep is actually quite an active state, and many life processes, especially mental processes connected with dreaming, are very much "turned on." Even organisms that hibernate over the winter do not turn off completely; like seeds and spores, they simply reduce their energy use. Why can't organisms just stop the flow of energy and materials through them completely for a while?

The answer is: *There is no way to stand still in this universe without doing work to maintain your position.* The reason is expressed in the same second law of thermodynamics which results in energy loss in food chains (Secs. 1-5, 1-10, and 2-1), rules out perpetual motion machines, and rules out doing work without creating heat. All the molecules in the biosphere are in constant motion. What is more, the constant motion of molecules, if not countered by some force, leads to greater and greater disorder. Another way of putting it is that every structure in the biosphere is always in a process of breaking down, of becoming disordered, of being dispersed. Unless energy is put into the system to rebuild, reorder, or recollect, things will continue to run downhill. Anyone who has ever attempted to keep a house or even a room clean and orderly should have an instinctive appreciation of the second law of thermodynamics.

From order to disorder The operation of this principle of one-way movement from order to disorder is easily observed in your everyday life. If you put a hot spoon in a dish of cold water, the spoon will cool and the water will be warmed until they are both the same temperature. The hot-cold ordering will disappear. Or if you put an ice cube in a glass of warm water, the cold cube–warm water ordering will be replaced by the less-ordered cool water. The lost order will not reappear by itself in either system. The spoon will not get hot when it is lying in a dish of cool water, nor will an ice cube reappear in the glass of cool water.

The spoon, of course, could be rewarmed; or a cube of ice could be made to re-form in the glass. A lot of *energy*, however, would be used to do any of these things, and a lot of heat would be created in the process.

An important principle to remember is that *energy can be used to create local order in some part of a system, but the disorder of the entire system will always be increased.* This is because heat and disorder are really different ways of looking at the same thing. When we say something is hot, we are saying that its molecules are moving faster than those of something we describe as cold. When we say a process releases heat, we are really saying it makes the molecules around it move faster. This will increase the disorder in the system just as shaking a pan full of ingredients for a cake will increase their disorder (mix them). Heat can be pumped away from the foods in a freezer, but the heat increase in the *entire system*—freezer plus the room around the freezer—will be *greater* than the heat loss from the food! The principle is the same as that of the room air conditioner (see Fig. 1-7).

To return to the question of why we sleep instead of "turning off"—all this means that if you just turned off instead of going to sleep, you would begin immediately to break down. The structure of your cells would start to become disordered. The material of your body would continue to disperse, but would not be replaced. You would very quickly lose any ability to "start up" again; you would be permanently dead.

Temperature and deterioration Even if you were frozen so that decomposers would not aid in your breakdown, you would still deteriorate, but at a slower rate (remember the relationship between heat and disorder). This, for example, is why frozen foods will not keep forever. Neither will frozen bodies. Recently there has been interest in freezing the bodies of people who die of presently incurable diseases, with the hope that they could be revived when, in the future, a cure is found for the disease which killed them. Now, it is true that frogs and insects may revive after being frozen for a short time in ice. And as noted in Sec. 8-12, semen, including human semen, can be frozen and used later to inseminate females. In fact, in 1973, a calf was born which had been frozen for six days as a blastula before it was reimplanted in the uterus of the cow which gave birth to it. Nevertheless, *eventually,* frozen things will deteriorate. As long as there is molecular motion, deterioration occurs. Even liquid nitrogen, which is used in these freezing processes, does not get to $-273°C$ (absolute zero), the temperature at which all molecular motion ceases. (Since temperature is a measure of molecular motion, there can be no temperature lower than $-273°C$.)

DOING THINGS

In addition to using energy for growth, reproduction, and maintenance, organisms also use energy to do things, to carry out movements. **Effectors** are systems or structures which "do things"—that is, they aid in the behavior of an organism or its response to the

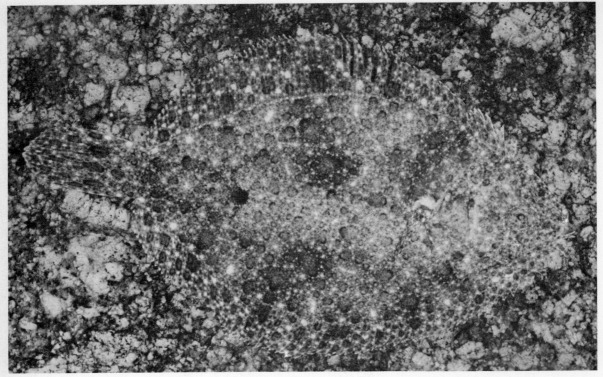

FIG. 13-9 Many fishes are able to change color. A flat fish can adjust its color pattern to match the background on which it lies. (*Courtesy of R. H. Noailles.*)

environment. For example, plants use energy to move their leaves and flowers and to open and close the stomata in their leaves. Movement in plants may be the result of movement of water across plant cell membranes, causing the cells to shrink or swell. This in turn causes the movements we observe. Animals have many kinds of effectors as well.

13-8 Effectors aid in maintaining a steady state in animals

Like the cells of many microorganisms, animal cells often have cilia. Cells that make up the walls of the tubes leading to lungs have such projections, which beat rhythmically. They move a sheet of mucus out of the lungs, carrying away foreign particles that have entered them.

The glands of animals are also effectors. For instance, when you smell delicious foods, your salivary glands do something—they produce saliva. Some animals, such as fireflies, have effectors that produce *light*. Others have special cells in which the distribution of pigments can change so that the animal can change color. Figure 13-9 shows color changes that can occur in a flatfish through the use of such effectors.

Organisms use energy in order to do things, such as moving around or changing their colors. They also use energy to maintain their bodies even when they do not seem to be doing anything. In order to carry out these processes, they must have synthesized the molecules that make up their muscles, glands, membranes, and enzymes. Thus energy is used for activities, for maintenance, and for synthesis in living things. Much of the energy is spent ensuring that the organism maintains a "steady state" suitable for carrying out its life processes (see Chap. 14). To help maintain a steady state, organisms may use their effectors to move to a more suitable environment.

13-9 Muscle tissue is a kind of effector in animals

Animals have specialized effectors, **muscles,** with which they perform movements of parts of the body. Muscles which move skeletal parts are generally arranged in pairs that work against each other. Figure 13-10 shows the opposed **biceps** and **triceps** muscles of the upper arm, which move the forearm. Muscles move things by exerting a pulling force when they shrink or **contract.** Muscles never move things by expanding and pushing. When the biceps muscle contracts and the triceps is relaxed (not exerting force), the forearm is moved toward the upper arm. When the triceps muscle is contracted and the biceps muscle is relaxed, the forearm is moved away from the upper arm.

Muscles are attached to bones by tough, noncontracting tissues called **tendons.** These cordlike tendons can be easily felt on the back of your hands if you spread your fingers as far as you can. This motion is accomplished by contraction of muscles on the back of the forearm which connect to tendons that go across the wrist and the back of the hand to attach to bones at the base of the fingers.
Voluntary and involuntary control If you engage in athletics, you are conscious of your **skeletal** muscles. You use your skeletal muscles for every movement of your body, from rotating your eyeballs to chewing your food or dancing. These are the muscles that are under **voluntary** control. You can contract a great many of them individually just by deciding to do so. These muscles are called **striated** because of their striped appearance under the microscope.

There is another group of muscles in your body that have a very different structure. They are not striated and are called **smooth muscles.** These muscles cannot contract as fast as striated muscles and are mostly associated with the digestive and respiratory tracts. Contractions of the smooth muscles of your digestive tract in waves move the food along. For many years Western scientists believed that almost all smooth muscles were **involuntary muscles;** that is, they could not be contracted by just deciding to do so. However, an obvious

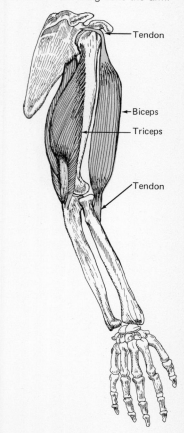

FIG. 13-10 Upper arm of a human showing the biceps muscle which draws the forearm up and the triceps muscle which straightens the arm.

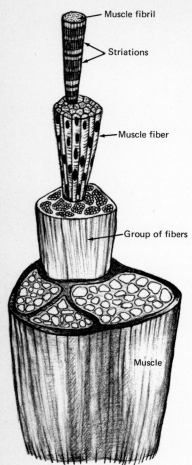

FIG. 13-11 Striated muscle can be subdivided into bundles of muscle fibers. Fibers are greatly elongated cells and have numerous nuclei. Muscle fibers contain fibrils composed of two proteins, actin and myosin (not shown), which are responsible for the contraction of the fibrils and thus of the fibers and muscle.

exception to the belief that smooth muscles are involuntary is the circular **sphincter** muscle that closes the anus.

Now it seems that many of the muscles considered involuntary by Western scientists are not really involuntary at all. Most people just never bother to learn to control them. In India dedicated yogis have been controlling their involuntary muscles for centuries. In fact, some of them have gone to such extremes as to completely reverse the direction of intestinal movement, with the expected result. In this country, it has been shown that many other "involuntary" functions such as heart rate, blood pressure, temperature, and even urine production can be controlled voluntarily in rats if they receive the proper feedback. There is still some question, however, whether there is a direct link between the voluntary and involuntary nervous system or whether the rats are using voluntary muscle contractions to accomplish these other controls. A promising beginning has been made in teaching people to control heart rate and blood pressure by the technique of **biofeedback.**

The heart itself is interesting since its muscle has a different structure from either striated or smooth muscle. It has striations which are more widely spaced than those of striated muscles and is capable of beating, at least briefly, without any outside nervous input. Control of heart rate is one of the more successful biofeedback experiments. Although most people never learn to control the contractions of their hearts, some people do, and they are able to control the rate of their heartbeat at will.

Muscle contraction The mechanisms of muscle contraction are partially understood. Figure 13-11 shows the details of the structure of a voluntary striated muscle, such as the biceps or triceps. Notice that it is made up of bundles of **fibers,** each of which is made up of many cells that have become united in the course of their development. These cells are richly supplied with mitochondria, which supply the energy for muscle contraction. Muscle fibers consist of subunits called **fibrils.** These in turn are made of filaments of two different proteins, one having thick strands (*myosin*) and one having thin strands (*actin*). Muscle contraction occurs when bundles of these filaments move past each other, shortening the muscle. The energy for muscle contraction comes from ATP. The exact method by which the filaments are pulled past each other, however, remains unknown. There are theories suggesting that there are cross-bridges between thick and thin strands that pull them together.

QUESTIONS FOR REVIEW

1 Why would life (at least as it is now constituted) be impossible without enzymes?

2 What do these terms mean: metabolism, synthesis, biofeedback, catalyst, repressor molecule, enzyme, effector system, structural protein, RNA, feedback loop, negative feedback loop, induction?

3 Organisms are said to be *open systems.* What does this mean? In what way is an ecosystem more closed than an individual organism? Is the biosphere more open or more closed than an ecosystem? Why? Can you think of any system, on whatever scale (of whatever size), that is completely closed?

4 What is the role of ribosomes?

5 Under what environmental conditions do certain organisms—land plants and amphibians, for instance—"slow down" so much that they seem to be almost "turned off"? During which of their developmental stages do seed plants seem very close to "off"?

6 When you pull against an object, your muscles are pulling (contracting). When you push against something, your muscles which are doing the work are also pulling. How can a pulling muscle enable you to exert a pushing force? Explain by describing how your forearm is moved relative to your upper arm by your opposing sets of arm muscles.

7 Why is the discovery that various *involuntary* muscles can be voluntarily controlled, a finding that is of considerable interest to physicians and others concerned with the advancement of human health?

READINGS

Helinski, D.: "Molecular Genetics," *Biocore,* unit VIII, McGraw-Hill, New York, 1974.

Lehninger, A. L.: *Bioenergetics,* 2d ed., W. A. Benjamin, Menlo Park, Calif., 1971.

Munger, B.: "Animal Diversity: Cells and Tissues," *Biocore,* unit X, McGraw-Hill, New York, 1974.

Nomura, M.: "Ribosomes," *Scientific American,* **221**(4):28–32, offprint 1157, 1969.

Peachey, L. D.: "Muscles and Motility," *Biocore,* unit XVIII, McGraw-Hill, New York, 1974.

Roller, A.: *Discovering the Basis of Life,* McGraw-Hill, New York, 1974.

Wolfe, S. L.: *Biology of the Cell,* Wadsworth, Belmont, Calif., 1972.

Wood, W. B.: "Molecular Design in Living Systems," *Biocore,* unit II, McGraw-Hill, New York, 1974.

———: "The Molecular Basis of Metabolism," *Biocore,* unit III, McGraw-Hill, New York, 1974.

Yanofsky, C.: "Gene Structure and Protein Structure," *Scientific American,* **217**(5):80–94, offprint 1074, 1967.

14

THE STEADY STATE

SOME LEARNING OBJECTIVES

After you have studied this chapter, you should be able to

1. Name three general *problems* of organisms which are discussed in this chapter, and state which type(s) of organisms—Monera, Protista, Metaphyta, Metazoa—must cope with each problem.

2. Demonstrate your knowledge of homeostasis and its mechanisms by doing the following:
   a. Distinguish between physiological and behavioral control of body temperature; give an example of each; state which of them are used by each of the following animal types—insects, reptiles, birds, mammals; and indicate which of these animal types can maintain a steady body temperature physiologically.
   b. Explain why water conservation is so important to land animals, name four ways in which the human animal loses water, and name three mechanisms by which humans reduce their water losses.
   c. Trace the process by which the body converts and excretes the metabolic waste product nitrogen, beginning with the prior process (name it) of which it is a by-product and ending with its passage from the body.
   d. Describe the mechanisms by which the body conserves (minimizes the loss of) water in the

process of eliminating nitrogenous wastes.
   e. Name at least three human metabolic disorders caused by blockages in the metabolic-breakdown pathway of the amino acid phenylalanine.

3. Name the two major coordination systems discussed in the chapter, say which of them provides high-speed coordination, and explain why most animals need a high-speed system as well as the other coordination system.

4. Demonstrate your knowledge of the human hormonal system by
   a. Stating how hormones are transported throughout the body.
   b. Naming the production site, the target(s), and at least one function of each of these human hormones: thyroxine, adrenaline, estrogen, insulin, testosterone.
   c. Naming the hormone which is called the "growth hormone,"

State its targets, and state the effects of its overproduction and underproduction in children.

5. Demonstrate your knowledge of how the nervous system functions by doing the following:
   a. Name the basic structural unit of the system, and describe how the nerve impulse is initiated and propagated (spread).
   b. State why the nerve impulse is termed an electrochemical response.
   c. Name the junction between two neurons, describe the process by which the nerve impulse crosses it, and explain how nerves are able to transmit nerve impulses in one direction only.
   d. State what is meant by calling a nerve impulse an "all-or-nothing" response to a stimulus, and describe the two processes by which this basic mechanism is employed to measure graded increases (or decreases) in the strength of a stimulus.
   e. State the function or purpose of reflexes, name the three major structural elements of reflexes, name one human reflex, and indicate one or more ways in which the behavior involved in this reflex has survival or other adaptive value for the human animal.

6. Further demonstrate your knowledge of nervous coordination by

a Describing what happens when you *see* an object—from the entry of light into the sensory organ to the visual image formed in the coordinating organ.
  b Name five sensory modes mentioned in the chapter.
  c Name three sensory modes other than vision which some nonhuman animals heavily depend on and, for each, name at least one animal that does so.
7 Name:
  a Four protective mechanisms shared by all vertebrates, including humans.
  b Two protective mechanisms, not covered in (a) above, which are both employed by some animals and some plants.
8 Demonstrate your knowledge of the immune response by
  a Stating what kinds of organisms employ it and what it protects against.
  b Describing its operation in terms of the functioning of antibodies, antigens, and interferon.
  c Saying what the process of artificially inducing it is called, and giving two examples of it (one mentioned in the chapter and one not mentioned).
9 Name, describe some symptoms of, and state some known causes of one disease of each of the following types: immune, autoimmune, environmentally induced, functional, and iatrogenic.
10 Explain why noninfectious diseases, such as those referred to in learning objective 9, are more prevalent today in this country than they used to be.

Organisms require a steady input of energy and materials, not just for growth and reproduction but also, as we saw in Sec. 13-7, for maintenance. Within limits, unicellular organisms can control the concentration of chemicals in their cells; some can also move to a more favorable location. Beyond this, unicellular organisms do well or poorly depending on their external environment. Multicellular organisms, however, maintain a relatively constant internal environment so that their cells are sheltered from some of the rigors and changes in the external environment. In a multicellular organism, each cell normally has sufficient oxygen provided in its environment. Many multicellular organisms are able to control their internal temperature so that it is neither too hot nor too cold for the enzymes in their cells to perform optimally. Multicellular organisms also have means for ensuring that the concentration of chemicals in their body fluids is just right to give their cells a suitable chemical environment in which to carry out the chemical reactions of metabolism properly. If the internal environment of an organism gets out of balance, then disordering processes outrun ordering processes, and the organism dies.

HOMEOSTASIS

Thus, much of the energy an organism spends on maintenance goes toward ensuring the stability of its internal environment. The steady state in which this environment is maintained is referred to as **homeostasis**. The mechanisms which produce it are called **homeostatic mechanisms**. The mechanism that speeds up your heart when you run and thus delivers more oxygen to your muscles is an example of a homeostatic mechanism.

14-1 Control of body temperature is a homeostatic mechanism

As you probably know, your body temperature is "normal" when it is 37°C (98.6°F). At about this temperature the chemical reactions of your metabolism run most efficiently. Suppose that you go outside in winter without sufficient clothing and your body begins to lose heat too rapidly. What happens? First, the blood vessels in your skin contract so that less blood gets near the surface to be cooled. Then you begin to shiver. This rapid muscular work produces heat as a side product, just as the second law of thermodynamics (discussed in Chaps. 1, 2, and 13) dictates. Both of these reactions warm your body.

Now, suppose that, like most people, you find shivering unpleasant and you run inside and settle down in front of a blazing fire. Your skin becomes flushed as the muscles of your blood vessel walls relax and let blood flow into your skin, where it can release heat. You begin to perspire, and the evaporation of the perspiration further cools you. These reactions to heat and cold are, of course, homeostatic.

Mammals and birds all have physiological mechanisms for maintaining body temperature at an optimal level. Other vertebrates and invertebrates generally lack such mechanisms. Their body temperature changes with the temperature of their environment. This does not mean, however, that they do nothing to regulate their body temperatures. They too have temperatures that are more or less optimal for their metabolic activities. When it is too cold, these animals are sluggish or inactive. Many of them, such as butterflies and lizards, will orient themselves to expose a maximum of their bodies to the rays of the sun on cool mornings (Fig. 14-1). This permits them to warm up to normal activity temperatures more rapidly. The high

FIG. 14-1 Lizards are able to regulate their temperature behaviorally. (*a*) In late afternoon this lizard lies parallel to the sun's rays keeping a body temperature appropriate for activity without overheating; (*b*) early in the morning it is buried in the sand with the head exposed to the sun, warming the blood in the head; (*c*) at noon the lizard avoids overheating from the sun by finding shade.

mountain *Colias* butterfly in Fig. 14-2 has dark pigments on the parts of the wing which lie directly over the body and come between the body and the sun. These pigments absorb more light energy from the sun than do light pigments (black absorbs almost all visible wavelengths; white absorbs almost none) and permit the butterflies to warm to "operating temperature" quickly.

14-2 The problems of wastes and water

One of the most difficult homeostatic problems faced by land animals is conserving water. Water is necessary for all metabolic reactions, and it, of course, is intimately involved with maintaining the correct concentration of various chemicals in the cells. For example, the sodium concentration in a cell would double if the amount of sodium in a cell remained constant while half the water was lost.

Mechanisms preventing water loss Terrestrial animals are always losing water. You lose water by evaporation from your skin and lungs. You lose water with your feces and urine. You must take in water to make up for these losses, or changes in the concentrations of substances will quickly make it impossible for vital reactions to take place. Your metabolism will break down, and you will die. If a man is lost in the woods and walks at night and rests during the day, he will die in seven days if he does not find water. This is assuming that the maximum daily temperature in the shade is 27°C (80°F). At 49°C (120°F) this man would survive only about one day. A person doing any work at all needs at least 2 liters of water a day to maintain water balance. In contrast, a person given water but no food can survive for quite a long time. Just how long depends on how much reserve fat is stored in the body, but it will be much longer than seven days. An overweight 34-year-old woman fasted under medical supervision for 236 days without apparent harm and reduced her weight from 281 to 184 pounds. In a protest hunger strike in 1920, Terrence MacSwinney, a man of normal weight, survived—without food—for 74 days before he died of starvation.

You would have to drink much more water than you do if it were not for various ways that your body has for limiting water loss. Your skin is relatively waterproof. Unless your sweat glands are active, you lose relatively little water through your skin. If the oxygen-exchange surface of your lungs were exposed rather than infolded as lungs, you would lose much more water through them than you do. But you do lose quite a bit, as blowing your breath against a cool window shows.

Excretion of nitrogen Another way in which your body controls water loss is by regulating the water content of your urine. Why should urine be produced at all? It is needed to allow the body to excrete certain wastes which are disposed of more easily when they are in solution. One of these is nitrogen. Often your diet may contain more amino acids than can be used immediately for protein synthesis. These extra amino acids are converted into carbohydrates or fats that can be stored or used for energy. Amino acids contain nitrogen, while

a

b

FIG. 14-2 Butterflies of the genus *Colias* occur in relatively warm lowland areas and in colder, alpine habitats. Those of alpine habitats (*a*) have dark pigment in the base of the wings and orient themselves perpendicular to the sun's rays, which warm the dark areas and the flight muscles behind. They are thus able to become active earlier in the day than would be possible if they were lightly pigmented like the lowland forms (*b*). (*After W. B. Watt,* Evolution, **22** (*3*): *437–458, 1968.*)

carbohydrates and fats do not. A major product of the conversion of amino acids to carbohydrates or fat is ammonia (NH_3). This nitrogen-containing substance is extremely poisonous and is immediately changed into a less toxic compound in your liver. This compound is called **urea.** From the liver, urea is released into the bloodstream. If too high a concentration built up in your blood, you would be poisoned; that is one of the things your kidneys prevent.

Reabsorption of molecules by the kidneys   How do your kidneys remove urea and other wastes from your blood? They filter the blood (Fig. 14-3) and produce a liquid medium, **urine,** which can be excreted with the waste products in it. There are three processes involved in the formation of urine: glomerular filtration, tubular reabsorption, and tubular secretion. These take place in the many thousands of structural units of the kidney, called *nephrons*. A nephron consists of a glomerulus (tiny knot of capillaries), Bowman's capsule (receiving chamber), and an extensive system of tubules. Much of the liquid part of the blood passes through the walls of the capillaries of the glomeruli and the Bowman's capsules (see Fig. 14-3). Left behind are the blood cells and protein molecules. This is glomerular filtration. Of the 24 percent of the heart's output that passes through the kidneys each minute, 125 milliliters (about half a pint) are filtered from the glomeruli to Bowman's capsules. This fluid contains waste molecules, molecules the body needs to save, and water the body needs to save. The second process is tubular reabsorption. Of the 125 milliliters of fluid filtered from the glomeruli, 124 milliliters are reabsorbed by the kidney tubules and go back into the bloodstream. Only 1 milliliter (0.03 ounces) goes on to become urine. The kidney tubules are very selective about what molecules they reabsorb with the fluid. They use both active and passive transport to ensure that needed molecules are not lost and that wastes are. Finally, the third process involved in urine production is tubular secretion. The kidney tubules also can use active transport to move substances from the vascular system into the urine. At the last portion of a tubule before it joins a collecting tube, hydrogen ions resulting from metabolic processes are actively secreted into the urine, helping to maintain the acid-base balance of the body. Some drugs, such as penicillin, can be secreted by kidney tubules. In the presence of a certain hormone, even more water can be absorbed from the urine. The amount of urine produced depends on water consumption and temperature and many other factors. Urine collects in tubes and in the ureters, which convey it to the bladder, where it is stored until it is voluntarily passed.

Your kidneys, and the kidneys and similar excretory organs of other animals, can be remarkably sensitive in what they excrete and what they reclaim. Kidneys are major devices for maintaining homeostasis. If the concentration of, say, glucose in the blood reaches a certain level, the kidneys will stop reabsorbing it. Normally *all* sugar is reabsorbed by the kidneys. Therefore, frequent, excessive amounts of sugar in the urine are a sign of an abnormally high concentration of sugar in the blood. When this occurs regularly, it is a sign of the

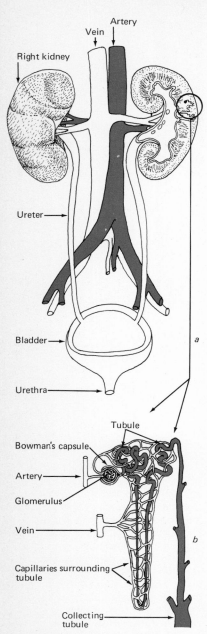

FIG. 14-3 Structure of human kidneys: (*a*) general anatomy of kidneys and associated structures (artery shown in color); (*b*) detailed view where color indicates the filtrate from the blood which becomes urine or is reabsorbed.

metabolic disorder known as *diabetes*. Diabetes results from failure of the pancreas to produce a hormone known as insulin, which plays a vital role in sugar metabolism. It especially affects the storage of carbohydrate by the liver and its distribution to the cells as fuel. Nowadays, a person with diabetes can live a nearly normal life if he or she follows a physician's instructions. Usually the diabetic must follow a proper diet and take insulin by injection. A diabetic usually checks his or her own urine sugar frequently to be sure that diet and insulin injections are properly adjusted. Diabetics should always tell their friends about their problem because they could go into a coma. A diabetic coma can result from either too much or too little insulin. A physician uses many symptoms to diagnose which it is; one symptom of lack of insulin is a smell of acetone (often mistaken for alcohol) on the individual's breath.

Excretion of ammonia Very few animals excrete urea as their nitrogen-containing waste. Aquatic animals excrete ammonia. As you would expect, fishes living in the sea continuously lose water because they have a higher concentration of water internally than does the sea around them. This is another way of saying that they have a lower salt concentration in them than the salt concentration of the sea. They use energy continually to move salt out of their bodies by way of specialized cells in their gills, and water flows out at the same time (Fig. 14-4). They must drink continuously to make up for this water loss. Since a great deal of outgoing water is available to dilute the toxic ammonia, they do not use energy to convert ammonia to less toxic urea. They simply excrete the ammonia through their gills.

Excretion of uric acid Vertebrates that lay shelled eggs have a special problem with excretion. While their embryos are developing inside their eggs, there is no way for toxic nitrogen wastes to be excreted. Both urea and ammonia are soluble in water; the embryo is floating inside the egg in a protective medium which is largely water. How can the embryo keep from being killed by its own waste products? It excretes its nitrogen wastes in an insoluble form—uric acid. Uric acid is stored as a solid inside the egg. Sometimes, after an egg has hatched, you can see a white powder left inside the shell; this is uric acid produced by the embryo during its development. Generally reptiles, birds, and insects use energy to convert their waste nitrogen to uric acid. Adults of these organisms, as well as embryos, use this same excretory pathway. The white part of a bird dropping is uric acid.

The human kidney If one of your kidneys is damaged by injury or disease, your other kidney can take over the entire job done by both kidneys previously. This allows you to donate a kidney for a transplant to someone else who has had both kidneys fail. Generally, kidney transplants are done between close relatives because of the problems of rejection (Sec. 14-13). Since kidneys are such important organs, it is indeed fortunate that we have two of them.

The kidney is an internal organ for which a somewhat satisfactory replacement has been devised. Artificial kidneys have been devised to carry out the cleansing of the blood using a system of coiled

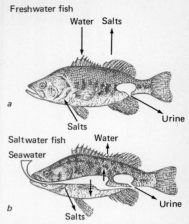

FIG. 14-4 Freshwater fishes and saltwater fishes face different problems since freshwater fishes tend to gain water by osmosis while saltwater fishes tend to lose water by osmosis. Freshwater fishes such as a largemouth bass (*a*) do not drink water and produce large quantities of urine. Saltwater fishes such as a blue rockfish (*b*) drink saltwater, excrete the salt by the gills, and do not produce large amounts of urine.

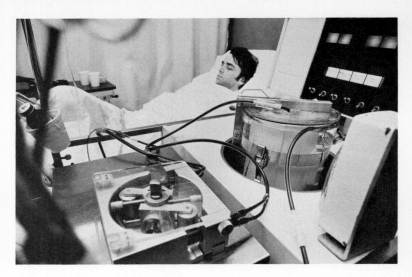

FIG. 14-5 Blood from this person is being diverted through a kidney machine which performs the same life-supporting functions his own kidneys can no longer do. (*Courtesy of Robert Goldstein.*)

membranous tubes in a tank of osmotically suitable fluids (Figs. 14-5 and 14-6). Numerous people with otherwise fatal kidney disease are kept alive today by being "plugged in" to a kidney machine several times a week. In 1974, there were 13,000 people in the United States being kept alive by kidney machines. Many of these were waiting for suitable kidneys so they could have a transplant operation. Children on kidney machines do not grow up physically, or mature; in order to grow, they must receive a suitable kidney transplant. Kid-

FIG. 14-6 Diagram to show how a kidney machine works. The blood flow is shown by arrows and color. Waste substances move through the walls of the coiled tubing into the tank.

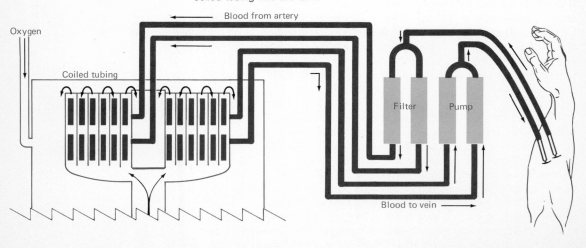

neys used in transplant operations often come from the bodies of people who die accidental deaths; in 1974 there was an approximately 20 percent reduction in deaths from highway accidents. This resulted in fewer kidneys to use in transplant operations and more people with a continuing need for kidney machines. Some of the social implications of transplants and artificial organs are discussed in Chap. 17.

14-3 The breakdown products of many reactions must be excreted

During metabolism, complex molecules of all kinds are put together or taken apart in series of reactions, one step at a time. Each reaction is catalyzed by a different enzyme produced at the direction of a different gene. Therefore, the absence or inactivity of an enzyme at any step in the sequence will affect the final outcome. An interesting example is phenylketonuria (PKU) and a series of related genetic disorders caused by the absence of enzymes essential to the breakdown and excretion of the amino acid phenylalanine. Phenylketonuria causes mental retardation unless treated (see Table 9-3). The other metabolic disorders caused by blocks in this pathway as diagrammed below are (1) albinism, a condition in which the skin, hair, and eyes lack pigment; (2) cretinism, a condition of thyroid deficiency leading to retarded physical and mental development; (3) tyrosinosis, a rare defect requiring no treatment, which allows hydroxyphenylpyruvic acid to accumulate in the urine; and (4) alkaptonuria, a condition with arthritislike symptoms in which the urine turns dark on exposure to air.

In the diagram on page 400, dashed lines indicate intermediate reactions which have been left out of the diagram. Tyrosine can be a direct breakdown product of protein since it is an amino acid; therefore, an absence of enzyme A does not lead to a deficiency of tyrosine and its breakdown products. Notice that three of these metabolic abnormalities result from the buildup of products that cannot be broken down and excreted because of a missing enzyme (PKU, tyrosinosis, and alkaptonuria), while the two others result from the absence of necessary products because the enzymes needed for their synthesis are missing (cretinism and albinism).

14-4 Homeostatic systems can be overloaded

A person under the influence of alcohol is more sensitive than when sober to drugs and other toxic agents. One explanation is that although alcohol is detoxified by the enzymes of the smooth endoplasmic reticulum, if this system is overloaded with alcohol, the smooth endoplasmic reticulum enzymes cannot detoxify *other* substances. These then may reach the brain and depress its function, sometimes to the point of death. Many drugs prescribed by physicians may be dangerous when used together with alcohol. The combination

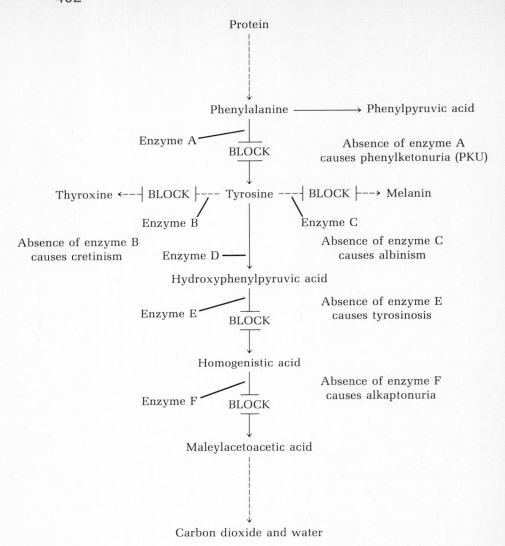

of alcohol and barbiturates or tranquilizers is particularly dangerous. Accidents when driving or using heavy machinery have occurred, as well as accidental deaths from taking sleeping pills and alcohol.

THE PROBLEM OF COORDINATION

It would do little good if, when your biceps contracted to bring an ice cream cone to your mouth, its contraction was so violent that the cone was mashed into your eye instead. Nor would things be very satisfactory if your pancreas squirted digestive enzymes into your small intestine when there was nothing to digest there. You would not last long if you shivered when you were hot and sweated when you were cold. In short, you, like all other animals, face the problem of coordinating your activities. If you are to maintain homeostasis, survive, grow, or reproduce, you have to do the right things at the right times.

14-5 Coordination generally involves feedback loops

You get hungry; you eat; you feel satisfied. That is, the action resulting from getting hungry produces a feedback which makes you less hungry. This is an example of negative feedback. If it were positive feedback, the more you ate, the hungrier you would get. We shall leave to your imagination the end result of such a system.

Digestion of fat is coordinated by a feedback system utilizing a chemical messenger—a hormone. When fats enter the small intestine, they cause the intestinal cells to release a hormone (cholecystokinin) into the bloodstream. That hormone, in turn, stimulates the gall bladder to contract and release bile into the intestine. The intestinal cells stop releasing cholecystokinin after the fat has been digested and passed on.

14-6 One major coordination system uses chemical messengers

Coordination may be accomplished by two major systems in animals. These are the **hormonal system** and the **nervous system.** Hormones are produced at specific sites called **endocrine glands** (glands without ducts) and carry their chemical messages through the circulatory system. Although all the cells in the body are thus exposed to hormones, a specific hormone affects only its **target cells.** For example, only the cells of the gall bladder respond to the hormone released by the intestine in response to the presence of fat.

Cyclic AMP, a form of adenosine monophosphate in which the phosphate group is in the form of a ring, plays an important role in the production of hormones by target cells. For example, thyroid cells have receptors that "recognize" thyrotropin, a pituitary-produced hormone that stimulates the thyroid cells to produce their characteristic hormone (thyroxine). The binding of thyrotropin to the thyroid cell receptor sites stimulates production of an enzyme in the cell membrane which produces cyclic AMP from the abundant supply of ATP in the cell. Cyclic AMP is then free to diffuse through the cell, where it acts as a "second messenger," instructing the thyroid cell to respond in its characteristic way by producing thyroxine. Cells of other glands also use cyclic AMP as a second messenger, but their receptor sites accept only their own first messengers and produce only their own products.

Some control systems involve both the hormonal and the nervous systems. The pituitary gland lies just under the brain and can release several hormones. This gland is controlled by nerves descending from various parts of the brain, nerves which secrete substances at their ends that control the hormonal secretions of the pituitary gland. This process is called **neurosecretion.** For example, when cells in the brain detect an increased concentration of chemicals (high osmotic pressure) in the blood, they stimulate the pituitary gland to release an increased amount of the hormone called antidiuretic hormone (ADH). This hormonal message circulates through the body and is "read" by the kidney tubules, which then increase their reabsorption of water from the tubule (Fig. 14-7). This adds more water to the blood plasma, which, in turn, decreases the concentration in the blood of the chemicals that started the process.

The many roles of hormones   There are many different hormones that play crucial roles in all aspects of life (Table 14-1). In men, the testes produce a hormone, **testosterone,** which, among other things, stimulates the growth of the penis at puberty. This hormone is also responsible for the development of male **secondary sexual characteristics:** distribution of pubic hair, beard, a deep voice, etc. In women, the ovaries release a number of hormones, for example, several kinds of **estrogen** and **progesterone.** These produce the changes undergone at puberty, such as the beginning of menstruation and the enlargement of the uterus. These hormones also stimulate the development of female secondary sexual characteristics: distribution of pubic hair, broader hips, and growth of the breasts.

Although most hormones have specific targets, some hormones have quite general activity. For example, there is one pituitary hormone that stimulates growth of the whole body. It is called **growth hormone** or **somatotropin** and is necessary for the regulation of growth. In children, too little results in their becoming dwarfs and too much results in giants. In adults, too little somatotropin results in a disease in which the body ages prematurely; too much results in acromegaly, a disease in which the feet and hands become enlarged and the bones of the face and skull become thickened.

The hormone **adrenaline** is produced in small glands that are attached to the kidneys and consist of an outer *cortex* and an inner *medulla*. Large amounts of adrenaline enter the bloodstream at times of fear or other stress. Adrenaline prepares the body for "fight or flight" reactions—to do battle or to flee. It speeds the heartbeat, stimulates the release of stored sugars into the blood, causes decreased blood flow to the digestive tract, stops the muscles of the digestive tract from moving food, and causes the hair on the head and body to "stand up." The functions of these preparations seem obvious. Fuel-rich blood goes to the muscles instead of the digestive tract, preparing for increased muscular activity. The matter of the hair standing on end seems to be an evolutionary remnant from our remote ancestors who had more hair. A reaction common in other mammals, it is believed to help in cooling their bodies during stress, and in many mammals it seems to be a threat signal. It makes a threatened (or threatening) animal look larger to its opponent. Often

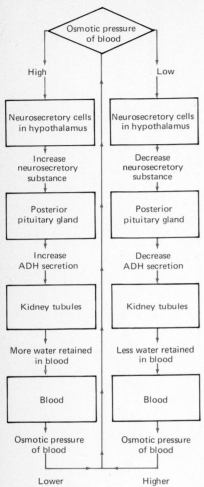

FIG. 14-7 One model of the way ADH (antidiuretic hormone) maintains the water content of the blood by a negative feedback mechanism. Cells in the hypothalamus of the brain sense the osmotic pressure of the blood and respond by "telling" the pituitary to make more or less ADH which circulates in the blood to the kidney tubules and "tells" them to conserve more or less water in the blood depending on whether the osmotic pressure of the blood is too high or too low.

TABLE 14-1 SOME HUMAN HORMONES OR HORMONELIKE SUBSTANCES

HORMONE	SITE OF PRODUCTION	TARGET	SOME FUNCTIONS
Antidiuretic (ADH)	Hypothalamus (stored in pituitary gland)	Distal kidney tubule	Water reabsorption
Oxytocin	Pituitary gland	Pregnant uterus	Muscle contraction
Adrenocorticotropin (ACTH)	Pituitary gland	Adrenal cortex	Influences adrenal cortex hormones
Somatotropin	Pituitary gland	Whole body	Influences growth of tissues
Thyrotropin (TSH)	Pituitary gland	Thyroid gland	Influences structure and secretion of thyroid gland
Gonadotropic hormones (FSH, LTH, LH, etc.)	Pituitary gland	Gonads (ovary or testes)	Influences gonadal functions
Thyroxine	Thyroid gland	Many parts of body	Influences metabolic rate and growth and development
Parathormone	Parathyroid glands	Special cells	Regulates blood calcium and phosphate and readiness of nerves and muscles to respond to stimuli
Unnamed hormone	Thymus	Unknown	Needed in youth if antibodies are to be produced
Glucocorticoids (cortisone, etc.) Mineralocorticoids (aldosterone, etc.)	Adrenal cortex	Many	Regulate ion and water balance; influence fat, carbohydrate, and protein metabolism; activity of lymphatic tissue and sex organs; permit response to stress
Adrenaline and noradrenaline	Adrenal medulla	Many	Reserve mechanism for stress responses
Angiotensin	Liver (in presence of renin from kidneys)	Kidney tubules	Regulates sodium and water in blood affecting blood pressure
Insulin and glucagon	Pancreas	Many	Regulate carbohydrate metabolism
Gastrin	Stomach	Stomach glands	Causes stomach to secrete acid
Enterogastrone	Duodenum	Stomach	Inhibits stomach secretions and movements
Secretin	Intestine	Pancreas	Stimulates secretion of water and ions
Pancreozymin	Intestine	Pancreas	Stimulates secretion of enzymes
Cholecystokinin	Intestine	Gall bladder	Causes gall bladder to empty itself of bile
Estrogens	Ovaries and adrenal cortex	Fallopian tubes, breasts, bones, pelvis, blood	Enhance activity of cilia in fallopian tubes; initiate growth of breasts at puberty; cause early ending of bone growth; cause widening of pelvis; control calcium and phosphate content of blood
Progesterone	Ovaries and adrenal cortex	Uterus, breasts, kidney	Prepares uterus for implantation of ovum; enlarges breasts and causes cells to secrete; enhances ion and water reabsorption in kidneys
Testosterone	Testes and adrenal cortex	Male sex organs, skeleton, muscles	Causes development of male organs; increases size and strength of bones; increases growth of muscles
Chorionic gonadotropin	Placenta	Same as estrogens, progesterone, and gonadotropins	Maintains pregnancy
Prostaglandins	Various tissues	Many	Many (see text)

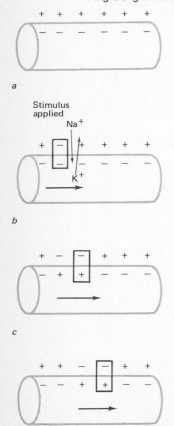

FIG. 14-8 A nerve impulse results from a change in the arrangement of positive (+) and negative (−) charges at the surface of a nerve cell. In the sequence a–d an impulse is shown moving along an axon.

the hair down the middle of the back is raised especially high, for example, in an angry domestic cat. The threat message is clear to any person who has previously dealt with an angry cat.

Substances which act like hormones Prostaglandins are hormonelike (not produced by endocrine glands) substances which affect a wide range of physiological processes. They were first discovered in the 1930s in semen, hence the name *prostaglandin,* referring to the prostate gland. Actually, as was later discovered, they occur in many body fluids and tissues. As of 1970, 14 of these substances had been identified in the human body. Prostaglandins have many different effects, and different ones may have opposite effects. They are known to stimulate smooth muscle to relax or contract, lower or raise blood pressure, slow down stomach secretions, and induce uterine contractions.

Prostaglandins are found in very small quantities in the body and are rapidly broken down; however, there is great interest in their use in medicine. They have been used to induce labor in pregnancy, and their possible use in inducing abortion is being investigated. They are also being investigated as therapeutic agents in relaxing bronchial spasms, regulating blood pressure, clearing nasal passages, and regulating metabolism.

14-7 The nervous system provides high-speed coordination

There are many necessary functions that hormonal systems cannot control with sufficient speed and precision. A mouse threatened by an angry cat cannot wait for a hormonal message moving slowly through the bloodstream to activate its muscles and allow it to escape the cat's pounce. Hormonal messages are also too slow for a small bird flitting from tree to tree or a person driving an automobile, who may have to respond very quickly to changing conditions. In human beings, the fastest-traveling blood moves at about 2.4 meters (8 feet) per second; however, it is much slowed down in capillary beds. Between 2 and 3 meters per second is too slow to transmit a message in an emergency. The nervous system supplies faster message and coordination service. The maximum speed of transmission in the nervous system is 120 meters (394 feet) per second!

The basic unit of the nervous system is the **neuron,** a usually elongated nerve cell. Neurons are hooked up into systems that pass information **electrochemically** from one part of the body to another. The ability of these cells to communicate with each other in this way has evolved from a general property of all cells. This property is the ability all cells have to maintain an **electrical potential** across their cell membranes. Many of the atoms and molecules dissolved in the fluids inside and outside of cells carry electrical charges. The distribution of these charged chemicals is usually such that there is a potential—a *difference in electric charge* between the inside and the outside of the cell.

The membrane of a neuron has the ability to respond to a **stimulus** (change) in its environment. The stimulus may be received from outside the organism by special nerve cells for detecting light, heat, etc., or it may be internal in origin. When a neuron receives a stimulus, its membrane alters the distribution of some of the electrically charged particles. The positive ions are Na^+ and K^+ and the negative ions are Cl^- and certain organic compounds. At first, Na^+ ions rush into the cell, then K^+ ions rush out as the electrical charge difference between inside and outside is briefly *reversed*. Then the original charge state is quickly restored.

The nerve impulse A membrane of a neuron accomplishes the shift in distribution of charged particles by both active and passive transport in a manner not completely understood. This shift in **membrane potential** is the beginning of a **nerve impulse** (Fig. 14-8). The shift first occurs at the point of stimulus; the change at this point in the neuron triggers a similar shift in the adjacent parts of the membrane. This then triggers another shift on either side; these shifts trigger further shifts beyond them, and so it goes down the neuron in *both* directions if the stimulus began in the middle. The nerve impulse is very much like a fuse lit in the middle and burning in two directions.

Transmitter substances In the nervous system, the two-directional membrane change can become a one-directional message. At a junction between two neurons, there is a very narrow gap called a **synapse** (Fig. 14-9). How does the nerve impulse cross the synapse? It triggers the release of chemical messengers called **transmitter substances** from the neuron on the "sending" side of the gap. These chemicals are stored in little organelles called **vesicles** next to the synapse and easily seen in electronmicrographs. The molecules of a transmitter substance diffuse across the synapse and act as a stimulus to start a nerve impulse in the "receiving" neuron. Thus, synapses make the nerves of the nervous system one-way transmitters. Transmitter substances can be released by only one of the two nerves meeting at a synapse. Thus, stimulation of a nerve "downstream" in the normal flow of nerve impulses will not result in an upstream message.

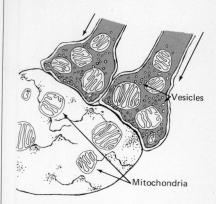

FIG. 14-9 Diagram of two synapses showing the vesicles of transmitter substance and mitochondria. At this magnification the actual gap is shown as a black line.

There is one slight problem. What happens to the transmitter substance once it is released into the gap? It is not chemically altered by the receiving neuron; therefore it should continuously stimulate the receiving neuron unless something were done about it. One of the most common transmitters is **acetylcholine.** A special enzyme called **cholinesterase** destroys acetycholine almost immediately after it is released so that the synapse quickly "recovers" and is able to transmit again.

There are compounds which you may take in from the environment that can interfere with this process. The best known are probably the organophosphate insecticides. They are cholinesterase inhibitors in both insects and mammals; in other words, they prevent the enzyme from functioning. Fortunately, most of them are more toxic to insects than to people. Some are so toxic, however, that a single drop on your skin will kill you! A number of farm workers and pesticide applicators are killed every year by these compounds, and many others are made ill. Low-level exposure results in headaches, dizziness, feelings of weakness, etc. The effect of long-term, very low-level exposure to these insecticides is not known. Unfortunately, because of the careless use of insecticides and devices that dispense them continuously into the environment, it is almost impossible to avoid exposure to organophosphate insecticides.

Effects of psychotropic drugs Adrenaline is a transmitter substance that acts at certain synapses. There appear to be several that act as transmitters in various parts of the brain. One of these is **serotonin.** A number of the drugs which alter consciousness have been thought to exert their effects on the mind by affecting the transmitter serotonin. Both reserpine (a tranquilizer) and LSD seem to affect it. Consciousness-altering drugs are called psychotropic drugs and seem to produce their effects by altering transmitter systems. A great deal more research needs to be done to determine how psychotropic drugs exert their effects, especially those drugs such as heroin which are addictive and cause great social problems. It is too early to assess the results of the various drug addiction treatments. At this point, we do not even know why alcohol (which is our most serious addiction problem) is addictive in some people and not in others. In fact, the details of how alcohol affects the nervous system are completely unknown.

The transmission of a nerve impulse A nerve cell either "fires" or it does not; there is no such thing as half a nerve impulse. Therefore, all individual neurons transmit only on an on-off basis. Once a nerve impulse has passed along a neuron, the neuron is quickly "reset" to transmit another impulse. The membrane has the original distribution of electrical charges reestablished and the cholinesterase resets the synapse.

As we have seen, a nerve impulse consists of a change in an electrical charge produced by the movement of charged particles across cell membranes. This is the reason it is called an electrochemical response. Nerve impulses cannot go as fast as telephone or radio messages, which travel at nearly the speed of light; they can, however, go much faster than hormonal messages, which depend on the blood circulation for their transportation. It is rather like calling a friend in a distant city on the telephone versus sending a letter via the U.S. Postal Service.

The speed of transmission of nerve impulses varies with the kind of nerve and the kind of animal. Nerves with coverings made of lipid-containing membranes called **myelin sheaths** transmit much faster than nerves without sheaths. Generally, cold-blooded animals have slower nerve transmission than warm-blooded animals. A comparison of how fast a nerve impulse can travel in relation to an electronic message in wire shows that if you pinch the ankle of a giraffe, its brain will get the pain message in almost exactly the length of time it takes for a telephone message to go from New York to San Francisco (assuming you have already made the telephone connection). The average nerve impulse in mammals travels at 100 meters (328 feet) per second, and every time it crosses a synapse there is a 0.0005-second delay. The delay at the synapse is caused by the time required for the transmitter chemical to diffuse across the gap between neurons.

14-8 Reflexes are examples of nervous coordination

If you pinched the ankle of a giraffe, it is highly likely that the giraffe would kick you before the pain message got to its brain. The reason for this is that the nervous system has built into it a shunting system which allows an immediate avoidance response to painful stimuli (Fig. 14-10). In the giraffe, the message would be picked up by receptors in the skin of the leg, would travel up an **afferent** ("leading to") nerve to the spinal cord, and would be shunted immediately, by way of a connecting neuron, to an **efferent** ("leading away") neuron, causing a response while the message was still on the way to the brain via spinal neurons. This is an example of a **reflex.** There are three elements involved in reflexes. One is a **receptor,** which receives information about a change in the environment; that is, it receives a stimulus. A second is an **effector,** which can do something about the change in the environment. A third element is, of course, one or more neurons connecting the receptor and the effector. In the giraffe kick, the effectors are muscles in the giraffe's leg.

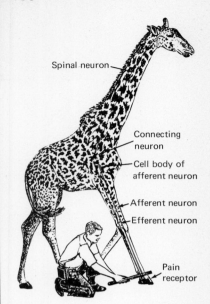

FIG. 14-10 Reflex arc in a giraffe.

Think of the advantage that reflexes give their possessors. If, for instance, you accidentally touch a hot pan with your finger, you pull it back *reflexively*—that is, you do not have to think about it. A pain receptor in your finger receives the stimulus and starts a nerve impulse in a nerve cell leading to your spinal column. In your spinal column, another nerve cell passes on the stimulus to a third nerve cell which leads to your biceps muscle. When this impulse reaches the biceps, the biceps contracts sharply, pulling your finger away from the stove.

Some reflex systems are even simpler than this, consisting of only two nerve cells, one leading to the spinal column and the other leading away. When you go for a physical examination, the doctor usually tests one of these reflexes—the stretch reflex controlled by a tendon in your knee. The doctor taps your knee with a little hammer. If your reflexes are functioning properly, your lower leg jerks. Failure of the reflex to occur would alert the doctor to possible damage in your nervous system.

14-9 The nervous system is highly organized

A nerve consists of many long extensions of neurons called **axons.** Each of the axons in a nerve is insulated from the others, just as the wires in a telephone cable are insulated from one another. Indeed, a single nerve may simultaneously be carrying sensory messages—in different axons—from one area of the body into the spinal cord and motor messages out from the spinal cord to effectors.

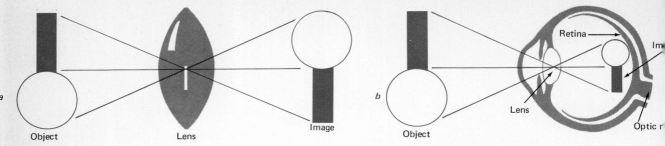

FIG. 14-11 Just as a camera lens (*a*) forms an inverted image of an object, so the lens of a vertebrate eye (*b*) forms an inverted image of an object on its receptor surface, the retina.

Control centers The nervous system in human beings and other vertebrates is extremely complex. It consists of numerous receptors to receive stimuli and, in human beings, many miles of nerves connecting them to effectors. There is, in most nervous systems, one other element that we have not yet mentioned—**control centers.** Not all stimuli and responses have a simple one-to-one relationship like that of the burnt finger and the contracted biceps. For instance, for most of us, the appearance of an attractive individual of the opposite sex is a stimulus, but ordinarily our response is more complex than a contraction of the biceps!

Messages pour into your brain from all parts of your body. Different parts of the brain are specialized for receiving and screening messages from different sets of receptors. Other parts of the brain are specialized for dispatching action orders to various effectors (see Fig. 14-13). Still other regions are centers for association, interpretation, decision making, and memory.

Visual centers What happens, for example, when you "see" a member of the opposite sex? Your eye functions much like a camera. It has an **iris diaphragm** to control the amount of light that enters and a **lens** (Fig. 14-11) to focus that light on a **receptor surface** (in the eye, the **retina**). A major, very important difference between the lens of your eye and that of a camera is that the lens of your eye focuses the image as muscles *change its shape.* A camera's lens focuses the image *by being moved closer or further* from the receptor surface.

The receptor surface of the camera is the film that is coated with light-sensitive chemicals. When the film is developed, a picture appears showing the pattern of chemicals struck by light of various intensities. The receptor surface of your eye also contains light-sensitive chemicals and is called the retina. The retina has light-sensitive cells that connect to neurons. In the light-sensitive cells, various pigments change their chemical structure when they are struck by light, and these changes trigger nerve impulses. A light stimulus starts many chains of nerve impulses traveling along the optic nerve toward the **visual centers** of the brain. The brain, in ways which are only beginning to be understood, integrates this wealth of information into the "picture" we "see." Much of our seeing is quite literally done by the brain, not the eye. We can see details that

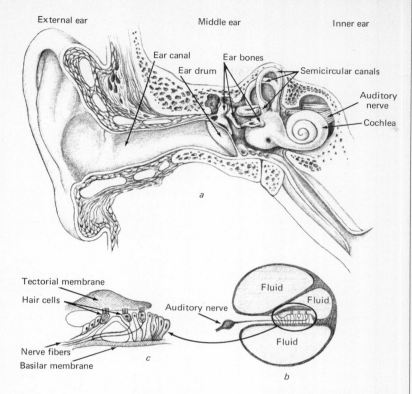

FIG. 14-12 (*a*) Vibrations of air molecules impinging on the eardrum are transmitted by three small ear bones to a membrane on the surface of a fluid-filled chamber which connects to the cochlea. These vibrations are transmitted through the coil of the lower chamber of the spiral, fluid-filled cochlea, a section of which is shown in (*b*), to the basilar membrane. Sensory cells called hair cells (*c*) on the basilar membrane are attached at their top to the tectorial membrane. The vibrating basilar membrane moves, pulling the hair cells and causing them to produce electrical impulses which are transmitted via the auditory nerve to the brain. Pressure on either side of the ear is equalized by the eustachian tubes which open into the back of the mouth. The semicircular canals are sensory structures controlling balance.

FIG. 14-13 The brain has areas of the cortex specialized for processing different sensory inputs. Motor areas govern muscular movement; sensory areas control sensation; association areas are concerned with the screening process, the putting together of inputs from different sensory pathways into a coherent picture, and with making decisions about inputs.

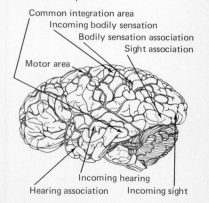

the eye is not physically capable of distinguishing, because our brain is capable of filling them in. For example, we can see a 6-millimeter- (one-fourth-inch-) thick wire four-tenths kilometer (one-fourth mile) away as a sharp, thin line, while a camera would show only a fuzzy dim line. The action of neurons in the brain produces this sharpening of the image. The human species is now slowly learning to imitate the superb eye-brain visual system produced by evolution. Increasingly, computers are being used to enhance (sharpen up) images that could not be resolved by the cameras on space probes.

Other sensory modes The appearance of a person of the opposite sex may register in other receptors than your eyes. If he or she speaks, your ear (Fig. 14-12) will transform the voice (vibrations of the air molecules) into chains of nerve impulses. These impulses go to **hearing centers** in the brain (Fig. 14-13). If the newly arrived person of the opposite sex is using shaving lotion or perfume, some molecules of these substances will land on **chemoreceptors** in your nose. These cells respond to certain molecules only, probably because of the shape of the molecules. Somehow, out of these diverse sensory inputs, your brain constructs the image of a person—perhaps one you learn to like very much indeed. If your encounter progresses from friendship to intimacy, other sensory modes, including touch and taste, may enter into your perception of this person.

Perception One of the great challenges of biology is to work out just how our brain processes the vast amount of information it receives.

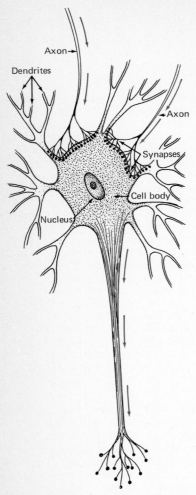

FIG. 14-14 A portion of a connecting neuron.

All this information arrives in the form of off-on nerve impulses. Somehow the brain converts them into a coherent picture of our environment. The study of perception is an extremely complex and controversial field. We do know that perception in human beings is not entirely genetically determined. Much evidence shows that normally functioning individuals of a culture generally perceive the environment in the same way. People from *different* cultures, however, perceive the *same* environment in *different* ways. In every culture, there are people who do not share the cultural norms of perception. Some have impaired sensory capacities, for example, blindness; others are defined as mentally ill or mentally retarded because they do not share the perceptions of the vast majority. Others are creative, ahead of their time, geniuses, or spiritual adepts. Even though very little is known about how perceptions are formed, many societies are quick to make life difficult for those who do not perceive "reality" according to the current standards of normality in that society.

Frequency of firing of nerve cells There are some simple events which are understood that can give one an indication of the kinds of building blocks on which the more complex thought processes are based. First of all, systems of neurons can "decide" whether or not to fire on the basis of the strength of stimuli. Nerve cells require a particular level of stimulus before a nerve impulse is fired. In other words, a **threshold** must be crossed. Not all nerve cells require the same level of stimulus before a nerve impulse is fired. Second, numerous sensory nerve cells often have all processes leading to a single connecting neuron (Fig. 14-14), and a certain percentage of them must release transmitter substances before the connecting nerve cell will fire. Both of these arrangements can be used to measure a stimulus.

Nerve cells can carry only an all-or-none nerve impulse, but they can carry such impulses repeatedly. Thus, once the threshold which fires the nerve cells is crossed, further increase in the stimulus can be transmitted as an increased **frequency** of firing. The frequency of an event is the number of times it occurs in a given period of time.

It seems likely that, in some way, the passage of nerve impulses over sequences of nerve cells in the brain makes future passages of nerve impulses easier. This is called **facilitation.** Such a system could provide the physical basis for learning and for some memory. There cannot be enough circuits constructed, however, even out of all 10 billion nerve cells in the human brain, to account for the amazing storage capacity of the human brain. There is now evidence that indicates at least some memory may consist of information stored in RNA molecules; however, storage in the form of protein molecules has not been completely ruled out. Goldfish that have learned simple responses have been shown to lose their "memory" when treated with substances which affect RNA synthesis. Experiments done a few years ago with the flatworm planaria seemed to show that a planarian could learn what another planarian "knew" by eating it. These experiments have not been confirmed in spite of efforts of many scientists. Perhaps it is fortunate that memory does not seem to work this way!

14-10 Perceptual modes vary in importance from organism to organism

Nervous coordination in animals other than human beings seems to differ little in principle from that in us, although the capacity for information processing of nonhuman coordinating centers is usually considerably smaller than ours. Nevertheless, the coordination of certain important tasks in nonhuman animals may be very impressive. For example, a praying mantis can catch an insect with a strike of its legs which takes 10 to 30 milliseconds (thousandths of a second).

We know that the "picture" of the world which many other animals put together in their brains must be quite different from the human world view. This is because the capabilities of sense organs and brains determines the world view. Humans are "sight animals"; in very large part, our world view is dominated by vision. Our ability to hear and smell is much less than that of, say, a dog. Many animals live in a rich world of odors and sounds in which vision plays a relatively minor role. Bats, winging through the night, can dodge wires and catch small insects even though many are virtually blind. They produce high-pitched sounds and use the returning echoes (reflected sound) to put together a picture of their environment. Whales and porpoises also use reflected sound to give them a picture of their environment. This mode of perception is called **echolocation.** Some fishes are nearly blind, but using special muscles, they generate electric fields around their bodies. This field changes in shape when it encounters objects in the environment of the fish, such as prey or predators. The fish can detect these changes in the field and interpret them into a coherent world view. Think of what it must be like to perceive the world as a series of echos or as deformations of an electric field. Is this any more or less "real" than perceiving it primarily by reflected light as we do?

It is interesting to note that *Homo sapiens* can *learn* to use senses other than vision very precisely. Blind persons, for example, may learn to read a special kind of printing called Braille in which letters are represented by raised dots. After practice, they can use the sense of touch in their fingertips to read accurately and rapidly. Experiments and practical experience also have shown that blind persons may keenly develop their sense of hearing. By the use of clicks or whistles in laboratory experiments, some subjects could locate and describe objects with astonishing accuracy. And the blind person's tapping cane is equivalent to the voice box of a bat, producing sounds that echo off buildings, trees, and automobiles, providing a sound picture of an environment. Recently a battery-powered sonar device has been developed. Its sound producer is worn on the forehead, and the echos produced by objects in the environment are fed to each ear. A blind ten-month-old boy has learned to identify his toys and play peekaboo using this device.

Perception is a fascinating subject leading to questions that have puzzled philosophers for thousands of years, such as "What is real?" and "If a tree fell in the forest and there was no one around, would there be any sound?" and "If there cannot be a heartbeat without a heart, can there be a thought without a brain?" These are fun to discuss, but they are the kinds of questions science cannot answer; they belong to the field of human learning called philosophy. Science studies *objectively verifiable reality*. That is, several different people must be able to agree on the result of an experiment and verify it by repeating the experiment. The sensory input that scientists use to determine the results of their experiments, however, is itself subjective. Therefore, science does not deal with absolute truth; it deals with data which more than one scientist agree they perceived. Philosophy is concerned with, among other things, the nature of truth and questions like those mentioned above.

THE PROBLEM OF PROTECTION

Nothing brings an end to homeostasis faster than being eaten. It is not surprising, then, that organisms have evolved diverse ways of avoiding this fate. Some of these are obvious to anyone. The spines and thorns of many plants (Fig. 14-15) discourage the attacks of herbivores. The horns of deer and antelope and the shells of turtles offer protection against carnivores.

The skin and mucus membranes are a first line of defense against invading disease organisms and foreign organisms of all sorts. In fact, substances inside the digestive tract are not technically inside the body at all. A parasite must breach the lining of the digestive tract of an organism and get in among its cells to be actually in its body. Most reflexes are behavioral defenses against the environment. The eye blink is one example. The entire body may participate in protective responses. When you are threatened, you may choose fight or flight.

The immune response, which will be discussed in Sec. 14-13, is also a mechanism for protection of the body. A great many other parts and functions of the body can also be considered as protective.

14-11 The blood clotting mechanism protects the body from blood loss

Such responses to serious injury as the closing of wounds are important protective mechanisms. In human beings and other vertebrates, one of the most critical of these responses involves the clotting of the blood. When a blood vessel is opened to the outside air by a

FIG. 14-15 Plants often are protected by spines and thorns. (*a*) An aloe from Zanzibar has spiny leaf margins; (*b*) a euphorbia from Africa has paired spines at points where leaves (which fall off very early) were attached; (*c*) a prickly pear cactus from New Mexico has clusters of tiny barbed hairs at points where leaves were attached on the much flattened stems.

wound, loss of blood starts to occur. If the vessel is a sizable artery, the loss can be rapid and even fatal. The mechanism operating to prevent such serious loss is **blood clotting** at the site of injury. Substances released into the blood plasma, the fluid portion of the blood, from damaged tissues or blood cells trigger a series of reactions among proteins resulting in the formation of a mesh of protein fibers. These fibers trap blood cells, and the mass, called a **clot,** blocks the leak in the blood vessel. The fibers and the clot shrink, squeezing out fluids, and the clot dries as a hard plug. Under certain circumstances a clot may form *within* a blood vessel; such a clot is called a **thrombus.** A clot within a blood vessel may break loose from the place to which it was attached and move in the bloodstream until it blocks a vital blood vessel.

Damage to the heart The human body is capable of repairing considerable damage. For instance, a fairly common occurrence among American men over forty is for a thrombus to block one of the arteries supplying blood to the heart muscle (Fig. 14-16). (It is somewhat less common in women and younger people.) The blockage is thought to occur in many cases because of fatty deposits which narrow the arteries. These deposits have been suggested to be a result of faulty diet, smoking, lack of exercise, emotional tension, or a combination of these. If such a block occurs, a portion of the heart muscle is deprived of oxygen by the block, and the cells die, causing a "heart attack" or "coronary." The term *coronary* refers to the coronary arteries, which are the ones that supply the heart muscle with blood. If the damage is not too severe, the person will survive and a portion of the heart muscle will be replaced by scar tissue. The heart repairs itself by enlarging the undamaged areas and increasing the size and number of the unblocked arteries to increase the blood supply; however, no new heart muscle tissue is formed.

The victim of a heart attack usually experiences severe pain below the breast bone, often with feelings of squeezing or pressing. The feeling of pressure in the chest may be followed by nausea, vomiting, and cold, moist skin. The coronary circulation may provide insufficient oxygen and cause the pain known as *angina pectoris* which consists of pain in the chest, left shoulder, and arm. Exercise, intense emotion, or a heavy meal may cause this in some people. Angina pectoris may be an early warning of impending heart attack, but it also commonly occurs in persons after an attack.

Of those who survive the first month after a heart attack, more than half will live 5 years and some 10 or 20 years. The fate of the patient is determined by age and other diseases that may be associated with the heart. A few years ago more than one-fourth of those taken to a hospital with heart attacks could be expected to die. Modern coronary treatment has now reduced this number to one-eighth. Many hospitals now have emergency coronary care units which are able to supply the very specialized treatment necessary to deal with stoppage of the heart, irregular rhythms of the heart, and the many complications which may develop in such patients.

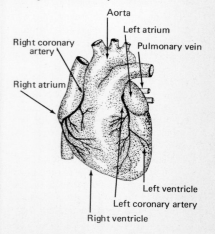

FIG. 14-16 The coronary arteries supply the heart muscle with nutrients and oxygen. Coronary arteries branch off the aorta close to the heart. About 10 percent of the heart output flows through the coronary arteries.

FIG. 14-17 Many kinds of plants produce substances which have biological effects on most organisms. For example (*a*) marijuana produces cannabidiol; (*b*) tobacco produces nicotine; (*c*) cinchona produces quinine (the bark was used as a source of the drug used to treat malaria); (*d*) coffee produces caffeine (as do a number of other plants); (*e*) the coca plant is the source of cocaine; (*f*) opium poppies produce opium (which can easily be made into heroin, morphine, and other substances); (*g*) the mescal cactus contains mescaline (which has effects similar to those of LSD); (*h*) the daisy pyrethrum produces a series of related pyrethrins (which are especially toxic to cold-blooded animals).

14-12 Some of the most effective defenses are not obvious

Many defenses of organisms are not obvious to the casual observer. They may be chemical in nature or involve patterns of coloration.

Poisonous plants Plants invest a lot of energy in manufacturing poisons, which they store in their tissues. The poisons were evolved for one purpose—to poison herbivores, especially herbivorous insects. We use some of these poisons for a variety of purposes. Some, like pyrethrins (Fig. 14-17), we use for their original purpose—to poison insects. Others supply the flavoring of what we call spices. Many we use for medicines: quinine to poison malarial parasites, digitalis to control heart disorders, and morphine as an anesthetic are examples.

FIG. 14-18 Larvae of the monarch butterfly are able to eat milkweed plants which contain poisons which make most other animals sick. The adult monarch (shown here) contains these poisons and is thus distasteful to predators, a fact it advertises with its bright orange and black colors.

Some people use caffeine from coffee and tea, and nicotine from tobacco, because they produce pleasant physiological effects (although they also produce unpleasant and sometimes dangerous side effects). Sometimes other plant extracts are used as drugs. The effects of these range from being a little "high" to being "spaced out." Marijuana from Indian hemp plants, opium and heroin from poppies, and mescaline from a cactus are used for their mind-altering properties (see Fig. 14-17). Can you imagine how these substances help ward off herbivore attack? Think of what an easy target for a lion a spaced-out zebra must be! These substances were evolved to poison animal nervous systems, and they do just that.

Poisonous animals Certain butterfly caterpillars are able to eat plants, poisonous to other species, without being harmed. They store the poisons made by the plant in their cells, and the poison is found in butterflies after metamorphosis. These poisons are vertebrate heart poisons and cause birds that try to eat the butterflies to vomit. Birds soon learn to leave that kind of insect alone, warned off by the bright colors announcing their poisonous nature (Fig. 14-18). Various fishes and other marine animals have defensive poisons, often injected by sharp spines. They, too, tend to advertise their poisonous nature with bright colors and contrasting patterns (Fig. 14-19). The male duck-billed platypus has poison spurs with which he jabs attackers.

Other animals, such as poisonous snakes, use poisons primarily to acquire and digest foods. But their poison can be used in defense also. Some poisonous snakes have found it useful to evolve warning mechanisms to announce their presence to large animals. A rattlesnake's rattle is a good example (Fig. 14-20). A rattlesnake does not have to waste poison on a buffalo it could never swallow when it can scare the buffalo away with a few shakes of its tail.

Camouflage The famous stone plants of the Africa desert are an outstanding example (Fig. 14-21) of inconspicuous plants. If they were conspicuous, they would be quickly devoured for their moisture and nutritive value in the dry, barren environment where they live.

Protective coloration and resemblance are so widespread and common in animals that whole books have been written that just scratch the surface of the subject. Animals often resemble their environments in both color and pattern. A great many look like leaves,

FIG. 14-19 Poisonous fishes are often brightly colored, as is this lion fish. (*Tom Myers, FPG.*)

some look like the droppings of other animals, and some look like animals that are dangerous and poisonous (Fig. 14-22). Biologists, especially those working in the tropics, are often surprised by the remarkable resemblances that have developed. For example, there are harmless bugs which have patterns on their backs which look like large ants that have nasty bites.

14-13 Multicellular organisms have internal as well as external defenses

In addition to protection from being eaten by predators from the outside, it is also desirable to have protection against being eaten by parasites from the inside. Since all organisms except photosynthesizers and a few chemosynthetic bacteria get their food from other organisms, it is not too surprising that many organisms have evolved modes of existence inside other organisms. Some of these organisms may cause damage to their hosts; others may live on or inside the host without killing or damaging it. Almost every adult human being in the world has mites living at the base of his or her eyelashes. These mites do no damage, and practically no one even knows they are there. On the other hand, some multicellular parasites, such as various parasitic worms, devour their victims slowly from within, and pathogenic protozoa and bacteria (Chap. 6) produce poisons which may cause illness or death when they multiply inside the body.

By convention, monerans that cause disease are usually studied by *microbiologists*, while protistans and animals that cause disease are usually studied by *parasitologists*. Both bacteria and parasites are often transmitted from individual to individual by other organisms called **vectors.** Many have complicated life cycles involving several hosts (see Figs. 6-41 and 6-59). Parasitic infections are most

FIG. 14-20 A rattlesnake warns of its presence by vibrating its tail rapidly, making a buzzing noise with the hollow segments of the rattle.

common and debilitating in the tropics. For example, about one-tenth of the 10 million people in the Volta River basin in Africa are afflicted with a worm transmitted by a black fly. The afflicted people are often blind and are greatly weakened. In some villages, most of the adults must depend on their children to lead them about, since the children have not yet acquired the disease and can still see. This disease, called *river blindness,* makes fishing and farming close to the river nearly impossible. In diseased populations, however, there are always some individuals who do not get the disease or who are less severely afflicted by it. Immune responses on the part of the potential host are responsible for this situation. Just as plants have coevolved with their herbivores, so have animals coevolved with parasites and disease. Animals, particularly vertebrates, have developed the immune system for coping with parasites, disease organisms, and other foreign materials that may enter the body.

Immune responses   In vertebrates, problems of defense from within are handled in part by **immune responses.** Substances foreign to the body, such as those on the surface of invading organisms, are called **antigens.** Certain cells in our bodies produce specific protein molecules, **antibodies,** which combine with these antigens and help defeat the invaders.

There are two kinds of cells involved in this process: *B cells* and *T cells.* B cells are derived from the marrow of certain bones, and T cells are derived from the cells of the thymus gland; the activities of both B cells and T cells decrease as a person ages. B cells differentiate into cells in the blood plasma that form antibodies. T cells form "killer" lymphocytes that kill foreign cells and probably some tumor cells; T cells also regulate immune responses, including those of B cells. T cells act as helpers for B cells; they are needed for B cell differentiation and to suppress B cell activity when necessary.

Antibodies can attach to the antigens on an invading bacterium, coating it and making it easier for a white blood cell to ingest and destroy. Antibodies can inactivate a key reaction site by attaching to antigens on bacteria or viruses, thus inactivating them. The immune system is able to "remember" past encounters with antigens and is often able to produce specific antibodies very rapidly when there is a second invasion. This is the basis of immunity to diseases caused by invading organisms. Sometimes, as with some of the classic childhood diseases like mumps and measles, a single encounter produces lifelong immunity. In other diseases, the "memory" may fade relatively rapidly.

Interferon   **Interferon** is a substance which is important in immune responses. It is a protein of low molecular weight which is produced by certain cells when they are stimulated by a variety of substances, for example, bacteria, some viruses, and nucleic acids. It nonspecifically inhibits many viral infections by blocking the replication of viruses, probably at the translational level. It also stimulates the immune responses of lymphocytes and encourages macrophages to engulf dead and dying cells. Interferon causes cells infected with a virus to die and thus be carried away by the macrophages, helping

FIG. 14-21 Some succulent plants from African deserts closely resemble the pebbles in which they grow and are commonly called stone plants. (*Courtesy of T. H. Everett.*)

to localize the infection. Furthermore, interferon can also stimulate the production of virus-specific antibodies and induce fever. Fever in a virus disease is beneficial to the patient because viruses are inhibited from carrying out their activities by body temperatures over 39°C (102°F).

Artificially induced immune responses Although people do not understand many things about immune responses, we have long used them to maintain our steady state. By injecting weakened or killed parasites into the body, an immune response is established with, at most, only a mild attack of the disease. This process is called **immunization,** and the material injected, a **vaccine.** A vaccine was first used by the English physician William Jenner in 1796. He noticed that people working with cows who had a mild disease called cowpox never got the often lethal, and always disfiguring, disease smallpox. By scratching some material from the skin of those with cowpox and placing it in scratches in the skin of healthy people, Jenner immunized them against smallpox. Thus, the immune reaction was used to protect people from a deadly disease long before anyone guessed what caused the disease; indeed, when viruses were completely unknown.

The immune response in organ transplantation Immune responses occur when tissues or organs are transplanted from one individual to another because different individuals differ in the details of their protein composition. An identical twin is the ideal donor for a patient needing an organ transplant because identical twins are genetically identical. Since most people do not have an identical twin, other donors must be found and tested for their degree of genetic similarity to the person needing the transplant. This involves matching tissue types much as blood types are matched for transfusions.

As long as the transplanted organ is not from an identical twin, there will be an immune response on the part of the host. This often leads to so-called *graft rejection* as the immune system of the host goes into action to destroy the foreign protein of the transplanted organ. The closer the match, the less the immune response will be. Drugs are used to suppress the action of the immune system and delay graft rejection; however, since the entire immune system is affected, a person taking such drugs becomes extremely susceptible to disease. This is the reason why many persons with "successful" transplants die of such diseases as pneumonia.

14-14 Some diseases are not infectious

Some diseases are not caused by invading organisms or viruses. These diseases come in a variety of types (defined below): allergic, autoimmune, functional, cancerous, degenerative, iatrogenic, and many others, including those caused by drugs and other environmental agents taken into the body. Many noninfectious diseases have a strong genetic component (Sec. 9-11). Between 1900 and 1970, infectious diseases, such as influenza and tuberculosis, lost out to noninfectious diseases as the leading causes of death in the United States (Fig. 14-23). In 1970, the front runners were cancer and diseases of the heart. This decline in infectious disease is due to improved health

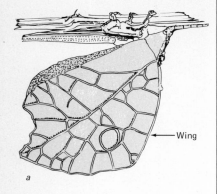
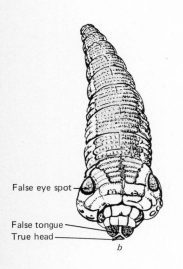
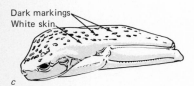

FIG. 14-22 Insects may mimic other things as a form of protection: (*a*) an insect mimicking a leaf, (*b*) a caterpillar mimicking a snake, (*c*) a tree frog mimicking an animal dropping.

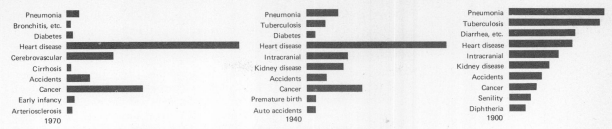

FIG. 14-23 Changes in patterns of the ten leading causes of death in the United States. Data from 1900, 1940, and 1970 are shown as relative lengths of dark bars, where the longest bar represents 360 deaths per 100,000 population. *Cerebrovascular* in the 1970 graph indicates disease of the blood vessels of the brain and is the same as *intracranial* in the 1940 and 1900 graphs. *Early infancy* indicates certain diseases of early infancy. *Pneumonia* includes other related respiratory disorders. During this period the average life expectancy in the United States increased from 47 in 1900 to 71 in 1970. (*Data from T. H. Dingle, "The Ills of Man,"* Scientific American, **229**(6):77–84.)

care, public health measures including water purification, sewage disposal, immunizations, and the development of many new medical techniques such as use of antibiotics. The decrease in infectious diseases, however, has led to increased longevity, which in turn has increased the incidence of functional (without clearcut physiological or structural cause) and degenerative diseases associated with aging.

Heart attacks have been the leading cause of death in the United States for the last 30 years. Hypertension (high blood pressure), often a functional disease, is present in about one out of seven adults in the United States. It is the most common chronic (lasting a long time) disease in the United States, and half the people who have it do not know they have it until they develop stroke, heart attack, or kidney failure. Often it takes 15 or 20 years of hypertension to produce these or other obvious symptoms (for example, severe headache) that something is wrong. A simple measurement of blood pressure can screen potential hypertensives from the population, and many types of hypertension can be treated if detected early.

Of the approximately 200 million Americans alive today, 50 million will develop cancer and 35 million will die from it—half before age 65—if current rates continue. The causes of many kinds of cancer are unknown; however, environmental factors, viruses, and hereditary factors all may be involved. The National Cancer Institute estimates that 80 to 90 percent of cancers are related to environmental factors; there is usually a 15- to 20-year delay between exposure to a carcinogen and the appearance of cancer.

It is known that 80 percent of lung cancer is attributable to smoking. Asbestos, however, is a widespread environmental pollutant also known to cause lung cancer. It enters the air from the demolition of old buildings, from insulation materials, from brake-lining wear in automobiles, and from industrial sources. When lung cancer has developed sufficiently to be detected by an x-ray picture, it is usually too late for surgical care. Only 5 percent of the cases discovered by x-ray survive for five years. Fortunately, it is now possible to detect lung cancer by a sputum test before it can be detected by x-rays.

There are many other known and suspected carcinogens: arsenic, vinyl chloride, aflatoxins, and nitrosamines are known to produce cancer in people or other mammals. Nitrosamines have been shown to be formed in the mammalian stomach from amino acids in foods and from nitrites. Nitrites are used as food preservatives and may be found in water supplies. Other food additives, cyclamates (artificial sweeteners), were removed from the market in the United States because they were shown to cause bladder cancer in rats. Most of the 1.8 million synthetic chemicals so far produced have never been tested for carcinogenesis even though many of them escape into the environment.

Radiation, both x-rays and ultraviolet rays, can cause cancer. Every year many people are exposed to unnecessary x-radiations, some recommended by careless physicians, some from TV sets, and some from unshielded electronic equipment. Only recently was screening for tuberculosis with mobile chest x-ray units abandoned; it was discovered that such screening was unnecessary and that these units give much higher doses of radiation than were produced by larger machines at hospitals and clinics. In the late 1930s fluoroscopes (a kind of x-ray machine which shows a picture instantly on a screen) were commonly placed in shoe stores so that customers could see whether their new shoes pinched their feet. Anyone, including children, could operate these machines as often as they wished. In the 1940s x-radiation of the face or neck was a common treatment for acne, tonsilitis, thymus problems, and other childhood diseases. Now, 35 and more years later, thyroid tumors are appearing in some people who had neck x-rays as children. Some doctors and hospitals are going back through their records and attempting to identify these patients; one recall of such patients has shown a 7 percent incidence of thyroid tumors in this group.

Allergies and autoimmune diseases involve the immune system. In allergies the system reacts to a foreign protein, such as ragweed pollen, by causing certain cells to release the substance histamine. Histamine causes the classic symptoms of hay fever, such as swollen mucous membranes and secretion of mucus. In autoimmune diseases, the system for some reason produces antibodies against cells or membranes of its own body. Rheumatoid arthritis seems to be an autoimmune disease.

Iatrogenic diseases are those caused by physicians in treating a patient. They are distressingly common. The fungus infections occurring after antibiotic administration, which were discussed in Sec. 6-10, are iatrogenic. Mercury poisoning (Sec. 1-11) and pesticide poisoning (Sec. 14-7) are examples of diseases caused by chemicals in the environment. There are more and more diseases like this appearing. Air pollution seems to be associated with the recent increase in bronchitis and emphysema, diseases of the respiratory tract and lungs, respectively (see Fig. 17-3).

14-15 Plants also have parasites and diseases

Plants are attacked by parasites and diseases, but they lack an immune-response defense system comparable to that of animals. In plants, as in animals, the invading organism may live compatibly with its host—for example, mistletoe on oak trees—or it may cause great damage, as do some of the fungal diseases of our crop plants, discussed in Chap. 6. Other plant parasites also can be very harmful to crop plants. For example, broom rape is a flowering plant that parasitizes tomatoes and other crops. Some fields can no longer be used profitably for growing tomatoes because broom rape seeds can lie dormant for years and tomato plants have no defense against broom rape. Nematode worms and insect larvae are also internal parasites of plants. Sometimes the plants produce abnormal cancer-like growths of tissue surrounding such invaders. These are called *galls*; oak trees commonly have galls they produce as a response to a small wasp which lays eggs in the stems. Plants also may have fungal, viral, and bacterial diseases. It appears that plants' most usual way of dealing with parasites and disease is to evolve resistant strains. The human species must continuously breed new strains of crop plants that have resistance to newly evolved disease strains. People treat some plant diseases by external applications of fungicides, and antibiotic injections are being used on some sick palm trees in Florida and elm trees in the Midwest.

14-16 Parasites and disease spread best when there are many suitable hosts close together

We have already seen that the monocultures of crop plants which people have created are perfect places for the spread of disease organisms. Not only are all the plants often genetically identical, but they are close together so that the disease organism can easily move from one to the other. Dense groups of individuals of the same species of animals are also good places for diseases to spread. A resident of a large modern city is exposed every day to many other people, some of whom may be carrying contagious diseases. Large modern cities also have airports so that there are always people who have newly arrived from distant parts of the world and are perhaps carrying exotic diseases. Cities thus are vulnerable to **epidemics** (rapid spread) of disease, as are boarding schools, hospitals, jails, and other places where people live close together in groups. In fact, a classroom full of students is a "monoculture" of people.

QUESTIONS FOR REVIEW

1 The text notes two ways in which single-celled organisms can partially control their internal environment. What are they?
2 The *normal* human body temperature is 37°C (98.6°F). Why is 37°C *normal* in a functional sense? That is, why—in terms of what goes on within the body—is 37°C "better" than, say, 29 or 39°C?
3 What do these terms mean: diabetes, target cell, neurosecretion, detoxify, homeostasis, echolocation, transplant operation, hormone, membrane poten-

tial, transmitter substance, vaccine, secondary sexual characteristic, neuron, autoimmunity, sensory mode, synapse, axon, epidemic, chemoreceptor, antigen, antibody?

4 An old saying has it that "One man's meat is another man's poison." As this chapter points out, some plants' poisons are some peoples' "meat" (things they eat or drink by choice). What are some of these poisons, how in general do they affect their human consumers, and how have they benefited their plant producers?

5 The immune-response mechanism generally works well for its producers (although it cannot always prevent serious illness or death). But as it affects one modern medical technique, it often works *too* well. How so?

6 "A classroom of students is a 'monoculture' of people." What does this mean, and what are its implications for human health?

7 What role is played by the kidneys in maintaining blood-sugar homeostasis; what is frequent, excessive blood-sugar in the urine a sign of; and what should be done if this sign appears?

8 Why do alcohol and tranquilizers or barbiturates make such a dangerous (and sometimes a deadly) combination?

9 Can you identify a control, or coordination, system that involves both the hormonal and the nervous systems?

READINGS

Barlow, G. W.: "Animal Behavior: I," *Biocore,* unit XIX, McGraw-Hill, New York, 1974.

Cannon, W. B.: *The Wisdom of the Body,* Norton, New York, 1932.

Dingle, J. H.: "The Ills of Man," *Scientific American,* **229**(3):77–84, 1973.

Dubos, R., M. Pines, et al.: *Health and Disease,* Time Inc., New York, 1965.

Edelman, G. N.: "The Structure and Function of Antibodies," *Scientific American,* **223**(2):34–42; offprint 1185, 1970.

Ehrlich, P. R., and R. W. Holm: "Evolution," *Biocore,* unit XX, McGraw-Hill, New York, 1974.

Galton, L.: *The Silent Disease: Hypertension,* New American Library, New York, 1973.

Gasner, D.: "Wanted: America's Top Ten Killers," *Family Health,* **7**(1):27–32, 1975.

Guyton, A. C.: *Function of the Human Body,* Saunders, Philadelphia, 1969.

Hilleman, M. R., and A. A. Tytell: "The Induction of Interferon," *Scientific American,* **225**(1):26–31, offprint 1226, 1971.

Jerne, N. K.: "The Immune System," *Scientific American,* **229**(1):52–60, offprint 1276, 1973.

Kennedy, D. (ed.): "Cellular and Organismal Biology," *Readings from Scientific American,* Freeman, San Francisco.

Kennedy, D.: "Integration," *Biocore,* unit XVI, McGraw-Hill, New York, 1974.

Longmore, D.: *The Heart,* McGraw-Hill, New York, 1971.

Mellon, D.: "The Nerve Impulses," *Biocore,* unit XVII, McGraw-Hill, New York, 1974.

Nourse, A. E., et al.: *The Body,* Time Inc., New York, 1964.

Pastan, I.: "Cyclic AMP," *Scientific American,* **227**(2):97–105, offprint 1256, 1972.

Pike, J. E.: "Prostaglandins," *Scientific American,* **225**(5):84–92, offprint 1235, 1971.

Wickler, W.: *Mimicry in Plants and Animals,* McGraw-Hill, New York, 1968.

Wilson, D.: *Body and Antibody: A Report of the New Immunology,* Knopf, New York, 1972.

15

BEHAVIOR AND SURVIVAL

SOME LEARNING OBJECTIVES

After you have studied this chapter, you should be able to

1. Demonstrate your knowledge of coevolutionary interactions by
   a. Explaining what *coevolutionary interaction* means, using an example that does not involve mimicry.
   b. Defining mimicry and giving three examples of it in animals and two examples of it in plants.
   c. Explaining, with examples, the difference between a model-mimic pair and a mimicry complex.
   d. Stating whether mimicry is (1) advantageous to the model, and why; (2) advantageous to the mimic, and why; (3) advantageous to the mimic when the mimic outnumbers the model, and why.

2. Demonstrate your knowledge of reproductive behavior by
   a. Stating three different functions, or roles, that courtship behavior may play in various animal species.
   b. Briefly describing some courtship behaviors of (1) a species of reptile; (2) birds; (3) nonhuman primates.
   c. Naming some different kinds of animals that show maternal care of the young.
   d. Describing some reproductive behaviors of sea gulls, seals, deer, wolves, and chimpanzees.

3. Distinguish between an aggregation and a social group; give an example of a type of animal whose typical group is (1) a hive, (2) a herd, (3) a troop, (4) a flock, (5) a mating pair, (6) an aggregation.

4. List at least four advantages of group life, and for each give an example of how a particular kind of animal benefits from it.

5. Demonstrate your knowledge of dominance by doing the following:
   a. State a major function of dominance relations in groups.
   b. Give an example of a dominance hierarchy in a nonhuman species, and explain why an army's organization, or chain of command, is an example of a linear dominance hierarchy.
   c. State some nonlinear dominance relations found in many animal species.
   d. Name three kinds of submissive behavior found among mammals.
   e. State what displacement activity is and give an example of it.

6. Demonstrate some of what you have learned about territoriality by
   a. Stating two important functions served by territoriality.
   b. Naming two kinds of territories and stating what they are defended against.
   c. Stating which of the following establishes territories: individuals, mated pairs, groups.
   d. Stating how dominance manifests itself in, or figures in, territoriality.

7. Show some of what you know about individual space and interpersonal distance by
   a. Stating the meaning of each.
   b. Stating two major functions that individual space, like territoriality, serves.
   c. Naming a kind of organism that exhibits each.

8. State one important way in which the division of labor within groups of certain kinds of social insects differs from that within social groups of other animal species, and name at least three kinds of such social insects.

9. Name four different modes of communication, and for each one name a kind of animal to which it is particularly important.

10. Name three principal regions of the vertebrate brain, and list some major functions each serves in (1)

mammals and (2) other vertebrates.
11 Name the part of the brain upon which conceptual behavior depends, say which of the brain's three principal regions it is a part of, and describe its role in (1) filtering sensory input and (2) cortical integration.
12 Show your understanding of certain aspects of language as a mode of communication by
   a Naming three important sensory modes employed in language use.
   b Explaining the nature of a symbol and how this human device differs from a signal in the animal world, and giving at least two examples of each.
   c Explaining, with examples from our own and from other languages, what it means to say that "The structure of a language molds our view of nature into a form easily handled by the language."
13 Demonstrate your knowledge of cultural evolution by
   a Stating at least three ways in which cultural evolution is similar to, and at least three ways in which it differs from, biological evolution.
   b Explaining the effects of cultural evolution on the brain and vice versa, and giving at least one example of how cultural evolution has affected selection pressures on (1) other organisms and (2) the human species.

If animals are to survive, they must carry out appropriate behavior. Behavior functions as a homeostatic mechanism, permits reproduction, and structures populations of animals.

BEHAVIOR AND COEVOLUTION

Behavior also is often an important component of coevolutionary interactions. The classic case of coevolution is the interaction of herbivores and the plants they attack. But there are many other well-known coevolutionary complexes: predators and prey, hosts and parasites, flowers and pollinators. In all such cases, each member of the pair is a selective agent relative to the other.

15-1 One of the most studied coevolutionary situations involves models and mimics

Certain animals have, like many plants, evolved poisons that protect them from predators. As was discussed in Chap. 14, they often have also evolved conspicuous warning colors, for example, a bold pattern of yellow or red and black (Fig. 15-1). Numerous experiments have shown that once a predator has experienced the consequences of tasting one of these organisms, it quickly learns to shun all those which are similar. One such experiment is shown in Fig. 15-2. A bumblebee stings a toad, thus impressing on the toad the significance of yellow-and-black beelike creatures.

It is not surprising that other organisms which are not poisonous but which resemble bees or monarch butterflies (see Chap. 14) are at a selective advantage. Natural selection improves any chance resemblances that occur. Thus a harmless bee fly has come to imitate the stinging bumblebee, and a relatively tasty butterfly of another genus, the viceroy, has evolved a resemblance to the monarch (Fig. 15-3). This situation is known as **mimicry**. In such pairs the harmful or distasteful member is called a **model**, and the harmless or palatable member is called a **mimic**. Model and mimic pairs in the butterflies can be so similar that species from different families can be distinguished only by very close examination (Fig. 15-4). In other cases, individuals held in the hand may look quite different, but because of behavioral mimicry, they may be virtually indistinguishable in flight.

The **mimicry complex** In some cases mimicry may not involve a model and mimic, but two or more distasteful species resembling each other. This presumably gives mutual protection by permitting a predator to learn to avoid all of the species by tasting a single individual of any one of them (Fig. 15-5). Such a group of species is called a **mimicry complex.**

One way in which such a complex may arise is for a previously tasty mimic to evolve distastefulness. Then a model-mimic pair is converted into two mutual mimics. Other members could join the complex via the same evolutionary route. One might then logically ask why all 15,000 species of butterflies are not part of one vast orange-and-red mutual complex of distasteful mimics. The answer is that not only do prey evolve in response to the selection pressure of predators but the predators also evolve in response to the selection pressure of the prey. When enough members join a mutual mimicry complex, it produces an enormous *selective advantage* for any predators that can evolve a mechanism for dealing with the protective device of the mimics. More and more kinds of predators will tend to become specialists feeding on the complex (there are, for instance, predators that eat bumblebees and monarch butterflies). Suddenly it no longer pays for butterflies to belong to the complex; the selective system changes and the complex starts to break up. The whole situation is controlled by a complicated series of interacting selection pressures.

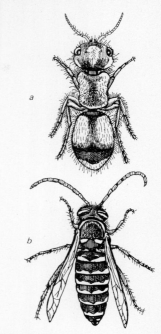

FIG. 15-1 The velvet ant (*a*) and the hornet (*b*) both have powerful stings and conspicuous warning patterns of black and red or yellow.

Think for a minute about a model-mimic pair. Is it an advantage for the model to have a mimic? What happens when the mimic becomes more common than the model? Obviously, at some point in the increasingly relative abundance of the model, *most* predator experiences with the complex will be *favorable*. That bodes ill for both model and mimic. However, experiments suggest that a model-mimic association will often be advantageous for the mimic even when it outnumbers the model. An occasional bad experience keeps the predators shunning the shared pattern. After all, if you found that one in ten bottles of beer of a particular brand contained turpentine, you would probably avoid that brand.

It has been estimated that more than one-half of all butterflies are involved in mimicry complexes. Mimicry is also widespread through other insect groups and, to a lesser extent, through most groups of animals.

Mimicry in fishes An interesting example of mimicry concerns two coevolutionary relationships. There is a coevolutionary complex among certain fishes that live on coral reefs and other fishes called **cleaner fishes.** Cleaner fishes scour other fishes, eating their parasites. Big coral reef fishes even open their mouths and gills to permit cleaners to work in those cavities. The small, brightly colored cleaner

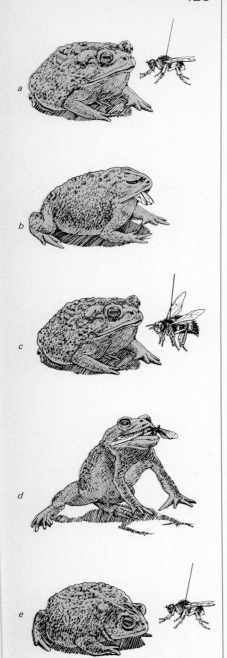

FIG. 15-2 (*a*, *b*) An inexperienced toad readily eats an insect, presented on a wire, which mimics a bee. (*c*, *d*) The same toad snaps up a bee and is stung. (*e*) Afterward, the toad refuses to eat both bees and bee mimics.

fishes establish "stations" on the reef, and other fishes regularly come to these stations to be cleaned (Fig. 15-6). Obviously the behavior of each member of the complex is advantageous to the other. Mimicry is also involved because another kind of fish has evolved into a perfect mimic of one of the species of cleaner fish. The mimic fish establishes a "station" and does an imitation of the characteristic dance of a cleaner "on station." The big fishes are attracted and go into their usual "customer" behavior. The mimic then takes a bite out of one of their fins! The big fish turns in surprise, but the little mimic stays right where it is, calmly chewing as if nothing had happened!

Mimicry in fireflies Females of the firefly *Photuris versicolor* prey on males of other species of fireflies by mimicking the flash responses of the prey's own females. Females of this species can attract males of four different species. The mimicry is quite effective; a female seldom flashed her light in answer to more than 10 males without catching one and feasting on it.

Mimicry in plants One well-known case involves flax plants, which serve as models, and several weeds which are the mimics. Farmers who grow flax winnow the seeds they harvest to separate the chaff. They toss the seeds up into a breeze and the flax seeds all travel a similar distance and collect in a pile. Other parts of the plant—and weed seeds—are lighter and blow farther away; thus they are sepa-

FIG. 15-3 The monarch butterfly (*a*) is mimicked by the viceroy butterfly (*b*). (*American Museum of Natural History.*)

rated out. Several weeds have evolved seeds which, although they do not *look* like flax seed, are shaped so that the wind carries them the same distance and they become mixed with the flax seed. They then are planted along with the flax. This is an example of what might be called *aerodynamic* mimicry.

15-2 Bats, moths, and mites make up an interesting coevolutionary complex

The principal predators of flying moths are bats. Bats hunt by "sonar," producing high-pitched pulses of sound and putting together a picture of their surroundings on the basis of returning echoes. Moths have evolved ears at the base of their abdomen which are designed to detect bat sonar. When a moth receives bat sonar pulses, it immediately plunges to the ground, usually avoiding being caught.

Many moths have colonies of tiny mites living in their ears. The mites are always found in only *one* ear of a moth. Since the mites destroy the hearing ability of the ear, the reason only single infections are found seemed obvious—moths that had mites in both ears got eaten by bats! Like many apparently obvious answers in science, this one turned out to be false. The mites have actually evolved a behavior pattern guaranteeing that only one ear per moth will have a mite colony. The first mite to climb on a moth (usually when the moth visits a flower) picks one ear or the other to settle in. *All subsequent mites follow a chemical trail left by the first mite to the same ear.* What selection pressures would have led to that behavior?

FIG. 15-4 The moth *Alcidis* (*a*) and the butterfly *Papilio* (*b*) are in different families but are very similar.

15-3 Pollination systems are usually coevolutionary systems

A plant pollinated by a specific insect or other animal has certain advantages over those plants that depend on wind for pollination. Generally animals tend to visit one or a few species of plants when they are foraging so that the pollen has a good chance to be delivered directly. Thus animal-pollinated plants generally do not need to produce as much pollen as those that depend on the unpredictability of wind to deliver their pollen. But they have the disadvantage that they must use some of the energy available to them to produce things that attract animals, usually colorful flowers and a supply of nectar.

Thus a plant usually pays for getting its pollen moved by supplying an attractive food source for an animal. The animals mostly benefit from this arrangement, although sometimes loads of pollen applied by the plant to its unsuspecting visitor may put it at a disadvantage. In some orchids, the flower resembles a female of the bee or wasp that is the pollinator (Fig. 15-7). Male bees or wasps visit these flowers, not for food but to copulate with them. In so doing, they remove tiny bags of pollen, which they carry to other flowers in repeat performances. This strange mode of pollination is called *pseudocopulation*. The females of the bees or wasps emerge after the flowers bloom, but in plenty of time for genuine copulation to occur.

FIG. 15-5 Two distasteful butterflies of a mimicry complex, (*a*) *Hirsutis megara* and (*b*) *Lycorea ceres,* are from different families but look like each other.

FIG. 15-6 The large fish is being cleaned by a tiny cleaner fish (on the larger fish's side) that will clean any fish that comes to its station in the coral reef. (*P. R. Ehrlich.*)

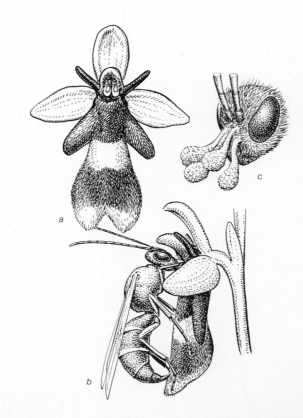

FIG. 15-7 This orchid flower (*a*) looks like a female bee; a male bee attempts to copulate with the flower (*b*) and in the process removes tiny bags of pollen attached to its head (*c*). (*From Wolfgang Wickler,* Mimicry in Plants and Animals, *translated from the German by R. D. Martin, World University Library, McGraw-Hill, 1968.*)

BEHAVIOR IN REPRODUCTION

Reproduction of animals commonly involves many kinds of behavior. For one thing, adult animals often are aggressive, even to other members of their own species. Reproductive behaviors include **courtship** behavior which, among other things, reduces aggression sufficiently for animals to mate. Actual mating behavior is often complex. Finally, there are behavior patterns associated with raising the young.

15-4 Courtship behavior announces readiness to mate

One of the problems faced in reproduction is the necessity for **synchrony** of readiness to mate or to produce gametes. Many marine invertebrates are synchronized by factors of the environment, such as day length, tides, or phase of the moon. Males and females release their gametes into the water simultaneously and external fertilization takes place.

In vertebrates and invertebrates with internal fertilization, factors of the environment, such as day length, are also important in setting in motion the processes leading to reproduction. Often they are associated with hormonal systems that directly affect the ovaries and testes (Fig. 15-8). These in turn produce hormones that alter the appearance and behavior of the organism. To other members of the species, these changes indicate readiness to mate.

Courtship behavior of invertebrates Approaching a large, hungry female spider with the intention of mating is a dangerous matter for a male. She may mistake him for a meal rather than a mate. Different kinds of spiders have evolved different kinds of courtship behaviors depending upon their way of life. You have probably seen the small, very active jumping spiders that do not make webs but roam about searching for prey. They have eight large eyes and are visually oriented, as you can see if you try to approach one. In jumping spiders, the male often has specialized appendages which are waved, like signal flags, in regular patterns, announcing readiness to mate (Fig. 15-9). The female receives this semaphore message and behaves accordingly. Different species have different patterns of movement, often of different appendages.

Web-weaving spiders, in contrast, are touch-oriented, being very sensitive to the movement of the strands of their web. Such movements indicate the size and position of a trapped prey insect. In these species, a male may indicate his readiness to mate and also that he is *not* trapped prey by carefully plucking or shaking the web so that the possibility of any misunderstanding is minimized. Even so, we must say in all honesty that after or during mating, the male often becomes a high-protein meal for the female, who must then synthesize egg materials.

Courtship in insects may also be complex. Butterflies may go through elaborate dances. In certain flies, males present females with a silken "bubble" as a "gift" in the course of their courtship dance.

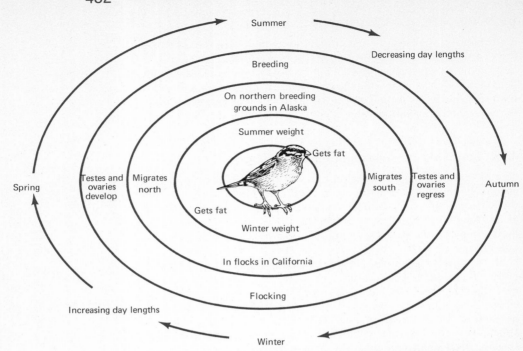

FIG. 15-8 North temperate zone birds, such as this subspecies of white-crowned sparrow, have hormonal systems that respond to changes in day length (hours of light per day). Lengthening days in spring cause the seasonal development of the ovaries and testes, and the deposition of body-fat stores which are utilized during the birds' spring migratory flight to their breeding ground in Alaska. They experience a molt (loss and regrowth of feathers) preceding their migration and breeding. These birds establish territories and breed in Alaska during the summer; as day lengths decrease, their testes and ovaries regress, and they migrate back to their winter home in California where they spend the winter in flocks.

As noted above, fireflies communicate with their mates at night by flashing their lights in a pattern characteristic for each species.

Courtship behavior of vertebrates Courtship behavior of vertebrates frequently is more complex than that of invertebrates. Lizards of the family Iguanidae have an elaborate courtship behavior. The male northern fence lizard recognizes a female of his species by her color pattern. He will rush toward her with a stiff-legged gait, his head and shoulders held high. As he approaches the female, he pauses several times to bob up and down. Meanwhile the female moves away in short jerky hops with arched back if she is receptive. (If she is not receptive, she simply runs away.) Finally, the male catches up with the seductively hopping female and grasps her neck in his mouth. Copulation follows, usually with the male retaining his grip on the female's neck. Crocodiles may have the least sophisticated courtship behavior of the reptiles. The male bellows, arches his neck,

FIG. 15-9 In this species, male jumping spiders court females by raising first the left and then the right foreleg.

lunges at the female, grabs her by the neck, and mates with her. In birds, it is common for males to sing species-specific songs which convey their readiness to mate and which indicate that they have established a territory (Sec. 15-8). Courtship often involves the display of patterns of colored feathers, fur, or skin. It may include posturing and dancing of various kinds (Fig. 15-10). In fishes and birds, nest building may be a part of the courtship routine.

The more complex patterns of behavior may, in fact, also have the function of inducing the female to ovulate. Many female fishes and birds cannot lay their eggs until the proper nest-building rituals have been followed (see Fig. 15-10). And the females of some mammals, for example, mice, rabbits, and mink, do not ovulate until courtship and mating take place. In some mammals, courtship and mating may be a violent, often bloody, affair. For example, in many kinds of cats that are solitary until mating (domestic cats and tigers), copulation often follows a vigorous fight. In birds, by contrast, the preliminaries to mating are often stately "dances" or humorous (to us) headbobbings. Sometimes the male presents the female with "gifts" such as nesting materials. Apparently some sort of ritualized behavior is essential in most vertebrates before reproduction can be carried out successfully.

In the primates, it is often the female who announces her readiness to mate. Males in a primate group sometimes drive away females who approach them closely. When the female is ready to mate, however, she adopts a submissive **presentation posture,** displaying her buttocks (often swollen and/or colored) to him. The presentation posture places the female in a position in which the male can easily copulate with her. After perhaps several mildly aggressive encoun-

FIG. 15-10 Gannets go through a courtship dance (a) followed by nest building (b) and laying eggs. [(a) Grant Haist, National Audubon Society; (b) Phyllis Greenberg, National Audubon Society.]

ters, the male accepts the female and mating takes place. Other mammals have special scents which may attract males from long distances, as anyone who has owned an unspayed female dog can testify. Male giraffes and some antelopes taste the urine of females for substances which indicate her reproductive state.

15-5 Parental care leads to group formation

Following mating, the female lays her eggs or the young develop within her body and are born some time later. In either case, the female often leaves her offspring, which are able to take care of themselves at once. This is true of many fishes, and most reptiles and amphibians. There are other groups, however, in which the female, and sometimes the male, remains with the young. Many birds, most mammals, and some fishes and reptiles fall into this category. Some invertebrates, especially arthropods, also show maternal care of the young.

Some birds, such as sea gulls, terns, gannets, and pelicans, form large groupings of individuals (Fig. 15-11). Nests are often side by side, and there may be hundreds of them. At each nest, one parent usually remains with the eggs or offspring while the other goes feeding. On returning from feeding, a bird must be able to find its own nest among the myriad others. Birds often have elaborate behavior for identifying their mates and for warning nearby birds when they come too close.

Superficially similar groups of breeding animals are found in seals, sea lions, and elephant seals (Fig. 15-12). In these marine mammals, the males arrive first at an island or other shared breeding place.

FIG. 15-11 A rookery of gannets showing large numbers of these sea birds crowded together in dense aggregations on a steep sea coast during the breeding season. (*Allan D. Cruickshank, National Audubon Society.*)

FIG. 15-12 Aggregations of seals on a rocky island during the breeding season. (*FPG.*)

They then establish defended territories, the size of which depends on the size and vigor of the male. Each male gathers about him a harem of females after they arrive. There is much fighting, as each male attempts to establish as large a harem as possible in his territory. Males learn to recognize females of their harem as well as competing males. After the young are born, each mother recognizes her own offspring by its smell.

Single family groups are rarely found among terrestrial mammals, although they do occur. For example, female bears and tigers remain with their young until they are weaned and able to fend for themselves. Gibbons (tree-dwelling apes) also commonly have groups consisting of male, female, and young of various ages. Most primates, however, occur in troops or bands of males, females, juveniles, and young. Males mate with any female that is receptive. Gorillas, chimpanzees, and baboons behave in this manner.

Elephants and some carnivores, such as lions, hyenas, and wolves, also have similar groups of animals of both sexes and all ages. In the carnivores, these groups often have rather complex hunting behavior. Ungulates (deer, antelope, etc.) have groups known as herds or flocks. In these groups, there is ordinarily only one sexually mature male during the breeding season.

SOCIAL BEHAVIOR

Groupings of individuals may result because of their common responses to certain environmental factors. A crowd of flies in and around a garbage can is an example of such a group called an **aggregation.** A different kind of group is formed as a result of the response of individual animals *to one another.* The responses of individuals to others of the same species are called **social responses.** The groups

FIG. 15-13 When threatened by wolves or other hunters, musk oxen form a defensive circle around their young. (*American Museum of Natural History.*)

that result are called **social groups.** All of the kinds of behavior in such groups are referred to as **social behavior.**

15-6 There are many advantages to group life

Among the more obvious advantages to group life are mutual protection and reproductive efficiency. Perhaps you have heard of the defensive behavior of musk oxen, which form a circle with the young in the center (Fig. 15-13). Baboon troops travel and feed in the open in groups arranged so that an approaching predator, such as a leopard, always meets vigorous adult males (Fig. 15-14). The advantages of group life for reproductive efficiency are clear. Both males and females are present and do not have to be attracted to one another over long distances. Their reproductive behavior can be correlated and structured.

There are other advantages of group life, however. It has been shown that for many animals there are physiological advantages. In both invertebrates and vertebrates, *groups* of individuals are more resistant to poisons or other stresses than are separate individuals. Groups of warm-blooded animals may conserve heat by snuggling together. Amphibia, such as frogs and toads, conserve moisture by the amphibian equivalent of snuggling. Individual birds, fishes, and other animals may be less susceptible to predation in large flocks or schools. Different species of reef fishes have coevolved similar color and behavioral patterns and form interspecific social groups called *heterotypic schools*, which are thought to function as predator-defense mechanisms. It is known also that animals in groups tend to observe and imitate one another. This facilitates learning and increases the general level of activity of the group.

FIG. 15-14 In the center of this troup of baboons is a group of adult females with their young. The adult males tend to be toward the outside of the group.

Finally, animals seem to feed more efficiently in groups. When one individual finds food, the others quickly find out about it. When a flock of ground-feeding birds crosses an area, its members seem to be more systematic and thorough than when alone. Their activities also stir up soil and plants and thus raise insects and other prey for the group. Of course, hunting groups, such as those of wolves, can bring down much larger prey than could an individual and, in such groups, prey is shared among all members.

Group structures We have already discussed several examples of how groups are structured. In primate groups, adult males often play an important role in the protection of the other individuals. In ungulate herds, the single adult male gathers around him a harem of females. Associations of seals are structured by the size and aggressiveness of the males. And associations of seabirds may be arranged according to the terrain and the amount of space available. Heterotypic schools of fishes are structured by size; fishes that are too large for the group move out and lead a solitary life. Nearly all aspects of structuring of social groups involve a kind of social behavior known as **dominance.**

Dominance relations When individuals of the same, or of different, groups come together, **aggressive behavior** commonly occurs. First encounters often involve active physical conflict—biting, pecking, or clawing. In later meetings of the same individuals, aggression may be displayed more by snarling, growling, or screaming; or by raised hackles, displays of canine teeth, horns, or antlers; or simply by a fixed, piercing stare. In other words, actual physical aggression is reduced. Such encounters establish one individual as socially **dominant** to another, more submissive individual. Dominance relations have been found in all groups of vertebrates, except amphibia, as well as in many invertebrates. In crickets, for example, the first individual to chirp in an encounter may become dominant to the other without actual fighting occurring.

In some birds an interesting form of dominance called **peck order** is found (Fig. 15-15). For example, a flock of chickens may develop a **linear dominance hierarchy.** Often one bird is dominant to almost all other birds in the flock and can peck any of them. At the lower end of the peck order (Fig. 15-15*h*) is the most submissive hen who can be pecked by every other bird in the sequence, but cannot peck any chicken in the flock.

Dominance relations are usually not so strictly linear. For example, in Fig. 15-15 individual *d* is submissive to *a* but dominant to *e*, which in turn is dominant to *a*. Dominance relations may change with the season, age, state of health, and sexual condition of individuals. In most animals, the larger and more vigorous animals are dominant to others. Males are usually dominant to females and older females dominant to younger ones.

In social groups where dominance relations strongly structure the group, submissive animals have various ways of weakening aggression on the part of the dominants. Slowly retreating with head turned aside and eyes averted is such a technique. In many carnivores, tucking the tail between the legs or rolling over on the back serves

FIG. 15-15 Three forms of peck order are illustrated. *a*, the dominant chicken, can peck all others but *e*. There are two linear hierarchies of dominance, $a \rightarrow b \rightarrow f \rightarrow h$ and $a \rightarrow c \rightarrow g \rightarrow h$. There is also a circle of dominance in which *a* dominates *d* and *d* dominates *e* which can peck *a*.

to indicate submission (Fig. 15-16). Grooming of an aggressive dominant individual by a subdominant one will often have the same effect. Thus licking the fur or, in the primates, combing and stroking it with the claws may reduce the tendency to violent aggression.

In social groups there is often a kind of behavior called **displacement activity.** This is a response to some stimulus in what appears to be an inappropriate manner. It is as if the organism has to do

FIG. 15-16 Submissive posture in coyotes is indicated by the facial expression and the position of the tail. In this sequence, a male and female encounter a foreign male coyote. The resident male approaches the foreign male with head high, ears forward, and tail extended horizontally—all dominance signals; the foreign male shows submissive behavior, lowering its tail and drooping its head (*a*). When the resident male attacks, the foreign male turns on its back, exposing its throat and underside, indicating his unwillingness to fight (*b*). The resident male and female then move on. (*After D. Hiser, in P. McMahon,* Natural History, **84**:*42–51, 1975.*)

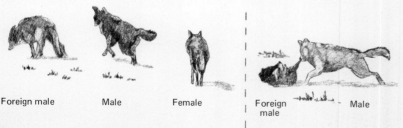

something, cannot do the appropriate thing, but must do something. For example, certain birds when approached by a dominant individual they dare not fight will preen their feathers, scratch, or stretch. Displacement activity also is found in human beings. People confronted with a problem they cannot solve will often scratch their heads, rub their noses, doodle, chew their pencils, etc.

It seems clear that dominance relations tend to substitute symbolic aggression for actual physical aggression. Each individual soon comes to know its position in the group and does not have to fight to maintain it. Dominance relationships in human beings serve essentially the same purpose—they reduce conflict. Many social groups, such as clubs, corporations, and the army, have dominance hierarchies, as do most groups of friends. Among your friends does one person make decisions or take the lead more often than others? There are various symbols of dominance used by group members to indicate their status. In groups such as armies, dominance symbols are quite obvious: insignia, uniforms, medals, etc. However, in American society as a whole there are more subtle dominance symbols. Many physicians drive Cadillacs not because they like Cadillacs but because it is expected of them in their position in society. Prestige, as indicated by a Cadillac, is a specifically human form of dominance. The posture of individuals (Fig. 15-17) may indicate their feeling of

FIG. 15-17 This photograph of a group of people in a meeting shows some kinds of body language. Dominant individuals tend to sprawl and appear "loose" while less dominant ones are more restrained. (*Marcia Weinstein.*)

social status or prestige. Some dominant ones tend to sprawl in a relaxed manner at meetings or in class while less dominant ones keep their hands in their laps, sit up straight, and generally present a more "closed" appearance. Are dominant or submissive individuals more likely to touch you, put their arms around your shoulders, hold you by the elbow, or take you by the hand? Of course, the details of such behavior are different from culture to culture.

15-7 Some organisms show territoriality

Another aspect of social behavior that is related to dominance is called **territoriality.** Individuals, mated pairs, or social groups often have a space or territory that is defended against others of the same species. Such territories may be breeding areas or feeding areas, and are usually proportional to the size of the animal or group.

Earlier we mentioned that the song of the male bird conveys the information that it is ready to mate. Such songbirds as robins, mockingbirds, and thrushes establish territories in which they nest and feed. Within its own territory, a male is dominant to other males of the same species. If it strays into another territory, it is less dominant and easily chased out.

Even when a territory is not established, individuals usually have an **individual space** within which other individuals are not permitted to remain. You may have seen a group of birds perching on a telephone wire. Usually they are evenly spaced. If another bird flies in and tries to settle down between two birds who are already at the minimum tolerable distance from each other, vocal protest occurs and feathers may fly. For this reason, flocks of birds are notoriously noisy when they are settling in a tree for the night.

Human individuals have an **interpersonal distance.** When others approach more closely than this distance, an individual feels threatened and uncomfortable. Interpersonal distance varies with the state of the individual. Courting individuals temporarily reduce their interpersonal distances, perhaps to zero. Interpersonal distances are, in part at least, learned. They vary from culture to culture. What is a comfortable distance to many people of the Middle East is considered threatening to people of Western cultures. And, on the other hand, what might be a proper, accustomed distance to an Englishman could appear as unfriendly and insulting to most Italians.

Territory and individual space act to arrange organisms more evenly over their environment, often resulting in more efficient use of the resources of that environment. By being restricted to certain areas (borders of territories), aggression in nonhuman animals is controlled and reduced. With changes in the environment and the availability of resources, territories and individual distances often shrink or expand accordingly.

Do territory and interpersonal distance play similar roles in human populations? In cities our interpersonal distance is often violated; many people feel uncomfortable in crowded buses and subways. There is considerable doubt as to whether individual human beings are territorial at all. Our closest living relatives, chimpanzees and gorillas, are not. Howler monkeys show group territoriality, but baboon troops coexist in the same areas. Some human hunting and gathering groups, such as some tribes of native Americans, had territories they defended from other tribes. This did result in efficient use of the resources of the environment. In fact, North America was fully populated at hunting-gathering and simple agricultural levels before Europeans arrived.

15-8 Social insects have very rigid groups

The social groups discussed up to this point all show some sort of division of labor among individuals. This division of labor is not rigid and unchanging, however. The role of an individual may change with time and its physiology. There are kinds of insect, however, in which division of labor is very rigid and where each individual's role is determined early in development for the rest of its life. These insects

FIG. 15-18 The social life of the honeybee centers about the hive. Honey is stored in cells in the honeycomb made of wax secreted by the workers. Each hive has a single, active queen who was fertilized by a male (drone) during a nuptial flight. The male soon dies, but he has provided the female with a store of sperm. Some cells of the honeycomb are used as brood cells for developing larvae. The queen lays an egg in them, and the workers feed and care for the larvae. They cap the cells when the larvae become pupae. From time to time new queens and drones are produced. A new queen either takes over the old colony or founds a new colony elsewhere.

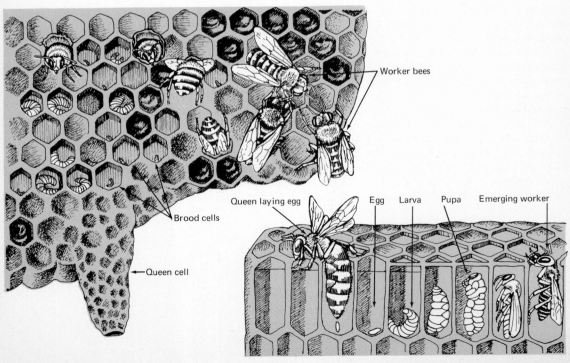

are known as **social insects.** Termites, ants, wasps, and bees all have species with rigid social groups; the best-known social groups are those of honeybees.

The social group of honeybees is called a **hive** (Fig. 15-18). In a hive there is only one reproductive female, the **queen.** The most numerous individuals are **workers,** which are sterile females. The workers constantly feed and clean the queen, and all the bees constantly exchange small amounts of food. The queen also passes to the workers, in such exchanges, a compound called queen substance, which she secretes from glands at the base of her jaws. The queen substance inhibits workers from constructing many of the special cells in the hive in which queens develop. She ensures that there will be only one queen by stinging and killing developing queens.

The workers gather nectar (which they make into honey) and pollen. They also construct the cells (or spaces) that make up the honeycomb, in which eggs develop into larvae, pupae, and finally adults (see Fig. 15-18). These cells are made of beeswax secreted by glands on the abdomen of workers. Finally, the workers feed the developing larvae in their cells; clean up the quarters of the colony, throwing out dead or dying bees; and defend the hive against predators. In its defense of the hive, workers may die since the stinger of a worker bee (but not that of a queen) can be used only once and in the process of stinging it is ripped out of the body.

When the workers become too crowded in the hive, or if something should happen to the queen, the workers are no longer inhibited by queen substance. They then begin to produce large cells called queen cells. Larvae that grow up in these cells are fed a special diet, including a substance the workers secrete called royal jelly. This diet permits the larvae to develop sexually, in other words to become queens. Reproductive males called **drones** are produced at certain times in the life of a hive. They are haploid—that is, unlike the queen and workers, they have only one set of chromosomes. A large overcrowded hive may produce several swarms, each containing a mated queen, which fly away to form new hives.

The three different kinds of bees in a hive—queen, workers, and drones—are called **castes.** In some social insects, there are even more castes. Termites sometimes have several warrior castes (Fig. 15-19). One type has powerful biting jaws. Another has jaws that are very much reduced in size, but the head is specialized into a kind of squirt gun from which a chemical repellant can be sprayed at enemies.
Communication among honeybees Worker honeybees forage widely for the nectar of flowers. When a successful forager returns to the hive, it is able to convey much information about its trip. By exchanging small amounts of the nectar with other workers, it informs them of the odor of the nectar. By performing a *dance* on the honeycombs, it also informs them of the distance of the source of nectar and, if it is relatively distant, the direction of the source from the hive. There are two kinds of bee dances. If the nectar source is nearby, the worker honeybee does a **round dance** (Fig. 15-20). The duration of the dance and its speed indicate how abundant the nectar source is.

If the nectar source is distant, the worker does a **waggle dance**

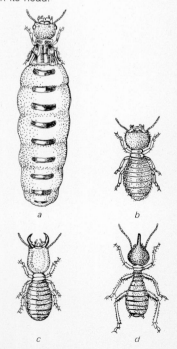

FIG. 15-19 Four castes of termites: (*a*) queen; (*b*) worker; (*c*) soldier, specialized for pinching, has large jawlike structures for use in combat; and (*d*) soldier, specialized for chemical warfare, can squirt a poison from glands in its head.

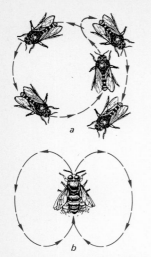

FIG. 15-20 The round dance of honeybees (*a*) is used to indicate the presence of a nearby food source. The waggle dance of honeybees (*b*) is used to indicate the distance and direction of a food source further than 90 meters from the hive.

(Fig. 15-20). In the waggle dance, the bee moves in a figure-eight pattern. The speed of the dance conveys the distance of the nectar, as does the number of tail waggings. The straight portion of the dance indicates the direction of the source from the hive. For instance, if the bee is dancing on a vertical comb and the straight portion is vertical—pointing upward—the nectar is in the direction of the sun. If the bee dances the straight portion pointing downward, the nectar is directly away from the sun. Should the nectar source be, say, 60° to the right of the sun, the straight portion of the dance will be pointed upward at an angle of 60° to the right of the vertical (Fig. 15-21).

Bees can use **polarized light** to determine the position of the sun because they (along with many other invertebrates) have an eye structure that enables them to detect patterns in the light of the sun which our eyes cannot detect. The polarization of the sun's light changes as the sun changes position. (Polarized light may be thought of as having its waves in a single plane.) By the use of this ability a bee is able to find the position of the sun even on a cloudy day, compare the flight direction to the nectar with the direction to the sun, and then transmit the information to other workers in the waggle dance. Bees also use their biological clocks (Sec. 4-11) to compensate for the changing position of the sun during the day.

15-9 Communication is carried out by many modes

A number of kinds of social behavior have now been discussed. All these require communication among individuals. You probably will have noticed that a number of different modes of communication can be used. Let us summarize the various modes of communication.

Chemical communication **Chemical communication** occurs because all animals secrete substances into the environment. If you have just finished playing an active sport, your friends may be able to determine this without your telling them! In the same way that sweat may communicate information, urine and other bodily secretions may also. Many animals produce substances whose major role is to act as *chemical messengers* outside the body. These substances are called **pheromones.** Ants, for example, deposit a trail substance which other ants can follow to a food source. Queen substance is also a pheromone. In some social insects there are so many pheromones that they amount to a kind of language in themselves.

There is growing evidence that pheromones are important in human beings as well as in other animals; that human beings may communicate with each other by odors without being fully aware of it. For instance, at times of stress the glands in the armpits produce a special, strong-smelling sweat, which is sometimes referred to in phrases such as, "You could smell his fear." Recent studies show that human females have substances in their vaginal secretions that are the same as those in other female primates which are known to possess sex-attractant properties. These potential human pheromones vary in production during the menstrual cycle, reaching a peak when ovulation occurs. The "pill" suppresses this rhythmic change in pheromone production. Other possibilities of pheromonal communication in human beings are also being investigated scientifically.

Tactile communication A second mode of communication is **tactile communication**—communication by touch. We have seen that tactile communication can be important to spiders. It also plays a role in reducing aggression in primates, as when a dominant male baboon is groomed by a female. The role of tactile communication in human societies varies from culture to culture. Most cultures encourage both adults and children to hold and cuddle babies. Many cultures do not permit tactile communication between adults other than lovers, except for brief ritualized exchanges such as handshakes. Recently, a new interest in the potentials of tactile communication has appeared in American society. Various types of encounter groups have experimented with tactile communication as a way of helping people relate more fully to life and their own feelings.

Sound communication Obviously, **sound communication,** a third mode, is important in a very large number of animals. The way sound is produced varies a great deal from group to group. Fishes can make sounds by grinding their teeth and crickets chirp by rubbing their wings together. In humans complex sounds are produced by our vocal cords and are called speech. It should be mentioned here that not all sounds which communicate information are designed to do so. Sounds often indicate the physiological or behavioral state of an organism; other individuals then interpret these sounds. Many animals, when disturbed, give a so-called warning cry. It is thought that the cry was not originally made for the purpose of warning others but perhaps served to startle predators. In the course of evolution and coevolution, it has come to have a warning function as other individuals have come to recognize it as indicating possible danger to themselves.

Recently it has been discovered that, in addition to the sounds they produce for echolocation, humpback whales produce underwater songs. These are repetitions of sequences of sounds, ranging from extremely low-pitched notes to extremely high-pitched notes, and lasting up to 30 minutes. Different whales have slightly different songs. Unfortunately the function of whale songs is not known; since sound can travel great distances underwater, the songs could be a form of communication.

Most sound communication in animals other than humans beings gives a more or less *instantaneous* indication of their state. For example, a courtship song indicates readiness to mate. In true language, however, there is usually a *delayed* transmission of information. This is why the dance communication of bees is sometimes called a language.

Visual communication A fourth mode of communication is **visual communication.** Visual communication is usually combined with sound communication. We have already discussed the visual displays of birds marking their territories and the warning colorations and patterns of distasteful or poisonous animals and the species that mimic them. Visual signs may also be used as alarm warnings. For example, when a deer or antelope is alarmed, it raises its tail, expos-

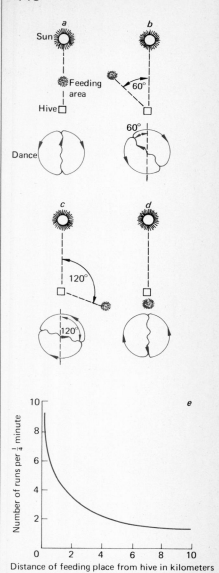

FIG. 15-21 The position of the sun with respect to the food source is indicated by the angle of the straight portion of the waggle dance (*a–d*). How the speed of the dance indicates the distance of the food source from the hive is illustrated in (*e*).

ing a patch of lighter fur on the rump. This acts as an alarm for other members of the herd.

The human species has tended to specialize in verbal sound communication and deemphasize visual signs and displays. There is abundant evidence, however, that visual communication still goes on in our species to a greater extent than most people realize. It is often called "body language" (see Fig. 15-17). Some persons are especially well versed in interpreting body language, for example, psychotherapists and used-car sellers!

CONCEPTUAL BEHAVIOR AND LANGUAGE

The development of language in human beings has had many important effects. The structure of a language may affect how a group of people think about the world. More importantly, language makes possible a kind of evolution essentially restricted to people. This sort of evolution is called **cultural evolution**, and it has wide-ranging effects on human beings and all other organisms.

15-10 Conceptual behavior depends upon the cerebral cortex

Many invertebrates and nonhuman vertebrates exhibit very complex behavior. Spiders construct complex webs; birds make intricate nests and carry out intricate courtship dances. All or many of these activities are inherited behavior patterns that are carried out, more or less automatically, under the right environmental stimuli, including those provided by other organisms. Usually such animals are not able to learn a new way of making a nest, nor are they able to invent new songs or web patterns. The human species is distinguished from other animals by the ability to learn and invent complex behaviors, such as speaking, writing, dancing, and athletic routines. These conscious activities are controlled by the largest, most conspicuous region of the brain, the **cerebral cortex** or cerebrum.

Figure 15-22 shows the principal regions of the vertebrate brain and the changes that have taken place in the course of evolution of the brain. The vertebrate brain consists of three principal regions: the **hindbrain,** the **midbrain,** and the **forebrain.** The hindbrain regulates such functions as heartbeat and breathing, and it contains the centers for taste and muscular coordination. The midbrain is the center for vision and carries out integrative functions in lower vertebrates, but some functions are taken over by the forebrain in mammals. The midbrain in people serves as a relay center. Figure 15-22 shows the great increase in the forebrain, including the cerebrum, in the course of evolution. Inputs from the various sense organs of the body, as well as inputs from the muscles, go to specific areas of the cerebral cortex, which becomes much folded or convoluted. There they are integrated with other inputs and stored information, and outputs to all parts of the body are produced (see Fig. 14-13).

The brain itself lacks sense organs and pain receptors. In people and other animals, it is possible to stimulate the brain electrically

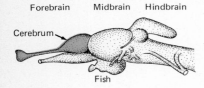
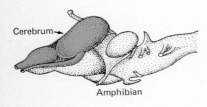
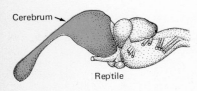
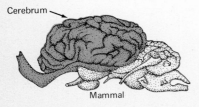

FIG. 15-22 The evolution of the brain in vertebrates using living forms to represent presumed evolutionary trends. The forebrain (color) is progressively larger and more complex in forms believed to have evolved more recently.

and observe the responses. Persons undergoing brain surgery, for example, have been studied in this way. Usually they are given a local anesthetic for such an operation and, since they are conscious, can be asked what they experience. When a very mild electrical stimulus is applied to various parts of the brain, patients have reported remembering past conversations or seeing places where they formerly lived. One patient heard an entire symphony when current was applied to one particular spot of the brain! As a result of such studies, the location of many integration centers in the cortex have been mapped. In Fig. 15-23, the amount of cortex on either side the median groove devoted to sensory and motor functions is shown by the relative sizes of the parts of the two figures.

It would appear that one important function of the cortex is to "disregard" a great deal of sensory input. That is, much of the information that our sense organs pick up is not brought to our awareness. If this were not so, we would be overwhelmed with sounds, sights, pressures, and odors and would be unable to function. The cortex somehow learns to filter, from the mass of stimuli presented by the environment, those inputs that need to be dealt with consciously. In the course of time, many behaviors that are at first conscious become unconscious. Think of the difficulty a child has in learning to tie a knot in shoelaces. Or think of the time and attention it takes to learn to drive a car. Once these skills are learned, however, they can be carried out automatically with little conscious attention.

Cortical integration of vision We tend to think that we see with our eyes. In fact, as discussed in Sec. 14-9, we see with the *help* of our eyes. The eyes are sensory organs that send input to the visual areas of the cerebral cortex. Along the way, and in the cortex itself, a great deal of integration takes place without which we would not see. There are many experiments that show this is true. For example, if you close your eyes and gently press on your eyelids, you will "see" light. Even though light energy does not reach your retina, your cortex tells you that you are seeing light because of the input it receives. The cortex knows only that it has received impulses from the retina, and it interprets these as light. Similarly, when you dream, as you do about six or so times every night, you "see" vividly (as most people report, in color) without there being any light stimulus.

It is not known exactly how the visual cortex of the human species accomplishes the amazing job of integrating the electrical stimuli it receives from the eye into "pictures" of the world around us. In frogs and cats, however, somewhat more is known. The frog's visual system is relatively simple. The frog's retina filters out, from all the light stimuli it receives, only four different kinds of stimuli to be sent to the brain. In the brain there are cells that respond to each of these kinds of input. The result is that a frog has a very limited visual world, but one which is quite suitable for a frog's life. The first kind of input detects contrasts in the environment that occur for long periods—it provides the overall view of the trees, rocks, pond, etc. of the habitat. The second type of input detects sudden moving shadows, such as a snake or other predator. The third input detects a sudden decrease in light, such as would occur if a large predator

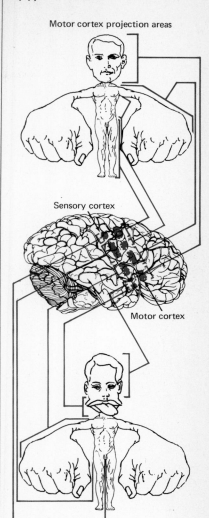

FIG. 15-23 In the human brain, along either side of a median groove in the cerebral cortex are areas called the motor cortex and the sensory cortex. Differing amounts of surface area of the cortex in these areas are associated with different parts of the body. The relative sizes of the parts of the two figures are proportional to the amount of cortex concerned with them. Thus the hands have enormously more cortical area than do feet, and the tongue and lips have much more surface area in the sensory cortex than in the motor cortex.

were attacking. The fourth system is most peculiarly frog-suited; it responds to small, wriggling or flying objects close to the eyes.

More complex animals have suitably complex systems with additional kinds of retinal detectors and more brain cells to receive them. In cats, for example, there are cells that detect corners and edges of objects as well as cells that respond to movement in various directions across the retina. Although we do not know as much about human visual systems, our retina and visual cortex and our ears and auditory cortex make possible, along with other areas of the cortex (for example, tactile), the development of written human language.

15-11 Language as a mode of communication is nearly restricted to the human species

Although some scientists call the bee dance language, others feel that true language is known only in the human species. Recent studies, however, have shown that chimpanzees and gorillas can learn various kinds of sign language. Some chimpanzees have learned to make sentences by putting colored plastic symbols in order. Both chimpanzees and gorillas have learned to use the sign language used by deaf people. These chimpanzees and gorillas are able to communicate meaningfully with their human teachers and to generalize the meaning of a word to a broader sense. For example, one chimpanzee learned the sign language for *open* with respect to a box of toys and later extended it to request that a water faucet be turned on.

Genuine language is generally considered to include the ability to delay the transmission of information. Gorillas and chimpanzees must be said to be able to learn language from people, although not apparently from one another *in nature*. People have the ability to remember things which happened in the past and to convey the information to other people at some later time. This may, of course, be done vocally or in writing. Writing may be preserved for thousands of years and the information it contains be transmitted. Animals, as we have seen, transmit **signals** that indicate the physiological (or psychological) state of the sending individual. It is, in a sense, incidental that the receiving individual can interpret these signals as warnings, courtship behavior, or the direction of a nectar source.

Symbols   In addition to the use of signals, people use **symbols** in the construction of language. Symbols are words or sounds we agree upon to represent something else. The word *fish*, for example, bears no special relationship to an actual fish—it does not look, sound, or smell like a fish. Nevertheless, persons using the English language have agreed, *and teach their young,* that these four letters put together in this order stands for a fish. Or, used as a verb, it stands for the even more abstract idea of "trying to catch a fish." As languages evolve, they often become more and more complex. For example, the adjective *fishy* may mean "fishlike." Or, when used in just the right way, it may convey the information that something is not quite right: "That sounds fishy to me."

By the use of symbols and groups of symbols, the human species has developed conceptual behavior. We classify the things we find important in the world around us and build concepts which are abstractions from the "real" world. Thought in people often involves the conception of a theoretical model of what goes on in the real world. Such a model may then be compared with the real world and modified as necessary or desired. Of course, it is possible to construct a purely imaginary model and deal with it as if it were the real world. Some *linguists* (students of language) believe language itself often has built into its structure assumptions about the real world.

Structure of language and cultural differences in perception In this view, once a language has been developed, the structure of the language molds our view of nature into a *form easily handled by the language.* Ever since the early Greeks, Western peoples have behaved as if every effect implies a cause and all things must be contained in something, and we, as **subjects,** perceive things in the real world as **objects** which are strictly separate from us. On the other hand, Oriental philosophies have usually emphasized the artificiality of the subject-object distinction. Our language has built into it a need to have a "doer" for everything that is "done." We are forced to describe objects and relationships among them. All this sounds so "natural" and "correct" that it is difficult to talk or write about—especially in our own language! The need for a doer for everything done puts us in the position of having to say, for example, "It is raining." A German would say, *"Es regnet,"* and a Mexican, *"Esta lloviendo."* What is the *"it"* that is raining? Merely a word we put in a sentence because our language requires such a construction.

It is broadening to consider how people of other cultures deal with objects in the real world. We find that they often do things very differently from what we consider the correct and natural way. Eskimos, for example, have no single, overall word to refer to water. They do, however, have a very detailed vocabulary for dealing with the many kinds of frozen and liquid water they encounter in their daily lives. The gauchos (cowboys) of Argentina have some 200 words for colors of horses. Clearly horses are important in their lives. In describing the vegetable world, on the other hand, they need only four words: *pasta,* for fodder; *paja,* for bedding; *cardo,* for woody materials; and *yuyos,* for all other plants!

A good way of understanding the relationship of language to behavior and to the description of nature is to compare our language with a very different language. One student of languages compared our language with that of the Hopi Indians, which had a quite different evolutionary history. Their language is so different from ours that the differences are difficult to discuss in English. We use numbers indicating quantity when we speak about either real or imaginary groups. We count 10 books and regard them as a group. We also, however, refer to 10 hours, or 10 days, or 10 years. With these units of time, of course, only *one* unit is experienced at a time. The others are remembered or predicted. Our language enables us to "know" that there was a day yesterday and that there will be a day tomorrow. Even though tomorrow has not yet arrived, we can divide it up quite precisely into hours, minutes, and seconds. It would not occur to a Hopi Indian to use numbers for things that do not make up a real, as opposed to a subjective, group. They recognize a group of five Indians. If they stay for a visit, their hosts would say that they "left after the tenth day," not that "they stayed 10 days." In the Hopi language there simply is no such thing as a group of days.

Other differences show up when the problems of dealing with physical quantity or time are considered. We have nouns that refer to homogeneous materials without definite outline or size. *Water* and *air* are examples of such words which imply indefiniteness. When we become more specific, we must say, "a body of water," "a dish of food," or "a bag of nuts." This shows that we must have some sort of container for the portion described. By using "of" we indicate that we are talking about the "contents" of something (body, dish, bag). In the Hopi language, general nouns imply indefiniteness but also a specific portion and size. No container is implied. The Hopi seem not to have a need for abstract, formless items such as our "time." There are no past, present, and future tenses of verbs. Nor could a Hopi construct sentences involving *imaginary space,* whereas we can say, "This chapter is *over my head!*"

When one turns from ordinary language to the language of science, there is no reason to think that language does not affect our perceptions of the world. "Scientific" concepts such as time, space, and matter are part of the structure of the language of the physicist. In the same way, much of what we think of as "real," "commonsense," or "beyond doubt" in biology comes directly from our language and culture. For example, we talk of natural selection as a "force" which "operates" on populations, when it is the description of what *has happened:* differential reproduction of genotypes. Similarly, we say "isolation" causes the evolution of two species from one when, obviously, "isolation" cannot itself cause anything. Biologists have a great deal to learn from the ways nature is viewed by cultures with very different languages.

Language is a means of conveying information about the real world and about worlds we construct in our minds. Unfortunately, language also provides us with the chance to fool ourselves. For example, people may tell themselves that they can control nature and they may believe it. In the real world, however, many things are happening which they have not yet perceived, making it impossible for them to control nature.

Language and cultural evolution Once spoken and written language had been acquired, our species had also acquired the capacity for a new kind of evolution. This new evolutionary process is called cultural evolution (Sec. 10-13). We have seen in earlier chapters that biological evolution is the transmission, with modification, of genetic information from generation to generation. This genetic information is contained in the DNA. Cultural evolution is the transmission, with modification, of nongenetic information from generation to generation. This nongenetic information is culture, especially in the form of language.

Cultural evolution is very limited in other animals. In a sense, cultural evolution goes on when a female lion teaches her cubs to hunt. It also occurs in monkeys and apes. Studies of Japanese macaques and chimpanzees, among several others, have shown that new behaviors may be acquired by an individual in a troop and then be passed on to other individuals. The way in which the new information is transmitted depends upon the species. In some, a male transmits a new food choice to others much more rapidly than a female. In others, younger members of the troop may develop a new pattern of behavior and transmit it to their peers, but adults never learn the behavior.

Cultural evolution has many similarities to biological evolution. Speaking in general terms, we might compare the acquisition of a new idea to a gene mutation. The change of a culture and its spread at the expense of others may be thought of as similar to natural selection. It is known, also, that people in small groups—for example, peoples on oceanic islands—sometimes develop their own specialized subcultures. This may be compared to genetic drift. And, of course, there are many examples of migration of ideas and cultures.

There are important differences between cultural and biological evolution. The transmission of genetic information is from parents to offspring in a biologically determined way. The transmission of cultural information may be among peers, between parents of offspring, between grandparents and grandchildren, etc. Cultural information may be transmitted by books, radios, television sets, and computers, etc. The information may be days, years, or hundreds of years old. Obviously, the transmission of cultural information is much more complex than the transmission of genetic information.

For at least 3.2 billion years, organisms have responded to changes in the environment with evolutionary changes. It is only in the last 100,000 years that cultural evolution has been possible. Rather than responding to environmental change with evolutionary change, the human species has usually responded by cultural evolution. From the time of the makers of simple stone tools to the present day with its awesome technology, people have remained, physically, rather generalized, unspecialized animals. The development of agriculture, of using metals, and of making machines occurred by cultural evolution. Each generation adds new elements to culture so that cultural evolution moves very rapidly compared with biological evolution. Vertebrates acquired the ability to fly over a period of some millions of years, by the transformation of the forelimbs into wings. But, people alive today can remember when human beings first began to "fly." Now our airplanes can fly faster and higher than any bird.

Interrelation of cultural and biological evolution The existence of cultural evolution does not mean that biological evolution has stopped for our species. Cultural and biological evolution do not take place independently of one another. The evolution of the brain and the cerebral cortex made possible an expansion and enrichment of culture. The existence of culture gave a selective advantage to certain types of brains, that is, those with extensive cerebral cortex development. A change in selection pressures has been a major effect of cultural evolution. The human species has eliminated most large predators, developed advanced medical techniques, increased the food supply locally in many parts of the world, and learned to control or anticipate change in the environment. This permits genotypes to survive and reproduce that would otherwise have been eliminated. Bad eyesight is corrected by glasses, and diabetes is controlled by insulin.

On the other hand, cultural evolutionary changes may bring about new selective pressures. Selective agents undoubtedly favor genotypes today that are relatively more immune to insecticides in food, to air pollution, and to nervous tension. Furthermore, since life span has generally increased, selection is occurring against diseases such as heart problems and cancer that were not selected against before. This is because more individuals are living to (and reproducing in) middle age, when differential susceptibility to these diseases begins to take its toll. Cultural evolution, by improving means of travel and by changing cultural barriers to mating, has changed the patterns of gene flow in human populations.

In general terms, biological evolution can be thought of as a genetic change in populations through time in response to environmental change, including other organisms. The processes of meiosis and the factors controlling the ways in which gametes come together determine the rate at which biological evolution can occur. The factors which control rate of evolution are themselves under genetic control and therefore subject to evolutionary change. It is to be expected that, in the course of time, the rate of evolutionary change will be slowed down or speeded up until it is proportional to the rate of environmental change.

Cultural evolution does not have the built-in regulators of rate that are found in biological evolution. Transmission of cultural information is not just from parent to offspring; it also may be transmitted by books or by radio and television to millions of people simultaneously. The rate of cultural evolution, therefore, can be immensely faster than that of biological evolution. Once started, cultural evolution appears to proceed at an ever-increasing rate—a positive feedback occurs in the system. Indeed, one of the major effects of cultural evolution is to develop improved means of transmitting information to more and more people.

Cultural evolution of this sort is available only to the human species. In responding to environmental change, we have changed the environment with our technology. Unfortunately, this has meant that the *rate of environmental change* has been speeded up enormously. In many parts of the world, the rate of environmental change is so fast that no organism, or only a very few, can respond by biological evolution. Only the human species by cultural evolutionary change has a chance to make an adequate response. And recently even cultural evolution has been sorely taxed to keep up. It is unfortunate also, from our point of view, that the organisms which *can* best respond to human-induced environmental changes are small, rapidly reproducing organisms. In other words, they are bacteria, fungi, insects, rodents, and the like, many of which are regarded as pests. Cultural evolution has permitted the human species to become the dominant organism on the earth. All too often this has been at the expense of the other organisms with which we share this planet (see Chaps. 3 and 5).

QUESTIONS FOR REVIEW

1 If the cleaner fish mimics prospered to the extent that they greatly outnumbered their models, what do you suppose would eventually be the effect on the cleaner fish–cleaned fish–mimic interrelationship?
2 We have seen that the males of some spider species are often consumed by the females—a situation which is clearly disadvantageous to individual males. Why, then, does this common fate not act as a kind of selection pressure that causes male spiders, over many generations, to evolve phenotypes for structures and behaviors that help them avoid it?
3 What do these terms mean: aggregation, dominance, waggle dance, presentation posture, social response, peck order, displacement activity, territoriality, pheromones, social insects, cortical integration, signal, symbol, filtration of stimuli?
4 How would you answer the question at the end of Sec. 15-2?
5 Can you give at least three reasons why the honeybee "language" is not a true language?
6 Which communication mode was first used for language: sound or vision? Which would you say is the more important mode today in the maintenance and transmission of culture?
7 What communication mode other than vision and hearing is sometimes employed for symbolic communication? What limitations does this mode have that precludes its being the primary mode of symbolic communication?
8 It is the brain, not the eye, that *sees*. Can you support this statement by explaining it?

READINGS

Barlow, G. W.: "Animal Behavior: I," *Biocore,* unit XIX, McGraw-Hill, New York, 1974.
———: "Animal Behavior: II," *Biocore,* unit XX, McGraw-Hill, New York, 1974.
Chambers, K. L. (ed.): *Biochemical Coevolution,* Oregon State University Press, Corvallis, 1970.
De Vore, I. (ed.): *Primate Behavior,* Holt, New York, 1965.
Doi, T.: *The Anatomy of Dependence,* Kodanshu, San Francisco, 1973.
Ehrlich, P. R., and R. W. Holm: "Evolution," *Biocore,* unit XXII, McGraw-Hill, New York, 1974.
———, ———, and D. R. Parnell: *The Process of Evolution,* 2d ed., McGraw-Hill, New York, 1974.
———, and P. H. Raven: "Butterflies and Plants," *Scientific American,* **216**(6):104–112, offprint 1076, 1967.
Fox, M. W.: *Concepts in Ethology,* University of Minnesota Press, 1974.
Hall, E. T.: *The Hidden Dimension,* Anchor Books, Doubleday, Garden City, New York, 1969.
Linden, E.: *Apes, Men, and Language,* Saturday Review/Dutton, New York, 1975.

Portman, A.: *Animals as Social Beings,* Viking, New York, 1961.
Roeder, K. D.: "Moths and Ultrasound," *Scientific American,* **212**(4):94–102, offprint 1009, 1965.
van Lawick-Goodall, J.: *In the Shadow of Man,* Houghton Mifflin, Boston, 1971.
von Frisch, K.: "Decoding the Language of the Bee," *Science,* **185**(4152):663–668, 1974.
Washburn, S. L., and I. De Vore: "The Social Life of Baboons," *Scientific American,* **204**(6):62–71, offprint 614, 1961.
Whorf, B. J.: *Language, Thought and Reality,* M.I.T., Cambridge, Mass., 1956.
Wickler, W.: *Mimicry in Plants and Animals,* McGraw-Hill, New York, 1968.
Wilson, E. O.: *The Insect Societies,* Harvard University Press, Cambridge, Mass., 1971.
———: *Sociobiology: The New Synthesis,* Harvard University Press, Cambridge, Mass., 1975.

16

CULTURE AND SURVIVAL

SOME LEARNING OBJECTIVES

After you have studied this chapter, you should be able to

1. Describe at least four impressive cultural achievements of various hunting and gathering societies which go far toward refuting the notion that these societies were and are "primitive," that is, culturally inferior to our own. Your list should include one achievement that has eluded our own "advanced" culture.

2. Demonstrate your knowledge of some aspects of the development and nature of agriculture by
   a Stating where, and about when, farming is believed to have appeared for the first time.
   b Explaining why the agricultural revolution was, as the text notes, probably the most important event in the history of our species.
   c Stating where and when the second major phase of the agricultural revolution began, citing two of its important developments, and saying what was its effect on the human population.
   d Listing three major impacts of

   agriculture which have taken place since the industrial revolution and have directly or indirectly been caused by it.
   e Stating a major negative impact of the agricultural revolution and describing some causes and consequences of this impact.

3. Demonstrate your knowledge of the interrelationship of the city and technological society by
   a Stating where the first cities appeared, and briefly explaining why the appearance and subsequent development of cities were dependent on the agricultural revolution.
   b Stating where the first true industrial cities arose, and tracing the interaction of landed estates, agriculture, mines, and mills as industrialization proceeded there.
   c Stating some reasons why industrial cities are so vulnerable to disruptions, and giving a reason why, besides the inconvenience and dangers they pose for individuals, this vulnerability and such disruptions are disturbing.

4. List three ways in which underdeveloped countries differ from overdeveloped countries and give an example of each type, say what is meant by "the revolution of rising expectations" and what increasingly negative environmental impacts this "revolution" is having, and explain why it is neither desirable nor possible for UDCs to become ODCs.

5. Explain what is meant by referring to our planet as "a unitary Spaceship Earth" and a "commons", explain what a "tragedy of the commons" is, and list three resources of our planetary "commons" which are now threatened with tragedy.

The genetic heritage of the human species may be thought of as a set of blueprints that are constantly being updated by natural selection. Without such up-to-date genetic blueprints, our species could not survive. But we also could not survive without cultural information appropriate to the environment in which we find ourselves. People can lead many different kinds of lives and have many different kinds of social orders, but only if they possess the appropriate cultural tools. Cultures have evolved as environments have changed; each individual born into a culture learns the information appropriate for that culture. Because of increased rate of environmental change, many individuals today live in several different cultures during their lives.

HUNTING AND FOOD GATHERING

The most important problem animal species face is obtaining food. Without food, individuals of a species cannot grow or reproduce. They cannot engage in searching for shelter or other behavior characteristic of their kind. The human species has for most of its time on this planet obtained its food by methods very similar to those used by other animals—hunting and gathering.

16-1 Hunting and food gathering was the human species' earliest way of life, and it still persists today

As we discussed in Chap. 3, our early ancestors ate fruits, berries, and roots, and captured fishes and small game, which they ate raw or cooked. When they lived near the seashore, they undoubtedly varied their diet with shellfishes and other animals from the tidal zone. Gradually human groups sharpened their hunting skills and began to take big game. The hunting capabilities of our ancestors tens of thousands of years ago are attested to by the existence of vast bone yards.

Some members of industrial societies tend to look down on human beings who still hunt and gather food. A closer examination, however, of the few hunting cultures that have survived until recently provides little support for such bigotry. This is becoming more and more widely understood by Americans whose ancestors heartlessly exterminated most native Americans as lesser beings. We now understand that native Americans had a culture with a high survival value, until the Europeans invaded. The native Americans, for instance, were able to adapt to life on the great prairie of the United States and live there without seriously disturbing its ecological balance for thousands of years. Europeans who moved to America, on the other hand, totally destroyed the prairie in less than a century. And it is not clear that the fragile agricultural ecosystem with which they replaced it will persist even another 50 years. Man for man, the native Americans made better light cavalry than Europeans, even though they had not seen a horse until the white people brought them. History may prove they were better people all around.

Similarly, no sensitive person who has carefully studied Eskimos, Ifaluk islanders, Australian aborigines, or South African bushmen can feel anything but admiration for the cultures these people have developed. Aborigines have developed a system for keeping track of relatives, a kinship system, which is vastly more detailed and intricate than that of our culture. The Eskimos show unsurpassed skill at hunting and killing big game under the most trying of conditions. The bushmen have developed a culture which permits them to survive where other human beings would quickly die of thirst. All these groups contain people with immense artistic talent, as even the most superficial examination of aboriginal bark paintings, bushman rock art, or Eskimo carvings will show (Fig. 16-1).

16-2 Probably no more than 5 million hunters and food gatherers ever walked the earth

Even the most skilled hunters and food gatherers require at least 100 times the land area to support one individual than do semiskilled agriculturalists. The productivity of a stable, natural ecosystem, in terms of food suitable for human consumption, is much lower than that of an unstable agricultural ecosystem designed with only human nutrition in mind. Based on the amount of land suitable for hunting and food gathering and the known population densities of modern nonagricultural people, the peak number of hunters and food gatherers was about 5 million people.

THE AGRICULTURAL REVOLUTION

The high point of human hunting and food gathering came around 10,000 B.C. By that time the human species had spread from its ancestral home in Africa to occupy virtually the entire land surface of the globe. Fossil remains in Asia indicate that the human species (or its ancestral species) has probably been there for at least 1 million years. Our arrival in the Western Hemisphere is much more recent, dating back perhaps only 25,000 years.

16-3 Human beings first practiced agriculture in the Middle East

On the basis of studies made in what is now the border area of Iraq and Iran, scientists are now convinced that village farming communities existed there between 7000 B.C. and 5500 B.C., and they estimate that the practice of farming began around 9000 B.C. to 7000 B.C.

One can only guess at the origins of agriculture. Food gatherers observed edible plants growing in the disturbed areas around their camps—areas where the soil was loosened by digging and which were rich in nitrogen from garbage and human wastes. They began to

FIG. 16-1 These drawings of Eskimo carvings show modern examples of an art thousands of years old. (a) Harpooner and (d) caribou came from Hudson Bay, Canada, and were done by Mosesee and Sywolee, respectively. (b) Musk ox and (c) mother and child come from Baffin Island, Canada, and were done by Sheeokjuk and Samulellee.

encourage these plants, perhaps by spreading their wastes around or by purposely loosening the soil. They also doubtless began to remove *weeds*, unwanted plants growing among the desirable ones. Thus, the process of plant domestication was begun and agriculture was underway. As more and more food became available near the campsite, the reasons to move in search of game became fewer and fewer. Staying in one place provided numerous advantages. Food could be stored against times of shortage. Arrangements for the defense of the group were easier to make. Implements too large to carry could be fashioned. And young animals could be raised in cages and pens—the beginnings of the domestication of animals.

16-4 Not all agricultural inventions originated in the Middle East

Although the evidence indicates that farming began in the Middle East, it seems very likely that farming developed in other areas shortly thereafter. Farming was an "idea whose time had come." Outside the Middle East, people living on lake shores or by the sea in tropical areas may have begun farming. A rich supply of fishes and shellfishes would have reduced the necessity for moving around, and their remains provided rich fertilizer. But whether the idea of farming arose once in the Middle East and spread like wildfire over the earth or whether, as seems more likely, farming began independently in several locations, it is certain that different plants and animals were domesticated in different areas (see Fig. 5-31). The horse was domesticated in western Asia and the donkey in East Africa. Native horses and donkeys, however, became extinct in North America long before domestic horses and donkeys were reintroduced by the Spaniards. Some of these escaped, and wild horses and donkeys still live in some areas of southwestern United States. Tobacco, corn, and potatoes originated in the Western Hemisphere. They were unknown in Europe until after the time of Columbus.

16-5 The agricultural revolution probably led to declining death rates

The agricultural revolution was probably the most important event in the history of our species. Farming is thought by some to have led to a life more secure from famine than hunting and food gathering, both because of increased yields of edible materials and because of the possibilities of food storage. Other hazards of life also may have been reduced by living in permanent settlements. Dangerous large animals could be exterminated in the vicinity of the settlement, and permanent defenses against various marauders could be constructed.

As a result of all this, the death rate in the human population seems to have begun a slow decline and, since the birthrate generally increased, the human population *began to grow*. Agriculture not only started the population growing, but it also, of course, provided the means for supporting a larger number of people per square kilometer. In their some 10,000 years of farming experience, people have continually sought to improve their techniques. In an early great spurt, they domesticated a great variety of plants and animals. *All* crops and farm animals important today were domesticated well before the time of Christ. People began to fashion tools out of copper and then iron. People developed wheeled carts in which they could move farm produce.

In the eighteenth century, another rapid series of improvements was made in European agriculture. Social changes led to bigger, more efficient farms. It was discovered that growing clover would help renew the soil. As we now know, bacteria that live in swellings in the roots of clover and other legumes are able to take nitrogen from the air and convert it to nitrogen compounds usable by higher plants. This ability to fix nitrogen permitted farmers to grow a clover crop to feed their animals and simultaneously to improve the soil. This made the process of leaving a field unfarmed every third year to save the soil unnecessary. Other improvements were made in methods of cultivation and animal breeding. Food production increased *and the population grew.*

With the industrial revolution came mechanical plowing and harvesting and the beginning of a fossil fuel "subsidy" for modern agriculture. With the use of gasoline and diesel-powered farm equipment, the energy stored by photosynthetic processes in the distant past is used to enhance photosynthesis for our benefit. In the twentieth century, great advances in yields have been made in agriculture in Western nations, largely through the process of artificial selection of crop plants. Strains have been developed which produce tremendous yields in the water-rich, heavily fertilized environments that people can provide. Techniques of fossil fuel–subsidized, high-yield agriculture are now being exported from Western nations to the poor countries of Asia, Africa, and Latin America in an attempt to provide food for rapidly growing populations there. Figure 2-8 shows the inputs required by a modern industrialized farm. Poor countries cannot afford these inputs. As fossil fuels increase in price as they become scarcer, the poorer countries will have to go back to traditional farming since they cannot afford fertilizer and fuel.

16-6 The agricultural revolution was a turning point in the human species' war against the environment

The agricultural revolution did more than just start the population explosion. It also marked the acceleration of large-scale destabilization and destruction of natural ecosystems by man. As people began to replace complex natural systems with simple agricultural systems, they began also to reap the consequences.

The world is covered with the evidence of the destructive impact of our agricultural activities. The once fertile region through which the Tigris and Euphrates Rivers flowed was a major agricultural area between 8000 B.C. and 1500 B.C. (Fig. 16-2). But the people who lived

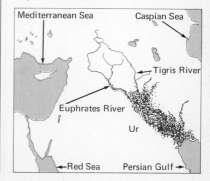

FIG. 16-2 A map of the Tigris and Euphrates Valleys. This area (color), now a desert, was once a fertile agricultural region with two annual wheat crops. The city of Ur, indicated on the map, flourished about 2100 B.C.

there did not reckon with the consequences of the extensive irrigation systems they built. Salts built up in the soil as water evaporated. Most crops could not be grown; a great civilization was wiped out, and the area was converted into a desert. Similarly, almost the entire Mediterranean basin, once wooded and rich, has been turned into a relative wasteland by overgrazing, especially by goats, incompetent agriculture, and clear cutting of forests. The great Sahara Desert was, at least partly, and perhaps entirely, created by the human species. Its southern border still advances southward up to 50 kilometers (30 miles) a year because of overgrazing combined with changing weather patterns.

Agriculture started us on our way both to planetary dominance and to the simplification and destruction of the ecosystems upon which we depend for much of our food, for the control of many pests, for the maintenance of the quality of the atmosphere, and for the disposal of our wastes. Remember that a critical portion of humanity's high-quality protein comes directly from the natural ecosystem of the sea. Furthermore, more than 95 percent of the potential pests of crops are controlled not by people but by predators that are part of natural ecosystems. And, of course, microorganisms in those systems play roles critical to mankind as decomposers and as agents in the cycling of elements. Modern agriculture produces more food than primitive agriculture, but it also produces more disruption of natural cycles. Because of this and because of its dependence on powerful biocides, it threatens to destroy the very life fabric of the earth.

THE CITY AND THE ORIGINS OF TECHNOLOGICAL SOCIETY

The evolution of cities was itself dependent on the development of agriculture. The very establishment of settlements in itself undoubtedly helped increase the efficiency of agriculture.

16-7 The first cities appeared several thousand years before the birth of Christ

The first cities appeared in broad river valleys in relatively temperate areas, such as the Fertile Crescent, which includes the Tigris and Euphrates Rivers. These areas were suitable for the cultivation of grains, relatively high-yield crops which permitted the accumulation of the required surpluses.

More than food surpluses are required for the growth of cities, however; one must also have a social organization that makes possible the gathering, storage, and distribution of those surpluses. Most early cities seem to have developed the same general form of government: a priest-king assisted by various administrative specialists. The appearance of written language at this time helped in the organization of cities. Records of production of grain and other financial accounts, taxes, laws, historical documents, and so on made the flow of commerce easy. Thus organization evolved that made possible the mobilizing of labor for the construction of irrigation canals, public buildings, and fortifications.

The earliest cities, such as Ur (Fig. 16-3), probably did not have more than 10,000 people in them; but, somewhat later, cities like Teotihuacán in Mexico may have had as many as 100,000 people. Such preindustrial cities were characterized by a ruling, literate elite and dependence at first on the power of human and animal muscles. In later preindustrial cities, both wind and water power were harnessed to some degree.

FIG. 16-3 A drawing of the famous Ziggurat (from the Assyrian word for pinnacle) at Ur. The people who built the Ziggurat had come from mountainous country and were used to worshiping at shrines on mountain tops. The Ziggurat provided them with an *artificial mountain* to continue this custom. It was built of brick to the glory of the moon god Nanna, and was composed of three tiers which rose 21 meters (70 feet) above shrines, storehouses, and homes of temple workers.

16-8 Industrial cities represent an entirely new stage in urban evolution

Industrialized cities are characterized today by mass literacy, a relatively fluid class structure, and a technology which permits the exploitation of sources of energy such as coal and oil. Industrial cities have elaborate transportation networks based on the use of these fossil fuels or on electricity. These transportation networks permit industrial cities to be larger than any preindustrial cities, often exceeding 1 million inhabitants. In 1960 there were 24 metropolitan areas on the earth with populations in excess of 3 million each. It is characteristic of these industrial cities that they tend to sprawl and blend together until it is almost impossible to determine where one leaves off and the next one begins.

Industrial cities evolved from preindustrial cities in many cases. The first true industrial cities arose in England, where the social system was considerably less rigid than that on the European continent. In England the Puritan work ethic was also strong, which further aided industrialization. Another important step toward industrialization was the inventiveness and ingenuity of English engineers in building engines that could pump deep mines dry (Fig. 16-4), carry goods and passengers on rails, and weave and spin fabrics. However, the early states of industrialization were still labor intensive; that is, a great number of workers were needed to load the trains, mine the coal and ore, and tend the machines. The feudal period had been ending in England over a number of centuries. Landlords were now viewing their estates as sources of cash revenues. Excluding the peasants and their animals, they enclosed the common grazing fields and farms in order to raise sheep for wool. Thus they created a landless peasantry, which became a labor force to serve the mines and the mills. Unfortunately, the transition was not a smooth one, and eighteenth-century England was plagued with the problem of the wandering poor, people dispossessed from their ties to land and lord, yet not part of the industrial work force. Poorhouses, often aptly called houses of terror, were established in some areas to care for them. But many chose to wander rather than subject themselves to the terrors of the poorhouse. Ultimately, however, these serious social dislocations led to the distinctions common today of *land, labor,* and *capital* and to the great industrial cities of England.

In England there followed a long period during which the workers were exploited; child labor was common, and living conditions for

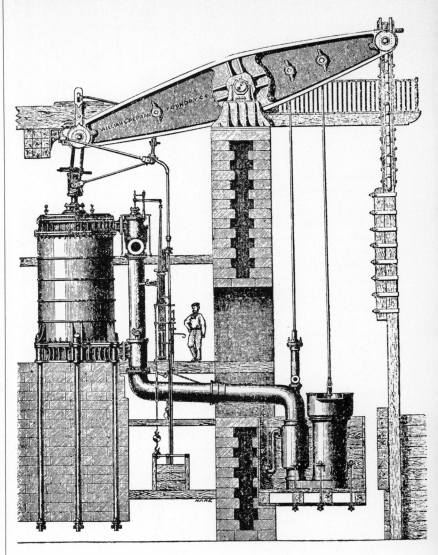

FIG. 16-4 This illustration shows a Cornish beam engine that was built about 1862. These massive steam engines were able to pump water from mines deeper than 610 meters (2,000 feet), making possible the mining of copper, tin, and other minerals from deposits which were previously inaccessible. The beam was attached at one end to the engine cylinder which sometimes had a diameter of 3 meters (10 feet) and at the other end to the pumping rod. The beams were often over 9 meters (30 feet) long, and many weighed more than 50 metric tons. Such engines were an important stage in the evolution of the machines of the industrial revolution. They provided engineers an opportunity to experiment with ways of increasing the efficiency of steam technology, and they provided needed materials for the developing factories. (*Courtesy of D. B. Barton,* The Cornish Beam Engine, *D. Bradford Barton Ltd., Truro, Cornwall, England, 1969.*)

most of the people were probably worse in the industrial city than in the preindustrial city. Eventually, trade unions and other groups corrected some of the abuses of the exploitation of the laboring class. In other countries, industrial cities soon emerged and the people from the countryside continued to move into them. Today, strikes and other worker discontents are common both in industrial cities and in supporting industries, such as mines. Nevertheless, cities continue to grow until the present-day megalopolis threatens to engulf all it encounters.

As cities in the United States grow larger, there is a general trend for formerly fashionable neighborhoods to be abandoned by the well-to-do as they move to the country. These neighborhoods then often go through an evolution of decay, as we have seen (Chap. 1).

The industrial city has spawned a new kind of society, the *consumer society*. In the United States, jobs are thought of as both a right and an obligation. However, to keep full employment, consumer demand for the products of workers must be great. Advertising and a continually growing economy have been the ways of balancing jobs and demand. Now that we are entering a period of real scarcity, however, many politicians and economists feel that other means must be found to create jobs. It might be possible to arrange other ways in which people in cities can meet their needs. We shall look at some possibilities in the next chapter.

16-9 Industrialized cities are vulnerable to disruptions

As we saw in Chap. 1, industrialized cities are dependent on inputs and outputs; without energy inputs, they die. In the consumer society of an industrialized city, each person is dependent on many others for the basic necessities of life. A strike, an epidemic, an energy failure can totally destabilize an industrial city. Industrial cities are dependent on fossil fuels for their very life. Without transportation of food and other products into a city, rioting and looting soon break out. In 1974, workers on the railroads entering Bombay in India went on strike. There were riots, and people searched the markets for whatever they could buy to feed their families. In fact, the great industrial cities are dependent on the entire planet for the energy and materials to feed their machines. They are products of what has been called the *cowboy economy,* the idea that "there is always plenty more of whatever is needed, so let's use it up." The industrial city is perhaps the most wasteful invention of the human species. Foods are brought from all over the world into industrial cities and end as sewage polluting the seas or rivers near by. Materials brought to the cities are made into products that are dispersed about the countryside, eventually ending in garbage, lost for further use.

The vulnerability of industrial cities is disturbing because they are now seats of most governments, centers of economic activities, publishing houses, book stores, libraries, universities, and TV and radio networks. Much of the important cultural heritage of humanity is stored in cities or concentrated there. Today, even cities such as Washington, D.C., and Brazilia, which function mainly as centers of government rather than industrial centers, are dependent on fossil fuels for transportation.

Terrorist organizations have taken advantage of the concentration of people and institutions of importance in cities in various bombing and kidnapping attempts. The nuclear "balance of terror" that hangs over the world today would be of much less consequence if people and the cultural institutions necessary for their survival were scattered around the countryside rather than being concentrated in industrial cities.

Sociologists equate civilization with the development of cities. The industrialized city has the cultural and other advantages of a pre-industrial center of civilization; however, the industrialized city has problems and a vulnerability that makes it a fragile vessel of civilization indeed. Nevertheless, cities continue to be major sources of new ideas, places where people of diverse cultures meet and create civilizations. Urbanization, the increase in size and number of cities, tends to be one of the most important social trends of this century.

16-10 Human beings have evolved a great diversity of cultures

Homo sapiens is, in comparison with most other animals, a physically variable species. But variability in such things as skin color, height, and blood type fades into insignificance in comparison with the cultural variability of humanity. Human beings speak some 2,800 different languages, describe their genetic relationships with each other with various complex kinship systems, worship a great diversity of gods and spirits, and are organized into groups that practice every degree and kind of government. People in all cultures fill their everyday lives with myriads of taboos concerning everything from forms of greeting, proper language for "ladies and gentlemen," and proper modes of dress, to which topics are appropriate for discussion or scientific investigation.

As noted in Chap. 15, cultural differences are far from superficial, and they may account for serious problems, especially within the "global village," which the world may be on the way to becoming (Sec. 16-11).

The world today is a world of contradictions. In some ways it is much like the world of 100 years ago, a world of separate countries each wishing to control its own destiny and mostly in competition with other countries for resources, territory, and political power. But in other ways today's world is utterly unlike even the world of 1945, let alone that of 100 years ago. One hundred years ago, communications among distant nations were slow, limited in most cases to telegraph contacts among small numbers of people and the circulation of written materials.

16-11 The world could be a global village

Just as almost everyone in a village tends to be pretty much aware of all the major activities of the village, so today almost all human beings could be aware of major global events. There are inexpensive transistor radios powered by batteries, the sun, or even heat from stoves. Satellites circle the globe, making possible instant television transmission around the earth. International TV, radio, telephone, and telegraph systems could bind all people in the world into a planetary communications network. However, how many Americans in 1973 knew about the famine in Africa? How many Africans know that the energy crisis in America may make it impossible for us to send surplus food to save them from starvation? How many Americans know that most of the people in the world do not sleep in beds, not because they are too poor but because they have always used mats on the floor or other arrangements? It is clear that there are many factors that keep the world from truly being a global village. One, of course, is human nature; are you *really* interested in what is happening in Sri Lanka (Ceylon)? Probably you are not, unless you have relatives there or come from there. Another factor is the restricted availability of radio and TV receivers because of poverty and government controls. There is also governmental control of the content of native-language programs in most countries.

In spite of all these inhibitions to communication, the communications revolution is still having a profound impact upon the human species. One of the most important effects is its encouragement of the so-called revolution of rising expectations. More and more people in the "poor," or *underdeveloped,* countries are being exposed to a distorted view of the affluent life of *some* of the people who live in the "rich" or *overdeveloped* countries.

The underdeveloped and overdeveloped countries Underdeveloped countries (UDCs) differ from overdeveloped countries (ODCs) in a number of ways. Underdeveloped countries tend to have subsistence farming or relatively inefficient food production and distribution systems. They have extremely low gross national products. Their rates of population growth and illiteracy are very high. The majority of people of UDCs live in poverty and misery which can only be made worse by their high rates of population growth. The overdeveloped countries, on the other hand, have very efficient food production and distribution systems, highly dependent on fossil fuel to run machines and fertilizers to produce high yields per hectare. They have low rates of population growth, and most of their people are literate. They have high gross national products and most of their people live relatively comfortable and well-fed lives. The United States is an example of an overdeveloped country. People in the United States use more of everything, especially energy, than they really need; this is characteristic of an ODC—wasteful consumption at the expense of the world's resources. The UDCs will never be developed to the level that the United States is developed simply because, given the rapid growth of their populations, there is not enough energy or material resources to make such development possible.

Economists, technicians, engineers, and government officials, in particular, in underdeveloped countries have been given the impression that the kind of development that occurred in the overdeveloped countries is both possible and desirable for them. Unfortunately, one cannot have an affluent society like ours without a great many unwanted side effects in the form of pollution, resource depletion, excessive urbanization, unemployment, and instability.

In the overdeveloped countries in the last hundred years, people's views of what constitutes poverty have completely changed. By today's standards the United States in 1870 was a poor country. The working and living conditions of most would seem intolerable to us today. No hot running water, electric lights, refrigerators, automobiles, washing machines, dryers, stereos, etc. How did people survive? They did not live as long on the average, but who is to say that a long life is necessarily desirable or that they did not enjoy what they did have just as much as we enjoy our lives today. It is unfortunate that just as we are entering an age of scarcity, when even the overdeveloped countries will have to cut back on their consumption, people in the underdeveloped countries feel that there is no reason they too should not have all the frills produced by the industrial revolution.

As we begin to feel the effects of shortages of energy, of food, of goods, of jobs in our own country, we may take another look at whether we want to label subsistence farmers as poor or whether we want to consider those few hunting and gathering cultures still left as underdeveloped. In the years that come, as the industrial revolution winds to its end with the depletion of the fossil fuels, we may find that we consider subsistence farmers very rich indeed. They can grow their own food, and they know how to cope with weather and insect problems without expensive pumps, fertilizers, or insecticides. They live where their food is produced, and so a transportation strike will not affect them. Poverty is a relative concept; the poor in America today, even if they are on welfare, often have appliances which the wealthy in some countries do not have and of which kings of the past never dreamed.

The real grinding, bitter poverty that people have experienced throughout their evolution has come from weather-caused famines, plagues, oppressive social systems, and overpopulation. Once, only isolated areas were overpopulated and people could migrate to new lands. Now it is the whole world that is overpopulated and there is no place for people to move on to.

It is not desirable or even possible for the underdeveloped countries to become overdeveloped countries. Although nations often like to consider themselves as separate units quite unaffected by the internal affairs of other nations, such independence is in fact nonexistent. The Arab oil embargo of 1974 made Americans very much aware of their dependence on other nations.

When a gardener in New York sprayed DDT on the rose bushes, part of that DDT entered the world ecosystem and some of it may now be part of the fat of an Eskimo, a New Zealander, a German, or a Masai warrior in Africa. The spray apparatus that the gardener used may contain steel made from iron ore mined in Canada, tin from Malaya, and plastics made from oil pumped from the ground in Iran. The process of converting the iron ore into steel may have produced the smog that helped to give a citizen of Gary, Indiana, lung cancer.

The resources of our planet, although they are very large, are also finite—that is, they are limited. Furthermore, they are unevenly distributed; some areas and nations control more than others. The capacity of the earth's ecosystem to absorb pollutants and to tolerate other abuse is also limited, and all parts of that ecosystem are interconnected so that disturbances are not just local affairs.

It is most important that the overdeveloped countries begin to stop using far more than their share of the world's resources and to greatly reduce the stress they place on the world's ecosystems. It is also important that ways be devised so that the underdeveloped countries can improve the lot of their people without creating the same disruption of social and ecological systems found in the overdeveloped countries. As we shall see in Chap. 18, many creative people are applying their talents to these problems. As long as present governments and ruling classes are in charge, however, there seems to be little chance that present trends can be reversed. Neither is it possible to make the people of the world aware of the global problems as long as communications are in the hands of government officials and businesses that stand to profit from increasing industrialization. Further, in the consumer societies, each person fears loss of his or her job if the economic system were to change drastically. Since consumers must buy everything, including food, it is unlikely that even though they get the message it will be heeded.

16-12 Our planet is like a spaceship

The only sensible way, then, to look at our global village is to consider it as a whole—as a unitary Spaceship Earth. That spaceship has life-support systems (ecosystems), and it also has concentrations of resources which, once they are destroyed or dispersed, will for practical purposes be gone forever. It is absolutely necessary that all people everywhere understand these basic facts: otherwise humanity may be destroyed in what biologist Garrett Hardin has called a "tragedy of the commons."

A **commons** is a resource which is not considered to be owned by anyone, or is owned jointly by a group, and which is used by a group. An example is a community pasture on which cows owned by a number of people are grazed. Suppose the pasture can support 100 cows without overgrazing, and suppose that 10 people graze cows on it. Since it does not cost a person any more to have 15 cows on the pasture than to have 1 cow, each individual says to himself or herself, "My best strategy is to have as many cows as I can buy. Then I'll get the biggest share of the grass." Each person thus buys all the cows he or she can. The individual strategy works just fine until the total number of cows reaches 101. Then the pasture is destroyed and everyone's cows die: a tragedy of the commons.

Destruction of the earth's commons Today each person, company, or nation tries to maximize his, her, or its share, rarely, if ever, considering the joint impact of the behavior of all those utilizing the commons. The overfishing of commercially important kinds of fishes is frequent because each fishing company or nation tries to get as many fishes as possible (see Table 3-1). The result is that not enough breeding fishes are left to replenish the stock.

Similar things happen with pollution, for the air we breathe is a commons. The capacity of the atmosphere to carry away poisons is limited, but it seems to be to the advantage of each polluter to continue using the atmosphere as a dump. (After all, who wants to walk when he or she can drive, and what corporation wants to spend part of its profits for pollution-control devices?)

Similarly, many people feel that the number of children they have is no concern of society's. Yet, if every family had four children, the population would nearly double every generation! Suppose the number of people who want to put cows on the commons doubles every generation. The commons would soon pass the 100-cow limit, and no one would have any place for his or her cows. What about increasing the size of the pasture, which is the commons? What about doubling it every generation? It would only take 26 generations (780 years) before a 200-hectare (494-acre) pasture increased to fill all the potentially arable land in the world, which is 3.18 billion hectares (7.86 billion acres). This is approximately the span of time since the signing of the Magna Charta in 1215! Problems of all kinds are created by this "solution." A critical point to remember is that problems in the functioning of a system rarely can be solved by increasing the size of the system; they usually can only be postponed—most likely until they are so huge as to be really unmanageable.

The difficulty in getting agreement on control of the commons Why is it so difficult to get people to agree on control of the commons? Control of the commons is a global issue, but people, especially in our culture, still have frontier traditions which make it difficult for them to face the limitations on using resources. Just as they once thought that there were inexhaustible numbers of buffalo, they now do not want to believe that the supply of oil or the pollution-carrying capacity of the air and water can be exhausted. The 1974 shortage in gasoline supplies in some states left most people convinced only that there was some conspiracy going on to raise prices. They resumed their old wasteful driving habits and many bought big new gas-guzzling cars as soon as the opportunity presented itself. People do not want to face the fact that fossil fuels are getting scarcer and that sooner or later supplies will run out. People also have the attitude that one person's (company's, nation's) activities cannot make all that much difference. One more candy wrapper in the street—"so what?"

16-13 Our future as a species depends on changing our attitudes toward the commons

Much of our destructive behavior springs from our attitudes toward commons. All people must come to regard the atmosphere, the oceans, and much of our mineral resources as commons. The exploitation of these commons cannot continue to be an every-man-for-himself scramble or we will have a tragedy which will dwarf all past human disasters. Any chance that we have of saving our own society from disruption and helping the really hungry and desperate people of the world depends upon finding ways of changing people's attitudes and consequently their behavior. Long ago people discovered how to graze a commons fairly; each family was allowed to put a certain number of cows on it, and that was that. If the commons began to be overgrazed, the number was decreased. Naturally, cheaters were punished. How can we do, at the global level with many resources, what was fairly easily accomplished at the village level with one resource?

QUESTIONS FOR REVIEW

1 The text makes some educated guesses about how agriculture got started. Can you summarize them?

2 Agriculture, unlike hunting or herding, requires that its practitioners stay in one place for awhile. What were four "side benefits" of staying put to those who, as farmers, stayed put?

3 In what geographic region were each of these organisms first domesticated: tobacco, horse, corn, potato, donkey? And about when were *all* of today's important crops and farm animals domesticated?

4 The text describes the evolution of cities and notes that the emergence as well as the subsequent development were dependent on agriculture. But what about the other side of this coin? How did the evolution of cities shape the evolution of agriculture?

5 What attitudes and practices have been jointly called "the cowboy economy"? Have you, in the past week, seen, smelled, heard, heard about, or read about any striking examples of the cowboy economy in operation? Who exactly are the "cowboys" in this meaning of the term?

6 Not everyone is blind to the grave dangers of potential "tragedies of the commons." Some governments, private groups (including at least a few corporations and other businesses), and individuals are making active efforts to prevent them. Make a list of at least six potential "tragedies of the commons," and turn it into a scoresheet by writing after each entry "substantial progress" or "little or no progress" to date in controlling or preventing it. Base each evaluation on what you have learned from this book and from other information sources such as newspapers. If you are not sure how to score an entry, do a little research on it. Does your completed score sheet appear to give you cause for optimism or pessimism? Did you think that your various information sources gave balanced pictures—neither too optimistic nor too pessimistic—of the ecological situations they were reporting on? Does your scorecard give you any ideas about how you might contribute to the preservation of our planetary commons?

7 The concept of *critical mass* is central to nuclear technology. Its essence is that the energy source must be packed at the proper density for the desired reaction to take place—a devastating explosion in the case of a nuclear bomb, the controlled generation of heat in the case of a nuclear-powered electricity generating station. The concept of critical mass can, by analogy, help one understand the development, continuity, and value of cities as well as their problem-causing aspect if one thinks of the "energy source" as human talent, creativity, perserverance, and so on. What are some kinds of desirable, beneficial human critical masses possible in cities that would be impossible or very difficult to achieve or maintain in more sparsely populated areas? Today it is possible, and indeed common, to find certain types of human critical masses far from cities—although in most cases their ability to reach and maintain the critical point depends on the goods and services generated by the industrial city. You may, for instance, now be one of a community of scholars who live in or near a small town from which the nearest large city is 50 or more miles away. Before the industrial revolution, this type of critical mass rarely coalesced very far from a city. Do you think that the rapid transportation and communication made possible by industrialization will eventually make the large city obsolete for the many kinds of human critical mass that it has traditionally supported? Why?

READINGS

Brown, L. R.: *In the Human Interest—A Strategy to Stabilize World Population,* Norton, New York, 1974.
———, and E. P. Eckholm: *By Bread Alone,* Praeger, New York, 1974.
Claiborne, R., et al.: *The First Americans,* Time-Life Books, Time Inc., New York, 1973.
Cloud, P.: *Resources and Man,* Freeman, San Francisco, 1969.
Editors of Time-Life Books, with text by B. Capps: *The Indians,* Time-Life Books, Time Inc., New York, 1973.
Ehrlich, P. R., and A. H. Ehrlich: *Population Resources Environment,* 2d ed., Freeman, San Francisco, 1972.
———, ———, and J. P. Holdren: *Human Ecology—Problems and Solutions,* Freeman, San Francisco, 1973.
———, and J. P. Holdren: "Impact of Population Growth," *Science,* **171**(3977):1212–1217, 1971.
———, ———, and R. W. Holm (eds.): *Man and the Ecosphere: Readings from Scientific American,* Freeman, San Francisco, 1971.
Hamblin, D. J., et al.: *The First Cities,* Time-Life Books, Time Inc., New York, 1973.
Hardin, G. (ed.): *Population, Evolution, and Birth Control,* Freeman, San Francisco, 1969.
———: "The Tragedy of the Commons," *Science,* **162**(3859):1244–1248, 1968.
———: *Exploring New Ethics for Survival; Voyage of the Spaceship Beagle,* Viking Press, New York, 1972.
Holdren, J. P., and P. R. Ehrlich (eds.): *Global Ecology,* Harcourt, Brace, Jovanovich, New York, 1971.
Howe, G. M.: *Man, Environment and Disease in Britain,* Barnes & Noble, New York, 1972.
Lamberg-Karlovsky, C. C., and M. Lamberg-Karlovsky: "An Early City in Iran," *Scientific American,* **224**(6):102–111, offprint no. 660, 1971.
Malthus, T. R.: *An Essay on the Principle of Population,* edited by A. Flew, Penguin, Baltimore, 1970.
Scott, R.: *Muscle and Blood,* Dutton, New York, 1974.
Study of Critical Environmental Problems (SCEP): *Man's Impact on the Global Environment,* M.I.T., Cambridge, Mass., 1970.
Swanson, C. P.: *The Natural History of Man,* Prentice-Hall, Englewood Cliffs, N.J., 1973.
Watt, K. E. F.: "Ecology," *Biocore,* unit XXI, McGraw-Hill, New York, 1974.

17

SOCIETY IN THE FUTURE

SOME LEARNING OBJECTIVES

After you have studied this chapter, you should be able to

1. Demonstrate your knowledge of some aspects of forecasting by
   a. Differentiating between the text's definition of a forecast and a speculation.
   b. Stating how the accuracy of a forecast is related to the amount of data on which it is based.
   c. Giving two reasons why any forecast of our species' future prospects cannot be considered a certainty.
2. List at least four forecasts about the future of human society, and demonstrate your grasp of the reasoning and/or the data or kinds of data on which *two* of them are based by putting each of the two in the "If _____, then _____" form described at the beginning of the chapter. (Example: "If the populations of crowded, highly polluted areas such as Los Angeles continue to grow, and if the pollution levels of these areas do not drop, then the populations will become less healthy than they are now.")
3. Explain, with examples, (1) why steps taken to prevent disease may lead to more disease; (2) why certain steps being taken to increase food production will probably create ever-greater health hazards; and (3) why in-

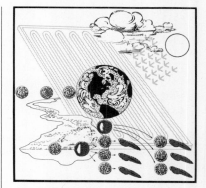

creased industrial activity may lead to increased health hazards.

4. Demonstrate your knowledge of the nature of biological engineering, and some of its problems and prospects, by doing the following:
   a. Define the term biological engineering.
   b. Give two examples of organ transplants, and name one serious medical problem that faces the transplant recipient and one serious medical-legal-ethical problem that can affect the donor.
   c. Name two general types of artificial organs, state which of them is a practical and ever-improving reality, and name at least five prosthetic devices that are in use today.
   d. Identify a social and legal problem that is likely to grow to major proportions as more and better prosthetic devices are de-

veloped for use by more and more people.

5. Explain what genetic screening is, give two examples of it, and state some of the difficult choices and ethical problems it poses.
6. Explain what genetic engineering is and describe one particularly hazardous type of genetic engineering experiment which many scientists hope will no longer be done until ways to control its dangers have been found.
7. State what the *Green Revolution* is and list at least three reasons why this program, which once seemed so promising, is now considered by many to be a failure.
8. Explain how the law of diminishing returns affects fertilizer use in agriculture and energy use in mining, and what the law's long-range implications are for both these activities if the world's population continues to grow and if countries continue to pursue *growth economies*.
9. State what is meant by ecological engineering, and list at least five general steps that an ecologically engineered system might take in pursuit of its goals.
10. List some physiological and psychological problems which some people think biofeedback techniques give promise of easing, curing, or preventing.
11. Demonstrate your knowledge of

the relationship of one's perceptions of reality to one's culture by
a Explaining what is meant by *the social fabric of reality,* and what it really means to say that someone has become *reality oriented.*
b Explaining why science cannot be completely objective, and why science itself can be viewed as an ongoing series of *reality adjustments.*
c Explaining how the declining use of the words *man, mankind,* and *he* when referring to women as well as men is both a cause and an effect of an important change that is taking place in our culture's *social fabric of reality.*

12 Give one largely negative historical example and one considerably more positive historical example of large-scale social engineering, and state one disadvantage of any possible worldwide scheme of social engineering which is so serious a drawback that any such scheme should probably never be implemented.

What sorts of societies will people live in in the future? It would be nice to have a crystal ball so that we could answer such questions with assurance, but lacking such a device, we are stuck with **forecasts** and **speculations.** By **forecasting,** we mean the formulation of "if-then" statements such as weather forecasters provide—for example, "If the hurricane continues to follow the same path, it will strike the coast of Florida near Miami at 6:00 P.M. Sunday." By **speculations,** we mean predictions or prophecies which do not consider the possibility that unforeseen change may alter the course of events—for example, "The human population will number more than 6 billion people in the year 2000." However, if we were to say, "If current growth rates continue, the human population will number more than 6 billion people in the year 2000," we have made a forecast. The speculation can be checked only by waiting until the year 2000 to see how many people there are. However, the forecast can be checked by looking up the current population size and growth rate and doing some arithmetic. Although forecasts can be checked immediately and speculations cannot, forecasts often do not come true. The "if" part of the statement may not come true and then the forecast does not come true either. In short, no one can know for certain what will happen in the future; however, one can make forecasts on the basis of present trends and then study the probability that those trends will continue. How likely such forecasts are to come true depends on how many data were considered. Unfortunately, in studying forecasts of the future of the human species, there are so many changing conditions that all can never be considered. Furthermore, conditions in the future may change in ways quite unexpected on the basis of our present knowledge. Nevertheless, it is important to consider some of the forecasts people have made about the future of society.

HUMAN PROSPECTS

It seems likely that the size of the human population will continue to grow. Such a larger population will probably be more crowded and less healthy. Steps taken to provide for such a large population will present significant hazards to human health. Nevertheless, advances in medical science may make it possible for people in some countries to live longer, healthier lives. Exciting experiments in organ transplantation, artificial organs, and genetic engineering may offer humankind possibilities for a healthier future. However, these also pose serious ethical and legal problems.

17-1 The size of the future human population will be an important determinant of the form of future societies

As we saw in the last chapter, the human species engaged in hunting and gathering to make a living for most of its past. There are only a few hunting and gathering societies left, however, and these are rapidly being eliminated. Thus it is easy to forecast a further dwindling of the remaining nonagricultural societies and a restriction on the possibility of returning to such an existence. This option is already closed to our species as a whole, since the planet will not support 4 billion hunters and food gatherers.

On the other hand, if the human population is dramatically reduced in size, as it could be by famine, plague, thermonuclear war, or widespread ecocatastrophe, then hunting and food gathering might well become the dominant social form. Indeed, if population size were reduced sufficiently, the maintenance of an industrial society might prove impossible. Relatively large numbers of people are required to permit the division of labor and far-flung supply networks of modern industrial society. In addition, if industrial society ceased to function, that mode of life might *never be possible again*. Very advanced technology is now required to obtain the raw materials necessary to run an industrial society. The high-grade ores and easily accessible fossil fuels on which industrialization was based are now exhausted. One must now drill deep to get oil and use complex techniques which consume large amounts of energy to separate many needed metals from their ores. It is highly unlikely that a post-industrial society could, in the future, start to industrialize again "from scratch."

Population forecasts With the size of the human population now doubling about every 35 years, it is easy to forecast a preposterously large future human population size. For instance, if growth continued at that rate, then the world population would exceed one billion billion people in only 1,000 years—1,700 persons for every square meter of the earth's surface, land and sea! No one would predict that such a forecast would be accurate; clearly, the rate of growth will change.

The United Nations in 1963 made a series of forecasts of the size of the human population in the year 2000 which include, on the "if" side, possible changes in growth rate due to changes in birth and death rates. Based on assumptions of three different rates of decline in birthrates, they forecast a low of 5.4 billion, a medium of 6.1 billion, and a high of 7.0 billion. Furthermore, they forecast a population of 7.5 billion if there were no decline in birthrates and death rates continued to decline. In 1968, the UN revised upward their forecasts for the underdeveloped regions of the world in a manner which would add 0.3 to 0.4 billion to those projections.

If these forecasts are accurate, barring disaster, human society around the turn of the century will contain roughly twice the number of people that it does now. The qualifying "barring disaster" is necessary since the UN forecasts are "surprise-free." They do not consider the possibility of rises in the death rate which could produce a population in the year 2,000 much smaller than the present one. A nuclear war could produce such rises in death rates; so could a worldwide epidemic disease. If droughts such as the recent one in Africa, which has led to famine in the region just south of the Sahara, should continue and crops should fail in other regions as well, growth in the human population could be stopped by lack of food. If these hazards are surmounted, it seems unlikely that world population growth can be halted before later than the middle of the next century, at a size of 15 billion or more. There is a momentum of population growth which is inherent in the age composition of the present population—over 37 percent of the world's people are under age fifteen. Based on extensions of present technology, we can forecast that such a large population would have fewer goods per capita, less food per capita, and by present-day standards a lower quality of life.

Effects of a larger population The trend toward urbanization, which we discussed in the last chapter, seems destined to continue. Cities will get larger, both from in-migration and from growth of their own populations. People will be crowded more closely together. Crowding will almost certainly make the population more subject to the spread of epidemic disease. The isolation of pockets of disease will be more difficult, and the possibility of the chance appearance of lethal mutant strains of disease will be increased.

The indirect effects of increased population size on health are already apparent. In crowded areas of high pollution, such as Los Angeles, it is estimated that people are losing about 10 years, on the average, of their lifespans. Diseases related to air pollution, such as emphysema, are more and more common in younger and younger people. Continued rapid growth of the human population will lead to a lower standard of health for everyone. More polluting cars will be needed for transport; more polluting factories will be needed to produce goods. More fertilizer and pesticide pollution will occur as people attempt to grow more and more food. Consider what a doubling of the human population in the next 35 years will mean. In rough terms, every facility for the support of human life that has been developed or constructed in the last 10,000 years will, in one sense or another, have to be duplicated if the per capita standard of living is to remain the same. Twice as many people would require twice as much of everything. Twice the present food supply would have to be made available as well as twice the freshwater supply. Double the number of housing units would be needed. Two doctors would have to be in practice for every one today, and twice the number of hospital beds would have to be available. The capacity of transport systems would have to be doubled, as would our cleverness at controlling pollution.

Will it be possible for humanity to double everything in the next

35 years? Consider that many large public works projects take more than a decade from planning to completion. Consider also that people are already facing an era of scarcity of both energy and materials. No competent scientist thinks that humanity will manage to double everything in the next 35 years; indeed, there is little sign that we will even try. Thus, on the average, a less well-fed, less well-housed, less well-clothed population than today's can be expected 35 years from now. In addition, it seems unlikely that it will have twice the medical facilities available to it as the present population. Decline in health due to crowding and lowered quality of living standards seems likely as the population skyrockets.

17-2 Many steps taken in attempts to care for increased numbers of people may lead to health hazards

Floods Increases in the population size will lead to more people living on flood plains. One forecast that can be made about low-lying areas near rivers, bays, or oceans is that these areas will be flooded every two or three years. Geologists can identify the flood plain of a river and forecast with reasonable certainty that once every 50 to 100 years that river will deeply overflow its flood plain and beyond (Fig. 17-1). As populations become larger and larger, more people must live on flood plains. There are now 20 million hectares (about 50 million acres) in the United States subject to flooding, and most of them are densely settled; in fact, 10 million people in the United States were living on defined flood plains in 1975. In 1936 and 1937 when the population was much lower, devastating floods occurred; at that time, instead of restricting building on flood plains, Congress encouraged expensive and subsequently inadequate projects of dikes and canals for flood prevention. Flood plains in many other parts of the world are densely populated. The greatest natural disaster of recorded history occurred in the low-lying delta region of Bangladesh in 1970. A devastating flood killed over 300,000 people when a storm drove a huge wave over the delta.

Diseases Steps taken to help prevent disease may ultimately lead to uncontrollable outbreaks of disease. There is a major worldwide effort to kill insects which are the vectors (carriers) of diseases with insecticides. As noted in Sec. 10-5, insects rapidly become resistant to insecticides. Figure 17-2 shows areas of the world in which the resistance of disease-carrying insects to insecticides is so severe as to make vector control of significant species difficult or impossible. About 77 percent of the rat populations in 40 cities, including Atlanta, Chicago, New Orleans, New York, Pittsburgh, and San Francisco, have been found to be immune to anticoagulant rodenticides, as are some of the mice. Exterminators may have to return to the highly toxic poisons the relatively safe anticoagulants replaced.

Poisons Steps taken to increase food production lead to health hazards. As attempts are made to grow more and more food on each acre of land, reliance on synthetic pesticides and inorganic fertilizers increases. These substances find their way into human beings through

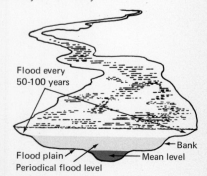

FIG. 17-1 Cross section of a river showing main channel and floodplain. Periodical flooding occurs every two or three years, while deep flooding takes place every 50 to 100 years.

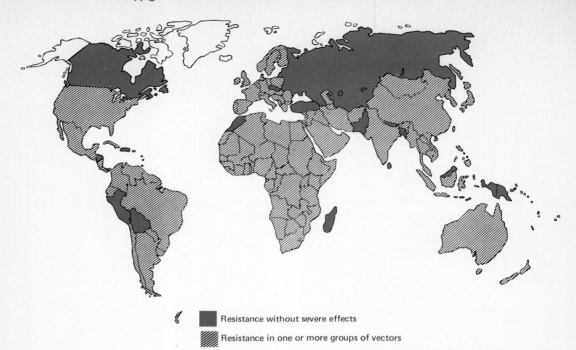

FIG. 17-2 The diagonal shading indicates countries in which vector control has been severely affected by insecticide resistance in disease-bearing insects. The colored areas indicate countries which have reported insecticide resistance without severe effects. Most of the remaining countries had not sent information at the time these data were accumulated. (*Data from WHO, 1971.*)

air, water, and food chains, and by direct contact. As insects have become resistant to the chlorinated hydrocarbons such as DDT, more use has been made of the organophosphate insecticides, many of which are extremely potent quick-acting poisons (Sec. 14-7). In California alone, in the early 1960s, approximately 1,000 farm workers annually reported illness from pesticides. Many more farm workers and their families are exposed to doses too low to cause immediate illness but which may have long-term effects. Most of these substances have not been studied at all for their long-term effects, but some have been banned for certain uses because they cause cancer or birth defects in animals (for example, Amitrole, an herbicide; 2, 4, 5 T, an herbicide; and chlordane and dieldrin, insecticides). Herbicides will become an even greater hazard to health as more desperate attempts are made to grow food for the ever-increasing population. Farmers have found a way to grow crops without plowing the soil by using herbicides, thus reducing labor and energy costs and erosion. Presently, 2.5 million acres of soil in the United States are being so treated. The long-term effects of such treatment on crops or ecosystems are completely unknown; however, the herbicides used are ones which are particularly dangerous to the person applying them. Herbicides are also used widely to clear brush from grazing lands.

Pesticides are put into ecosystems for many purposes other than vector control or increased food production. The cut-flower industry is a heavy user of pesticides. Rugs are often treated with pesticides. Buildings are treated or sprayed to prevent or kill termites or household insects; lawns and flower gardens around houses are often sprayed with pesticides, sometimes for no valid reason at all. Cities use herbicides on vacant lots and street edges; highway departments use them along highway edges (wild berries picked from roadsides may be contaminated). Even nations use them as agents of war to clear vegetation where the enemy may hide. Recently, the United States used herbicides to remove the leaves from trees along the border between the United States and Canada so that people crossing the border illegally could be seen from airplanes.

In March 1974, F. W. Kutz gave the first report of the Environmental Protection Agency's National Human Monitoring Program. On the basis of a sample of 17,000 persons, he stated that *every person* in the United States has pesticides stored in his or her body. This is not too surprising since the fatty tissues of all people tested for DDT (or its breakdown product DDE) show its presence. Nobody knows what effect, if any, these pesticides are having on the people who have them stored in their bodies. If, in fact, everybody in the country has them, how would it be possible to find out if they are having any effect? We have already seen the deleterious effect of chlorinated hydrocarbon pesticides on the eggshells of predatory birds (Chap. 2). Other pesticides are believed to have caused death of estuarine organisms important as food for people. Societies of the future may be able to determine the long-term effects of persistent pesticide use by comparing their impoverished environments with the diversity shown in books about present environments.

Another consequence of intensive agriculture is fertilizer pollution. Fertilizer runoff is implicated in speeding up the eutrophication of lakes, as we saw in Chap. 2. Inorganic fertilizers turn up in drinking water supplies when they are heavily used. *Nitrates* are not poisonous in small amounts to adults. In the digestive tracts of infants, however, they may be converted by bacteria to *nitrites* which reduce the capacity of hemoglobin to carry oxygen. Severe anemia or death may be the result. Several regions of California have sufficiently high nitrates in their water supply that parents of small children are told by physicians to buy bottled water for babies. In California and several other states of the United States, infants have become ill and died from excess nitrate in their drinking water or in foods, such as beets and spinach, which contain much nitrate when they are overfertilized. Many farm animals have become ill and many have died from eating plants fertilized with excess nitrogen. Ultimately, the excessive use of inorganic nitrogen fertilizers will be self-limiting. The increase in yield diminishes as more and more fertilizer is applied. For example, 25 times as much nitrate was applied to soils in Illinois in 1969 as in 1940, but the yield did not increase by anywhere near 25 times. In Holland some soils have been overnitrated to the point where nothing will grow in them. Fortunately, these soils have recovered when organic fertilizers are added.

17-3 Increased industrial and building activity may also increase health hazards

As the population has increased, industrial activity has increased. Since the industrial revolution, most people no longer live on the land or provide their own food and shelter. Therefore it has been necessary to provide housing, transportation, and jobs for people in ever-increasing numbers. Whether these activities are carried out by private enterprise or by governments, they lead to increasing amounts of land covered by structures. Greater flows of waste products from industry enter the environment, and nonrenewable resources, such as fossil fuels, are being used up faster than ever.

Industrial pollution Japan exhibits many of the tragic side effects of rapid industrial growth. We already noted, in Chap. 1, the tragic deaths and illness resulting from mercury pollution in Minamata Bay. In other areas of Japan, cadmium poisoning (Itai-Itai disease) has afflicted many. In fact, in some Japanese streams, photographic film can be developed without the use of further chemicals! Smog is so bad that there is concern over a widespread deterioration of the respiratory health of the urban populations. Oxygen is available from coin-operated machines on the street, and traffic police officers at busy intersections must be relieved frequently. Japan has been called the "miner's canary" of the industrialized nations. Like the canaries miners took into mines to detect poison gases, Japan will show the rest of us just how far we can go in polluting the environment before people start becoming ill and dying in unacceptable numbers. Already we know that the lives of people in crowded, polluted areas of American cities are shorter on the average than those of country dwellers. We do not know how serious such pollution must get before people decide to take action. We do know that death rates from emphysema and respiratory cancer have been rising for about the last 20 years (Fig. 17-3).

Reduced controls As fossil fuels become scarcer and more expensive, many things are happening. Pollution controls are being relaxed in obtaining, transporting, processing, and using fossil fuels. For example, drilling for oil in offshore areas vulnerable to spills is being increased, and the Alaskan oil pipeline is being stretched across fragile tundra. Acid rain (Sec. 1-11) is so severe that in May 1975 the First International Symposium on Acid Precipitation and the Forest Ecosystem met at Ohio State University to consider the problem. In power plants more attention is paid to efficiency than to low emissions of pollutants, and pollution standards for automobile engines have been delayed.

Radioactive pollution There also has been an increasing rush to install nuclear fission reactors as sources of electric power. With present technology, these reactors may carry with them high risks of radioactive pollution of the environment. Low-level escape of radioactivity occurs from all these power plants; many informed nuclear physicists fear a catastrophic accident which could release high levels of radiation. A number of physicists, biologists, and other people fear the possible diversion of reactor fuel or waste by terrorist

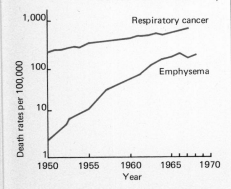

FIG. 17-3 The trends in respiratory cancer and emphysema in California. (*After K. E. F. Watt,* Principles of Environmental Science, *McGraw-Hill, New York, 1973; data from California State Bureau of Vital Statistics.*)

groups. Finally, the waste products of nuclear power plants must be stored away from living things for hundreds of thousands of years if people and other organisms are to be protected from their radiation. Ionizing radiation can lead to alterations of the genetic material and many qualified physicists and biologists believe that there is no safe dose of such radiation. They believe that higher levels of radiation will bring higher incidence of birth defects (from mutations) and cancers. Certain kinds of radioactive elements concentrate in food chains; and, largely because of the testing of nuclear bombs in the 1950s and 1960s, all people have strontium 90 in their bones in much higher concentrations than were found before the tests. Strontium 90 behaves much like calcium in the vertebrate body and is taken in mostly in milk. An interesting study in the Arctic showed the concentration of another radioactive element produced by the tests in a simple food chain (Fig. 17-4). Migratory caribou feed on lichens at certain seasons, and Eskimos feed on the caribou when they are available during their migrations. The lichens accumulate cesium, and when radioactive cesium 137 became available after the tests, they concentrated it—and so it went, up the food chain. No one has

FIG. 17-4 Concentration of cesium 137 in an arctic ecosystem. A computer simulation model (*B. E. Vaughan,* Proc. 6th Berkeley Symp., **6:***506, 1972*) based on extensive observation (*W. C. Hanson,* Health Physics, **13:***383–389, 1967*).

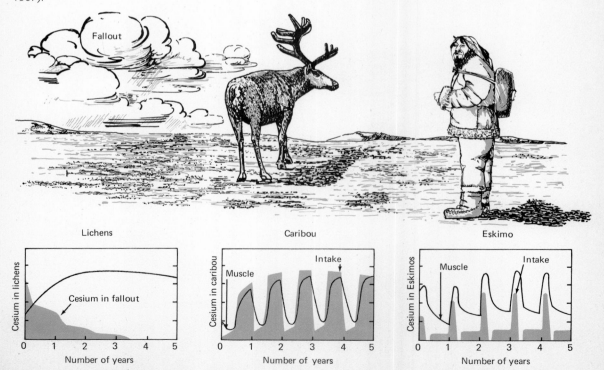

found any harmful effects on the Eskimos—yet—but follow-up studies have not been reported, to our knowledge.

Alternatives The former Atomic Energy Commission long promoted the development of nuclear power. On January 19, 1975, its functions were split and turned over to two new agencies. The Nuclear Regulatory Commission (NRC) will function to regulate the development of nuclear power and its safety problems. The Energy Research and Development Commission (ERDA) will promote development of nuclear power and other power technologies. There are many other potential sources of energy such as solar, geothermal, and wind and tide power. These have been low on research budgets in the past because there was no government agency likely to get prestige from their development and no private industry apt to make huge profits

FIG. 17-5 A house designed to make the maximum possible use of the sun's energy. Water is heated as it passes through coils in a collector unit on the roof. Provision is made for emergency heating and use of a stove, fireplace, or furnace during cloudy periods. Air is simply and inexpensively heated in collectors attached to windows. Cool air from a room flows into the collector, is heated by the sun, and reenters the room as much as 15°C hotter.

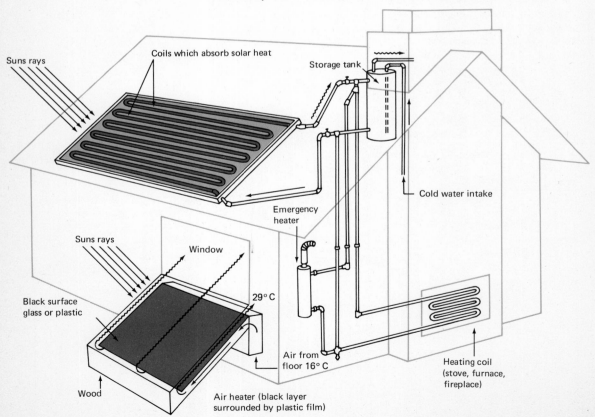

from them. In spite of this, some especially promising attempts are being made to utilize solar energy (Fig. 17-5).

The most logical alternative, reducing energy consumption, has had only minor promotion. The speed limit on all highways in the United States was reduced, by presidential order, to 88 kilometers (55 miles) per hour in the winter of 1973 to 1974. Both the effort to reduce gasoline consumption and the reduced death toll from highway accidents led to Congress enacting a law setting the speed limit at 88 kilometers per hour. Local power and light companies have produced ads and brochures on how to save energy. Often in the same brochure they advertise the *need* for more nuclear power plants so that their proposed energy conservation measures will be only temporary!

The vicious circle Thus, with increased population size, the human species is faced with increased risk of health problems associated with industrial society. Lung cancer, heart disease, emphysema, and heavy-metal (e.g., lead, mercury, cadmium) poisoning appear to be associated with industrial societies and their pollution. At the same time, the sprawl of the structures of human society has covered valuable agricultural land. The drift of smog over farmland has decreased the productivity of plants and even made it impossible to farm in some areas. The flow of industrial pollutants into the sea may well reduce the harvest of marine fishes. For example, recently, fisheries in Japan's inland sea were severely affected by an oil spill. Thus, in a vicious circle, humanity's attempts to care for a growing population not only directly assault human health but add to our vulnerability to disease through malnutrition.

17-4 Health in future societies may be improved by biological engineering

As you know by now, the great increase in the size of the human population is due to a decrease in death rates, not to an increase in birthrates. As we noted in Chap. 14 medical science has made tremendous advances in treating and preventing the diseases caused by bacteria, viruses, and parasites which plagued our ancestors. This success has allowed people to live longer, and thus more people are subject to the degenerative diseases of aging. Comparatively little research has gone into the problem of the health of the aged. After saving many people from the diseases of childhood and middle age, medical science has then abandoned them to whatever degenerative diseases may set in with advanced age. Beyond glasses and hearing aids, often little can be done to help the aged ill. More research on prevention and treatment of stroke, emphysema, arthritis, and the mental problems of the aged is badly needed. Instead, as we noted in Chap. 7, the aged ill are often left to die in understaffed nursing homes. Between 1963 and 1972 the number of nursing homes in the United States increased from 13,000 to 22,000; no doubt this number will continue to increase with the growing population of the aged ill.

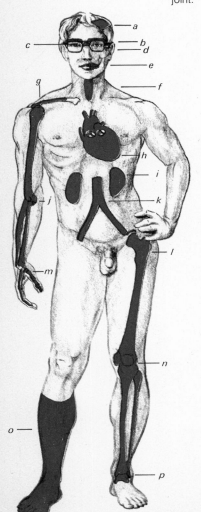

FIG. 17-6 Some of the replacement parts now under development for use in the human body: (*a*) skull plate, (*b*) artificial eye and brain stimulation or sonar sight, (*c*) cornea, (*d*) implanted hearing aid, (*e*) implanted tooth, (*f*) speech device to replace vocal cords, (*g*) shoulder joint, (*h*) artificial heart and valves, (*i*) artificial kidney, (*j*) elbow joint, (*k*) blood vessels, (*l*) hip joint, (*m*) finger joints, (*n*) knee joint, (*o*) leg and foot, (*p*) ankle joint.

Transplanted organs Biological engineering is the ability to modify, repair, and substitute parts of living things. It can range from putting a new gene in a bacterium, to transplanting a human heart, to providing a wholly artificial chemical or mechanical substitute for some part of the body or some necessary function. In the last few decades, enormous progress has been made in the ability to understand, modify, and repair the human body. As noted in Chap. 14, numerous people have had other people's kidneys transplanted to replace their own failing ones, and years have been added to the lives of some people by this technique. The immune system unfortunately attempts to reject such transplants, and powerful drugs must be used to suppress it, often causing death from inability to resist infections. Most physicians seem to feel that the rejection problem will be solved and that donated organs will be widely used. Nevertheless, there will remain the problem of finding donors for transplants. If transplants ever become really common, this problem could be immense. Recently the heart of a murder victim was transplanted to a patient whose own heart was beyond repair. The attorneys defending the alleged murderer used as a defense the argument that the doctors who transplanted the heart, and not their client, were the real murderers. It was shown, however, that the victim's *brain* was dead when the heart was removed and the doctors were vindicated. In 1975 as legal attempts were being made to define death as a cessation of brain wave activity, a heart attack victim revived after 12 hours with no detectable brain activity. His body was being maintained by a machine which took over the function of his heart and lungs so that his organs would be in good condition for transplantation; his wife had just signed the permission slip for his organs to be used as transplants when his eyelids fluttered and he regained consciousness. Defining death is a major problem in organ transplantation.

Artificial organs and parts There could be two types of artificial organs: those made of wires and tubes and mechanical parts and those structured by culturing the individual's own cells to produce an organ in tissue culture which can then be implanted. Little research has been done on the second type since researchers seem to feel that the problems connected with transplantation between individuals will be solved. Great advances have been made with the first type. Some artificial organs are small and can be implanted in the patient; others are large, and the patient must be connected to them while the machine remains in place. Artificial organs or parts of the mechanical-electric type are known as *prostheses*. They include plastic or steel hip joints, plastic heart valves, false teeth, dacron arteries, artificial legs and arms, and heart-lung and kidney machines. Figure 17-6 shows some of the devices now under development which can be implanted or worn by an individual. Artificial legs and arms which are operated by the movement of small muscles in some remaining part of the body are now in use. Some have small electric motors to operate them. Artificial kidneys are becoming smaller and cheaper; a portable one is under trial but is not yet available for purchase. Numerous people have heart pacers implanted in their chests; these regulate irregular heart beats. The heart-lung machine is now used during surgery to take over the functions of both heart and lungs while the surgeon makes repairs. Reports have appeared in the medical journals that a completely artificial heart small enough to be implanted in the chest is under development. An artificial pancreas to aid the diabetic is also under development.

Other aids A great deal of research has gone into devices to help the blind. There is a new device which helps those who have difficulty seeing in dim light. Another device called an Optacon translates printed letters from a page into tactile stimuli which a blind person can "read." Hearing aids are common and are being improved and made smaller all the time. A method has just been devised for implanting new teeth into gum sockets after tooth extraction.

As more and more advances are made in the development of medicine and of artificial organs, more and more people will be living longer, yet little thought seems to have been given to the problems of the aged population which will result. What shall be done for a ninety-year-old person with a healthy, functioning heart (natural, artificial, or transplant) and a fuzzy mind which makes it impossible for him or her to live in society? Will we fill up even more nursing homes with people with artificial organs? Some futurists (people who attempt to predict and study the future), for example, Arthur Clarke, have suggested that human life will become obsolete as machines are able to take over more and more of our functions. Perhaps machines *are* slowly evolving toward dominance and after us, evolution will continue by means of self-reproducing computers.

Ethical problems of cloning A clone is a group of genetically identical individuals or cells (Sec. 9-8). Plants and animals that can reproduce asexually often form clones. The nucleus of an egg of a frog can be replaced with a nucleus taken from a cell of a late blastula of another frog. The egg will subsequently grow to be an adult frog identical genetically to the donor of the nucleus. In this way, a great number of genetically identical frogs can be produced (Fig. 17-7). Cloning has not yet been carried out in mammals. It has been predicted, however, that if current research continues, a clonal human being will appear on earth sometime in the next 20 to 50 years. Such a being could be implanted in a woman's uterus early in development or could even be nourished by a completely artificial womb. He or she would be genetically identical to the donor. Who is to decide who should be the donor? Should we use cloning as livestock breeders hope to do, to preserve and perpetuate the finest genotypes that arise in our species? Since we still know so little about how much of a person's ability and character is formed by environment and how much by heredity, is such an experiment worth the effort? Totalitarian governments might delight in turning out assembly-line people—people like those in power to rule and people all like each other (and docile) to do the work.

Ethical problems in research Even more ethically puzzling are the consequences which could result from current experiments in which cells of species as different as mice and human beings are allowed to fuse in tissue culture and then divide. These mouse-human hybrid cells continue to divide for a number of generations and successively

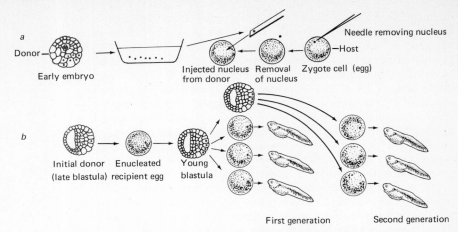

FIG. 17-7 (a) The nucleus of a zygote (enucleated egg stimulated to develop without fertilization) is removed and replaced with one of many nuclei extracted from an early embryo. (b) The same experiment repeated for two generations. Color indicates the initial donor or a nucleus therefrom. (After J. Gurdon.)

eliminate the human chromosomes. For this reason, they have been used to localize certain genes controlling easily observed biochemical reactions on particular human chromosomes. They have also been used to show that viruses do indeed attach themselves to chromosomes. It is almost certain that someone soon will hybridize human cells with chimpanzee or gorilla cells. It is just as certain that, if cloning techniques were available to allow development of an organism from a cell from tissue culture, somebody would attempt to do it. Already, some people are concerned about whether such an organism should be called a "person" and what the legal rights of a creature half human and half something else should be.

Until July 1, 1975, all live virus vaccines were contaminated with bacteriophage (see Sec. 6-15). On that date federal regulations went into effect that should eliminate phage from human vaccines. While there is no conclusive evidence that any of the millions of people who have received live virus vaccines for polio, measles, mumps, or rubella have suffered harm, some National Institutes of Health biologists are deeply worried that such contaminated vaccines were ever used. Some human diseases are caused by bacteria infected with a phage that causes them to secrete a toxin, for example, scarlet fever and diptheria. Presumably, harmless forms of these bacteria could be converted into toxic forms by phage contaminants of vaccines. Phages can also transmit genes into human cells in tissue culture; could this happen in a vaccinated person?

The transmission of genes into genetically defective individuals by use of viruses has seemed a likely way of correcting genetic diseases. But how could a researcher be sure that a virus carrying a gene would not infect persons other than the intended recipient? How could the researcher be sure that such a virus would not have new and unpredictable infectious properties? In July 1974, a letter signed by 11 scientists working in this field was published in two major scientific journals and many newspapers and magazines. This letter asked that two types of experiments with great potential hazards be postponed until methods are devised for doing them safely. First, the scientists proposed that introduction of genetic material into bacteria not be carried out if it would introduce antibiotic resistance or toxic characteristics not found in that group of bacteria in nature. An example would be introduction of penicillin resistance into a strain of streptococcal bacteria which are not naturally resistant to penicillin. Second, they proposed that construction of new DNA molecules which contain DNA from animal viruses be stopped. In particular they wanted to avoid constructing new DNA molecules which contained animal tumor viruses. *Escherichia coli,* intestinal bacteria commonly used in research, could be the host for these recombinant DNA molecules. These bacteria could then infect research workers. From the research workers potentially hazardous DNA could spread throughout the population by the usually harmless but infectious *E. coli,* which would exchange genetic material with other *E. coli.*

These 11 scientists presented their proposals in the form of a letter so that the seriousness of the situation could be known quickly to other scientists. A breakthrough in genetic engineering techniques made such experiments so easy that a high school student could set up a lab and carry out such experiments. Not all scientists have abided by the moratorium on the two kinds of hazardous research until guidelines for safety could be agreed upon. Such research has such a great potential that it could not be totally abandoned. It could lead to greater understanding of the genetic material and to great practical results, for example, cure of genetic diseases or production of new and useful bacteria which could synthesize insulin or chlorophyll or serve as living fertilizer in agriculture. Therefore, 150 scientists met in February 1975 and worked out guidelines for deciding how such experiments can be done safely.

Ethical problems of genetic screening **Genetic screening** is another technique that poses serious ethical problems. Already 36 states of the United States have some type of genetic-screening program. It is now possible to do prenatal screening for an ever-increasing number of genetic defects by **amniocentesis** (Fig. 17-8). A needle is inserted into the amniotic fluid surrounding the fetus and a sample of the fluid is withdrawn. The fluid contains cells from the fetus, and these can be studied. If a genetic abnormality is found, the parents can decide whether to have a therapeutic abortion. Unless a person is opposed unconditionally to abortion, this seems a beneficial and nonthreatening process. However, problems arise in *who* decides what defect is serious enough to warrant abortion. Should fetuses with abnormal sex chromosomes be aborted? There is some ambiguous evidence that XYY males tend to criminal behavior. Is it sufficient to justify their elimination before birth? It is also possible that children, determined at birth or before to be XYY, would go through life stigmatized by this label. Parents, teachers, counselors, etc., would find it difficult, if not impossible, to treat such a child

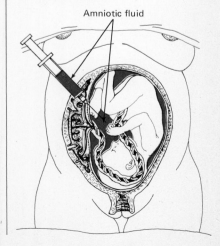

FIG. 17-8 Amniocentesis. The fluid which surrounds a developing fetus contains cells from the skin of the fetus and its respiratory tract. A needle can be inserted into the pregnant woman's abdomen and into the fluid-filled amnion. The material withdrawn contains both cells and fluid which can be tested for genetic abnormalities.

the same way they did his peers. This alone could affect his behavior. Since the sex of the infant can be determined by amniocentesis, genetic screening could be used as an instrument of control of sex of offspring. In cultures that favor male children, females could be aborted before birth. Obviously, such a practice would result in a drastic alteration of sex ratios in that society and a decrease in population size. In America it would probably result in a high proportion of guilt feelings among parents who had embryos aborted just because they "didn't want another boy or girl."

As we noted in Chap. 9, it is also possible to do genetic screening for some genetic conditions by detecting heterozygote carriers of deleterious recessive genes. Sickle cell disease (Chap. 10) is one in which the heterozygotes can be identified; however, it is not yet possible to detect a fetus which is homozygous for sickle cell disease by amniocentesis (research *is* underway to provide a safe test). Thus, two heterozygotes who marry and have children take the risk that one-fourth of their children will be homozygous for this very serious disease. Tay-Sachs disease, on the other hand, can be detected in heterozygote carriers *and* by amniocentesis; thus heterozygotes who marry each other can have their prospective offspring tested while it still can be safely aborted. The availability of these tests will reduce human suffering greatly; however, it will also give people new and difficult choices which will have many emotional and social consequences. How will the parents feel who knowingly take the risk of having a child with a serious illness? How will parents carrying Tay-Sachs gene feel when they are told that their second or third fetus is afflicted and should be aborted?

With all these attempts to change, control, improve, or correct human genetic traits, problems arise: Who will make the decisions? Will the decisions be made wisely? Will genetic engineering create horrible new diseases or cure genetic diseases and create useful organisms? Will genetic screening laws be used to discriminate against certain groups of people such as certain "races," minority groups, or the offspring of the so-called nonproductive members of society? Or will genetic screening laws be used intelligently to prevent human suffering while preserving human diversity? As the size of the human population continues to increase faster than the ability of the earth to feed it, there will be increasing pressure to make parenthood the right of a selected few instead of the right of all, as it is now. Will the techniques of genetic screening be used to determine the selected few who will contribute genes to the next generation?

Ethical problems created by medical advances We have already seen that falling death rates, not rising birthrates, have led to the current exponential growth of the human population. Medicine and public health played a large part in reducing the death rate. Neither medicine nor society as a whole foresaw that reduced death rates without comparably reduced birthrates would lead to the enormous population problem now faced by humankind. We have also seen how medical advances that make organ transplants possible lead to ethical and legal problems in deciding when the donor is truly dead. Medical science can now maintain a person in a state of semiexistence—keeping the body "alive" when the brain is dead. With accident victims and the aged, the problem of when to "pull the plug" of life-preserving machines is becoming a very serious one. The high cost of medical care can bankrupt a patient's family while the patient has no chance of ever becoming a functioning person again.

For most people in the world, medical care is still substandard or nonexistent. Under these conditions, how can we justify the vast amount of medical effort put into heart transplants which benefit only a few people? Why is there not more medical effort made in preventive medicine rather than intricate patching up of patients after a crisis has occurred? Cardiovascular problems and certain kinds of cancer can be treated or even cured if they are detected in their early stages. Why is not more attention devoted to environmental medicine? What effects are pesticides, food additives, deodorants, hairsprays, patent medicines, and air and water pollution having on human health?

As noted in Chap. 1, an extremely rare form of liver cancer has begun to appear at an unexpectedly high rate in workers who work with vinyl chloride, a substance used in the production of polyvinyl chloride commonly used in the plastics industry. It is highly likely that many more cases will appear since environmentally caused cancers take 20 or 30 years to develop and there are 19 companies in the United States alone making vinyl chloride-type compounds and the annual production is in the billions of kilograms. Setting safety standards for the exposure of workers and the general public to such substances is only one of the pressing problems of environmental medicine.

Ethical problems of medical research Medical advances result from research. Often such research must be done on human beings. Ideally, such research subjects should be *fully informed* volunteers. The past history of behavior of some research scientists is an indication that safeguards are needed against abuses. In the early 1970s, Americans were shocked to learn that some blacks in the Southern states had been used as unwilling guinea pigs by the U.S. Public Health Service in an "experiment" on syphilis. Several hundred black men participated in this "study," which began in 1932 and lasted until 1972, when the conditions under which it was conducted were brought to light. Many of the men were given no treatment so that the experimenters could observe the course of the disease; the men, however, believed they were being treated. In 1975 the U.S. Attorney General approved an out-of-court settlement for the 100 men still surviving. Those who went without treatment and are still alive got $37,500 each; the survivors of those who went without treatment and died received $15,000. No amount of money can compensate for the needless suffering which resulted from this "experiment." Such behavior on the part of some physicians makes it clear that society as a whole must have greater awareness and control of biomedical activities.

A great number of experiments are reportedly being carried out on prisoners today in attempts to modify their behavior. At one institution, inmates who lie or swear are injected with a drug that makes them vomit uncontrollably for up to an hour. Elsewhere, drugs, hypnosis, electroconvulsive shocks, and psychosurgery (destruction or removal of part of the brain) have reportedly been used in attempts to modify the behavior of prisoners.

There are important biomedical decisions which will affect the whole of human society. Even though most of the people in the world will have no chance whatever of getting a heart transplant, decisions on procedures relating to genetic screening, the availability of life-prolonging drugs, and the availability of birth control devices will affect great numbers of people. How much time and energy a society should devote to medical care for the various age groups in its population is a serious question when for most societies the most obviously pressing problem has been, and will continue to be, obtaining food.

ECOLOGICAL PROSPECTS

The future of all life on earth now depends on the activities of human beings. The future of *our species* also depends not only on our activities but on the rest of the earth's components: the air, the oceans, the natural geological processes, the ice caps, and, of course the rest of the biosphere. The biosphere must maintain its flow of energy and cycling of materials if life is to continue on earth. The onslaughts of the industrial revolution and the exponentially increasing human population and its activities are already putting significant stress on the functioning of the biosphere. They also may affect the climate of the earth itself.

17-5 It seems likely that the world ecosystem will suffer further deterioration in the decades ahead

As we have seen, the size of the human population seems likely to increase greatly. As a result of the rising expectations of people in the developing countries, the demands each individual places on the ecological systems of the planet may also increase. As humanity struggles to provide more food, better clothing and housing, and other necessities for a growing population, natural ecosystems will inevitably suffer. Simplification of ecosystems will occur as populations and species go extinct under the pressure of hunting, poisons, subdivisions, highways, and simply lack of habitat. Our species is already exploiting some 40 percent of the earth's land surface and has reduced the amount of terrestrial vegetation by an estimated one-third. Roughly 5 percent of net, global, photosynthetic productivity now occurs in agricultural ecosystems. Our species now moves many materials around the surface of the planet more rapidly than do natural geological processes. For instance, human input of oil into the oceans is estimated to be some 20 times that which enters from natural seepage.

Meteorologists have shown that climatic change can occur rapidly in response to relatively small changes of a variety of variables. As we noted in Chap. 2, the release of CO_2 into the atmosphere from the burning of fossil fuels should increase the greenhouse effect and warm the earth. The CO_2 concentration of the earth is increasing by about 1 part per million anually. Since 1940, however, the mean global surface temperature has *dropped*. Meteorologists believe the drop is due to increased particulate matter in the atmosphere from agriculture, industrial processes, and an increase in volcanic activity. They believe that changes in CO_2 *may* be capable of producing shifts in the distribution of rainfall. The earth is presently in a period of rapid climatic change, and no one as yet can predict what the change will produce. It is impossible to know whether areas which have been subject to years of drought will continue to suffer; we can only wait and see. It does seem likely that rapid climatic change will cause decreased agricultural yields since most of our crop plants were selected for maximum yield during the so-called normal period from 1930 to 1960. In terms of the last thousand years, this period was actually not normal at all (Fig. 17-9).

By a wide variety of measures, *Homo sapiens* is already a major global, geological, and biological force, and our impact seems bound to increase rapidly. Such increase bodes ill for the functioning of natural ecosystems.

Fewer options in the future Although it is difficult to forecast exactly what will happen to the human environment, some trends seem clear. For instance, as more and more land is occupied by more and more people, attractive recreation areas will become scarcer. And there will be more people wanting access to shrinking parks, seashores, and the like. Already the National Park Service of the United States is experimenting with rationing the time people can spend in national parks. In California, people are advised to reserve their campsite in a state park months in advance if they want to be sure of a place to vacation. One of the prices paid for the freedom of unrestricted reproduction is restriction in recreational options.

More serious, as population grows, will be restriction in dietary options. Because of the loss of energy in transfers in food chains (Chaps. 1 and 2), many more people can be supported on a vegetarian diet than on meat. Meat is already scarce in the diet of most human beings and is becoming scarcer and more expensive. Pollution of the oceans and overfishing are likely to reduce the fisheries' yield of the sea so that, in the face of sharply rising demand, per capita supply will be reduced. Some countries that get their protein largely from the sea will be directly affected. Other countries will be affected indirectly because much of the fisheries' catch has been used to feed poultry and other animals used for meat. A decline in the catch of Peruvian anchovies has led to fewer and fewer people being able to afford fried chicken in America. Similarly, crop failures and human needs for grains have made supplies of feed grains for pigs and cattle scarcer and have caused meat prices to go up.

Massive starvation While people in the United States worried about increasing meat prices, people in some other parts of the world did not even have grain to eat. In the area of Africa known as the Sahel, just south of the Sahara Desert, hundreds of thousands of people

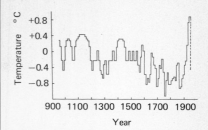

FIG. 17-9 The mean annual temperature in Iceland during the past thousand years. The period 1930–1960 which has been considered normal by international agreement is actually the most unusual 30 years of the thousand. Dashed line indicates the rate of recent decline in temperature. (*After Bryson and Bergthorsson.*)

and their livestock died of starvation during a prolonged drought. These peoples depended on seasonal monsoon rainfall to grow their crops and feed their livestock. The monsoons failed for six years in a row in the late 1960s and early 1970s; starving people clustered together at refugee camps, and what little food they got came from donations from other countries. People who remained in remote areas suffered greatly since relief agencies had great difficulty transporting food across rugged terrain with poor or no roads. In 1974, the rains returned but were too little and too late for planting crops. The overgrazed and wind-eroded land eroded further.

In 1973, world grain reserves were at a record low and it was clear that food production cannot long keep up with human population growth. Presently perhaps one-half of the human species receives an inadequate diet and millions are dying of starvation annually.

The food situation is most threatening in the underdeveloped countries where population growth is also the most rapid. Since many of these nations could double their populations every 25 years at current growth rates, they are faced with the impossible task of doubling their food supplies in that period just to stay even.

Failure of the Green Revolution Much hope was placed in the so-called Green Revolution, an attempt to greatly increase agricultural productivity in underdeveloped countries by introducing high-yield "miracle" strains of grain. For example, one strain of dwarf rice, IR-8, is capable of producing more than twice the harvest of traditional rice strains grown in the same area. In spite of this, the Green Revolution is now widely considered a failure. These high-yielding strains of grain require intensive fertilization, high pesticide use, and a great deal of water. Since the energy "crisis" of 1974, the price of fertilizer has more than doubled. Fertilizer factories need natural gas, which is expected to be used up worldwide in a decade or so. Producing pesticides and drilling wells require energy from fossil fuels. Plastic pipe and pesticides are made from oil derivatives. Transporting fertilizers and running farm machinery require energy also. Many farmers have been forced to plant their old strains of grains because they cannot afford the expense of growing the high-yield miracle grains.

Even before the energy shortage, the Green Revolution had many problems. In countries such as India, the cost of high-yield agriculture made the use of miracle grains easier for relatively wealthy farmers than for relatively poor farmers. The better-off profited more than the most poverty-stricken. This increased already sharp class differences and led, in India, to mob violence, including the murder of landlords. Food riots involving protests over high food prices also occurred. Of course, the people of a country must be able to pay for food. Otherwise, increased harvests simply drive prices down and farmers find themselves with little or no reward for their increased expenses and effort. Shortage of capital (money and property) is a major problem of underdevelopment, and this shortage ties the agricultural situation to the entire problem of development.

Another unfortunate aspect of the Green Revolution was that it tended to encourage the mechanization of agriculture. In underdeveloped countries, the large-scale replacement of farm labor with machines was a great disaster. These countries *already* suffered staggering unemployment problems and their cities are growing at rates of up to 7 percent a year, doubling in size every 10 years. Mechanizing agriculture has put more people out of work and sped migration into cities like Bombay (Fig. 17-10), which are already disaster areas.

Perhaps the most serious biological problem associated with the Green Revolution is the loss of genetic variability of crops. Genetic variability is the raw material of selection (Chap. 10), and it is that variability that has permitted plant geneticists to produce high-yield strains of crops as well as strains which are resistant to various diseases and pests. Modern agriculture is built on a base of plant genetics.

It is absolutely essential that the genetic variability of crops be maintained. Every type of resistance which can be bred into a crop can be countered by the pest evolving a new mode of attack. This intertwining, reciprocal evolution of plant and pest is a classic case

FIG. 17-10 Many people of Bombay have no homes and live and die in the streets. In spite of the hard life in Bombay, people continue to migrate there from farms and villages where life may be even harder. (*Marilyn Silverstone, Magnum Photos.*)

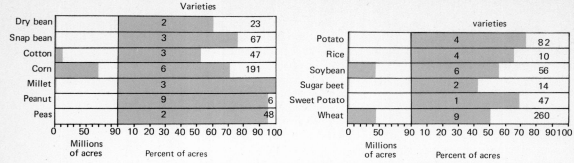

FIG. 17-11 Major U.S. crops: Most of the acreage planted to each crop (*left*) was given over to only a small number of the varieties available (numbers in boxes on *right*). (After *Genetic Vulnerability of Major Crops*, National Academy of Sciences, 1972.)

of coevolution (Chap. 10). The plants and the pests can be thought of as being in coevolutionary races. The price of staying in the race is maintaining genetic variability so that selection can occur; the price of losing the race is extinction. If humanity allows crops to lose their genetic variability, it will mean the end of high-yield agriculture. Even in the United States we have taken great risks of losing crops to disease by planting most of the acreage of our major crops in only a few varieties (Fig. 17-11).

The originators of the Green Revolution intended only to buy time while population controls were carried out. Unfortunately, when people have more food, they are not likely to understand the need for controlling their rate of reproduction. The Green Revolution may even have made the food-population crisis worse by giving people the illusion that some miracle could allow the population to expand indefinitely.

17-6 Food is not the only resource which is likely to be scarce in future societies

It is difficult to estimate the future resource situation because it depends on such a large number of variables. One is the number of people to be supplied with the resources. Another is the amount each individual will use. A third is the extent of the reserves of the resource, and a fourth is the potential for substitution. Finally, it is possible that many resources can be recycled. Population size seems destined to grow, for at least a while, as does the per capita demand. But known reserves of most mineral resources also will probably expand as shortages drive up prices and encourage exploration. Higher prices may make previously uneconomical substitutions possible. Higher prices may also encourage recycling of materials.

Table 17-1 shows future supplies of several minerals under two different assumptions. The static assumption shows the number of years required to use up the resource reserves known in 1970 at the 1970 rate of consumption. The growth assumption shows the number

TABLE 17-1 WORLD MINERAL DEPLETION

	Number of years supply	
MINERAL	STATIC ASSUMPTION	GROWTH ASSUMPTION
Coal	2,300	150
Petroleum (liquid)	31	50
Chromium	420	154
Nickel	150	96
Tungsten	40	72
Manganese	97	94

SOURCE: D. H. Meadows, et al., *The Limits to Growth*, Universe Books, New York, 1972, pp. 56–59.

of years required to use up 5 times the 1970 known reserves assuming that the rate of consumption increases at the rate projected by the U.S. Bureau of Mines. The actual time at which a mineral will become so scarce as to be exhausted for practical purposes probably lies somewhere between the two figures.

Notice that, under either assumption, our supply of petroleum seems likely to give out in the not-too-distant future. The 1974 energy minicrisis foreshadowed events which will occur when this resource really does run out. Many people became aware for the first time how much their life styles depend on petroleum. For some uses, power from nuclear fission or fusion can be used as a substitute. However, this power is electrical, and only about 25 percent of American energy consumption is now in the form of electricity. Petroleum is the direct or indirect source of over 40 percent of our energy use. Therefore, dramatic changes in the kinds of vehicles and machinery that are in use will be required if electricity is to carry the major load. A satisfactory long-range electric car has yet to be developed. Thought has been given to using electricity to split water into hydrogen and oxygen and then using the hydrogen as fuel for cars. Since hydrogen is a highly volatile, explosive gas, many safety features would be needed with such a system.

It is possible to get oil by liquefying coal, but that would shorten the life of the coal reserves. It is possible to get oil from rocks called oil shales, but there is dispute as to whether it will ever be economically feasible. The ecological costs of using oil shales will be very high. Huge areas will be chewed up in mining shales, and enormous amounts of waste material will remain afterward. Most importantly, energy and materials must be put into any of these projects. Many coal reserves are most conveniently strip-mined. The machines that mine them must be built of materials extracted using energy, and energy must go into their construction and use. Oil shales also require elaborate machinery for their extraction. Both strip mining and oil shale mining need large amounts of water. Water must be transported and pumped by the use of energy. Somewhere in this process of extracting fuels from less concentrated sources the **law of diminishing returns** will begin to apply, and people will find that they are putting more energy into the operation than they are getting out of it.

For a number of reasons, it has proved profitable to develop increasingly larger tank ships, which cannot use the Suez Canal, to convey crude oil from the Middle East to other parts of the world by passing around the tip of South Africa (see Fig. 1-14). These ships are so large that their responses to the stress and strain of weather, water, and their load of oil (up to 550,000 tons) are not entirely understood. Often called superships because they may be 400 meters long, they are not easily maneuvered and are subject to power and electrical breakdowns which leave them at the mercy of the elements. They have been involved in innumerable collisions, have broken up at sea (they travel in the roughest waters of the oceans), and have exploded when empty and burned when loaded. Tremendous quantities of oil have been released into the ocean and the effects of this in the long term are not known. In the short term it is clear that millions of birds and other animals have been killed, with one species of penguin perhaps brought to extinction. The effects of oil spills on phytoplankton—the base of oceanic food chains—could, of course, be disastrous. Because of the distribution of currents in the most dangerous areas off Africa (see Fig. 5-2), oil is being spread to some of the most ecologically productive regions of the ocean.

The availability of energy to mine and transport other resources may impose limits on their availability. Even if nuclear fusion could be made to work and we could use this source of energy to extract needed minerals from common rock or seawater, we face another limit—that of *waste heat*. Even at the present rate of growth of energy consumption, it would probably be less than 100 years before the disposal of heat from human activities seriously upsets the climatic system of the earth and caused widespread disaster. The second law of thermodynamics places a rather firm lid on our species' hopes for energy use.

Other ecocatastrophes Both food and nonfood resources of the earth are used disproportionately by the relatively small segment (about one-quarter) of humanity that lives in the affluent nations. Indeed, the United States, with some 6 percent of the population of the world, is estimated to consume about 30 percent of the annual resource flow. This disparity between the rich and the poor nations has been increasing over recent decades and, if current trends continue, may increase in the future. However, the ability of the Arab states to control the majority of the world's oil may change this trend rapidly. One possible consequence in either case could be political instability leading to a thermonuclear war.

The destructive effects of a nuclear explosion are caused by the sudden release of enormous amounts of energy in the vicinity of the bomb. At the center of the blast, the temperature rises to millions of degrees Celsius, and this high temperature causes a shock wave with pressures as high as 7,000 times normal atmospheric pressure.

A 1-megaton nuclear bomb (equivalent to 1 million tons of TNT) will destroy all houses in an area of 50 square kilometers (19 square miles) around its center. The fireball formed by the explosion attains temperatures of tens of millions of degrees Celsius and radiates much like the sun. The heat from a 1-megaton bomb causes paper to burst into flames as far as 14 kilometers (9 miles) from its center. Buildings and people are destroyed by the shock wave and by direct thermal radiation, as well as by the fires and firestorms created and the nuclear radiation released. The nuclear radiation spreads over a wide area immediately with the flash, and radioactive elements spread over an even wider area as fallout from the radioactive cloud. Worldwide thermonuclear war would disrupt terrestrial ecosystems seriously; radioactive materials would enter food chains, leading to deaths and mutations; firestorms would sweep vast areas bare of vegetation; and weather modification would result from the huge amounts of dust entering the atmosphere. Marine ecosystems would suffer from the huge quantities of silt and chemicals washed into the oceans from burned-off land and destroyed cities. Surviving people would face degraded ecosystems, ruined agriculture, communication and transportation breakdowns, loss of key parts of technological civilization, and uncertain weather conditions. Even if civilization could be restored, there would be a high incidence of cancer and mutation-induced defects in people and other living things for many generations.

Another horrible possibility is a worldwide plague, starting perhaps in the ever-more-crowded, ever-weaker populations of the poorer nations. The probability of such a disease killing off a substantial portion of humanity seems to be increasing. Indeed, as we noted in Chap. 6, there already have been some rather frightening near misses. In 1967, 30 people in a laboratory in Germany were infected with a hitherto unknown disease carried by a shipment of vervet monkeys that had passed through London airport on the way to Marburg, Germany. Of the 30, 7 died, even though they were well-fed laboratory workers and had the best of medical care. If this Marburg virus had infected personnel at the airport, it could have spread *all over the world* before anyone realized what had happened. Poorly fed, crowded populations would be especially vulnerable to such an epidemic. Another frightening possibility is that a disease designed to be used in biological warfare might escape from a laboratory. Presumably the United States has renounced biological warfare and destroyed its stocks of biological weapons. Some other countries are no doubt continuing research into biological warfare, and as long as such weapons are available in any part of the world, the possibility remains of an accidental escape or a terrorist diversion.

It is not possible to list all the ecocatastrophes which may be caused by the actions of the human species because we simply do not know what they are. We have no idea of the long-term effects of most of our technology on the earth's systems. A number of things that seem harmless may turn out to do great harm. Some scientists believe that substances from aerosol cans (the fluorocarbon propellants) or from nitrogen fertilizers could destroy the ozone layer which protects us from the deadly ultraviolet rays of the sun. Other scientists believe that attempts to modify the weather for our benefit could backfire and disrupt the whole atmospheric system. In fact, we may have already set in motion some unsuspected ecocatastrophe whose cause we may not be able to identify even when it occurs.

17-7 Ecological engineering could help the human species to avoid some of these grim possibilities

Humanity, of course, does not have to sit around like the proverbial bump on a log waiting for disaster. Many steps could be taken to help *Homo sapiens* operate within the limits of the earth's life-support capacity and to adjust the scale and kind of human activities so that they do not inevitably lead to the end of civilization.

One step is to do everything possible to halt the growth of the human population in a humane manner, that is, by limiting births rather than increasing deaths. Then, by keeping the birthrate slightly below the death rate for a long time, the size of the population could be gradually reduced until it was well within the carrying capacity of an ecological system which, unlike the present one, did not exist at the expense of continually mortgaging the future.

The design of such a system would be in the domain of what we might call **ecological engineering.** The general outlines of a possible system have already been developed by those concerned with the future of society. It would include an economic system which puts great emphasis on high-quality goods that last, rather than on continuous replacement of things used for a short time. Limits would be put on the rate of depletion of resources. This, in turn, would make many kinds of recycling economically feasible, and would make such things as large automobiles, which are produced and used at great cost to resources and environment, prohibitively expensive.

Great effort would be put into eliminating much of today's wasteful energy use by, for instance, requiring that all new structures be carefully insulated so that less power would be required to heat and air-condition them. Research into increasing the efficiency of fossil-fuel power plants, fusion power (which is relatively free of pollution hazards), and the use of solar energy would get top priority. So would the development of an ecologically sound technology for high-yield agriculture. For instance, emphasis in pest control would continue a trend already started away from the broadcast use of synthetic pesticides to integrated control. Integrated control involves, among other things, the use of natural enemies to reduce pest populations, the enhancement of the natural defenses of the plants, and, when necessary, the judicious use of nonpersistent pesticides in a manner which minimizes the chances of resistant pests evolving (Chap. 10). The goal of integrated control is the *management* of pest populations so that their population size remains below the level at which damage is done economically. It is not the *extermination* of pests, a goal which has never been reached by any chemical control program.

A new occupation, that of **ecosystem manager,** should be created.

People as skilled in the workings of ecosystems as physicians are in the workings of the human body should make the decisions about environmental modifications. When and where people have become the dominant species in an ecosystem, they have traditionally managed the ecosystem without really being aware that they were indeed managing it. In these people-dominated ecosystems, people have acted in such a way as to maximize their own short-term benefit and convenience. They have consistently tried to dominate nature instead of working with it. It is clear that ecosystems provide essential services for people which are easily disrupted when management takes into account only the short-term benefit and convenience for people. For example, the recycling of sewage and the natural control of many pest species are functions of ecosystems which have been seriously harmed by people. Replacement of these functions is accomplished by wasteful expenditures of energy. The ecosystem manager would treat such a system as a sick patient and attempt to restore its ecological health.

17-8 A redistribution of wealth seems a critical part of any global program of ecological engineering

As already noted, the growing gap between haves and have-nots in a world full of rising expectations poses an especially serious problem for humanity. When the developed countries propose population control for the underdeveloped countries, the underdeveloped countries point to the wasteful use of resources by the developed countries as the real problem. Recent conferences on environment and food have not been very productive because of this problem. Restrictions on population growth are often seen by underdeveloped nations as being a case of the rich saying, "We've got ours; now you restrict your population growth in order to save the world." The obvious answer is that *two* things must happen: The developed countries should cut back on their wasteful use of energy (Fig. 17-12) and resources while restricting their own population growth. Second, the underdeveloped countries should attempt to develop in ways which are less wasteful than those of the developed countries *and* control their population growth. The underdeveloped countries cannot hope for successful development and improvement of their people's lives as long as their populations are growing so rapidly they cannot even be fed adequately. Neither can the overdeveloped countries hope to continue their industrial growth. In 1973 the Arab oil embargo showed that peoples of the Third World are ready to withhold resources from the developed countries. The Organization of Petroleum Exporting Countries may be just the first of a series of organizations of countries that produce raw materials needed by the industrialized countries.

CULTURAL PROSPECTS

It is largely the culture-transmitting aspects of our species that have led to the rather grim human and ecological prospects we have just

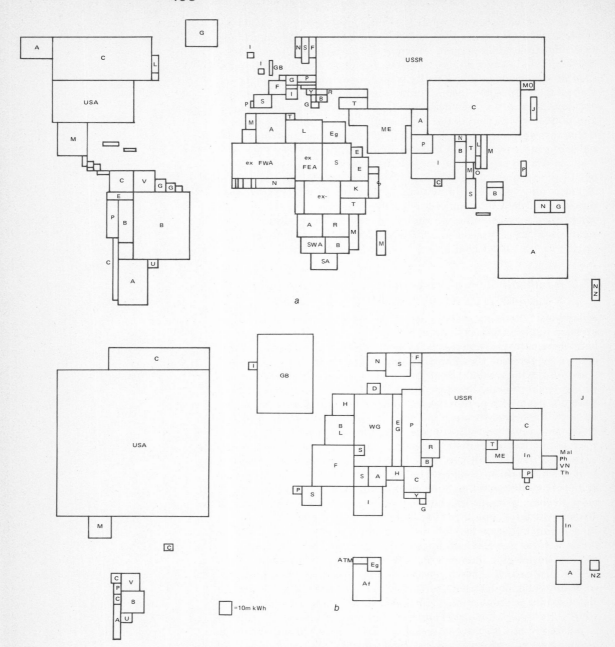

FIG. 17-12 Energy consumption (1965) of countries of the world. (*a*) Countries represented as rectangles of size proportional to surface area; (*b*) countries represented as rectangles proportional to the consumption of energy. (*After J. W. Burton,* World Society, *Cambridge University Press, England, 1972; based upon data from* Oxford Economic Atlas of the World, *3d ed., Oxford University Press, England, 1965.*)

forecast for the future of earth. Cultural evolution is responsible for the transmission and improvement of agricultural techniques, industrial inventions, social organizations, and economic arrangements that made possible the population explosion, pollution, environmental deterioration, and the gap between the rich and poor nations. Culture also consists of art, music, literature, and the languages and religions of peoples. What are the chances for conservation of diversity of cultures in the world of the future? Can cultural evolution occur fast enough to invent new social forms in time to respond to environmental crises?

17-9 Can human nature be different in societies of the future?

Many people today believe that only a fundamental change in human nature can ensure the survival of civilization. They see the shortages appearing on the horizon as heralding the end of the industrial era. A new socioeconomic structure will be necessary to cope with the change to the postindustrial era. Only people who are fundamentally different from the people who are presently leading the governments and businesses of the world can deal creatively with such dramatic change. Only individuals who are more identified with human needs and less with material possessions can make the transition.

It appears that, in the past, human behavior *has* changed in response to differing conditions. In the Middle Ages, people willingly spent their lives working on cathedrals whose completion they could not live long enough to see. Today, few people are interested in devoting their energy to long-term projects. They are too impatient even to build rapid transit systems because they may take five or ten years to complete. In the present, life is comparatively easy for most people in the United States (compared with that in many other countries). Yet virtually no persons say they are satisfied with their lives. Perhaps such dissatisfaction can lead to fundamental changes. It does seem unlikely that people will voluntarily undertake a reduction of their energy consumption and wasteful use of resources. Perhaps they will adapt readily if such changes are gradually forced on them by higher prices.

The role of human aggression A whole school of human behaviorists has emphasized the idea that most of human behavior is innate. They say, for instance, that the conflict of nations over territory is the result of the same kind of genetically determined "drive" that produces territoriality in songbirds (see Chap. 15). Some of these people believe that the only answer to humanity's innate aggression which leads to territorial conflicts is to channel it into relatively harmless activities such as sports.

This is an unfortunate oversimplification of the nature of human beings. For instance, careful investigation of territoriality in a wide range of animals shows it to be a very complex phenomenon, virtually always subject to modification by the environment. It quite likely evolved numerous times in response to diverse selection pressures. Species, for instance, may be territorial in some environments and not territorial in others. The ayu, a fish related to the salmon, is territorial in shallow streams but forms schools in deep water. Vervet monkeys may be territorial in some environments and not defend territories in others. Furthermore, the nearest living relatives of *Homo sapiens,* gorillas and chimpanzees, do not show clear-cut territorial behavior, and there are no data which must be interpreted as showing a compelling drive to territoriality in *Homo sapiens*. Many human groups show what may be considered territorial behavior in a broad sense, but this does not mean that territoriality is a genetically fixed element of the human species' behavioral repertory. Unquestionably, the human genotype (the genetic inheritance common to all *Homo sapiens*) will interact with appropriate environmental stimuli to produce behavior patterns which we describe as territoriality and aggression. There is also no doubt that in other environments such behavior will not be elicited.

Another indication of the important environmental component of aggression is the variation from society to society. African bushmen are very nonaggressive, both among themselves and toward other groups. Anger is very rare, and harmony has high cultural value. The Eskimos, another group of hunters and gatherers, also are nonaggressive. On the other hand, the Yanomama, who live by gardening and hunting in the tropical rain forest, are extremely aggressive. The men pride themselves on their fierceness, beat their wives, and continually argue and fight with each other. The women often kill female babies. Yanomama villages fight with one another; there are raids, feuds, and even massacres. Interestingly, the Yanomama are not born aggressive. They are carefully trained in aggression, and some are better pupils than others.

It may be possible that a tendency to interpersonal aggression in our society may have the same origin as the often stereotyped combat which makes up the "wars" of primitive societies. The Danae of New Guinea are caught up in a never-ending feud between groups in which one side kills a member of the other, which then kills in revenge. Warring is a way of life, but it is highly ritualized. Casualties are low, with a day's battle (several episodes of less than an hour) often leading to wounds but no deaths. No attempt is made to occupy enemy territory or to subjugate or exterminate the enemy group.

Can holocausts such as World War II simply be extensions of this kind of behavior? Those who would extrapolate from defense of territories by birds to the behavior of nations would doubtless say yes. But the question is still being investigated in detail, and at the moment the answer is unclear. Indeed, it is possible we may never know. We do know that overpopulation, that is, too many people for the perceived resource needs of a nation, may be a cause of modern wars. But the relationships are complex and are just beginning to be untangled in careful statistical studies by political scientists.

Some very good work has been done on what makes some people more aggressive than others. For instance, it has been shown that heavy punishment of aggression in young children is likely to be less successful at controlling their aggression than a home environment where aggression is regulated by family rules. Parents who use praise and displays of love counterbalanced with the withdrawal of same when rules against aggression are broken appear to raise children with strong consciences and little tendency to strike out against other children or their parents.

There is also some general understanding of factors which tend to produce aggression in individuals. Assassins tend to have been lonely, often having poor relationships with their parents and distrusting them. There is considerable evidence that watching aggressive behavior in TV shows may promote aggressive behavior in children. And there are also data which indicate that an environment in which there are large numbers of firearms stimulates aggression. As one psychologist said, "The finger pulls the trigger, but the trigger may also be pulling the finger."

17-10 We are gradually learning something about control of the human mind and body

Now that more and more is being learned about the workings of the human nervous system, it has been suggested that aggression and other undesirable behavioral patterns may be controlled by controlling the mind. The psychiatrist today has an arsenal of drugs to sedate, tranquilize, or lift the emotions of patients. These have literally revolutionized the treatment of the mentally ill. California reduced the number of mental patients in its state hospitals by 80 percent between 1961 and 1973. Drugs *have* enabled seriously ill patients to recover and have been particularly valuable in treating anxiety in patients who might otherwise have been hospitalized.

But the use of some of these drugs and other "downers" and "uppers" has moved out of the hospitals to the streets and school yards. Illegal drugs, such as heroin and cocaine, are in wide use also. For a time in the late 1960s and early 1970s there was a drug subculture with its own life style, music, and cultural heroes. Several of the finest young musicians of this popular culture died of overdoses of drugs.

Meanwhile the use of the most ancient narcotic drug, alcohol, has also increased. Between 1960 and 1970 the per capita consumption of alcohol in the United States increased 26 percent. Per capita consumption in the United States is now at an all-time high. After heart disease and cancer, alcoholism is the country's biggest health problem. Unfortunately, some people cannot drink alcohol without becoming addicted to it; this may have a strong genetic component, but heavy drinking may also be learned. An alcoholic's life span is shortened by 10 to 12 years, and alcoholics create untold suffering among the members of their families and very high costs to society in the form of traffic accidents and crimes related to alcohol use. Liver damage results from heavy use of alcohol and some researchers believe that the brain is damaged permanently by long continued use of alcohol even in a nonalcoholic. Most tragic is the fact that the use of alcohol and other drugs is spreading to younger and younger members of our society so that one now finds drug abuse, including alcohol use, among junior high school and even elementary school children.

The use of drugs has been part of our culture since the first person drank fermented fruit juice. There seems little possibility that drugs in some form will not be used by people. It is certainly hoped that the use of drugs by children may be one of those passing fads of American culture. Alcoholics and other drug addicts *are* receiving more attention and treatment than ever before, and so their chances for recovery are good if they take advantage of the help available.

The possibility that a totalitarian government would use drugs for mind control has been an issue of continuing concern. In individual cases of brainwashing, drugs already have been used. It seems unlikely that there will be deliberate attempts to use drugs to control the minds of masses of poeple in the near future.

Other techniques of mind control Throughout history certain charismatic (**charisma** is extraordinary power to influence others) leaders have had an intuitive understanding of how to control the minds of the masses. No doubt charismatic leaders of the future will possess similar abilities; whether they will use them for the good or ill of the species remains to be seen. Recent research has shown regions of the brain that when stimulated electrically, produce exquisite pleasure or excruciating pain, depending on the location. Electrical stimulation of pleasure centers has been used successfully in the treatment of certain mental problems. Because of the difficulty of implanting electrodes in exactly the right position, it seems unlikely that this technique will be used to control the masses in the near future. On the other hand, experiments, especially in Russia, have shown the possibility of inducing sleep with gentle electrical stimulation of the head. This technique, thought to be harmless, may make it possible to ensure a pleasant, relaxed sleep under circumstances where this might otherwise be impossible. It might also reduce drug use!

Other experiments in which the *corpus callosum* (the tract joining the left and right cerebral hemispheres) was cut show that there are apparently two minds in our brain—or two brains in our mind! Such operations have been done to treat severe epilepsy. After the operation, the patients function normally in almost every way. However, when they are tested carefully, it is shown that the so-called dominant hemisphere (the left in a right-handed person) deals with verbal, visual, and "intellectual" phenomena. The other deals with touch and emotional states. Both hemispheres seem to be intelligent, but the right one cannot direct speech. It *can* direct the hand it controls (the left) to select objects shown in photographs to the patients. In an intact brain, the two hemispheres are in communication. It has been suggested, however, that a person is artistic or intellectual depending on which hemisphere is dominant. For those of us who are strongly one way or the other, these studies suggest that it might be possible for us to "get into" the other half of our brain and get to know our other "self." In fact, it is quite possible that mystics have been doing this for thousands of years already.

An interesting new development is the discovery that human beings and other animals can learn to control activities of the body formerly thought to be completely involuntary (Chap. 13). The technique used is called biofeedback, which usually requires expensive and complex instruments. Animals who are rewarded by brain stimulation can learn to change their blood flow and blood pressure, the functioning of their kidneys, the contractions of their intestines, and the electrical activity of their brains. If rewarded for maintaining a difference in blood flow to its two ears, an animal can learn to produce one warm ear and one cold ear. In people, the widest use so far of biofeedback is in relieving psychosomatic problems. For example, learning to relax by receiving feedback from their muscles can be tremendously beneficial to some patients. Migraine headaches (caused by changes in the blood flow in the head) can often be stopped short by people who can learn to warm their fingers by increasing blood flow through them. Biofeedback may also hold promise for such problems as asthma, insomnia, ulcers, and high blood pressure of some kinds. Many scientists think that biofeedback can be tremendously useful for healthy people in helping them to get in better touch with their internal world. The kind of dream state induced by certain kinds of biofeedback training may lead to creative insights.

Finally, we should mention the growing interest in control of the mind without drugs and without instrumentation of any kind. More and more people are learning techniques of deep relaxation, self-hypnosis, and meditation. Remarkable changes for the better in physiology, as well as psychology, of many people using these methods have been described. In our overcrowded, noisy, technological society, many people use such techniques to make life endurable. Indeed, those who are adept at self-hypnosis can experience all the effects various drugs are known to produce.

17-11 Your world view is a product of your culture

As we noted in Chaps. 9 and 15, people are born with a genetic potential which is then modified by their environment. Culture is a part of the environment of all human beings; and it is possible that many cultures may stifle the inborn capacities of people born into that culture. For example, a Yanomama infant might have a great genetic potential for tenderness and compassion, but such tendencies would have little chance for expression in the harsh and vindictive Yanomama culture. Obviously every human individual must grow up in a culture; however, comparative studies of culture have revealed vast differences in how they determine which aspects of an individual's potential will be developed and which suppressed.

It appears to many people that our view of reality itself is a product of our cultural background. They feel that our way of representing the world to ourselves arises out of the whole society in which we live and grow. The work of some scientists suggests that the "real world" each social group perceives is based in large part on the language structure of the group. No one looks at the world with a purely innocent eye. When we say that a child has become "reality-adjusted," we just mean he or she has become another strand in the social web of his or her culture. But that social web, like all others, has a world view that is to some degree arbitrary and at the same time flexible.

The social fabric of "reality" is apparently maintained by social pressure arising from people who fear chaos when their view of reality is questioned. The Inquisition, during which Galileo was forced to renounce his views of the structure of the solar system, is an extreme example of such social pressure. In all ages and places some degree of social enforcement of views of reality occurs. Cultural leaders such as politicians, religious leaders, prominent scientists, and judges use their power and influence to reinforce their particular views of reality.

Contact and communication of Western cultures with other cultures are much greater than they were in the past, and this has resulted in increased tolerance of divergent views of reality. To the extent that we are flexible enough to accept the existence of other ways of viewing the world, we have some measure of direction in selecting our own world view. To the extent that we recognize the effects of our own culture on our world view, we can attempt to filter and select from the myriad stimuli to which we are continually exposed. For example, we can turn off the TV set when we choose, ignore the urgings of a politician whose world view we do not accept, or choose a profession which will lead to a world view compatible with our potential.

17-12 How does culture evolve?

The evolution of science is an example of cultural evolution. Science (like culture, society, and life itself) is a process, not a thing. The process of science occurs when individual men and women question nature and carry out observations and experiments which will answer their questions. Preliminary answers to questions are called **hypotheses** and are rejected or accepted as the process of science continues. Scientists write up their study and submit it to a scientific journal; before it is published in a journal, it is reviewed by other scientists familiar with that area of science. These reviewers decide whether the study meets the standards of science and should be published. Most often, a published study leads to further studies, often by other scientists, which may support, refute, or expand the original study. Thus, in a series of studies, some hypotheses become **theories,** and some theories become **laws.** Presumably, at all times in this process everything is open to question; nothing is sacred except absolute honesty in the presentation of research results. (In practice some scientists hold their own hypotheses sacred, and sometimes science as a whole will create and cling to dogmas. But sooner or later dogmas in science are challenged and if found wanting, discredited.)

The body of knowledge presented in science books such as this one results from the process of science—it is *not* science. It is a slice in time through the accumulation of data, hypotheses, theories, and laws which are currently considered to be verified objectively by scientists. So many people have contributed that it would be impossible to mention them all; therefore, we have mentioned only a few by name. These people contributed not only to changing the body of knowledge but also to changing the process of science itself. As we noted in the Preface and Chap. 15, there is an inevitable subjectivity in the supposed objectivity of science. Science is a product of our minds and thus is subject to our perceptual and cultural biases; we *cannot* stand outside the process of our perception because to some unknowable extent we *are* the process itself. Nevertheless we can look back and view science as process evolving. In Sec. 4-4 we noted that it took $2\frac{1}{2}$ centuries and many astronomers to "discover" the solar system which we all accept as reality today. During these hundreds of years new data were gathered and new instruments were invented, but most important, new views of reality were accepted. You can pick up a series of biology texts from years past and observe not only changes in content as new things were discovered but changes in emphasis as science and culture evolved. You can even find different descriptions of "the scientific method."

Traditional scientific method has been criticized because it never provided a source for hypotheses or even suggested which questions were important to ask. Some scientists have approached science as a massive collection of facts which are to be added to like bricks in a wall. Our most creative scientists, however, have always realized that no scientific theory is just a collection of facts. Albert Einstein said, "But on principle it is quite wrong to try founding a theory on observable magnitudes alone. In reality the very opposite happens. It is the theory which decides what one can observe." In our opinion, science as it is known today has as its domain only a small portion of potential views of reality. We expect science to continue to evolve and quite possibly to encompass more potential views of reality; but we do not know what views or how the evolution will occur. Science and its offspring, modern technology, are powerful tools used by modern civilization; they are too valuable to lose, but they must change their destructive impact on the biosphere if society is to survive.

Apparently culture in general evolves in a manner similar to that of science; the chief difference appears to be that new ideas (hypotheses in science) are tested by the whole of the society, particularly the cultural leaders, not for some sort of "objective" verification but for general social usefulness in that place at that time. There is an example of the cultural evolution of language in this book. You have probably noticed by now that the words, *man, mankind,* and *he,* as used to refer to the human species, do not occur in the text but only in the titles of supplementary reading. Concepts relating to the status of women in society are changing. Presumably the result of this change, our grammatical usage, will result in changed perceptions of the structure of society, which may then result in more changed concepts, and so on.

Some people believe that a really major change in a culture can occur only when there is a crisis and a restructuring of a whole block of ideas. Fundamental to this restructuring appear to be (1) an intense preoccupation with that discipline which includes the block of ideas and (2) the asking of an ultimately serious question. This is precisely the state in which people who are concerned with the question of the fate of humankind now find themselves. We are faced with a search which will it is hoped reshape our concepts in favor of the kinds of perceptions necessary to "see" answers. This is similar in a sense to the familiar self-fulfilling prophecy; if you go to an exam convinced you will do well, the chances are you will do much better than if you enter in a state of terror, convinced you will do poorly. The dimensions of restructuring of the mind, that is, the activity of the brain as manifested in behavior, have been explored by many cultures. Restructuring of the mind seems to allow "mutations" in what the anthropologist Levi-Strauss has called a culture's "word-built world."

The remarkable achievements of fire walkers under the most rigid scientific observations seem to be one example of a suspension of "reality." Others are perhaps related to the miracle cures at various shrines and the extraordinary powers of certain people of uncommon charisma. It seems possible that we are about to experience a new perception of "reality." Since almost everyone agrees that a fundamental attitude change must occur before any significant progress can be made toward solving the population-environment-resource crisis, this could be the breakthrough that makes attitude and behavioral changes possible.

Social engineering Although almost all nonhuman animals must depend in part on slow, genetic evolution to adapt them to many environmental changes, human beings alone can adapt through evolution of a complex culture (Chap. 15). At the moment, cultural evolution is largely undirected. There is a possibility that **social engineering** could direct cultural evolution in the future. Biological and ecological engineering will have major effects on evolution if they ever are used to any great extent. Social engineering has already been used on a limited scale and has been remarkably successful in the short term. In only a few years, Adolph Hitler was able to alter the culture of one of the most scientifically advanced and literate countries in the world so that genocide on an unprecedented scale was committed and the entire future of world history was changed by a devastating war. Many other negative examples of social engineering will appear if you thumb through the pages of any history book. On the more positive side, the People's Republic of China has used social engineering to change the culture and habits of its people—apparently to their benefit and that of the environment. People in today's mainland China are mostly well fed, have adequate medical care, and have been able to slow their population growth. Even so, not every person has approved of the methods used, and many have moved to Hong Kong and other areas.

Social engineering, then, is quite capable of modifying the behavior of masses of people in a comparatively short time. It is certain, however, that the methods used must be culturally specific. What worked in Germany probably would not work in China, although there are similarities which the reader can find by pursuing the subject. Because of this, designing a culture for the whole world is quite different from designing one for certain countries. Moreover, one would have to think in terms of designing a culture for the *future,* yet we do not *know* what the world of the future will be like in spite of our forecasts and speculations. It seems, therefore, that it would be the rashest foolishness to design a culture for the future. We do not even know what kind of society can best survive the coming age of scarcity and crises which we have forecast.

It appears that one thing to do is see that local or *worldwide* social engineering is *never* implemented. If civilization survives, it will have passed through a period of selection of cultural attributes which are successful. It would appear wise to apply to human cultures the same principles we should apply to plant genetics. Let us allow the survival of *diversity* everywhere possible so that there is a greater chance of some civilization somewhere in the world being able to adapt to conditions of the future. Let us let diversity of cultures survive because each has different viewpoints about the universe and is a treasure house of ideas and possibilities.

A culture represents the interaction of the genetic potential of people with their environment and their history. We do not expect Australian aborigines to become pianists—their environment lacks pianos. We do not expect Joni Mitchell to know her exact kinship relationship to more than 100 people—she was not raised in an aboriginal environment where kinship is of extreme importance. This does not mean that an aborigine could not, if raised in a different environment, become a fine pianist. And it does not mean that Joni Mitchell, if she had been reared as an aborigine, would not know the kinship system of her group backward and forward. Nevertheless, we do not have the faintest idea why aborigines place so much emphasis on kinship or Europeans and Americans on piano playing. In fact, we can only speculate on how any culture evolved. Perhaps the restructuring of the mind of which we have spoken is a chance occurrence like a mutation. There may be some natural selection for cultural attributes of a practical nature. For religious and artistic characteristics, however, there may be simply some kind of cultural momentum similar to gene fixation in genetic drift.

Just as the pressure of human populations is leading to loss of diversity in natural ecosystems, so is the pressure of overpopulation leading to the destruction of cultures and cultural diversity. Technological society with mass transportation and communication which threatens to turn the world into a global village will also lead to increasing loss of cultural diversity. However, hybridization of ideas from various cultures may lead to creative new approaches to the problems of humanity. Rather than designing a culture, it would seem preferable to keep lines of communication open so that cultures can hybridize instead of being lost forever. If we cannot preserve a culture intact, at least we can save some of its ideas and art.

17-13 Many people are attempting to forecast the future

At the beginning of this chapter, we indicated that one reason most forecasts are uncertain is that they do not include all the variables. Recently people have begun to list variables which will affect the future of society and to attempt to make forecasts based on a number of variables. Table 17-2 lists some of the more important variables that will affect the future of human society. Perhaps the most famous study which attempted to forecast the future is *The Limits to Growth* study done at MIT (Fig. 17-13). This study used only the first five of the variables listed in Table 17-2. Even so, a very complex computer model was necessary. Another study by Kenneth Watt asked only four basic questions and came up with 16 possible scenarios for the future.

Almost everybody agrees that these systems are still *too* simple. Technological optimists say that is why the studies all imply that both economic and population growth must cease if civilization is to survive. If they were more complex, however, there is no certainty

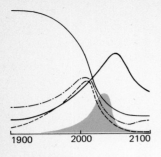

FIG. 17-13 A graph resulting from a computer simulation done at MIT. In this model, nonrenewable resources fail about the year 2000. Following this, food per capita, industrial output per capita (related to capital in Table 17-2), and pollution fall rapidly. Population continues to increase for about 50 years and then crashes. Vertical axis shows relative level of variables. (*After D. H. Meadows, et al.:* The Limits to Growth, *Universe Books, New York, 1972.*)

TABLE 17-2 MAJOR TRENDS IN VARIABLES AFFECTING THE FUTURE

VARIABLE	TREND IN 1975
Population	Increasing exponentially in most countries
Capital	Declining due to inflation
Food	Reserves lowest in 30 years
Nonrenewable resources	Exponential increase in use
Pollution	Exponential increase
Market saturation	Occurring in developed countries
Differences between developed and underdeveloped countries	Increasing
Rising expectations in undeveloped countries	Increasing
Energy use	Increasing exponentially
Economic interdependence of nations	Increasing
Nuclear, biological, and conventional arms races	Balance of terror
Weather trends	Changing rapidly
Epidemic disease	Appearing in many underdeveloped countries
Famines	Increasing in 1975
Urbanization	Increasing exponentially
Species extinction	Holding steady at one to two major species per year
Ecosystem simplification	Increasing exponentially, especially in tropics
Success of international conferences	Dubious
Technological breakthroughs	Possible fusion power in 20 to 30 years

that the results would be any different. You can make your own projection just by looking at the variables in Table 17-2. It is quite obvious that these trends, if taken all together, indicate a high probability of such things as more famines, widespread disease, and breakdown of the life support systems of our planet. All these will be accompanied by human strife, including riots, looting, and perhaps conventional and nuclear wars. These can be regarded as corrective effects by which an overextended system could regain some steady state. Nevertheless, from our personal point of view they are undesirable and should be avoided if possible. People are trying to alter these trends. There is increased interest and publicity about pollution and population problems. International conferences are being held to study these questions. Individuals are writing to their representatives in Congress, boycotting polluters, running for office, and working for candidates for office. In the next chapter, we explore what your role in the future may be.

QUESTIONS FOR REVIEW

1 As the human population becomes larger and larger, more and more people must live on floodplains despite the inevitable danger of doing so. What is one major reason why today, as in the past, many people *choose* to live on floodplains?
2 What is vector control, and why has it become difficult or impossible in some areas of the world?
3 What do these terms mean: ecocatastrophe, hypothesis, limit of waste heat, amniocentesis, amniotic fluid, "normal period of 1930–1960," variable, megaton bomb, charisma, "global village"?
4 What is fertilizer pollution, and what are two dangerous examples of it?
5 What disturbing biological harvest is now being reaped that was sown in the atmospheric testing of nuclear bombs two decades ago?
6 The atmospheric concentration of CO_2 is increasing slightly each year, and more atmospheric CO_2 should mean a greater greenhouse effect. The greenhouse effect raises temperatures—yet the mean surface temperature of our planet has been *dropping* in recent years. To what do some meteorologists attribute this drop in temperature? Why are some experts worried about both the increased CO_2 level and the temperature drop?
7 What is some of the evidence indicating that human aggression is not *innate* and that explanations of human conflict in terms of *human territoriality* must be viewed with suspicion?
8 What cultural changes do you foresee in American society by the year 2000? Are these ideas speculations or forecasts? How many of them can you put into the forecasting ("If _____, then _____") form?

READINGS

Bronowski, J.: *The Ascent of Man*, Little, Brown, Boston, 1973.
———: *Science and Human Values*, rev. ed., Harper & Row, New York, 1965.
Bryson, R.: "A Perspective on Climatic Change," *Science*, **184**(4138):753–759, 1974.
Clarke, A.: *Profiles of the Future*, Bantam, New York, 1971.

Daniels, D. N., M. F. Gilula, and F. M. Ochberg: *Violence and the Struggle for Existence,* Little, Brown, Boston, 1970.
Ellul, J.: *The Technological Society,* Knopf, New York, 1964.
Farvar, T., and J. Milton (eds.): *The Careless Technology,* Natural History Press, Garden City, New York, 1972.
Gofman, J. W., and A. R. Tamplin: *Poisoned Power, The Case Against Nuclear Power Plants,* New American Library, W. W. Norton, New York, 1974.
Halacy, D. S.: *Genetic Revolution—Shaping Life for Tomorrow,* Harper & Row, New York, 1974.
Heilbroner, R. L.: *An Inquiry into the Human Prospect,* Norton, New York, 1974.
Henry; J.: *Culture against Man,* Vintage Books, Random House, New York, 1963.
Holdren, J. P., and P. Herrera: *Energy,* Sierra Club Books, San Francisco, 1972.
Judson, H. F.: "Fearful of Science," *Harpers,* **250**(1498):32–41, 1975.
Kuhn, T. S.: *The Structure of Scientific Revolutions,* 2d ed., University of Chicago Press, 1970.
Lunde, D. T.: *Murder and Madness,* Freeman, San Francisco, 1975.
McClary, A.: *Biology and Society—The Evolution of Man and His Technology,* Macmillan, New York, 1975.
McPhee, J.: *The Curve of Binding Energy—A Journey into the Awesome and Alarming World of Theodore B. Taylor,* Farrar, Straus & Giroux, New York, 1974.
Meadows, D. H., et al.: *The Limits to Growth,* Universe Books, New York, 1972.
Mesarovic, M., and E. Pestel: *Mankind at the Turning Point—The Second Report to the Club of Rome,* Dutton, New York, 1974.
Mostert, N.: *Supership,* Knopf, New York, 1974.
Odum, H. T.: *Environment, Power and Society,* Wiley-Interscience, New York, 1971.
Ornstein, R. E.: *The Nature of Human Consciousness: A Book of Readings,* Freeman, San Francisco, 1973.
———: *The Psychology of Consciousness,* Freeman, San Francisco, 1972.
Pearce, J. C.: *The Crack in the Cosmic Egg—Challenging Constructs of Mind and Reality,* Julian Press, 1971.
———: *Exploring the Crack in the Cosmic Egg,* Julian, New York, 1974.
Pirsig, R. M.: *Zen and the Art of Motorcycle Maintenance—An Inquiry into Values,* Morrow, New York, 1974.
Ruddle, F. H. and R. S. Kucherlapati: "Hybrid Cells and Human Genes," *Scientific American,* **231**(1):36–44, 1974.
Sage, W.: "Choosing the Good Death," *Human Behavior,* **3**(6):16–23, 1974.
Samples, R. E.: "Learning with the Whole Brain," *Human Behavior,* **4**(2):16–23, 1975.
Singer, S. F., (ed.): *The Changing Global Environment,* D. Reidel Publishing Co., Boston, 1975.
Sui, R. G. H.: *The Tao of Science—An Essay on Western Knowledge and Eastern Wisdom,* M.I.T., Cambridge, Mass., 1957.
Uetz, G., and D. L. Johnson: "Breaking the Web," *Environment,* **16**(10):31–39, 1974.
Watson, J. D.: "Moving toward the Clonal Man," *Atlantic,* **227**:50–53, 1971.
Watt, K. E.: *The Titanic Effect—Planning for the Unthinkable,* Sinauer Associates, Stamford, Conn., 1974.
Wylie, P.: "Cultural Evolution: The Fatal Fallacy," *Bioscience,* **21**(13):729–731, 1971.

18

YOUR ROLE IN THE FUTURE

SOME LEARNING OBJECTIVES

After you have studied this chapter, you should be able to

1. State one yardstick which sociologists use to measure the impact of individuals on their environment, and show why Japan—whose population, area, and natural resources are all much less than China's—by this yardstick nevertheless has a total environmental impact that is more than twice China's.

2. Explain how the life style one chooses affects one's environmental impacts; give a real or hypothetical example of a life style with a relatively great, negative environmental impact, one with a moderately negative impact, and one which has a moderately negative impact in some respects and a rel-

atively great, positive impact in other respects.

3. Explain, with examples, why it is that up to a certain point an increase in population means a decrease in per capita energy con-

sumption, but beyond that point further population growth means an increase in energy use per capita; and then state the practical implications of this relationship for (a) increased population growth, (b) zero population growth, and (c) negative population growth, that is, declining population.

4. List at least five ways in which you can reduce your negative environmental impact.

5. Explain why planned, gradual change is often better than sudden, unplanned change; and why group action, not just individual action, is often necessary for effective change.

6. List some of the things that you, as a citizen, can do to strengthen the *ecological point of view* in politics and government.

CHAPTER EIGHTEEN

By now you should be able to see that humanity's future depends a great deal on the actions of individuals today. If we continue to let our population increase unchecked and to destroy the ecosystems upon which our lives depend, the most grim of the forecasts of the future of our species cannot help but come true. The laws of biology and physics cannot be revoked by *any* technological advance. We may even say, "Who cares about posterity; it never did anything for me." But time is now so short for *Homo sapiens* that people born today will live most of their lives in the twenty-first century. That is, our children will live their days in that world of over 8 billion, and perhaps 15 billion, people with greatly impaired freedoms and impoverished environment (if we are so lucky as to avoid famine, plague, or nuclear war). Besides, if you were twenty in 1975, you will be forty-five in the year 2000, so *it is your future too.*

YOU AS AN INDIVIDUAL

Perhaps you think there is nothing you can do about the future. *Your* individual actions, however, *will* help to determine the future of human society. Society is made up of individuals and it is their combined actions which lead to the conditions society as a whole experiences. Society is also given to following its leaders; you may have an opportunity for leadership.

18-1 Individuals are genetically unique

There is no question that people's genetic endowment plays a major role in shaping human behavior. People seem to vary in their genetic capacity to solve mathematical problems, express themselves verbally with facility, coordinate the movements of their bodies, compose music, or create aesthetically pleasing paintings or sculpture. Because *you* are genetically unique (unless you have an identical twin), *you* are uniquely valuable to society. The combination of abilities and energies which you have may never appear on earth again. Everywhere on earth there are people whose uniqueness harbors the creative answers that will make it possible for civilization to continue. A fundamental change in human consciousness is apt to involve agriculture, building construction, child care, cooking and cleaning, etc., as well as formal disciplines such as science, art, religion, politics, literature, and economics in a total reshaping of values and relationships. Therefore, no matter what you do in life, you will have an opportunity to affect the future. Even if you decide to sit back and enjoy yourself while being part of the problem, your action still affects the future. As one ecologist has said, "We can never do nothing."

Homo sapiens is a "sight" animal and so tends to focus more on the color of individuals than on their voice pattern or odor. Nonetheless, we are not condemned to be enslaved by our genetic heritage. Many of us have learned that biologically there are no separate races, and we treat people accordingly. You can respect the cultural diversity of people who are different from you.

Men and women are perhaps genetically different in their ability to express love, hate, envy, and compassion. In different cultures the roles of men and women have differed widely. While there is every reason to give the sexes equal opportunity, there is no reason not to utilize any special talents one sex or the other possesses. For example, it has been shown that, in American society, female police do better at resolving family fights than male officers. In many societies today, the potentials of one sex to do jobs ordinarily assigned to the other sex are unknown. They have never had the opportunity to try.

Certainly it is time to let people try different roles for which they are appropriately trained. Better job opportunities for women could lead to a drop in population growth. Women who have careers have fewer children than women who stay home. The presence of more women in government could have an important influence on aggression among governments. Studies have shown that when mixed groups of men and women are crowded together, the group members are significantly less aggressive than in an all-male group (Sec. 1-12). Presently, our nations are being ruled almost entirely by men and postmenopausal women. More young women in governments, armies, and businesses might lead to positive changes in the behavior of such groups. Presently the leadership and creative potential of almost half the human race is being wasted. If you are a man, you can actively support the rights of women. And if you are a woman, you can aspire to jobs which, although traditionally done by men, offer you a chance to make a difference as an individual.

18-2 The life style you choose affects your environmental impact

The deterioration of the earth's environment is the result of the impact of each individual on the environment multiplied by the number of individuals. Suppose you could choose any life style you wished. You could, for example, choose to be a Buddhist monk or nun of certain sects. You would be forbidden to have money, and the only food you would be allowed to eat is that placed in your begging bowl by the pious. You would own your begging bowl, a robe, a razor, a needle and thread, and that is all. Your presence on earth would cause very little deterioration of the environment. On the other hand, you could choose to be an American millionaire. You would own several houses and have numerous automobiles and perhaps a jet plane. You might own several polluting factories and have a half dozen children—all eventually with their own houses, cars, and polluting factories. It is obvious that you would contribute much more to environmental deterioration than the Buddhist monk or nun.

How can we measure the impact of individuals on their environment? Ecologists think that a reasonable approximation of the environmental impact of an individual is the amount of energy he or she consumes. Energy is used in manufacturing. Energy is used in lighting and transportation. Energy is used in agriculture. Energy is used in mining and construction. There is no doubt that the amount of energy an individual uses is closely related to the impact of that individual on society and the ecosystems of the world.

Calculating energy consumption Let us look at how much energy an American used on the average in a recent year. In 1970, an American used, on the average, the amount of energy equivalent to that which could be generated by the burning of 12.3 tons of coal. Look at Table 18-1, which shows the environmental impact of various nations. The first column gives the 1972 population (N), in millions, of a representative series of countries. The second column is an approximate measure of the environmental impact of the average individual in that country (I), which is an estimate of energy consumption of each individual (in tons of coal equivalents). This is a

TABLE 18-1 NATIONAL CONTRIBUTIONS TO ENVIRONMENTAL DETERIORATION

COUNTRY	N	I	E $(N \times I)$*
United States	209	12.3	2,564
Sweden	8.2	6.9	57
United Kingdom	57	5.9	334
Netherlands	13	5.6	74
Germany (West)	59	5.6	333
Soviet Union	248	4.9	1,213
Japan	106	3.5	374
People's Republic of China	786	0.22†	173
India	585	0.21	123
Burma	29	0.07	2
Egypt	36	0.29	10
Burundi	3.8	0.01	0.04
Kenya	11.6	0.17	2
Uganda	9.1	0.08	0.7
Zambia	4.6	0.06	0.3
South Africa	21.1	3.0	63
Colombia	22.9	0.64	15
Chile	10.2	1.3	14
Argentina	25	1.9	46
Costa Rica	1.9	0.38	0.7
Cuba	8.7	1.1	10
Haiti	5.5	0.05	0.3

*Small variations caused by rounding.
†Rough estimate.
SOURCE: UN Statistics.

very rough measure since, among other things, it does not consider the differing impacts of various energy-consuming technologies. But for our purposes it is sufficiently accurate. The third column is our index of the amount of global environmental deterioration (E) caused by that country. It is calculated by multiplying together the first two columns so that $E = N \times I$.

Notice, for instance, from column E, that although the United States has less than one-half the population size of India, it contributes (by this measure) more than 20 times as much to global environmental deterioration. Sweden and Cuba are about the same size, but Sweden applies almost six times as great a destructive force to the ecosystems of the planet. Column I can also be very instructive. I implies, for example, that the birth of each child in the United States represents roughly 60 times more ecological threat than an Indian baby, 175 times that of a Burmese baby, and well over 1,000 times that of an infant born in Burundi.

But that makes sense, doesn't it? We are a good example of an *effluent* (waste-producing) society, with almost 120 million motor vehicles polluting the atmosphere and burning precious fossil fuels, and huge plants turning out billions of plastic containers, aluminum beer cans, and assorted other items which quickly turn into rubbish. We make most of the world's biocides and loose large amounts of mercury into the environment. The Indians do not even have a start at most of these activities, and even European nations are still far behind us.

"But wait a minute," you may say, "look at the Dutch. The Netherlands consumes nearly half as much energy per capita as the United States and yet the Dutch have such a neat and clean country, with few environmental problems." Well, if you could see Amsterdam, you might change your mind about the "neat and clean." Even if it were the cleanest city in the world, this would only indicate that the Netherlands reaps the benefits of industrialization while shoving many of the costs off on others. For instance, the Netherlands imports much of its food (it is the second largest per capita importer of protein), most of its fiber, almost all its industrial minerals, and all its chemical fertilizer. Therefore, all the environmental deterioration associated with the agricultural, fishing, and mining operations necessary to produce and transport those imported materials must be chalked up to the Netherlands' account. Furthermore, the Netherlands also imports about half its energy, so that the environmental costs of producing it are borne elsewhere!

Effect of population size on per capita power consumption Too few people realize that reduction of population size will have a disproportionate effect on environmental deterioration. This is because *per capita energy consumption is itself related to population size*. Up to a certain point, a larger population will mean less energy consumption per capita. More people in a society permits specialization, which is energetically efficient. Assembly-line production of cars by thousands of people, for instance, requires less energy per car than construction of each car singly by a small group of people. Such savings of energy are called **economies of scale.** Beyond a certain population size, however, an opposite effect takes over. More and more energy per capita is required as the population grows. A simple (and imaginary) example will show you how this works.

Suppose two cities, each with a population of 1 million people, are located 10 miles apart and are connected by 10 miles of highways (Fig. 18-1a). Now, what happens if a third city of 1 million grows up 10 miles from the first two (Fig. 18-1b)? The total population has grown 50 percent, from 2 million to 3 million. But the additional amount of energy required will be much more than 50 percent. One reason is that 20 miles of new highways will have to be built, an increase of 200 percent (10 miles to 30 miles). Building highways requires energy. Another is that when the first two cities were built,

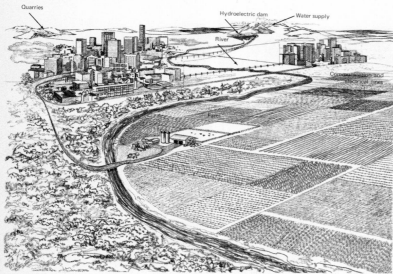

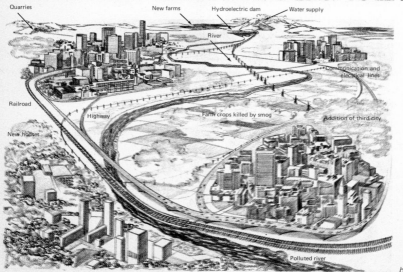

FIG. 18-1 Per capita impact is itself dependent on population size. In this drawing, two cities (a) have an impact I on their environment. In (b) the population of the area has increased by 50 percent and a new city has been founded: the impact on the environment is much greater than I plus 50 percent of I; thus per capita impact has increased.

most of the rock needed could be mined nearby. But the two cities exhausted the local quarries, and now rock must be moved from a great distance. Moving rocks takes energy. And, of course, the flow of the local river is enough to supply the first two cities, but not the third. A dam must be constructed in distant mountains, pipes constructed, and water pumped to the new city. Construction and pumping require energy.

All three cities now dump their sewage into the river. Its natural ecological system could degrade the wastes of two cities, but not three. Sewage treatment must be started, and sewage treatment requires energy. The 50 percent increase in smog makes the air so poisonous that local farmers lose their crops. They must move, and distant, less fertile soil must be cultivated to make up for the loss. Moving and establishing new farms require energy, and the new farms on less fertile soil require a higher level of energy input than the old ones did.

Thus you can see that a 50 percent increase in population can have much more than a 50 percent greater impact on the environment (we did not even count in the farms paved over to make the city and highways). Similarly, a 50 percent reduction in population can result in more than a 50 percent reduction of environmental impact.

18-3 Your decision on your family size affects the size of the population as a whole

In light of the impact of increasing population size which we have just explored, you have a personal decision to make: How many children should you have? If you have more than two, you will be more than replacing yourself and your spouse. If all couples have just three children—no more, no less—the population of the United States will continue to grow. That growth of population is likely to adversely affect the quality of your children's lives. It may even affect their chances for survival. In the United States in 1974, people were choosing to have slightly fewer than two children per couple. This does not mean that the population has stopped growing. As mentioned in Chap. 17, however, there is a momentum to population growth because many children already born will grow up and have children of their own before their parents die. Even if this low reproductive rate continues (which it may not), the population of the United States will continue to grow well into the twenty-first century (Fig. 18-2). Some people are choosing not to have children at all; this is another possible reproductive decision you may want to make. Whatever you choose to do, it will have an effect on the future, not only directly in the impact your children have on the earth but also indirectly in the influence your decision has on other people. People *are* influenced by the decisions of their friends. You may want to join a national organization such as ZPG (Zero Population Growth), which is dedicated to halting growth of the American population as soon as possible.

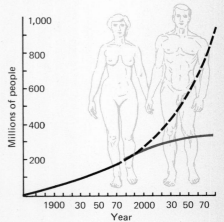

FIG. 18-2 United States population growth assuming the average family size is two children (colored line) or the average family size is three children (black line).

What if you came from a large family and always wanted a large family? You can adjust your ideals and have a small family or you can go ahead and have a large family. If you do the latter, however, you must remember that someone else is bearing the environmental cost of your reproductive decisions. If you really enjoy large families, you can take care of foster children or adopt children. Or you can choose one of the communal life styles in which all the adults in the group look after all the children as needed. In one of these groups, you can be surrounded by children without having *any* of your own Whatever you decide, you can be sure your decision will make a difference for the future.

18-4 You can try to reduce your environmental impact

There are many simple ways you can reduce your own environmental impact without greatly inconveniencing yourself. In fact, you will probably be healthier and happier if you follow some of the following suggestions. Do you drive to work or college? How about taking a bus or train, or bicycling, or even walking if it's close enough. Do you buy highly processed, overpackaged foods? How about resolving to buy nothing processed, canned, or frozen for a week. See if what happens to your diet is not an improvement. Do you keep the TV or radio on when you are not watching or listening? Turn it off. Do you leave lights on that you do not need? Turn them off. Do you buy fancy cosmetics and cleaning products in aerosol cans and plastic bottles? Try soap and water. Do you throw away your aluminum beer and soft-drink cans, your old newspapers, your nonreturnable bottles? Recycle them or, still better, buy only returnable bottles and share a newspaper with a friend. We could go on and on, but you get the idea. Now, what would happen if everybody suddenly stopped using energy-wasting products? Environmental impact would be reduced greatly. But the entire American economic system would also be disrupted since many people make their livelihood at jobs which produce the wasteful consumer goods we have just described. Therefore, this must be a gradual transition, with these people finding other ways to make a living.

Farming Those people who would be put out of work by the proposed reduction of impact will eventually be put out of work *anyway* by the shortage of energy. In fact, the market is already saturated for many of their products. What can the technologically unemployed do? One way to become independent of the industrial-technological society is to return to subsistence farming. Many people are now attempting to reverse the trend to the city, which we noted in the first chapter, by going back to the farm. This has resulted in many books being published on skills of country life. We have lost touch with the knowledge of our ancestors so that now we need to read and study how to grow plants, how to butcher a hog, how to assist a cow during a difficult calving, how to manage a hive of bees, and many other subjects that our ancestors knew as part of their folk wisdom.

The success of city people who return to the country is varied. Usually they must start with poor land because all the good land belongs to large farmers. If they want to farm organically, they often have to fight the prejudices of their neighbors who may be dyed-in-the-wool pesticide users. In spite of the problems, many people are attempting to return to farming. Also, many people who live in cities or suburban areas are attempting to reduce their consumption. People are planting vegetable gardens in such numbers that there have been shortages of seeds. Even if you live in an apartment, you can have vegetables (for example, carrots and tomatoes) for house plants, and many people are doing just that. People in suburbia are learning about composting, mulching, and rabbit raising and are asking their neighbors if they mind a few chickens in the neighborhood.

18-5 Individuals are concerned with surviving the crises they foresee

In America, the energy crisis of 1973 and 1974 was the first taste of shortage for many people. People too young to remember the Second World War had never experienced shortages, and many of them were shocked into awareness by the energy crisis. People are also concerned with inflation, rising prices, and dwindling supplies of paper, plastics, metals, and other materials. Again, books are being written about how to survive various kinds of crises from monetary ones to the breakdown of civilization. Financial advisors are getting rich by telling people how to invest in whatever the latest fad is, for example, old coins, silver, gold, or art objects, in order to protect the value of their money. Unfortunately, you cannot eat any of these things, and some people are investing instead in a year's supply of food. As an individual, the best thing you can do to prepare for the coming time of shortage and crises is to increase your abilities and knowledge of basic survival techniques. If farming is not your thing, it would be wise to make sure you have some skill which can be traded for food. If a real crisis comes, there will be little market for advertising executives or automobile salespeople. Nearly everyone has or can learn a useful skill. There will be plenty of work for people who can sew, make shoes, fix a plow, or for that matter do anything *useful* or *entertaining* with simple materials and tools available close at hand.

18-6 Knowledge from a diversity of other cultures is available to individuals today

People today are not only living through a time of unprecedented crises, they are also living in a time of unprecedented opportunity. People in the United States can learn about almost any other culture they wish. It is now possible to travel almost anywhere in the world and see first hand how people live there. It is also possible to observe what is happening on the other side of the world, as it happens, on television. Translations of contemporary books are available from many cultures of the world. There are also translations and reprints of historical and religious writings going back to the beginning of written culture of our species. For the first time it is possible to see the human species in perspective: all our various cultural mores and wars and superstitions and sciences. We can discover that the history of our species actually has been one of danger and uncertainty. Other people in other times have faced many of the problems we face today. Famine, plague, and war have always haunted us. The only difference today is the scale on which events may occur: we have gotten ourselves into a position in which we may destroy the entire biosphere. Of all the things we might call ourselves—greedy, stupid, lacking foresight, selfish—perhaps the most useful is adventurous. We are adventurous animals living in what may prove to be the most adventurous time of all. We live, as the philosopher Ortega y Gasset has put it, with "that insecurity which is essential to all forms of life, that anxiety both dolorous and delicious contained in every moment, if we know how to live it to its innermost core, right down to its palpitating vitals." By studying writings of the past and present of other cultures as well as our own, we can gain a perspective on our place as individuals in the universe.

For example, the study of different world views may lead us to that fundamental restructuring of our ideas, which was discussed in the last chapter, that seems to occur when cultures evolve. We find that some ideas from ancient cultures fit today's circumstances very well. Consider Isaiah 5:8 from the King James Bible: "Woe unto them that join house to house, that lay field to field, till there be no place, that they may be placed alone in the midst of the earth!" That could have been said by a modern environmentalist bewailing a new subdivision. On the other hand, world views of other cultures may seem very foreign to us, yet may express the essence of what modern science understands in another way. For example, the theory of "interdependent origination" has many expressions in Buddhist philosophy. The Tibetan view binds it up with the instantaneousness and impermanency of all phenomena, consisting of discontinuous flashes of energy. Alexandra David-Neel writes: ". . . nothing is produced by one single cause; the combination of several causes is always necessary to bring about a result. The seed without the cooperation of earth, dampness, light, etc., will never become a tree." This is the essence of modern ecology, yet it also contains a grain of that other world view. Knowledge of other world views may help to transform the world view of individuals of the industrial age into the necessarily changed viewpoint which would enable them to cope with the coming postindustrial era and the crises between now and then.

How can you tell an authentic philosophy from a phony soft sell by a get-rich-quick spiritual teacher? The interest shown by many people in American society in Eastern philosophies and religions has led to numerous groups and individuals setting themselves up as offering *the* unique insight into truth. It is now possible to find almost any religion or spiritual discipline of the world being practiced somewhere in America. You can even take yoga at many colleges. Among all this diversity of world views how can one choose? One *can* consider the kind of people who belong to a particular group. Are they your kind of people? One can ask what consequences will arise from the world view being taught. Will the consequences be like the negative aspects of the drug subculture? Are these consequences that you would want to be responsible for?

There is no easy way to choose between useful and useless world views. It is important to remember that there are *many*. Therefore, any one taken too seriously can lead to an end of your ability to be receptive to new ideas. Some of the popular movements today do not allow for this perspective or for change. The history of science seems to show that it is *always* an error to assume the existence of absolutes. Throughout this book we have stated current theories and laws of science. Within our present world view, they seem unlikely to be repealed. They are not absolute, however, because we cannot ever study them in all places at all times in all possible meanings they may take. In *our* time, for people with *our* world view, they *seem* to be applicable and useful.

YOU AS A CITIZEN

You may have adopted a low-impact life style and developed a new nontechnological world view. None of this will necessarily change society. In fact, as we have seen, if all people in our society suddenly decided to change their habits of consumption overnight, there would be chaos.

18-7 Individual action is often insufficient for change

Consider the automobile, for example. As it is now manufactured and used, the automobile causes some of our most serious problems. Automobiles pollute the air with their exhausts, with particles and gases from the friction of their tires against the road, and with particles of asbestos (which can cause lung cancer) from their brakes. They are a major drain on our precious fossil fuels, and they pour large quantities of heat into the environment. Much valuable land has been paved over for highways and parking lots. A great deal of energy is used in obtaining the raw materials from which automobiles are constructed and in the process of their manufacture. A substantial societal effort is required to care for the roughly 5 million Americans injured by automobiles annually. And, of course, the "cost" of the 55,000 annual deaths attributable to cars is incalculable.

Suppose, suddenly you and everybody else decided not to buy or drive cars. The problems we have just enumerated would be translated into other social problems. In places such as Los Angeles, many people would starve to death because the food distribution system depends on cars and trucks. People in Los Angeles could not get to their jobs because there is no suitable mass transit system. During the recent gasoline crisis, people there were afraid of not being able to get gas to get to work. Indeed, many people everywhere would find that their jobs no longer existed because they depend directly or indirectly on cars and trucks. In fact it is hard to picture an area of our economy which would not be severely damaged. Such sudden individual action would not accomplish a worthwhile goal. People will have to get together and make plans, before petroleum supplies run out, to phase out automobiles gradually and replace them with other modes of transportation. Those groups of people who would be out of work will have to have help to find other things to do.

18-8 Fruitful interaction among people is necessary for societal action

In order to prevent the demise of the automobile from creating social chaos, people must work together to plan alternatives. Individuals can work together for laws, which might start out by limiting the size and horsepower of automobiles. They might buy only the smallest, least polluting, most economical cars. Meanwhile they can vote for the use of gasoline taxes to provide mass transit. The people of California did just this in the spring election of 1974. Automobile factories could be awarded contracts to design mass transit facilities. People who presently service cars will find other jobs—some perhaps servicing mass transit. If many people work together over a period of time, problems can be solved. If we refuse to work with each other, problems will be solved by default and only with much suffering. To solve the problems of the automobile and the eventual end of petroleum supplies, citizens must work with governments, industries, oil companies, and unions. As long as each of these groups continues trying to get the most money or power from the other, no constructive changes will come about. Oil companies will raise prices; citizens will be unable to afford the big cars built by industry; unions will demand higher wages; prices will go even higher. Government will pretend to regulate the whole situation by appointing to energy study groups oil company executives who contribute heavily to political campaigns—and so on, until the earth is thoroughly polluted and the last accessible resources are used up.

Cooperation is going to be necessary between all kinds of people. There are blacks who think, wrongly, that the ecological crisis is an invention to draw attention from the need for social equality. There are whites who think, wrongly, that our population problems are due to poor blacks. It will not be easy for such people to work together, but if we cannot find ways of working together, everyone will suffer. Think of yourself as being stuck on a spaceship that is in deep trouble. You are going to have to solve its problems with the crew you have. Do you think solutions will be easy to come by if many of the passengers are constantly mistreated by others? Do you think that people who have never had a chance to get a fair share of the good things of life are going to worry about the overall problems of running the ship until some attention is paid to their problems? By ourselves none of us as individuals can right the inequities on our spaceship. We *can* join organizations of others who care and are trying to help equalize the distribution of wealth.

We can also work to learn to cooperate with all kinds of people. If we are to keep our spaceship functioning, everybody is going to have to talk to everybody else, and everybody is going to have to listen carefully. If you are young, how long has it been since you *really* talked with a person older than yourself—even your parents? If you are middle-aged, do you ever *really* talk with young people? Have you ever talked to a very old person? Older people remember when the world was very different and you can learn a great deal from them and they from you. Do you know any children? Are children today any different than when you were a child? The children of today will very shortly be tomorrow's adults; you are going to have to work with them. If you are liberal, meet some conservatives—you are going to have to work with them too. If you seek out variety in your acquaintances, you will not only learn a lot about other kinds of people but also learn how to talk and listen to different viewpoints.

18-9 Fruitful political action requires more than just voting

Certainly you should vote and encourage others to vote. Remember when it comes time to vote the biological principles which we have presented in this book. You can ask candidates about food chains, diversity, and other subjects and very quickly find out if they know enough about biology to make intelligent decisions. Would you want to vote for a representative who had never heard of ecological succession which explains eutrophication? Almost every day when Congress is in session, votes are taken on the floor or in committee on measures, such as strip mining, which will return whole areas of the earth to primary stages of succession. No society can escape the fact that it is dependent on a functioning biosphere. Therefore, it is not very wise to elect officials who have no knowledge of or interest in how the biosphere functions.

Political candidates need many people to work for them if they are to be elected. Telephones must be staffed, envelopes must be addressed, and campaign literature must be folded and stuffed into them. Someone must go door to door to persuade people to vote for the candidate. Often others volunteer to drive voters to the polls on election day. There is a lot you can do and lots you can learn by doing it. Just pick a candidate, show up at his or her headquarters, and say, "I want to work." Recently many important issues have been brought directly to the people on the ballot. The processes of initia-

TABLE 18-2 SOME SELECTED ENVIRONMENTAL ORGANIZATIONS

ORGANIZATION	ADDRESS	PUBLICATION
Audubon Society	950 Third Avenue New York, N.Y. 10022	*Audubon Magazine*
Defenders of Wildlife	2000 N Street N.W. Washington, D.C. 20036	*Defenders of Wildlife*
Environmental Action	Room 731 1346 Connecticut Avenue N.W. Washington, D.C. 20036	*Environmental Action*
Environmental Defense Fund	162 Old Town Road East Setauket, N.Y. 11733	*EDF Newsletter*
Environmental Fund	1302 18th Street N.W. Washington, D.C. 20036	*World Population Reference Sheet*
Federation of American Scientists	203 C Street N.E. Washington, D.C. 20002	*F.A.S. Newsletter*
Friends of the Earth	529 Commercial Street San Francisco, Calif. 94111	*Not Man Apart*
League of Conservation Voters	324 C Street S.E. Washington, D.C. 20003	Publishes voting records on environmental issues
National Parks and Conservation Association	1701 18th Street N.W. Washington, D.C. 20009	*National Parks and Conservation Magazine*
National Resources Defense Council	15 West 44th Street New York, N.Y. 10036	No publication
Planned Parenthood Federation of America	810 Seventh Avenue New York, N.Y. 10019	Annual report and newsletters of local affiliates
Sierra Club	1050 Mills Tower 220 Bush Street San Francisco, Calif. 94104	*Sierra Club Bulletin* and newsletters of local chapters
The Wilderness Society	1901 Pennsylvania Avenue N.W. Washington, D.C. 20006	*The Living Wilderness*
Zero Population Growth	1346 Connecticut Avenue N.W. Washington, D.C. 20036	*ZPG National Reporter*

tive and referendum allow measures to be placed directly on the ballot. Many important environmental proposals have become law this way. So have a number of political reforms. Investigate the laws of your state and local area; with a group of like-minded citizens, you may help to get an issue on the ballot.

Increasingly, the courts are being used to protect the environment. Presently, federal law requires an **environmental impact statement** to be prepared for important projects. Many states also have environmental impact laws. Individuals and environmental groups have sued various branches of the government when they felt environmental laws had been violated. Such lawsuits require volunteers to round up expert witnesses, make phone calls, gather information, inspect government records, and finally raise funds to pay for court costs and legal fees. It is probable that an environmental law suit of some sort is going on in your area right now. One lawyer has suggested that, in the future, laws may change so that an individual or group may be appointed legal guardian of a natural area. Such guardians would have rights of inspection and be able to represent their "wards" at administrative and legislative hearings.

There are now local, state, national, and even international environmental organizations of every shape, size, and variety. (There are also organizations which pretend to be saving the environment while actually protecting special interests.) Whatever your interest or tastes, you can find an environmental organization to join. Some are content with your membership fee. Others expect you to write letters, talk to people, circulate petitions, or even join in demonstrations. Table 18-2 lists a sample of environmental groups. There are now so many that it is not possible to list them all.

The quality of governmental decisions should be in large part a reflection of the citizens who make up the society. That is where you come in. Your vote, the example you set by your behavior, your persuasiveness with your friends, and your degree of political involvement will help shape your future and that of humanity.

YOU AND YOUR CAREER

As was noted earlier in the chapter, *you can never do nothing*. Unfortunately, many people must take whatever job is available. They may find themselves in a polluting factory making throw-away plastic wrappings or in a mill which is dumping wastes into a river. They can do little to change these situations without quitting or being fired. People in such positions *can* realize that the economic and social system has made them into unwilling parts of the problem. As fossil fuels disappear, it is hoped that new jobs will appear that will enable people to make useful and nonpolluting products.

18-10 Whatever career you choose, you will have an influence on the future

However, it is possible for many people to use their jobs to have a positive impact on environment. In 1970 an airline pilot was fired for refusing to carry out a polluting, fuel-dumping procedure. He received a lot of publicity and the airline decided to change the procedure and rehire him. A housewife in our nation's capital has been carrying on a one-woman crusade against more freeways. A San Francisco clothing designer has been financing campaigns to halt thoughtless development in the San Francisco Bay area. The class speaker at a college graduation made national headlines with her dramatic statement that the best way *she* could help an overpopulated world would be by having no children. Two physicists at the Lawrence Radiation Laboratory, funded by the Atomic Energy Commission, had their jobs threatened when they pointed out that the AEC standards for permissible levels of radiation pollution were set much too high. But eventually the AEC was forced to admit they were right. And there was the government employee who lost his job for exposing the way the Air Force and a large aircraft manufacturer were running up huge bills at the taxpayers' expense.

Many people have risked fines or imprisonment in their attempts to help. Some have thrown themselves down in front of bulldozers, chained themselves to trees, plugged up polluting smoke stacks, and dumped outflow from a polluter's plant on the rug of the main office. For the most part, however, people are quietly working away, writing their representatives in Congress and the heads of government agencies when they see environmental abuses in the course of their jobs. It is not necessary to lose your job to make a difference.

Choosing a career specifically related to solving population-resource-environment problems Some careers will put you right in the midst of opportunities to work for solutions to the problems of the future of the human species. Doctors and lawyers can easily move into environmental medicine or environmental law. Engineers often find themselves writing environmental impact reports and can often contribute to environmental engineering. Geologists, biologists, and physicists can use their special knowledge directly in teaching about the workings of nature, in researching new environmentally sound ways of doing things, in influencing politicians, and in acting as expert witnesses in environmental lawsuits. Chemists can spend time devising safe and environmentally acceptable ways of producing useful chemicals. Microbiologists can study biological reactions that lead to more efficient disposal of waste and production of novel foods and ways of coping with new strains of disease organisms. People in agriculture can attempt to return to ecologically sound methods of producing foods.

Special careers will undoubtedly appear in the future which will give people opportunities to work on solutions to the population-resource-environment crisis. We mentioned in the last chapter the possibility that there will be ecosystem managers in the future with knowledge and responsibility equivalent to that of physicians but in relation to the environment. Environmental law has already become a recognized specialty. Some people are now professional administrators of environmental organizations; whole institutes may be devoted to ecological research in the future. There may be careers in areas as yet undreamed of as research continues in biology.

Although a number of colleges have specialties in environmental education, there are at present few jobs open to graduates of such programs. Many of these programs are not sufficiently biologically oriented and deal with the so-called human environment only. There may be possibilities that you can help by choosing to study biology as your major. The future possibilities for biology, however, may be either very great or very poor. These possibilities depend on whether our society wishes to put a sincere effort into solving environmental problems, and if society chooses, biologists will be in great demand. They will be needed to identify problems and find solutions and, of course, to teach. One reason for the present problems of humanity is that society too long ignored the biological constraints on human activity.

Being aware of the human and biological prospects We have shown how our species has come from small groups of hunters and gatherers roaming a few isolated sections of the earth to the present, when people are everywhere on earth and threaten to damage the working of the biosphere on which they depend. There is no indication that the long journey was either easy or pleasant for our fellows who participated in it. In fact, there is much evidence to the contrary. Our ancestors lived with a very high death rate—they knew most of their children would die before reaching maturity. They had little protection against epidemic diseases and parasites. They had no means of transportation to evacuate areas of natural disaster. Famines have been common throughout human history. People have always lived with uncertainty, fear, and insecurity.

Today values are such that many want only security and an easy life. Evolutionarily, however, we are not adapted to the "easy life" which the industrial revolution has made possible for us. We substitute activities such as skydiving, desert motorcycling, and watching sporting events for fighting wild beasts, seeking food, and battling storms. People seem to need adventure and in the world of the future they will have plenty of it. There is no time for despair or gloom. There is too much that should be done to improve the human prospect, to reduce human suffering during the period of transition to the postindustrial age, and to save some of the environment for our descendants to enjoy. A crisis is a dangerous opportunity. The human prospect improves as people gain knowledge about the biological prospect. Whatever your career, you can refuse to despair, reject apathy, and help people past difficult times by your knowledge, stability, and cheerfulness.

QUESTIONS FOR REVIEW

1 Country A, an industrial nation, imports most of its fuel, fertilizer, food, and industrial materials. Assume that it pays fair world-market prices for all these things. Why, then, must it be said that Country A reaps the benefits of industrialization while shoving many of its costs off on others?

2 "You can never do nothing." What does this mean in ecological terms?

3 Can you list some factors which cause certain nations to contribute more than other nations to environmental deterioration?

4 Compare the environmental impact of three cities, each of which has a population of 1 million, and is 10 miles from each of the other two.

5 As it is now manufactured, used, and provided with fuel, the automobile causes some of our most troublesome environmental problems (and it will continue to do so even if cars' gasoline mileages continue to improve). What are several serious, negative environmental impacts of the automobile and its supporting industries in this country?

The auto is presently a necessity for many people as well as a luxury that is abused by many others, and it is currently a mainstay of our national economy. Could the negative environmental impacts of the automobile and its supporting industries be substantially reduced without inflicting serious hardship on many people and without seriously disrupting the national economy? What are some of the steps that could be taken to minimize the social and economic problems that would almost surely result from any major effort to greatly reduce this country's dependence on automobiles?

6 Explain, using Table 18-1:
  a What would happen to the index of environmental impact on the United States if everyone used only half as much energy?
  b What would happen to the United States impact if we cut our population size in half and did not change our energy consumption?
  c What would happen to the United States impact if we doubled our population size and halved our per capita power use?
  d What would happen to the United States impact if we halved both our population size and our per capita energy consumption?

7 Make a list of as many things as possible you as an individual can do to reduce your negative impact on the environment.

8 List as many positive things as possible you can do to influence your friends and society.

READINGS

Angier, B.: *Survival with Style,* Vintage Books, Random House, New York, 1974.

Brown, L. R.: *In the Human Interest: A Strategy to Stabilize World Population,* Norton, New York, 1974.

Daly, H.: *Toward a Steady State Economy,* Freeman, San Francisco, 1973.

David-Neel, A. and Lama Yongden: *The Secret Oral Teachings in Tibetan Buddhist Sects,* City Lights Books, San Francisco, 1967.

Editors of the *Ecologist: Blueprint for Survival,* Signet, New American Library, New York, 1972.

Ehrlich, P. R.: *The Population Bomb,* rev. ed., Sierra Club/Ballantine Books, New York, 1971.

―――, and A. H. Ehrlich: *The End of Affluence—A Blueprint for Your Future,* Ballantine Books, New York, 1974.

―――, and R. L. Harriman: *How to Be a Survivor: A Plan to Save Spaceship Earth,* Ballantine Books, New York, 1971.

Environmental Protection Agency: *Working Toward A Better Environment—Some Career Choices,* E.P.A., Washington, (in press).

Fanning, O.: *Opportunities in Environmental Careers,* Vocational Guidance Manuals, Louisville, Ky., 1972.

Fritsch, A. J.: *The Contrasumers: A Citizen's Guide to Resource Conservation,* Praeger, New York, 1974.

Koberg, D., and J. Bagnell: *The Universal Traveler. A Soft-Systems Guide to: Creativity, Problem-Solving, and the Process of Design,* rev. ed., William Kaufmann, Los Altos, Calif., 1974.

McHarg, I. L.: *Design with Nature,* Doubleday, Garden City, N.Y., 1971.

Olkowski, H., and W. Olkowski: *The City Peoples' Book of Raising Food,* Rodale Press, Emmaus, Pennsylvania, 1975.

Pirages, D. C., and P. R. Ehrlich: *Ark II: Social Response to Environmental Imperatives,* Freeman, San Francisco, 1974.

Schumaker, E. F.: *Small Is Beautiful,* Harper Torchbooks, Harper & Row, New York, 1973.

Stone, C. D.: *Should Trees Have Standing? Toward Legal Rights for Natural Objects,* William Kaufman, Los Altos, Calif., 1974.

Terry, M.: *Teaching for Survival,* FOE/Ballantine Books, an Intext Publisher, New York, 1971.

APPENDIX

METRIC TABLE

UNIT	NUMERICAL VALUE	SYMBOL	ENGLISH EQUIVALENT
Area			
square meter		m^2	10.764 feet
hectare	10,000 m^2	ha	2.471 acres
square kilometer	1,000,000 m^2	km^2	0.386 square miles
Length			
meter		m	39.37 inches
kilometer	1,000 m	km	0.621 miles
centimeter	0.01 m	cm	0.394 inches
millimeter	0.001 m	mm	
micrometer	0.000001 m	μm	
nanometer	0.000000001 m	nm	
angstrom	0.0000000001 m	Å	
Mass			
gram		g	0.035 ounces
metric ton	1,000,000 g	t	2,205 pounds
kilogram	1,000 g	kg	2.205 pounds
milligram	0.001 g	mg	
microgram	0.000001 g	μg	
Time			
second		sec	
millisecond	0.001 sec	msec	
microsecond	0.000001 sec	μsec	
Volume (solids)			
cubic meter		m^3	1.308 cubic yards
cubic centimeter	0.000001 m^3	cm^3	0.061 cubic inches
cubic millimeter	0.000000001 m^3	mm^3	
Volume (liquids)			
liter		l	1.05 quarts
milliliter	0.001 liter	ml	
microliter	0.000001 liter	μl	

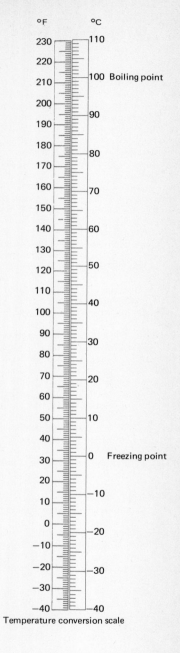

Temperature conversion scale

GLOSSARY

Aborigine An individual belonging to the earliest known group of human beings to inhabit a country; often refers specifically to an Australian aborigine.

Abortion Spontaneous or induced expulsion of the fetus from the uterus during the first two-thirds of the period of pregnancy.

Acetylcholine A chemical produced at the junctions between nerves and muscles and at certain synapses in the brain; serves to transmit nerve impulses to the muscles and to areas of the brain, in mammals, involving alertness.

Active transport The energy-expending process by which cells move substances across cell membranes, often from lower concentrations to higher concentrations against the diffusion gradient.

Adenosine triphosphate (ATP) The major source of usable chemical energy in metabolic processes. When ATP loses one phosphate group to become ADP, usable energy is released.

Aerobic Occurring in the presence of air, specifically, oxygen. The complete respiratory breakdown of compounds to carbon dioxide and water occurs only under aerobic conditions.

Afterbirth The placenta and membranes connected to a fetus which are expelled after a fetus is born.

Alga (plural, **algae**) A photosynthetic eukaryotic organism which lacks multicellular reproductive structures. (The blue-green algae are photosynthetic prokaryotes.)

Algal bloom A population explosion of algae often due to a sudden influx of nutrients (for example, sewage) accompanied by other favorable environmental conditions.

Allele One of the two or more alternate states of a gene that occupy the same position (locus) on homologous chromosomes.

Allergy A disease resulting from overproduction of antibodies against foreign substances, for example, pollen or dust.

Amino acids The subunits from which protein molecules are built; amino acids are nitrogen-containing organic molecules.

Amniocentesis The removal of a sample of the fluid inside the amnion surrounding the fetus by inserting a needle through a pregnant woman's abdominal wall.

Amnion The embryonic membrane which immediately surrounds the fetus.

Anaerobic Able to carry out life activities without oxygen. Some microorganisms not needing air for growth and reproduction are called anaerobes; strict anaerobes cannot live in the presence of oxygen. Also used to refer to the first steps of aerobic respiration, which do not require oxygen.

Anemia A condition characterized by decreased oxygen-carrying capacity of the blood caused by too few red blood cells, too little hemoglobin, or faulty hemoglobin.

Annual plant A plant which completes its entire life cycle in a single growing season.

Antibiotic A natural organic substance which retards or prevents the growth of organisms; generally, substances formed by microorganisms which prevent growth of other microorganisms.

Antibody A protein carried in the blood which is synthesized in response to a foreign substance (antigen) in the body; it helps to inactivate the foreign material.

Antigen A chemical which can cause synthesis of a specific antibody.

Arthropod A member of the phylum Arthropoda, distinguished by jointed appendages, an exoskeleton, and a ventral nerve cord. Examples: insects, spiders, crabs.

Artificial selection Selection done by human beings for particular genotypes of a species to "improve" domestic animals or plants or as an experiment. See also **Selection**.

Asexual A reproductive process that does not involve the union of gametes.

Asteroid One of many comparatively small, planetlike bodies, most of which remain in solar orbit between Mars and Jupiter.

Atoll An island or group of islands consisting of a coral reef surrounding a lagoon.

Atom The smallest unit into which a chemical element can be divided and still retain its characteristic properties.

Autoimmune An immune response of the body against itself; may lead to a disease where, for example, the body damages itself by producing great numbers of antibodies against some of its own proteins.

Axon The long extension of a nerve cell which serves to carry information to other cells.

Bacteriophage A virus that infects bacteria.

Bacterium (plural, **bacteria**) A unicellular, prokaryotic organism lacking chlorophyll a (a type of chlorophyll found in blue-green algae but not in photosynthetic bacteria).

Bee language A form of transmission of information used by honeybees to inform other members of their hive of the distance and location of a food source.

Biocide A substance used to kill living things. Biocides are classified by their target organism, for example, insecticides, fungicides, herbicides, yet they are rarely or never completely harmless to nontarget organisms.

Biodegradable Capable of being broken down by living things.

Biofeedback A technique by which so-called involuntary processes may be learned to be controlled by an individual when that individual is able to observe or sense them by means of special instruments, for example, to control heart rate when a stimulus (sound) is given at the desired rate.

Biogeographic region A region of the earth which contains a taxonomically distinctive group of organisms, for example, the North American biogeographic region.

Biological amplification Concentration of a chemical by successive members of a food chain so that the species at the top of the food chain comes to have the highest concentration of the chemical.

Biological clock An internal system which has the capacity to "measure" time and control the innate rhythms in plants and animals.
Biological control A method of pest control by which pests are controlled by other biological agents, for example, predators, parasites, and diseases.
Biological engineering The ability to modify, repair, or substitute parts of living things.
Biological oxygen demand (BOD) A measure of the organic material in water based on the amount of oxygen consumed by bacteria in water during a specified time under standard conditions.
Biology The study of all living things and their interdependencies.
Biome A group of communities characterized by distinctive vegetation and climate; for example, all grassland areas taken together form the grassland biome.
Biosphere The zone of air, land, and water occupied by living things.
Blight A disease or injury of green plants leading to withering of leaves and cessation of growth.
Bonds, chemical The forces that hold atoms together in molecules. Energy is sometimes released in reactions when bonds are broken and new bonds are formed; thus, bonds are often said to "store energy."
Calorie The amount of energy in the form of heat required to raise the temperature of one gram of water one degree Celsius. The caloric value of food is given in kilocalories (Calories). One Calorie is equal to one thousand calories.
Canopy The top layer of a forest, consisting of the crowns (which often touch) of the tallest trees.
Carbohydrate An organic compound consisting of carbon atoms connected to each other and having hydrogen and oxygen attached in a 2:1 ratio, for example, sugar, starch, glycogen, cellulose.
Carbon dioxide A compound consisting of one carbon atom attached to two oxygen atoms. Carbon dioxide is used by plants in photosynthesis and produced by animals and plants in respiration as well as by the burning of fossil fuels.
Carbon monoxide A compound consisting of one carbon atom attached to one oxygen atom. Carbon monoxide is produced by incomplete combustion of fossil fuels and can bind irreversibly with hemoglobin, blocking its ability to carry oxygen.
Carcinogenic Having the ability to produce cancer.
Carnivore An animal that feeds on other animals.
Catalyst A substance that can accelerate the rate of a chemical reaction but which is not used up in the reaction. Enzymes are catalysts.
Cell The structural unit of organisms.
Cell wall The outermost rigid layer of plant cells as well as of some protista and prokaryotes.
Cellulose The chief component of cell walls in most plants, a complex carbohydrate.
Celsius The temperature scale invented by the Swedish astronomer Celsius which has the freezing point of water at 0°C and the boiling point at 100°C, and is thus a centigrade scale.
Cervix (adjective, **cervical**) The lower portion of the uterus which connects with the vagina; also any necklike portion of an animal.
Chlorinated hydrocarbon A synthetic carbon compound which contains chlorine; examples are DDT, DDE, PCBs. Most are not readily biodegradable and concentrate in food chains.
Chlorophylls A group of green pigments necessary for photosynthesis. Chlorophylls are found in green plants, some protista and some prokaryotes.
Chloroplast An organelle in green plant cells which consists of chlorophyll-containing and other membranes; chloroplasts are the site of photosynthesis.
Cholinesterase The enzyme which destroys acetylcholine at a synapse or neuromuscular junction, thus allowing another nerve transmission to cross the synapse or junction.
Chorion The membrane which encloses the fetus and the amnion in pregnant mammals.
Chromatid One of the two daughter strands of a duplicated chromosome.
Chromosome A body in the cell nucleus which contains the genetic material. Chromosomes are most easily seen during mitosis or meiosis.
Cilia (singular, **cilium**) Short, hairlike structures present on the surface of certain cells.
Circadian rhythm A rhythm which has a period or cycle of about 24 hours.
Class The taxonomic category between phylum and order.
Classification The systematic arrangement of organisms into groups based upon some definite scheme.
Clone A population of genetically identical individuals descended by mitotic division from a single ancestor.
Cloud seeding A process in which silver iodide crystals are introduced into clouds in order to produce condensation of water vapor and, eventually, rain.
Coal A fossil fuel, solid and usually black, resulting from partial decomposition of organic materials not exposed to oxygen during long periods under the earth's surface.
Coevolution The patterns of evolutionary interaction between taxonomically distinct organisms with a close and evident ecological relationship, for example, plants and herbivores.
Colostrum The first "milk" secreted by a female mammal after giving birth. Human colostrum contains antibodies from the mother and is higher in protein and vitamin A and lower in fat and carbohydrate than human milk; colostrum is also a mild laxative.
Comet One of a number of bodies in extreme elliptical solar orbits. They periodically approach the sun exhibiting a glowing tail and then recede great distances from the sun.
Community The group of organisms found in a particular place.
Complementary strand One of the strands of a DNA helix which has its bases so arranged that it can pair with the other strand.
Compound A combination of atoms in definite ratio held together by chemical bonds.
Condom A contraceptive device consisting of a sheath worn over the penis during copulation. Condoms may also prevent transmission of venereal disease.

Conservation Preserving, guarding, or protecting natural resources for the purpose of maintaining them for enjoyment and use of present and future generations.
Contagious disease An infectious disease which can be passed from individual to individual by direct or indirect contact. Many infectious diseases are not contagious since they require a special agent (vector) of transmission, for example, malaria.
Contraception Prevention of pregnancy.
Copulation The process by which sperm are transferred from the male to the female's reproductive tract. During copulation in mammals, the penis of the male is inserted into the vagina of the female.
Cowboy economy An economy based on the idea that there are endless resources and that waste is justified because there are plenty more "wide-open spaces" in which to find riches.
Crossing over The exchange of corresponding segments of genetic material between chromatids of paired homologous chromosomes at meiosis.
Crystal A body formed when a chemical element or compound solidifies in such a way that it is bounded by plane surfaces.
Cultural evolution The transmission with modification of nongenetic information; change in a culture over time.
Culture Nongenetic information, especially in the form of language.
Cycle A series of changes which lead back to its starting point, for example, cycles of elements during which elements enter into different compounds but are not lost as they go through the cycle again and again.
Cytochrome Any of several iron-containing compounds of the electron transport chain.
Cytoplasm The material inside the cell membrane exclusive of the nucleus.
Daughter cell The two cells formed from division of one cell.
DDE See **DDT**.
DDT Dichlorodiphenyltrichloroethane, a persistent, chlorinated hydrocarbon insecticide which concentrates in food chains mostly in the form of DDE, a breakdown product which has one less chlorine atom in its molecule.
Decomposers Organisms, especially bacteria, fungi, and tiny animals, that break down organic material into smaller molecules that can be reused in the biosphere.
Decomposition The process of breaking down organic materials to simpler molecules.
Deciduous Of green plants, shedding all leaves at a certain season.
Deoxyribonucleic acid (DNA) The carrier of genetic information, composed of chains of phosphate and sugar molecules, and nitrogen bases. DNA is capable of directing the synthesis of copies of itself and determining RNA synthesis.
Destabilization The process of breaking down a stable or steady condition, often leading to wild fluctuations or drastic changes.
Detritus Decaying organic material.
Diabetes Diabetes mellitus is a disease caused by a lack or deficiency of the hormone insulin.
Dialysis The separation of substances by their unequal diffusion across a membrane.

Diaphragm (1) A contraceptive device used to cap the uterus and prevent sperm from entering during copulation. (2) A sheet of muscle between the thoracic and abdominal cavities which is the major muscle used in breathing in mammals.
Differentiation A process by which an unspecialized cell undergoes a progressive change to a more specialized cell; also the specialization of cells and tissues during development.
Diffusion The movement of dissolved or suspended particles from a region of greater concentration to a region of lesser concentration (diffusion gradient) due to the random motion of individual molecules. Diffusion tends to equalize the concentration of the particles throughout the medium.
Diminishing returns See **Law of diminishing returns**.
Diploid Having two sets of homologous chromosomes.
Dispersal The act of scattering from a center point; in organisms, the property of moving into areas other than that of their origin.
Diurnal Relating to the daytime, as opposed to nocturnal.
Diversity Complexity or variety. In biology an ecosystem with many organisms of many kinds with many interrelationships among them is more diverse than one with only a few organisms and few interrelationships. See also **Stability**.
DNA See **Deoxyribonucleic acid**.
Dominance A social status of power or authority over others.
Dominant In genetics, a gene that exerts the same phenotypic effect regardless of its allelic partner; that is, the phenotype is the same whether the locus is homozygous or heterozygous.
Dormancy A period of suspended activity or development in an organism.
Ecological engineering Planning, designing, constructing, and managing ecosystems based on biological and ecological knowledge; it is hoped that its purpose is maintaining the stability of the ecosystem necessary for continued well-being of its inhabitants, including human beings.
Ecological succession An orderly, often slow progression of changes in the composition of a community from the initial colonization of an area to the attainment of the most complex community the area can support.
Ecology The study of the interrelationships among organisms and between organisms and their environment.
Economy of scale A condition which exists when the price per unit of a product can be reduced by a large-scale production operation, such as an assembly line where many workers each do a small part of the job of producing a mass-produced item.
Ecosystem A system which comprises the interactions among a group of living organisms and their nonliving environment.
Effectors Organs of response, for example, glands or muscles.
Ejaculation The process in which semen is ejected from the penis.
Electron A subatomic particle with a negative electric charge equal in magnitude to the positive charge of the proton but much less in mass. The electrons surround the

nucleus of an atom and determine its chemical properties.
Electron transport chain A series of compounds along which electrons from glycolysis and the Krebs cycle flow to the ultimate acceptor, oxygen; during this transport ATP is produced.
Element A substance composed of only one kind of atom. One of about 100 types of matter which together make up all the materials of the universe.
Ellipse A closed curve (as a circle) which is elongated in one direction; *ellipse* has a precise mathematical definition but is often used to refer to any egg-shaped curve.
Embryo In plants, the young plant in the seed before germination. In animals, the young animal formed by the mitotic divisions of the zygote.
Emigrate To leave a region for residence in another region.
Endocrine Tissues or organs (without ducts) which secrete hormones into the blood stream.
Endometrium The lining of the uterus.
Endoplasmic reticulum An extensive system of membranes present in most cells which divides the cytoplasm into compartments and channels. When endoplasmic reticulum is coated with ribosomes, as it often is, it is called rough endoplasmic reticulum.
Endosperm A plant tissue found only in angiosperms which is usually triploid and contains stored food for the developing embryo.
Energy The ability or capacity to do work.
Environment The total of all external conditions which affect an organism, population, society, ecosystem, etc.
Environmental impact A term used in the National Environmental Policy Act of 1969 and hence subject to continuing legal interpretations; generally taken to mean a major negative or positive change in the environment, often including the economic environment of human beings.
Enzyme A complex protein which speeds up the rate of a chemical reaction; an organic catalyst.
Epidemic (1) A disease or other factor common to many individuals in a community. (2) The rapid spread of something, for example, a disease through a community.
Epidermis The outermost layer (plants) or layers (animal) of cells.
Equatorial Of or pertaining to the equator, the imaginary line which is an equal distance from the North and South Poles of the earth; used especially in reference to conditions near the equator.
Equinox The date on which the vertical rays of the sun are on the equator and the days and nights are approximately equal everywhere. The vernal equinox is on or about March 22 and the autumnal equinox is on or about September 22.
Erosion The process by which land is worn away by the action of wind, water, ice, and human activities.
Estrus The portion of the estrus cycle (sometimes called "heat") during which the female is receptive to the male.
Estrus cycle A cycle of nonprimate female mammals similar to the menstrual cycle of primates. The estrus cycle differs from the menstrual cycle largely in having a definite period of heightened sexual receptiveness and lacking significant vaginal bleeding at the end of the cycle.
Estuary An arm of the sea where tides of the oceans meet fresh water from rivers and streams.

Eugenics In its positive sense, genetic manipulations aimed at the overall improvement of the human population through selective breeding for what some person or group decides to be "desirable" characteristics. In its negative sense, genetic manipulations aimed at preventing births of defective individuals or perpetuation of defective genes such as those causing serious diseases.
Eukaryote (adjective, **eukaryotic**) An organism with cells which have membrane-bound nuclei, mitochondria, and other organelles.
Euphenics Phenotypic manipulations aimed at the treatment of genetic diseases in such a way as to improve the functioning of the afflicted individuals.
Eurasian Pertaining to Europe and Asia as a whole.
Eutrophication Natural or human-enhanced addition of nutrients to bodies of water resulting in succession from fewer organisms to many, and a change in species composition.
Evolution Change. Organic evolution is inherited change through time in populations.
Exoskeleton A skeleton on the outside of the body of an animal, for example, the shell of a crab.
Exponential growth Growth of a quantity by a constant percentage of the whole during any given period. Each increase produces additional increases as in compound interest on money. The number of years it takes for an exponentially growing quantity to double is approximately the percent yearly increase divided into 70.
Fallopian tube See **Oviduct.**
Family (1) A taxonomic category between order and genus. (2) A group of organisms which are biologically or socially related, for example, a man, a woman, and their children.
Fats Organic compounds containing carbon, hydrogen, and oxygen. The proportion of oxygen to carbon is much less in fats than it is in carbohydrates.
Fauna The animals characteristic of a geographic area.
Feces Digestive and metabolic waste material discharged from the anus.
Feedback A means of self-regulation by which an end product of a process controls the process itself. See **Biofeedback.**
Feedlot A place where cattle are crowded together and fattened by giving them grain and other relatively high-protein food. The average steer uses 21 kilograms of protein to create 1 kilogram of meat while being "finished" on a feedlot.
Feral Referring to a formerly domestic animal which has gone wild.
Fermentation Use of energy-containing compounds by cells in the absence of oxygen. In yeast a by-product of fermentation is alcohol; in muscle tissue a by-product is lactic acid. Fermentation is often used synonymously with anaerobic respiration.
Fertilization The fusion of two gametes, especially fusion of their nuclei, to form a diploid zygote.
Fertilizer A substance added to soil to increase plant growth.
Fish meal Ground-up dried fish waste and fish used as fertilizer and animal food.
Fission power Power produced by the splitting of the nuclei of heavy atoms to produce two nuclei and some parti-

cles whose combined mass is slightly less than that of the original nucleus. The remaining mass is converted to energy.

Flagellum (plural, **flagella**) A fine, threadlike structure protruding from a cell body, used in locomotion and longer than a cilium.

Flora The characteristic plants of a geographic area.

Food chain A chain of organisms each of which feeds on the one below it and is eaten by the one above it.

Food web A set of interconnected food chains.

Forecast An "if-then" statement which can be checked by repeating the forecaster's calculations. The forecast may not come to pass because the "if" part of the statement may change or because other important factors were not taken into consideration.

Fossil The remains of an organism that lived in the past, or direct evidence of its presence, such as chemicals, tracks, or imprints.

Fossil fuel Materials such as coal, oil, natural gas, peat, and their by-products which came originally from the remains of ancient living things and are presently used as sources of energy and organic molecules.

Fungicide A biocide designed chemically to be especially toxic to fungi.

Fungus (plural, **fungi**) An unicellular or multicellular protist which obtains food by absorption from other organisms or organic matter.

Fusion power Power produced by the fusion of the nuclei of atoms to produce one atom slightly lighter than the sum of the mass of the two original atoms. The remaining mass is converted to energy. The energy of the sun comes from fusion reactions.

Galaxy One of the millions of groups of stars which make up the universe. Each galaxy contains billions of stars; our sun is a star near the edge of the galaxy called the Milky Way.

Gamete A mature functional reproductive cell (haploid) which can fuse with another gamete of the opposite sex to form a zygote.

Gene A unit of hereditary information transmitted in a chromosome. A gene is capable of self-replication and controls the development of a trait in interaction with the environment.

Genetic drift Changes, unpredictable in direction, occurring in the genetic composition of a population due to chance events. When the population is small, drift leads to a decay of variability in which genes are lost.

Genetic screening The process by which individuals are tested to see if they carry deleterious recessive alleles or alleles which may cause disease in later life.

Genetics The study of the mechanisms of heredity by which traits are passed from generation to generation.

Genital organ One of the organs involved in reproduction; the external parts of the reproductive tract are often called external genitalia.

Genotype The genetic information of an individual; may refer to only one or a few characteristics.

Genus (plural, **genera**) The taxonomic category between family and species. Usually contains several species which share some common characteristics.

Gestation The period of time during which a female mammal carries her young in her uterus.

Ghetto An area of a city in which only certain types of people reside and from which they usually cannot easily migrate. This segregation of types of people may be based on race, religion, country of origin, economic status, or other factors.

Global village A combination of words meant to suggest that people around the globe today are not only as interdependent as people in a village but, because of technology, are all in close communication.

Glucose A six-carbon sugar.

Glycogen A complex carbohydrate made up of glucose subunits which is used for energy storage in animals.

Goiter An enlargement of the thyroid gland.

Golgi material (Golgi apparatus) A kind of organelle made up of flat sacs and tubules which functions as a collection center for materials made by the cell; often called a dictyosome, especially in plants.

Grain (1) A simple, dry, one-seeded fruit characteristic of the grasses; (2) A domesticated plant which produces food grains.

Gravitation The property of matter which leads material bodies to be attracted to each other. The weight of objects is caused by gravity, the attraction of objects to the earth or to other bodies in the universe.

Green Revolution An attempt to increase the food supply of some less developed countries by transforming their traditional agriculture by the introduction of high-yielding "miracle" grains.

Greenhouse effect The heating of the atmosphere caused by the presence of gases such as carbon dioxide and water vapor which allow light energy to pass in freely but retard the outward passage of heat energy.

Gross national product (GNP) The value of all the goods and services produced in a country in any one year.

Haploid Having a single set of chromosomes.

Hardpan A cementlike layer formed in soil which makes cultivation difficult.

Heartwood The nonliving wood in the central portion of a tree trunk through which water transport does not occur.

Heat balance Pertaining to the earth, the difference between the incoming solar radiation and the outgoing earth radiation. For the earth as a whole the two must be equal or the temperature will rise or fall.

Heat energy Energy in the form of heat, that is, in the form of increased random motions of molecules.

Hemoglobin The oxygen-carrying substance in red blood cells.

Herbaceous Pertaining to any nonwoody plant.

Herbicide A biocide designed chemically to be more toxic to plants than other kinds of life.

Herbivore An animal that eats plants.

Heredity The pattern in which inheritance of genetic traits takes place.

Heterozygote An organism which has two different alleles at the same locus on a pair of chromosomes.

Hibernate To spend the winter in a sleeping or torpid state.

Hierarchy A series of items classified by ranks, each of which includes or rules over lower ranks.

Homeostasis The maintenance of a stable, internal environment in an organism, population, or ecosystem.
Homozygote An individual which has identical alleles at the same locus on a pair of chromosomes.
Hormone A chemical messenger produced in minute amounts in one part (endocrine gland or tissue) of an organism and having an effect on another part (or all) of that organism.
Host In biology, any animal or plant infested with one or more parasites.
Humus The organic part of soil consisting of partially broken-down plant and animal materials.
Hydroelectric power Electric power produced by turbines turned by the energy of flowing water trapped behind a dam built for that purpose.
Iatrogenic Of diseases, caused by a physician.
Immigrate To come into a region from another native region for the purpose of permanent residence.
Immune response The response of the body to foreign materials, especially those produced by disease organisms.
Implantation The process by which the embryo of a mammal becomes buried in the lining of the uterus early in its course of development.
Incubate To sit upon eggs to hatch them or to maintain eggs or young in a condition favorable for development.
Independent assortment The inheritance of one pair of characteristics independently of the inheritance of other characteristics; that is, characteristics assort as though there were no other characteristics present. See **Linkage** for exception.
Individual space An area around an individual which it protects from invasion by other individuals except under special circumstances; for example, it allows its young and its potential mate to enter the area.
Inducer A substance which will cause the specific differentiation of an embryonic tissue.
Induction (1) In animal embryos, the process by which one tissue causes the differentiation of another type of tissue. (2) The process by which the synthesis of a particular enzyme is caused by the presence of a substrate.
Industrial revolution A social and technological process made possible by the harnessing of inanimate energy (for example, wind, water, and fossil fuels) which led to increased division of labor and accelerated production and consumption of material goods.
Infanticide The killing of an infant; a practice common in some cultures, especially in times of famine.
Infectious Any disease caused by the entrance, growth, and multiplication of organisms such as bacteria and protozoa or of virus particles.
Infrared Rays just beyond the visible red of the electromagnetic spectrum; heat waves, that is, those with wavelengths longer than visible light and shorter than radio waves.
Inheritance In biology, the transmission of genetic characteristics from generation to generation.
Input Any matter, energy, or information which is put into a system.
Insecticide A biocide chemically designed to be more toxic to insects than to other kinds of life.

Insulin A hormone produced by endocrine cells of the pancreas which decreases the level of sugar in the blood.
Integrated control A pest population management system that utilizes all suitable techniques and information to keep pest populations from causing economic damage to crops. It maximizes the use of existing regulating factors in the ecosystem, thus maintaining environmental quality.
Interpersonal distance A culturally determined distance at which human beings feel comfortable during various kinds of social relationships. See also **Individual space**.
Intertidal zone The zone along an ocean shore between the lowest low tides and the highest high tides.
Intrauterine device (IUD) A coil, loop, or other shaped device placed in the uterus to prevent pregnancy.
Invertebrate An animal which lacks a backbone.
Involuntary muscle Muscles such as those of the stomach which are not usually under voluntary control.
Irrigation In agriculture, the artificial watering of crops by ditches, canals, flooding, etc.
Itai-Itai disease A serious and painful disease caused by poisoning with the element cadmium. (*Itai* means "ouch" in Japanese.)
IUD See **Intrauterine device**.
Kelp Brown algae which grow as seaweeds, often becoming extremely large.
Kilowatt-hour The amount of energy used when a 1,000-watt appliance is used for an hour.
Kingdom The highest taxonomic category.
Kinship system Relationship in human groups by marriage or by degree of genetic relatedness, for example, mother, sister, cousin, nephew, mother-in-law, second cousin.
Krebs cycle A cycle of cellular reactions during which a three-carbon breakdown product of glucose is oxidized to electrons, hydrogen atoms, and carbon dioxide.
Kwashiorkor A protein-deficiency disease found in children.
Lactation Secretion of milk by the mammary gland.
Language A form of communication which usually involves symbols and delayed transmission of information.
Larva (plural, **larvae**) The immature form of any animal which is so unlike its parent that it undergoes metamorphosis before becoming adult.
Laterite Red soil containing iron compounds which become rocklike when repeatedly dried as when their vegetation is removed or they are used in agriculture. The process of forming the rocklike material is called laterization.
Law of diminishing returns A "law" which applies to a situation in which more and more input is required to produce less and less output. For example, when a species is being overfished, the effort put into catching the species increases while the size of the catch decreases.
Leaching The process by which a liquid such as water dissolves and carries away the soluble parts of a substance such as soil while filtering through it.
Legume A green plant of the pea or bean family; the fruit or seed of such a plant.
Lichen A fungus which can live in nature only when it has a particular alga within its tissues.
Life form The overall shape, size, and way of life of an organism.

Light energy Radiant energy in the visible or nearly visible part of the electromagnetic spectrum.

Linkage The tendency of genes to be inherited together because they are located on the same chromosome.

Locus (plural **loci**) In genetics, the place on a chromosome where the alleles of a gene for a specific trait are located.

Lymphocyte A white or colorless blood cell derived from lymphatic tissues and having the ability to move by extensions of its cytoplasm like an amoeba.

Mammary gland A tissue in a female mammal which secretes milk after she has given birth. Male mammals have undeveloped mammary glands.

Manure A fertilizer consisting of the urine, feces, and often the stable litter of domestic animals or of the droppings of colonies of seabirds.

Marsh An area of soft, wet land often exposed to flooding from tides, rains, or overflow from rivers or lakes. An ecosystem adapted to such conditions.

Marsupial A mammal in which the female carries the young in a pouch during most of its period of development.

Mass The quantity of matter in a body.

Materials Things consisting of matter; the substance of which anything is composed.

Matrix That in which something else is imbedded or takes place.

Matter Anything that takes up space or has weight. There are three main states of matter: solid, liquid, and gas.

Medieval Referring to the Middle Ages, the years in European history between the fall of Rome (A.D. 500) and an acceleration in the revival of learning (about 1400). The first part of this period is often called the Dark Ages because of the lack of intellectual development.

Megalopolis A very large city.

Meiosis A sequence of two divisions of a cell, but only one division of its chromosomes, by which a diploid cell produces four (potentially) haploid gametes or spores.

Membrane Any thin, soft sheet or layer. A cell membrane is a layer of protein and phospholipid molecules which forms the outermost membrane of plant and animal cells.

Menstrual cycle A monthly cycle in human females and the females of certain primates. During the cycle hormonal changes lead to the build-up of the endometrium, the release of an ovum into the fallopian tube, and if fertilization does not occur, menstruation, the discharge of blood and uterine tissue from the vagina at the end of the cycle.

Menstruation See **Menstrual cycle.**

Meristem A growing point or undifferentiated tissue in plants from which new cells arise.

Messenger RNA (mRNA) RNA that takes genetic information from DNA to ribosomes where it is translated into the order of amino acids in a protein.

Metabolism The sum total of all the chemical reactions occurring within a living cell or organism.

Metamorphosis The transformation by which a larva becomes an adult.

Meteor A meteoroid that enters the earth's atmosphere and is visible as it burns from friction with the atmosphere.

Meteorite A meteoroid which has fallen to earth and survived the journey.

Meteoroid A stony or metallic body smaller than an asteroid in orbit around the sun.

Meteorologist A person who specializes in the study of the earth's atmosphere, especially changes in the atmosphere which result in climate and weather.

Methane A gaseous hydrocarbon, CH_4, which is odorless and burns readily. It is produced by the decomposition of organic materials under certain circumstances.

Methyl mercury A compound produced from mercury which is highly toxic to living things. It may be formed by microorganisms in lakes, steams, or bays which are polluted with mercury and concentrate in food chains. See also **Minamata disease.**

Microorganism A very small organism usually visible only under magnification such as by a microscope.

Migration In biology, a regular seasonal movement of an individual or species from a breeding location to a wintering location; also used to mean any movements of groups of particles or individuals.

Minamata disease A disease caused by eating foods contaminated with methyl mercury. Named after a bay in Japan where it was first observed.

Mitochondrion (plural, **mitochondria**) An organelle surrounded by a double membrane which contains the enzymes of Krebs cycle and electron transport chain. Mitochondria are found only in eukaryotic cells.

Mitosis Nuclear division usually accompanied by cell division in which the chromosomes divide longitudinally, separate from each other, and form genetically identical daughter nuclei.

Molecule The smallest possible unit of a compound, consisting of two or more atoms.

Monoculture A large area planted with a single crop.

Monsoons Winds which reverse themselves seasonally, especially the summer monsoons in Asia and Africa which bring quantities of rainfall from the inflow of air from the sea over the continents.

Motile Moving or having the ability to move.

Multiple alleles Three or more alleles, any one of which may occur at a particular locus on a chromosome.

Mutagen Any agent that can cause a mutation.

Mutation An inheritable change from one form of an allele to another or a change in the structure of a chromosome.

Mutation rate The number of mutations that occur in a given unit of time or per generation.

Mycorrhiza The combination of certain fungi with the roots of vascular plants.

Natural fertilizer A fertilizer such as manure or phosphate rock which is not synthesized by people in fertilizer plants.

Natural gas Gas associated with petroleum deposits made up mainly of methane but often containing quantities of carbon dioxide, nitrogen, helium, and other gases. Natural gas is used as a fossil fuel and as a raw material in the synthesis of many chemicals by industrial processes.

Natural selection Differential (nonrandom) reproduction of genotypes without the conscious interference of human beings.

Neutron A particle in the nucleus of atoms which has a mass about that of the proton but is neither negatively nor positively charged.

Nitrate An ion composed of one nitrogen atom and three oxygen atoms. Also a compound containing such an combination, for example, sodium nitrate. Nitrates are usable

nutrients for green plants. Nitrites have one nitrogen atom combined with two oxygen atoms. Nitrates and nitrites may be found as pollutants in drinking water and as food additives or contaminants. In excess they lead to serious illness, especially in young children. Nitrites combine with some protein breakdown products to form carcinogens.
Nocturnal Occurring or pertaining to the night; active at night.
Nomadic Referring to a group of people that wanders from place to place.
Nuclear energy Energy which is produced from the nuclei of atoms when part of their mass is converted to energy. Nuclei may produce energy by being split apart (fission) or put together (fusion).
Nucleotide Organic molecules consisting of a nitrogen-containing base, a five-carbon sugar, and a phosphate group.
Nucleus (1) In biology, the specialized body in eukaryotic cells which contains the genetic material and is bounded by a double membrane. (2) In physics, the central core of an atom composed of protons and neutrons.
Nutrient (1) Elements necessary for the growth of photosynthetic organisms. (2) Elements or compounds necessary in the diet of organisms which absorb or ingest their food.
Nymph The immature form of an organism with incomplete metamorphosis.
Omnivore An animal which eats both plants and other animals.
Order A taxonomic category between class and family.
Organ A structural part of an organism made up of different tissues cooperating functionally to carry out one or more specific functions.
Organ system A group of organs which work together to accomplish a major function, for example, the digestive system.
Organelle Small, membrane-bound structures in the cytoplasm and specialized for various tasks.
Organic Referring to materials and compounds synthesized by living beings as part of their normal metabolic processes. (Chemists often refer to any carbon compound as "organic" regardless of its source.)
Organic food Food grown without the use of synthetic pesticides or fertilizers.
Orgasm A contraction of muscles in the genital tract leading to a release of sexual tension; in males, ejaculation occurs during orgasm.
Osmosis The movement of water across a membrane as a response to a concentration gradient.
Outputs Materials, energy, or information which are produced by or given off from a system.
Ovary (1) The part of a plant which develops into a fruit and which contains the ovules. (2) The egg-producing organ of animals.
Overdeveloped country (ODC) A country which is highly industrialized and in which food production is highly mechanized. ODCs are characterized by a high GNP, high per capita income, and a low rate of population growth. They also have high levels of pollution and use much more than their share of the world's resources, especially energy.
Oviduct The tube which conveys eggs from the ovary to the uterus; fallopian tubes in humans.

Ovulation The release of an egg from an ovary.
Ovule The structure in seed plants which contains the female gametophyte with the egg cell and surrounding structures and which develops into a seed.
Ovum (plural, **ova**) The female gamete or egg.
Oxidation (1) Combination with oxygen, as in burning. (2) A chemical reaction in which a molecule loses electrons or hydrogen atoms, that is, becomes oxidized.
Ozone A highly reactive gas containing three oxygen atoms in each molecule (O_3) which has the capacity to absorb ultraviolet radiation.
Paleontology The science that deals with the life of past geological periods based on the study of fossils.
Parasite An organism which lives in or on another organism, its host. The relationship benefits the parasite but harms the individual host.
Parthenogenesis The development of an unfertilized egg into a new individual.
Peat moss Mosses of the genus *Sphagnum*, etc., which grow in moist places and often partially decompose to form peat, a smoky-burning fuel.
Peck order A dominance relationship in which individuals form a hierarchy of dominance.
Pedigree A table or chart presenting genetic relationships among individuals having ancestors in common.
Penis The male organ used for copulation.
Perennial plant A green plant which persists from year to year and which reproduces in at least two different years.
Pesticide A substance designed for killing pests, that is, destructive or unwanted organisms.
Phage See **Bacteriophage.**
Phenotype The result of the interaction of the genetic information with the environment in the development of an individual. The total appearance of an individual or of one characteristic.
Pheromone A chemical, commonly emitted externally, used for communication among members of the same species.
Phloem The tissue in plants which conducts organic molecules from their place of synthesis to their place of storage or utilization.
Phosphate An ion containing one phosphorus atom combined with four oxygen atoms; also a compound having such a combination in it, for example, calcium phosphate.
Photosynthesis The conversion of light energy to energy stored in chemical compounds by the production of carbohydrates from carbon dioxide, using light energy in the presence of chlorophylls.
Phylum The taxonomic category between kingdom and class.
Phytoplankton Photosynthetic microorganisms which float in the upper layers of oceans and bodies of fresh water.
Placenta An organ formed in the wall of the uterus during pregnancy; it is made up of tissue from both the embryo and the uterus and serves to transport food to and waste materials from the embryo.
Plankton Organisms, mostly microscopic, which float freely in water.
Plutonium A highly toxic, radioactive element produced in nuclear reactors and used for making nuclear bombs or as fuel in other reactors.
Pollen The mature male gametophytes of seed plants.

Pollination Transfer of pollen from its site of formation to a surface where it can form a pollen tube which carries the male gametes into the ovules.

Pollutant Any substance harmful or unpleasant to living things in air, water, soil, food, etc. Pollutants may consist of synthetic substances or of natural substances present in excessive amounts.

Pollution The sum total of the pollutants present in a given area at a given time.

Polychlorinated biphenyl (PCB) A chlorinated hydrocarbon used in industrial processes, etc. PCBs have escaped into the environment and are concentrated in food chains.

Polymer A large molecule composed of many similar molecular subunits.

Polypeptide A polymer made up of amino acid subunits joined by peptide bonds.

Polyploid Possessing more than two sets of chromosomes.

Polyunsaturated fats A fat molecule which has two or more double bonds between carbons and thus does not have every carbon attached to the maximum number of hydrogens. See also **Saturated fats**.

Population All the organisms of a given kind in an area at a specific time; also may refer to all species in an area.

Population dynamics The patterns of change in population size.

Power The rate of doing work; the rate of transfer of energy in an electrical system.

Preindustrial city A city existing before the industrial revolution or presently in existence but lacking industrial developments.

Productivity In ecology, the rate at which energy is trapped and stored in the biomass of an ecosystem or other biological system.

Prokaryote (adjective, **prokaryotic**) A unicellular organism with unicellular reproductive structures which lacks a membrane-bound nucleus and well-defined organelles.

Protein malnourishment Lack of sufficient protein in the diet. Leads to disease and permanently retarded growth and development in young children. In adults results in increased susceptibility to disease.

Proteins Long polymers made up of amino acid subunits.

Proton A positively charged particle found in the nucleus of atoms; the nucleus of a hydrogen atom. The proton has a mass 1,845 times that of the electron.

Puberty The period during which sexual maturity is reached; the period during which an individual is first capable of producing eggs or sperm.

Pupa The stage between the larva and the adult in insects with complete metamorphosis.

Radiant Emitting heat or light.

Radiation Energy in the form of waves (for example, light) or in the form of particles such as may be given off by radioactive elements.

Radioactive element An element which has the property of spontaneously releasing energy from its nucleus in the form of waves or particles.

Rate The quantity of something measured per unit of time.

Receptor An organ or simple nerve ending capable of detecting energy or change in the environment; a sense organ.

Recessive Used in reference to an allele which is expressed only in the homozygous state, that is, when paired with an identical allele.

Recombination Formation of combinations of alleles or genes not present in the parental types.

Recycle To return a substance or element to use, as, for example, elements are recycled in ecosystems.

Replication The production of a copy; the production of a second DNA molecule like the first; the production of a sister chromatid.

Respiration (1) Intracellular process during which food is oxidized with the release of useful energy. (2) Breathing.

Ribonucleic acid (RNA) (1) A polymer of nucleotides which functions in the transcription of DNA and its translation into proteins. (2) The genetic material of many viruses.

Ribosomal RNA (rRNA) The RNA which occurs in ribosomes.

Ribosome A tiny organelle made up of rRNA and protein; ribosomes are the sites of protein synthesis.

Rickets A childhood disease caused by deficiency of vitamin D in which the bones fail to develop properly.

Rickettsia A kind of microscopic rod-shaped parasite smaller than most bacteria.

RNA See **Ribonucleic acid**.

Rodenticide A biocide, chemically designed to kill rodents. Some rodenticides, for example, 1080 (sodium monofluoroacetate), concentrate in food chains and are toxic to other organisms.

Rural ecology The study of living things, their relations with each other, and their relationship with their nonliving environment in the country or on farms.

Sahel region The region of Africa just south of the Sahara Desert.

Saturated fat A fat molecule which has every possible carbon bond attached to a hydrogen.

Scavengers Animals that eat dead and decaying animals.

Second growth Trees in forests where the original trees have been cut down.

Second law of thermodynamics See **Thermodynamics**.

Selection A situation in a population in which some genotypes have more offspring than others. The "choice" of which individuals have the most offspring may be either natural as a consequence of the interaction of individuals and their environment or artificial as a consequence of human intervention.

Semen The fluid ejaculated during orgasm by the male. Semen is composed of sperm and fluids from the genital tract and associated glands.

Sewage Waterborne waste matter (feces, urine, etc.) carried away from areas of human habitation by ditches or sewers.

Sex The characteristic of being male or female. Sexual reproduction in animals involves fusion of gametes produced by a previous meiotic process.

Sex chromosomes The pair of chromosomes that participates in determination of the sex of an individual.

Sexuality The properties of the life of an individual which revolve around the exercise of his or her sexual functions, interests, and being.

Signal A sign, sound, or action which is part of the anatomy or behavior of an individual member of a species and

which may lead to response from other individuals of the same or other species.
Silting The process by which bodies of water are gradually filled in with soil particles which settle out of water.
Slash and burn A practice of farming in which a clearing cut and burned in a forest is farmed until it no longer produces a satisfactory crop.
Social engineering Planning, designing, construction, or management of human societies.
Social insect A species of insect which lives in societies; an individual of such a species.
Society (1) In biology, an enduring group of individuals functioning to maintain itself and the species. (2) A human group within which individuals have ties of interdependence and homogeneity. (3) The human species as a whole.
Soil A material found in relatively thin layers on parts of the earth's surface consisting of fragments of rock and decaying organic materials in which plants grow and many different kinds of organisms live.
Solar power Power produced by using the sun's radiation to produce heat or electricity directly.
Solar system The sun and the bodies which revolve around it.
Spaceship economy (spaceman economy) An economy which would recognize that the earth is like a spaceship because there are not unlimited reserves of materials or unlimited spaces to absorb pollution.
Speciation The splitting process of evolution which results in the formation of species.
Species (plural, **species**) A group of similar individuals judged to be sufficiently different from members of other groups to be recognized as a distinct kind of organism.
Speculation Used in this text in the special context of a prediction or prophecy which does not consider the possibility that unforeseen change may alter the course of events. Also means predicting on the basis of reasoning rather than on experience or experiment.
Spore A reproductive cell capable of developing into a new organism without fusing with another cell.
Solstice (from Latin for "sun's stance") The longest (on or about June 21) or shortest (on or about December 21) days of the year.
Stability The property of a system which enables it to return to an equilibrium or steady state when it is disturbed. In ecosystems, the capacity which allows them to persist.
Stamen The organ of a flower which produces pollen.
Starch The chief food storage substance of plants. Starch is a complex carbohydrate, insoluble in water, which is a polymer of several hundred glucose subunits.
Steady state (stationary state) A dynamic condition in which the major characteristics of the system do not change through time. For example, a population in steady state would have birthrate and death rate equal.
Sterilization (1) The process of making an individual incapable of reproduction. (2) The process of killing or eliminating microorganisms.
Stigma The surface in a plant upon which pollen grains germinate.
Stoma (plural, **stomata**) A tiny opening in the epidermis of leaves and stems through which gases pass in and out.
Stratosphere The layer of the earth's atmosphere, 16 to 24 kilometers (10 to 15 miles) thick, above the tropopause.
Subphylum A taxonomic category below phylum and above class.
Subsistence farming Farming which produces just enough food for the people who practice it.
Sugar (1) A carbohydrate with the formula $(CH_2O)_n$, where n is usually 3 (triose), 5 (pentose), or 6 (hexose); a compound made up of two or more such units. (2) Sucrose, a sugar made up of one glucose subunit and one fructose subunit.
Swidden agriculture The practice of growing crops in temporary forest clearings until the trees which grow up naturally get large enough to interfere with the crops.
Symbiosis Living in close association. In a symbiotic relationship one of the partners may benefit while the other is harmed (parasitism) or they may both profit (mutualism).
Symbol Something which by common agreement or custom has come to mean or stand for something else.
Synapsis The process in which homologous chromosomes synapse (come together) early in meiosis.
Synthesis The process of building up a compound by combining its parts or elements under the proper conditions.
Synthetic Referring to materials and compounds synthesized in laboratories, factories, or other human-created situations.
Synthetic fertilizer Fertilizers synthesized in laboratories or factories.
System A group of objects, ideas, or other elements united by some form of regular interaction or interdependence, for example, an ecosystem.
Taboo A restriction on use of certain objects or participation in certain acts established by social custom or by religious or tribal authority.
Taxonomy The science of classification of organisms.
Technology (1) A collection of techniques, information, and tools by which human beings utilize the resources of their environment to satisfy their needs and desires. (2) The modern technology which has developed during and since the industrial revolution.
Temperate zones The areas on the earth's surface between the tropics and the polar circles; that is, the areas bounded by about 24° latitude and about 66° latitude.
Template A pattern or mold used as a guide in forming a new structure.
Teratogen Any agent capable of causing a defect in an organism during its development.
Terrestrial Living on the land as distinct from in trees, oceans, etc.
Territoriality The complex of behavior and physiology associated with marking and defending a territory.
Territory In biology, an area of space defended by an individual or group of individuals.
Testis (plural, **testes**) The sperm-producing organs in animals.
Thermal pollution Pollution by heat; for example, water used for cooling by industries is often returned to a lake or river hot enough to kill or harm organisms that live there.
Thermodynamics The science of heat, its transfers and relationships. The first law of thermodynamics states that

energy can be neither created or destroyed. The second law of thermodynamics is difficult to define simply, but many examples can be given: (1) a loss of useful energy always occurs when energy is changed from one form to another as in a machine; (2) heat cannot flow from a cold body to a hot one; (3) order tends to become disorder; (4) all energy is eventually degraded to heat.

Thorax The chest region of an animal; the part between the neck or head and the abdomen.

Threshold A critical level above which a system will respond.

Tissue A group of cells of similar structure organized to perform a specific function or functions.

Torpid Sluggish or inactive and, in birds and mammals, often with lowered body temperature.

Toxic Poisonous.

Trace element Elements which are needed in only tiny amounts by living things.

Tragedy of the commons A phrase coined by Garrett Hardin to refer to a situation in which each individual benefits from maximizing his or her use of a collectively used resource to the point at which the resource is overused and ruined for all users.

Tranquilizer A medicine or technique designed to calm a nervous or excited patient. Tranquilizers often reduce anxiety and encourage sleep or relaxation.

Transcription The process by which DNA acts as a template for messenger RNA.

Transfer RNA (tRNA) A molecule of RNA which brings an amino acid to the place where it is to be attached to the chain of amino acid residues being united during polypeptide synthesis.

Transformation A change in the form or composition of a substance or quantity.

Translation The process by which messenger RNA (mRNA) is translated into polypeptides. Each group of three mRNA subunits makes up a "word" which codes for an amino acid. Specific tRNAs carrying amino acids attach themselves to the appropriate "words" while their amino acids are connected by peptide bonds to the growing protein chain.

Transpiration Loss of water from plants by evaporation through their stomata.

Tropical zones The areas of the earth between about 24° north latitude and 24° south latitude, including the equator.

Tropopause The boundary between the troposphere and the stratosphere which varies in height with season and latitude, being higher in summer than in winter and higher near the equator than near the poles.

Troposphere The lower layer of the earth's atmosphere in which weather takes place, varying in thickness from 6 to 8 kilometers (4 to 5 miles) to 17 kilometers (11 miles).

Tubal ligation Sterilization of the female by cutting and tying the fallopian tubes (oviducts).

Turgor The pressure within a cell which results from movement of water into that cell by osmosis.

Turpentine An oil or resin derived from substances produced by various coniferous trees.

Umbilical cord The cord, containing blood vessels, that connects a mammalian embryo to its placenta.

Underdeveloped country (UDC) A country which is not heavily industrialized and in which food production is not mechanized to any great extent. UDCs are characterized by a low GNP, low per capita income, many people who cannot read, and a very high rate of population growth.

Upwelling A current of water which rises from deep areas of the oceans to the surface near certain coasts, bringing nutrients with it.

Uranium A radioactive element used in nuclear bombs and reactors.

Urban ecology The study of living things, their relations with each other and their nonliving environment in cities.

Urban-industrial city A city which contains industries and is dependent on fossil fuel energy and machines.

Urbanization The increase in the size of cities due largely to migration of people to them from the countryside.

Uterus The internal organ in a mammal in which an embryo develops and is nourished until its birth. The uterus is sometimes called a womb.

Vaccine A preparation made from a killed or weakened disease agent which can produce immunity to the disease when introduced into the body of an organism that has not had the disease.

Vagina The passage from the uterus to the exterior in female mammals.

Vas deferens (plural, **vasa deferentia**) A duct through which sperm pass on their way from the testes to the urethra.

Vascular tissue (1) Tissue which contains many blood vessels. (2) Xylem and phloem in plants.

Vasectomy Sterilization of the male by cutting and tying the vasa deferentia.

Vector An organism which carries and transmits disease organisms or viruses, for example, mosquitos that carry malaria.

Vertebrate An animal with a backbone.

Vinyl A carbon compound containing a vinyl group ($H_2C=CH-$); often a long-chain polymer of such groups used in plastics, for example, polyvinyl chloride.

Vitamins Organic molecules required in small amounts in the diet of organisms which cannot synthesize them.

Voluntary muscles Muscles such as those of arms or legs which are readily under voluntary control.

Waste (1) In nature, by-products of metabolism or decomposition which are readily broken down and recycled unless in excess, for example, urine and feces. (2) In society, material which is thrown away as useless, such as packaging material, broken machines, sewage, garbage, refuse from building construction and destruction, and by-products of industrial processes which are difficult or impossible to recycle.

Watt A measure of power, the unit of electrical power; watts equal the pressure of the electrons (volts) times the amount of electrons passing a point per unit of time (amperes).

Wavelength The distance between the successive points at which a wave is in the same phase.

Weaning The process of withdrawing a child or other mammal from mother's milk.

Work Activities of organisms or machines which involve

transfers of energy (moving things, growing, maintenance, etc.).

World view A person's total view of the nature of the world and his or her place in it.

Xylem The vascular tissue in plants which conducts water and minerals; in trees popularly known as wood.

Yolk A store of nutrients in an egg for use by a developing embryo.

Zooplankton Small animals which float near the surface of water and eat phytoplankton.

Zygote The cell resulting from the fusion of gametes; a fertilized egg.

INDEX

Page numbers in *italic* refer to pages on which there are illustrations.

Aboriginal bark painting, *60*
Aborigines, *59*
   variations in blood-group genes of, *304*
Abortions, 242—243
Absolute age of fossils and rocks, 316
Absorption:
   energy obtained from, 360—368
   by fungi, 191—194
Acetic acid, as product of fermentation, 358
Acetylcholine, 407
Acorn woodpeckers, individual differences in head markings of, *265*
Active transport, energy required for, 351
Adenosine diphosphate (ADP), 352
Adenosine triphosphate (ATP), 352
   bonds of, *353*
   production of, 356
ADP (adenosine diphosphate), 352
Adrenal glands, 404
Adrenaline, 404
   table, 405
AEC (Atomic Energy Commission), 480, 525
Aerodynamic mimicry, 429
Afferent neurons, 409
Africa, fauna of, 135
Age of fossils and rocks, 316
Aggregation, defined, 436
Aggression:
   dominance and, 438
   role of, 499—501
Aging process, 223—227
Agouti rabbits, coat color inheritance of, *270*

Agribusiness, defined, 42
   (See also Industrialized farms)
Agricultural ecosystems, instability of, 74—76
   (See also Rural life)
Agricultural revolution, 457—460
Agriculture:
   absence of, among early ancestors, 57—59
   failure of Green Revolution, 490—492
   first practiced in Middle East, 457—458
   organically grown food, 326
   slash and burn, 41
   (See also Rural life)
Air circulation, 99—100
Air conditioners, 9
Air pollution, urban, 27
   effects on inhabitants, 29—30
Albino rabbits, coat color inheritance of, *270*
Alcohol, mind control and, 501
Alcoholic fermentation, 358, 360
Algae:
   as aquatic Protista, 186, *188*, *191*
   green, life cycle of, *215*
   kinds of, *189*
   photosynthesis by, 320
   blue green, as photosynthetic Monera, 197
Algal blooms, 50
Alleles:
   defined, 263
   multiple, 269—271
   recombination of, 292
Alligator nests, 246

Altitudinal gradients, as environmental gradients, 117—118
Amino acids:
   apparatus used to form, *289*
   defined, 277
   residues of, defined, 349
Ammonia, excretion of, 399
Amniocentesis, *485*
Amnion (membrane), 236
Amoebas, *152*
Amphibians, 169, *171*
   evolution of, 328, *329*
   reproductive strategy in, *249*
Anaerobic microorganisms, 359
Anaerobic respiration, 358
Anaphase:
   of meiosis, *213*
   of mitosis, *208*
Angina pectoris, 415
Angkor Wat, *122*
Animal kingdom, classification of, table, 163
Animals:
   browsing, 336, *337*
   Carboniferous forest, 325
   composition of, 95—96
   diversity of, in forests, 71
   domestication of, *143*
   early land, 328—335
   early sea, 319—322
   earth rhythms and, 100—101
   effects of day and night lengths on, 104, *105*
   forest dwelling, 72
   major kinds of, 161—180
   periodic behavior of, 105—108
   poisonous, 417, *418*
   predatory, 35

542

Animals:
   reproduction of, other than human beings, 244—250
   (See also Distribution of life; and specific phyla, subphyla, and classes)
Annual plants, earth rhythms and, 100
Antelopes, *167*
Anthracite, 326
Antibiotics, 192
   microorganisms resistant to, 296—297
Antibodies, function of, 419
Antigens, defined, 419
Anus, 363
Aorta, *364*
Apical meristems, 219
Aquatic Protista, algae as, 186, 188, 191
Aquatic reptiles, 333
Arable land, defined, 39—40
Archaeopteryx, 333, *334*
Armadillos, *135*
Arterial blood, 365
Artery(ies), 364
   coronary, *415*
Arthropods (trilobites), 170—172, 320, *321*
   diversity of, *174*
   evolution of, 336—337
   movement in, *173*
Artificial organs (and parts), *482, 483*
Artificial selection:
   breeding by, 301
   defined, 293
Artificially induced immunity, 420
Artificially produced mutations, 292—293
Asexual reproduction, 207—209
   cloning as, 208—209, 273—274, *484*
   human, 483
Asia, fauna of, 135, *136*
Asteroids, 89—90

Atmosphere:
   composition of, 95
   rain and snow patterns, *101*
Atmospheric movements, *101*
Atoll(s):
   formation of, *65*
   Ifaluk, *63, 64*
Atom, defined, 5
Atomic Energy Commission (AEC), 480, 525
ATP *(see* Adenosine triphosphate)
Attitudes toward death, 224—225
Australia, fauna of, 132, *133, 134*
Australopithecus africanus, 341, 342
Automobiles, as urban problem, 21
Auxin, 219
Axis, earth, 98—99
Axons, described, 409

B cells, 419
Baboons, *341*
   social behavior of, *437*
Bacteria:
   kinds of, 197, *198,* 199—201
   petri dish with colonies of, *199*
   resistant to antibiotics, 296—297
Bacterial diseases, 199—201
   table, 200
Bacteriophages, 204
Barnacles, *177*
Bats, *167*
   in coevolutionary complex, *429*
Bean plants, sprouting seeds of, *180*
Bees *(see* Honeybees)
Behavior:
   periodic, animal and plant, 105—108
   survival and, 425—454
      conceptual behavior, 446—453

Behavior:
   survival and,
      reproductive behavior, 431—436
      social behavior, 436—446
   (See also Coevolution)
Biceps, 391
Bile, 363
Binocular vision, 340
Biodegradable, defined, 27
Biofeedback, 392, 503
Biogeographic regions, *131*
   faunas of [*see* Fauna(s)]
Biogeography, 131—147
   defined, 131—132
Biological amplification, 51
Biological clocks, 106
   of humans, 108
Biological communities, as living parts of ecosystems, 68
Biological control, 144—146
Biological engineering, health improved by, 481—488
Biological evolution (*see* Evolution)
Biological research, ethical problems of, 483—485
Biomass, defined, 37
Biomes:
   defined, 118—119
   desert, 124—125
   of evergreen trees, *126, 127*
   freshwater habitats of, 128—130
   grassland, 122—123
   major, *119*
   temperate forest, 125—126
   as natural ecosystem, 70—76
   tropical rain forest, 119, *120, 121,* 122
   tundra, *127,* 128
Biosphere, 111
Bird-hipped dinosaurs, 329, *330*
Birds, 169, *170*
   digestive tract of, *361*
   eggs of, 246—247
   evolution of, 335—337

Birds:
    peck order of, 438, 439
    readiness to mate, 432
    reptiles give rise to, 333—335
Birth:
    human, 237
    of mammals (other than human), 245
    year-round sexuality and single, 340
Birth control, methods of, 240—244
    table of failure rates, 242
Birth rate, 250—252
    and change in population size, 250—251
    human control of, as unique, 255
    natural growth rate and, 251—252
    table, 252
Bituminous coal, 326
Blood, 299, 364—366
    arterial and venous, 365—366
    regulation of osmotic pressure of, 404
Blood cells:
    human, 222
    normalized, 300
Blood clotting mechanism, 414—416
Blood-group genes in aborigines, variations in, 304
Blood groups in humans, table, 271
Blood vessels, 364
Blue whales, 79
Body control, 501—503
Body language, 440, 446
Body processes, aging and slowing of, 224
Body temperature, control of, as homeostatic mechanism, 395—397
Bombay, India, 491
Bone-forming cells, 222
Bony fishes, 322
Botulism, 199—200

Bracket fungi, 190
Brain:
    hearing center of, 411
    mind control, 501—503
    perception and, 412—413
    sensory inputs processed by, 411
    size of: culture and, 343—344
        of first human fossils, 341—342
    vertebrate, evolution of, 446
    visual center of, 410, 411
Breast feeding, 237—239
Breeding, 301
    (*See also* Reproduction)
Brewer's yeast cells, 192
Brontosaurus, 331
Browsing animals, 336, 337
Budding, as asexual reproduction, 208—209
Buffalo hunting by train, 61
Butterflies:
    body temperature control of *Colias*, 397
    defense mechanisms of monarch, 417
    development of checkerspot, 175
    mimicry in, 429

C_3 pathways, 353
C_4 plants, 354
Calorie, defined, 359
Cameras, eyes compared with, 410
Camouflage:
    as defense mechanism, 417—418
    of fishes, 390
Cancer, trends in respiratory, 428
Canines, 335
Capillaries, 364
Carbohydrates:
    described, 349
    digestion of, 362
    as energy source, 370—372
    starch and glycogen as, 350
Carbon cycle, 70

Carbon dioxide molecules, transport of, 365—367
Carbon fixation (dark reactions) of photosynthesis, 353, 354, 355
Carbon monoxide, ill effects of, 21
Carboniferous forests, 324—325, 327
Career, using one's, to affect the future, 525—527
Carnivore, defined, 35
Carrier molecules, carrier-molecule-hydrogen combination, bonds of ATP and bonds of, 353, 354
Carriers, defined, 201
Castes:
    honeybee, 443
    termite, 443
Casts, fossils as, 314
Catalysts, enzymes as, 380, 381
Cell membranes, chemical structures of, 351
Cells, 151—153
    aging related to specialization of, 223—224
    B and T, 419
    blood, 222
        normalized, 300
    brewer's yeast, 192
    differentiated, 222
    electrical potential of, 406
    energy and structure of, 346—351
    mitotic divisions of, 215—216
    molecule transport to and from, 363—365, 366, 367
    nerve, 406
        firing frequency of, 412—413
    organ formation and movement of tissues, 216—217
    sex, 209—214
    target, 403

Cells:
    vary in number of chromosome sets, 214–215
Cellulose, described, 349, *350*
Cephalopods, 172
    cephalopod mollusk, *322*
Cerebral cortex (cerebrum), conceptual behavior and, 446–448
Cervix:
    described, 234
    diaphragm insertion and, 241
Cesium 137, concentration of, 479
Chambered nautilus, *322*
Change:
    environmental, 290–291, 452
    role of citizen in bringing about, 521–523
Charisma, 502
Checkerspot butterflies, development of, *174*
Chemical communication, 444
Chemical compounds, defined, 6
Chemical elements:
    essential minerals, table, 45
    major cycles of, *38*
    trace, 44
Chemical fossils, defined, 314
Chemical messengers of coordination system, 403–406
Chemical structure(s):
    of cells, 348–351
    of deoxyribose sugar, *382*
    indicating, *7*
Chemical systems, origins of life in nonliving, 288–289
Chemoreceptors, 411
Chicken embryos, development of, *218*
Chinchilla rabbits, coat color inheritance of, *270*
Chlamydomonas, life cycle of, *215*

Chlorinated hydrocarbons, 50–51
Chlorophyll pigments in photosynthesis, 351–353
Chloroplasts, *348*
Cholinesterase, 407
Chorion (membrane), 236
Chromatids, defined, 213
Chromosomes:
    in asexual reproduction, 207
    cells vary in number of sets of, 214–215
    crossing over of strands of, during synapsis, *214*
    halving, 210, 212
    haploid number of, *213*
    human female, *212*
    human male, *211*
    independent assortment of, *214*
    independent behavior of pairs of, 212–213
    location of genes on, 265–268
    (*See also* Genes)
    X chromosome inheritance, *268*
Cilia, defined, 152
Circadian rhythm, 105–106
Circulatory system:
    of grasshoppers, *370*
    invertebrate, absence of high-pressure in, 368
    lymphatic system as, 367–368, *369*
    of multicellular animals, 363–365, *366, 367*
Cities, 1–33
    as ecosystems, 2–13
    industrial: as new stage in urban evolution, 461–463
    as vulnerable to disruptions, 463–464
    interaction between ecosystems of farm and, *68, 69*
    origin of, 19–20

Cities:
    technological society and, 460–469
    urban ecology, 2, 13–19
    urbanization: defined, 2
    impact of, 19–33
Civic responsibilities in the future, 521–525
Civilization, temperate zone forests and, 73–74
[*See also* Culture(s)]
Clarke, Arthur, 483
Class(es):
    defined, 158
    examples of vertebrate, *164*
Classification of organisms, 153–161
    hierarchical, *157,* 158
    how classified, *156*
    table, 163
Cleaner fishes, 427–428, *430*
Clitoris, described, 231
Cloning, 208–209, 273–274, 484
    human, 483
Clots, defined, 415
Club mosses, 187
Clutch, defined, 247
Coal, formation of, 325–326
Coat color inheritance:
    in guinea pigs, *258, 262*
    incomplete dominance in, *269*
    multiple alleles and, *270*
Cobra, *168*
Coevolution:
    behavior and, 426–430
    coevolutionary interactions with flowering plants, 328
    defined, 305
Colias, body temperature control of, *397*
Colonies of organisms, 153
Colostrum, defined, 237
Comb jelly, *178*
Comets, 89–90
Commons:
    defined, 467

Commons:
  destruction and control of, 467—468
  need to change attitudes toward, 468—469
Communication:
  among honeybees, 443—444
  language as, 448—453
  modes of, 444—446
Communities, biological, 68
Compass directions, *107*
Complete proteins, 369
Compounds:
  chemical, defined, 6
  organic, of cells, 348—349
Conceptual behavior, language and, 446—453
Condoms, 241
Cone-bearing plants, 182
Coniferous trees, 126
Conjugation process, 209
Connecting neurons, portion of, *412*
Consciousness-altering (psychotropic) drugs, 408
Constipation, defined, 363
Consumer society, 463
Consumers in food chain, defined, 35
Contagious diseases, 199
Continental drift, 136
  postulated stages of, *137*
Continuously varying traits, genes controlling inheritance of, 271—277
Contour plowing, 48
Contraception, defined, 240—241
Contractions:
  labor, 237
  muscle, 391, 392
Control centers of nervous system, 410
Cooperative action, citizens involved in, 522—523
Coordination, steady state and, 402—414
Copulation:
  defined, 230

Copulation:
  estrus and, 245
  premarital, 240
Copulatory organs, defined, 230
  (*See also* Sex organs)
Cornish beam engine, 462
Coronary, defined, 415
Coronary artery, function of, *415*
Corpus luteum, defined, 235
Cortex:
  of adrenal glands, 404
  cerebral, 446—448
Courtship behavior, readiness to mate and, 431—435
Cowboy economy, 463
Coyotes, *160*
  dominance relations among, *439*
Crabs, rhythm of color change of, *105*
Cro-Magnon people, 342—343
Crops, major U.S., 492
Crossing over, described, 213—214
Crowding, effects of urban, 30-31
Cultural differences in perception, 449—450
Cultural evolution, 305—308
  defined, 446
  interrelation of biological and, 451—453
  language and, 450—451
  process of, 504—507
  urban areas as centers of, 31
Cultural interaction, cities as centers of, 31—32
Culture(s):
  brain size and, 343—344
  future, 497—509
  knowledge available to an individual from another, 519—521
  Netsilik Eskimo, 60—61, *62*, *63*

Culture(s):
  sexual activities and, 239—240
  survival, 455—470
    agricultural revolution and, 457—460
    by hunting and food gathering (*see* Hunting and food gathering cultures)
    technological culture, 460—469
Cuticles of vascular plants, 323
Cytoplasm, defined, 152

Daisies, *181, 183*
Dark reactions (carbon fixation) of photosynthesis, 353, *354, 355*
Dating, fossil, 316—317
Daughter cell, defined, 207
David-Neel, Alexandra, 520
DDT (dichloro-diphenyl-trichloroethane), 50—51, *52*
  resistance to, selection for, *294*
Death:
  changes in patterns of ten leading causes of, table, 421
  extinction of animals, 139, *140*
  of organisms, 223—227
Death rate:
  and change in population size, 250—251
  differential mortality in water snakes, *299*
  effects of agricultural revolution on, 458—459
  natural growth rate and, 251—252
  table, 252
Decay of variability, 302
Deciduous forests, 71, 120—121, 125—126

Decomposer in food chain, defined, 35
Defense mechanisms (protection), 414–424
Deoxyribonucleic acid (see DNA)
Deoxyribose sugar, chemical structure of, 382
Dependency:
  of the aged, 225–226
  of cities, 15
Deserts:
  as biome, 124–125
  increase of, 140–141
Detritus-based food chain, defined, 27–28
Diabetes, 399
Diaphragms (contraception), insertion of, 241
Diarrhea, defined, 363
Diatoms, 189
Diceros bicornis, 136
Diet, 368–377
Differential heating, 98
Differential reproduction of genotypes, 301–302
Differentiated cells, examples of, 222
Differentiation, 153
  in asexual reproduction, 208
  of populations, 303–311
Diffusion, defined, 115, 116
Digestive enzymes, table, 362
Digestive tract, 361, 362, 363
Dihybrid cross, defined, 262
Diminishing returns, law of, search for new fuel sources and, 494
Dinichthys, 324
Dinosaurs (ruling reptiles), 329, 330, 331, 332, 333
  fossilized footprints of, 316
Dioecious plant, defined, 209
Diploid cell, defined, 210
Discontinuous traits:
  Mendelian study of, 260–262

Discontinuous traits:
  problems in study of, 269–271
Discontinuous variation, continuously varying traits and, 272
Disease(s):
  bacterial, 199–201
    table, 200
  control of, increased health hazards and, 475
  fungal, 193
  genetic, 282
    table, 283
  noninfectious, 420–422
  plant, 423
  venereal, table, 244
  viral, 201–204
    table, 202
  (See also Epidemics; and specific diseases)
Dispersal, 135–147
  barriers to, 136–137
  impact of structures on, 141–142
  people as agents of, 142–147
Displacement activity, 439–440
Distribution of life, 111–135
  east and west, 130–135
  north and south, 118–128
  vertical, 111, 112, 113–118
  (See also Biogeography; Biomes; Dispersal)
Diurnal activity, 104–105
Diversity, 287–312
  of arthropods, 174
  differentiation of populations and, 303–311
  of forest animals, 71
  natural selection and, 293–302
  origins of, 288–289, 290, 291
  plant: in cities, 19
    in forests, 71
  in populations, 291–293

DNA (deoxyribonucleic acid):
  genes as sections of, 277, 278, 279–280
  transcription of, to RNA, 381, 382
Dogs:
  artificially bred, 301
  feral, 4
  jaws and teeth of, 165
Doldrums, 99
Domestication, places of, 143
Dominance, genetic: defined, 261
  incomplete, 269
Dominance behavior, 438–441
Dominant life form(s):
  insects as, 325
  mammals as dominant land vertebrates, 335–336
  reptiles as dominant land vertebrates, 329–333
Dominant traits, defined, 261
Dormancy of forests, 71
Double helix of DNA molecules, 277, 278, 279
Doubling time, relationship between exponential growth and, 252
Drives, 499
Drones, 443
Drosophila, 253
Drugs:
  antibiotic, 192
    microorganisms resistant to, 296–297
  mind control with, 501–502
  psychotropic, 408

Earth, 102, 103–108
  commons of: defined, 467
    destruction and control of, 467–468
    need to change attitudes toward, 468–469
  cross section of, 95
  distribution of life on (see Distribution of life)
  revolution of, 99
  (See also Sun)

Earth:
    in solar system, 91—92
    as spaceship, 467—468
Earthquakes, 136
Earthworms, *178*
Echolocation, defined, 413
Ecocatastrophes, examples of, 494—496
Ecological engineering, purpose of, 496—497
Ecological relationships, classification based on, 155
Ecological roles:
    of bacteria, 197—199
    distribution of life and, 131
Ecology:
    defined, 2
    ecological succession in forests, *74*
    future of, 488—497
    rural, 43—48
    (See also Cities; Ecosystems; Environment; Rural life)
Economies of scale, defined, 516
Economy:
    cowboy, 463
    economic transformations in cities, 10
    manufactured goods, 31
    (See also Industrial cities; Industrialized farms)
Ecosystems:
    agricultural revolution and destruction of, 459—460
    cities as, 2—13
    defined, 2
    farms as, 39—43
        interaction between cities and, 68, *69*
    future world, 488—492
    managers of, 496
    natural, 56—82
        components of, 67—70
        examples of, 70—80
        self-sufficient life, 57—67
        stability of, 80—81

Ecosystems:
    principles of, for understanding rural ecosystems, 35—39
Edema, defined, 367
Effectors:
    energy used by, 389—392
    of reflexes, 409
Efferent neurons, 409
Egg(s):
    chicken, *246*
    defined, 209
    fertilization of sea urchin, *216*
    fertilized, 209
    formation of hollow-ball stage after division of, *217*
    human, *210*
        implanation of, 234—235
        release of, *233*
        production of, 230
    of reptiles and birds, 246—247
Einstein, Albert, 87
EIS (environmental impact statement), 19, 524
Ejaculation, defined, 231
Electric charge differential, 406
Electrical appliances, energy consumption by, 23
Electrical potential of cells, 406
Electromagnetic spectrum, *88*
Electrons:
    defined, 5
    transport chain of, 357, *358*
Elements:
    defined, 4
    important, in living material, 5
    (See also Chemical elements)
Elephas maximus, 136
Embryo(s):
    defined, 209—210
    development of chicken, *218*

Embryo(s):
    human, 211, *236*
    three-layered, 216
Emphysema, trends in, *478*
Endocrine glands, 403
Endometrium, 234
Endoplasmic reticulum, *347*, 382, *383*
    smooth, 386
Endosperm, plant, 219
Energy:
    alternative sources of, 480—481
    of cities, 15—17
    as input to city, 8—9
    obtaining, *15*, 345—378
        by absorption and ingestion, 360—368
        cell structure and, 346—351
        by nutrition, 368—377
        by photosynthesis, 351, *352*, 353—354, *355*
        by respiration, 355—358, *359*, 360
    transfer of, photosynthesis and, 352—353
    (See also Food; Fossil fuels; Sun)
Energy consumption, 379—393, *498*
    of cities, 22—25
    by effectors, 389—392
    of electrical appliances, table, 23
    future of, table, 508
    individual, 514—517
        table, *514*
    for maintenance, 386—389
    oil, 326
    per capita: population size and, 515, *516*, 517
        U.S. (1850—1970), *24*
    in synthesis, 380—386
Energy losses in food chains, 37—38
Energy Research and Development Commission (ERDA), 480

Environment:
  agricultural revolution and war against, 459–460
  beneficial effects of urban, 31–33
  deterioration of, national contributions to, table, 14
    (See also Pollution)
  development may be upset by unfavorable, 219–221, 223
  diversity of, 289–290
  effects of, on IQs of twins, 276
  effects of individual life style on, 513–517
  inorganic, for photosynthesis, 44–46
  intelligence and, 310
  interaction of genotypes with, 217
  population-resource-environment problems, careers related to, 526–527
    (See also Cities; Ecology; Natural resources; Rural life)
Environmental change, 290–291, 452
  cultural evolution and, 452
Environmental gradients, altitudinal gradients as, 117–118
Environmental impact statement (EIS), 19, 524
Environmental organizations, table, 524
Environmental Protection Agency (EPA), National Human Monitoring Program of, 477
Enzymes:
  as catalysts, 380, 381
  defined, 277, 348–349
    (See also specific enzymes)
Epidemics:
  conditions favoring, 423

Epidemics:
  worldwide plague, 495
Epiphytes, defined, 120
Epistatic, defined, 271
Equus asinus, 161
Equus burchelli, 161
ERDA (Energy Research and Development Commission), 480
Erie, Lake, eutrophication of, 131
Erosion:
  reducing soil, 46–48
  wind, 49
Escherichia coli, cells of, 151
Eskimo carvings, 457
Esophagus, 362
Estrogens, 235, 404
  table, 405
Estrus, defined, 245
Estrus cycle, 245
Estuaries, distribution of life in, 115, 117
Ethical problems, 483–488
  of biological research, 483–485
  of cloning, 483
  of genetic screening, 485–486
  of medical research, 486–488
Eugenics, defined, 284
Euglena, 153
Eukaryotic cells, 346–347
  animal, 151
  genetic material of, in chromosomes, 207
  plant, 152
Euphenics, defined, 284
Euphydryas chalcedona, 174
Eurasia:
  fauna of, 133, 135
  land bridge between North America and, 135
Eurypterids, 321
Eutrophication, 50
  of Lake Erie, 131
Evergreen trees, biomes of, 126, 127

Evolution, 301–308, 323–344
  of early land animals, 328–335
  factors other than selection affecting, 302
  human, 337–344
    genetic evolution, 305–308
  industrial melanism as form of, 297, *298*
  interrelation of cultural and biological, 451–453
    (See also Cultural evolution)
  of land plants, 323–328
  of mammals, birds, arthropods, flowering plants, and insects, 335–337
  organic, defined, 291
  of species, *303*
    splitting process of, resulting in new species, 304–305
  of vertebrate brain, 446
  of vertebrates into land-dwelling forms, 328–329
    (See also Coevolution)
Excretion, 399–401
Exponential growth:
  defined, 251
  relationship between doubling time and, table, 252
External defense mechanisms of multicellular organisms, 418–420
External fertilization, 230, 247–248
Extinction of animals, 139, 140
Eyes, *410*
  of frogs, *220*
  transplant of, in salamanders, *221*

F_1 generation, defined, 261
F_2 generation, defined, 261

Fallopian tubes:
  described, 231
  and ovulation, 233
  tubal ligation, 243
Family (classification system), defined, 158
Family size, population growth and, 517–518
Farming, return to individual, 518–519
  (See also Rural life)
Fat-soluble vitamins, 372
Fats:
  as energy source in diet, 370–372
  in respiration, 360
Fatty acids, 371
Fauna(s):
  African, 135
  Asian, 135, *136*
  Australian, 132, *133, 134*
  defined, 132
  Eurasian and North American, 133, 135
  impact of people on, 139–147
  South American, 132–133, *135*
  (See also Animals)
FDA (Food and Drug Administration), 372, 376
Feather color inheritance, multiple loci and, *272*
Feces, 363
Feedback loops, coordination and, 403, *404*
Feral dogs, 4
Fermentation:
  alcoholic, 358, *360*
  products of, 358–359
Ferns, 185–186, *187, 326,* 327–328
Fertile land, 40
Fertilization:
  external, 230, 247–248
  human, *210*
  internal, 230
  of sea urchin eggs, *216*
  of seed plants, 181–182

Fertilization membrane, 216
Fertilizers:
  as major output of farms, 50–52
  natural and synthetic, 40
Fetus, defined, 237
  [*See also* Embryo(s)]
Fibers, muscle, 392
Fibrils, muscle, 392
Fiddler crabs, *176*
Filter feeders, 361
Fireflies, mimicry in, 428
Firing frequency of nerve cells, 412–413
Fishes, 169, *172*
  camouflage of, *390*
  cleaner, 427–428, *430*
  as first vertebrates, 321–322
  flow through gills of, *388*
  with light producing organs, *114*
  lobe-finned, 322, *325*
  lungfishes, 322, *324*
  mimicry in, 427–428
  nests of, *173*
  overfishing, table, 79
  poisonous, *418*
  ray-finned, 322
  water loss by, *399*
Flagella, defined, 152
Flight patterns of bees, *106*
Flood control, 475
Floras:
  defined, 139
  impact of people on, 139–147
  storied vegetation, 119–120
  (See also Plants)
Flowering plants:
  evolution of, 335–337
  life cycle of, *224*
  as most recent seed plants, 328
  pollinating insects and, 337
Flowers, longitudinal section through, *180*
Flying reptiles, *333*
Follicle stimulating hormone (FSH), 234–235

Food:
  absorption and ingestion of, 360–368
  as energy source, *15*
  four basic groups of, 376–377
  in Ifaluk society, 67
  staple foods, *66*
  molecules of, transported by circulatory system, 364
  nutrition, 368–377
  organically grown, 326
  respiration as controlled oxidation of, 356–358
  scarcity of: failure of Green Revolution and, 390–492
  massive future starvation, 489–490
  table, 508
Food chains, *36*
  defined, 14, 35
  detritus-based, 27–28
  oceanic, *78*
  as part of ecosystems, 35–38
  in temperate forests, 75
Food and Drug Administration (FDA), 376
Food webs, 35, *36*
  described, *15, 16*
Foraminiferans, *320*
Forebrain, 446
Forecasting:
  defined, 472
  of the future, 508–509
  of population growth, 473–474
Forest fires, aftermath of, *73*
Forests:
  biome of tropical rain, 119, *120, 121, 122*
  Carboniferous, 324–325, *327*
  deciduous, 71, 120–121, 125–126
  ecological succession in, 74
  fossilized, *315*

Forests:
  scrub, 122
  second growth, 70
  temperate: biome of, 125–126
    as natural ecosystem, 70–76
Fossil fuels:
  defined, 16
  depletion of, 493–494
  formation of, 325–326
  for harvesting, 43
Fossils:
  dating of, 316–317
  defined, 16
  as records of past organisms, 314–316
Fraternal twins, 274–275
Freshwater habitats of biomes, 128–130
Frogs:
  eyes of, 220
  formation of three-layered stage in, 219
  metamorphosis of tadpoles into, 171
Fruit-bearing plants, 180–182
Fruit flies:
  population cycle of, 252, 253
  studies of heredity on, 264
FSH (follicle stimulating hormone), 234–235
Fungi:
  absorption process of, 191–194
  associations of, in soil, 46
  diseases caused by, 193
  reproductive structure of three kinds of, 190

Galaxy(ies), 84–86
  composition of, 84–85
  Milky Way, 86
  (See also Solar system)
Galileo, 504
Gall bladder, 362, 363

Gametes, sexual reproduction begins with production of, 209–214
Gannets:
  courtship behavior of, 434
  group formation of, 435
Gas exchange, 387
Gastropods, 172
Genes, 263–265
  behavior of, on X chromosome, 268
  blood group, 304
  controlling inheritance of continuously varying traits, 271–277
  defined, 213
  frequencies of, 300
  location of, on chromosomes, 265–268
    (See also Chromosomes)
  as sections of DNA molecules, 277, 278, 279–280
  sickle-cell, 298–300
  (See also Mutation)
Genetic code, 277
  language of, 279–280
  table, 281
Genetic diseases as causes of mutations, 282
  table, 283
Genetic drift, defined, 302
Genetic evolution (see Evolution)
Genetic information, transmission of, 307
Genetic IQ range, race and, 310
Genetic material:
  exchange of, in asexual reproduction, 209
  halving, 210, 212
Genetic screening, ethical problems of, 485–486
Genetic uniqueness of individuals, 512–513
Genetics:
  defined, 258
  Mendelian, 258–265

Genetics:
  (See also Chromosomes; Genes; Heredity; and under Genetic)
Genotypes:
  in asexual reproduction, 208
  differential reproduction of, 301–302
  interaction of, with environment, 217
Genus, defined, 158
Geographic variation in populations, 303
Geological time table, 317
Gestation period, 245
Giant mammals, 336
Gigantism, defined, 336
Giraffe, reflex arc of, 409
Glands (see specific glands)
Global ecosystems, 69–70
Global village, world as, 464–467
Glycogen, 350
Glycolysis, reactions of, 356
Glyptodon, 135
Goiter, 375
Gonorrhea, incidence of, table, 244
Gorillas, skull of, 342
Graft rejection, 420
Grains:
  harvests of, 43
  production of, 40
  proteins in, 369
Grapes, cloning of, 274
Grass anole, 168
Grasses, 123
Grasshoppers:
  circulatory system of, 370
  life cycle of, 223
Grasslands:
  biome of, 122–123
  as natural ecosystems, 76–77
  people and increase of, 140–141
Grazing animals, 336
Green Revolution, failure of, 490–492

Greenhouse effect, defined, 97–98
Group formation, parental care leading to, 435–436
Group life, advantages of, 437–441
Growth hormone (somatotropin), 404
Guinea pigs, 259
  coat color inheritance in, 258, 262

Haploid cells, defined, 210
Hardin, Garrett, 467
Hardpan, 141
Harmless protozoa, 196
Hatching, 247
Haversian system, 371
Health, improved by biological engineering, 481–488
Health hazards:
  future, 475–477
  from industrial and building activity, 478–481
  posed by anaerobic microorganisms, 359
Hearing aids, 482, 483
Hearing center in brain, 411
Heart, 364, 365
  muscle fibers of, 222
Heart attacks, 415
Heat:
  defined, 97
  in thermodynamics, 8
  waste, 494
Heat balance of earth, 98
Hectare, defined, 15
Heights:
  distribution of male, in U.S., 272
  mountain, 112
Helmont, Jan Baptista van, 14
Hemoglobin:
  described, 299
  transport of, 365
Hemophilia, occurrence of, in European royal families, 267

Hens, chicks and, 248
Herbaceous plants, 76
Herbivore, defined, 35
Heredity, 257–286
  complexities of, 265–277
  Mendelian genetics, 259–265
  pedigrees and, 258–259
  workings of, 277–285
  (See also Chromosomes; Genes; Mutation)
Hermaphroditic species, defined, 209
Hermit crab, 177
Heterotypic schools, defined, 437
Heterozygosity, defined, 262
Hibernation, 100
Hierarchy:
  in classification system, 156, 158
  linear dominance, 438
High-pressure circulatory system, absence of, in invertebrates, 368
High-speed (nervous) coordination, 406–409
Himalayan rabbits, coat color inheritance of, 270
Hindbrain, 446
Hives, bee, 442, 443
Holdfasts, 104
Hollow-ball stage following division of zygote, 217
Homo erectus, 162
  brain size of, 342–343
Homo sapiens:
  appearance of, 337, 342–343
  cultural evolution of, 305–308
  jaws and teeth of, 165
  skull of, 341, 342
  tree-dwelling ancestors of, 339–340
  (See also Human beings)
Homeostasis, steady state and, 395–402
Homozygosity, defined, 264

Honeybees:
  as social insects, 442, 443
  sun position and flight patterns of, 106
Hormonal system, role of, in coordination, 403
Hormones:
  birth control, 241
  role of, 404, 405
  table, 406
Horsetail, 187
Hosts, defined, 37
Housefinch nests, 247
Human beings:
  as all belonging to single species, 305–308
  biological clocks of, 108
  blood-cell types in, 222
  blood groups in, table, 271
  chromosomes in: female, 211
    male, 212
  circulatory system of, 363–365, 366, 367
    lymphatic system as, 367–368, 369
  digestive tract of, 361, 362, 363
  diseases affecting (see Diseases)
  embroyo of, 211, 236
  evolution of, 337–344
    genetic evolution, 305–308
  female: menstrual cycle in, 232–233, 234, 235–236
    puberty, 238
  in future society, 472–488
  impact of: on desert biome, 124–125
    extinction of other species and, 150
    on floras and faunas, 139–147
    on grassland biome, 123
    and increase in deserts and grasslands, 140–141
    on oceans, 78–80

Human beings:
  impact of:
    on succession in lakes and ponds, 130
    on taiga biome, 127
    on temperate forest biome, 125—126
    on tundra biome, 128
  language as mode of communication restricted to, 448—453
  life cycle of, 215
  male: distribution of male heights in U.S., 272
    puberty, 238
    sperm, 209, 232
  nervous system of (see Nervous system)
  nutrition of, 368—377
  races as arbitrary divisions of, 308—311
  reproduction in, 210, 230—244
    reproduction in other animals compared with, 245—246
    [See also Egg(s), human; Sexual reproduction]
  sex organs of: female, 231—232, 233, 234—236
    male, 210, 230, 231, 232
  social conditions and life expectancy of, 225—226
  subject to natural selection, 298—300
  (See also Cultural evolution; Culture; Evolution; Heredity; Homo erectus; Homo sapiens; Individual, the; Neanderthal man; Population growth; Survival)
Human development, environmental conditions affecting, 219—221, 223
Human fossils, brain size of first, 341—342
Human nature, societal structure and, 499—501

Hummingbirds, feeding young, 249
Humus, defined, 46
Hunting and food gathering cultures, 456—457
  examples of, 59—60
  methods used by, 57—59
  of North American peoples, 60—63
  (See also Ifaluk society)
Hydrocarbons, 50—51
Hydroelectric plant, diagram of, 18
Hydroelectric power, described, 16—17
Hydrogen bonds, 277, 278
Hypotheses, defined, 504

Ichthyosaurus, 333
Ideas, cities as sources of new, 31
Identical twins, 274—277
Ifaluk Atoll, 63, 64
Ifaluk society, 65
  foods of, 67
  staple foods, 66
Immunity:
  artificially induced, 420
  as internal defense mechanism, 419
Implantation, human, 234—235, 236
Incisors, 335
Incomplete dominance, 269
Incomplete metamorphosis, 218
Incomplete proteins, 369
Incubation, egg, 247
Independent assortment of chromosomes, 213, 214, 265
Individual, the, 512—528
  civic responsibilities of, 521—525
  effects of life style of, 513—517
  family size, population growth and, 517—518

Individual, the:
  genetic uniqueness of, 512—513
  knowledge available to, from other cultures, 519—521
  means of minimizing effects of, on environment, 518—519
Individual space, territoriality and, 441
Inducer, defined, 216
Induction, 216
  protein synthesis controlled by, 385—386
Industrial cities:
  as new stage in urban evolution, 461—463
  as vulnerable to disruptions, 463—464
Industrial health hazards, 478—481
Industrial melanism in moths, 297, 298
Industrial pollution (see Pollution)
Industrialized farms:
  estranged from resource base, 42
  impact of, 48—55
  spread of, 42—43
  transition to, 41—42
Infancy, 237—239
Infanticide, 65
Inflorescence in seed plants, 180
Influenza virus, 203
Infrared radiation, defined, 97
Ingestion, energy derived from, 360—368
Inheritance (see Heredity)
Inland genotype plants growing in seashore habitats, 274
Inorganic substances, defined, 6
Inputs:
  city, 12

Inputs:
  city:
    city as dependent on, 11, 13
    defined, 4
    energy, 8—9
    matter, 4—8
    measuring, 13
  to industrialized farms, 42—43
  sensory, processed by brain, 411
Insecticides, effects of, 296
  (See also DDT)
Insects, 170—176
  defenses of, 427
    mimicking as, 420, 428
  digestive tract of, 361
  as dominant life form, 325
  evolution of, as interrelated with evolution of birds, mammals, and flowering plants, 336—337
  as pollinators, 48, 337
  social, 442—444
Integrated control, defined, 296
Intelligence, environment and, 310
Interferon, described, 419—420
Internal defense mechanisms of multicellular organisms, 418—420
Internal fertilization, 230—232
Interpersonal distance, territoriality and, 441
Intertidal zone, defined, 115
Intestines, surface area of, 366
Intrauterine devices (IUDs), 241, 242
Invertebrates, 170—180
  circulatory system of, 368
  courtship behavior of, 431—432
  of early seas, 320—321
  types of, 173—178

Involuntary control of muscles, 391—392
Involuntary muscles, 391
Ions, defined, 6
IQ (intelligence quotient):
  genetic IQ range, 310
  as measure of school performance, 311
  of twins, effects of environment on, 276
Iris of the eye, 410
Irrigation, defined, 44
IUDs (intrauterine devices), 241, 242

Jaws:
  dog and *Homo sapiens*, 165
  in early fishes, 322, 323
Jellyfishes, *178*
Jenner, William, 420
Jet lag, 108
Jupiter, 92—93

Kakabekia, 319
Kangaroos, *133*
Kelp, 188
Kidney machines, *400*
Kidneys, reabsorption of molecules by, *398*, 399
Kilowatt-hour, defined, 22
Kingdoms:
  defined, 158
  four, 161
    classification table, 163
Kinship structure, defined, 60
Koala bears, *134*
Krebs cycle, *357*
Kutz, F. W., 477

Labor (birth), defined, 237
Lactation:
  in animals other than man, 245—246
  defined, 237
Lactic acid, as product of fermentation, 359

Lakes:
  eutrophication of, *131*
  as habitats, 129
  succession in, 129, *130*
Lampreys, *323*
Land, farm ecosystems dependent on, 39—40
  (See also Soil)
Land animals, evolution of, 328—335
Land plants (see Plants)
Land vertebrates, 328—336
  evolution of, 328—329
  mammals as dominant, 335—336
    (See also Mammals)
  reptiles as dominant, 329—333
    (See also Reptiles)
Language:
  body, *440*, 446
  conceptual behavior and, 446—453
  of genetic code, 279—280
Large intestine, *362*, 363
Larvae, 218
Lascaux paintings, *343*
Laterite, defined, 41
Laterization, *122*
Latimeria, *325*
Latitudinal gradients, 118—128
Lawrence Radiation Laboratory, 525
Laws, defined, 504
Leaching, 50
Leaf fossil, *315*
Legumes, proteins in, 369
Lemurs, *339*
Lens of the eye, 410
Lichens, *194*
Life:
  first signs of, 314—319
  origins of, in nonliving chemical systems, 288—289
  self-sufficient, 57—67
    (See also Ecosystems, natural)

Life:
   (See also Distribution of life; Diversity; Evolution; Life cycles; and specific life forms)
Life cycles, 207–215
   of flowering plants, 224
   of grasshoppers, 223
   of green algae, 215
   of human beings, 215
Life form, defined, 117
   [See also Dominant life form(s); and specific life forms]
Life style, individual, effects on environment of, 513–517
Light, honeybees' use of polarized, 444, 445
Light energy, photosynthesis and, 351–352
Light reactions of photosynthesis, 353, 355
Limits of Growth, The (MIT study), 508
Linear dominance hierarchy, 438
Linkage of genes, defined, 265–266
Lipid metabolism, 360
Lipids, described, 349
Liver, 155, 362, 363
   cells of, 347
Lizards, 170
   body temperature, control of, 396
   jaws and teeth of, 165
Lobe-finned fishes, 322, 325
Locus (loci):
   defined, 213
   multiple, 271
      feather color inheritance and, 272
Loxodonta africana, 136
Lung circulation (pulmonary circulation), 367
Lungfishes, 322, 324
Lungs, 368
   inflation of, 387

Luteinizing hormone, 235
Lymph nodes, 367
Lymphatic system, 367–368, 369
Lymphocytes, 367

Macromolecules, defined, 8
Macrophages, 367
Magnolia, 181
Maintenance, energy for, 386–389
Malaria, 300
Malarial parasites, life cycle of, 195
Malnourishment, protein, 369–370
Mammals, 162, 163, 166
   evolution of, 335–337
   giant, 336
   placental, 166, 167
   reproduction in (see Reproduction)
   reptiles give rise to, 333–335
   (See also specific mammals)
Manhattan (New York City), 3
Mankind (see Human beings)
Manufactured goods, urban areas as sources of, 31
Mars, 92
   composite photograph of, 93
Marshes, importance of, 117
Marsupials, 133, 134
Mating:
   in asexual reproduction, 209
   courtship behavior and, 431–435
   (See also Reproduction)
Matter, 4–8
   cycles of, through ecosystems, 38–39
Mean annual temperature in Iceland, 489
Medical research, ethical problems of, 486–488
Medulla of adrenal glands, 404
Meiosis:
   defined, 210

Melosis:
   process of, 210, 212–214
   stages of, 213
Melanic, defined, 297
Melanism in moths, 297, 298
Membrane potential, as beginning of nerve impulse, 406, 407
Mendel, Gregor, 259, 260–265, 291
Mendelian genetics, 259–265
Mendel's laws, 264
Menopause, defined, 235
Menstrual cycle, 232–233, 234, 235–236
Menstruation, defined, 234
Mercury (planet), 90–91
Messenger RNA (mRNA), 381
Metabolism:
   energy for, 349–351
   lipid, 360
Metamorphosis, 171
   complete and incomplete, 218
Metaphase, 208
Metaphyta, classification of, table, 163, 164
Metazoa, classification of, table, 163, 165
Metric table, 529
Microorganisms:
   anaerobic, 359
   resistant to antibiotics, 296–297
   (See also specific microorganisms)
Midbrain, 446
Middle East, agriculture first practiced in, 457–458
Migrations, human, and extinction of animals, 140
Milfoil, variation in size of, 275
Milky Way galaxy, 86
Mimicry:
   defined, 426–427
   in fishes, 427–428
   in plants, 428–429, 430

Mimicry complex, 427
Mind control, 501–503
Minerals, 374–375
   essential, table, 45
   trace, defined, 44
   worldwide depletion of, 492
      table, 493
Mines, Bureau of, 493
Mites in coevolutionary complex, 429
Mitochondria, *347, 348*
Mitosis:
   in asexual reproduction, 207
   stages of, *208*
Models in coevolution, 426–430
Molars, *335*
Molds (fungi), *190*
Mole, defined, 360
Molecules:
   carrier, 353, *354*
   control of protein synthesis and repressor, 385
   defined, 6
   DNA, genes as sections of, 277, *278, 279*–280
   macromolecules, 8
   reabsorbed by kidneys, *398, 399*
   transport of, to and from cells, 363–365, *366, 367*
Mollusks, 172–173, 176, 178
   early sea, 321, *322*
   examples of, *177*
Monarch butterflies, defense mechanisms of, *417*
Monera, 196–201
   classification of, table, 163–165
Monoculture, 48
   and agricultural instability, 75
   growth of, 81
Monoecious plants, defined, 209
Moon, 89–92
   earth rhythms and revolution of, *103*, 104
Moss plants, *188*

Moths:
   in coevolutionary complex, 429
   distribution of peppered, *304*
   industrial melanism in, 297, *298*
Motor cortex, 447
Mountains:
   gradients of vegetation on tropical, *117*
   heights of, 112
Mouth, the, in digestive process, 361, *362*
mRNA, 381
Mules, *161*
Multicellular, defined, 151
Multicellular organisms, 152–153
   circulatory system of, 363–365, *366, 367*
   defense mechanisms of, 418–420
   development of, 215–223
Multiple alleles, 269, 271
   coat color inheritance and, *270*
Multiple loci, 271
   feather color inheritance and, *272*
Muscles:
   cells of smooth, *222*
   fibers of, *222*
   tissue of, as effector, *391, 392*
Mushrooms, *190*
Musk oxen, social behavior of, *437*
Mutagen, defined, 282
Mutation rate, defined, 292
Mutations, 280–285
   artificially produced, 292–293
   diversity in population and, 291–292
   genetic diseases as causes of, 282
      table, 283
   predicting probability of, 292

Mycorrhiza, defined, 46
Myelin sheaths, 408
Myths, racist, 309–311

National Cancer Institute, 421
National Environmental Policy Act (NEPA; 1969), 19
National Human Monitoring Program (EPA), 477
Natural ecosystems (*see* Ecosystems, natural)
Natural fertilizers, 40
Natural gas, formation of, 326
Natural growth rate, 251–252
   table, 252
Natural resources:
   failure of nonrenewable, *508*
   population-resource-environment problems, careers related to, 526–527
   worldwide depletion of, 492
      table, 493
Natural selection, diversity and, 293–302
Nature-nurture controversy, 273, 276, 310
Neanderthal man, 342, *343*
Negative feedback loops in control of protein synthesis, 385
NEPA (National Environmental Policy Act; 1969), 19
Nephrons, described, 398
Neptune, 93–94
Nerve cells, *222, 406*
   firing frequency of, 412–413
Nerve impulses, *406*, 407
   facilitation of, 412
   transmission of, 408–409
Nervous coordination (high-speed coordination), 406–409
Nervous system:
   coordination and, 403, 406–409

INDEX

Nervous system:
  as highly organized, 409—413
  (See also Brain)
Nests:
  alligator, 246
  bird, 247, 248, 249
  fish, 173
Netsilik Eskimos, as hunters and gatherers, 60—61, 62, 63
Neurons:
  afferent and efferent, 409
  as basic units of nervous system, 406
  membranes of, responding to stimuli, 406—407
  portion of connecting, 412
Neurosecretion, defined, 404
Neutrons, defined, 5
Nitrates, 477
Nitrogen, excretion of, 397—398
Nocturnal activity, 105
Nomenclature of classification system, 158—159
Nongenetic information, transmission of, 307
Noninfectious diseases, 420—422
Nonliving chemical systems, origins of life in, 288—289
Nonprimates, reproduction of, 245
Nonsolar energy, described, 17
North America:
  fauna of, 133, 135
  land bridge between Eurasia and, 135
Northern solstice, 98
Nuclear Regulatory Commission (NRC), 480
Nuclear war, 494—495
Nuclei of cells, 151, 347
Nucleotides, 279
Nursing (breast feeding), 237—239

Nutrition:
  human, 368—377
  oceans and poor supply of, 77—78
  soil nutrients, 44—46
  (See also Food)
Nymphs, 218

Obesity, causes of, 370
Objectively verifiable reality, science as study of, 414
Oceans:
  currents of, 113
  depths of, 111—112
  food chains in, 78
  habitats of, 113
  habitats provided by, 112—117
  inhabitants of rocky shores of, 115
  inland genotype plants growing in seashore habitats, 274
  as natural ecosystems, 77—80
  tides of, 103, 104
ODCs (overdeveloped countries), 465—467
Oil:
  formation of, 326, 327
  world movement of, 26
Omnivore, defined, 37
One part per million, diagram of, 53
OPEC (Organization of Petroleum Exporting Countries), 497
Order (classification), defined, 158
Orders of magnitude, relative sizes in terms of, 94
Organ systems:
  defined, 153
  major, 156
Organelles of cells, 151
Organic compounds, major, of cells, 348—349
Organic evolution, defined, 291

Organic substances, defined, 6
Organically grown food, 326
Organization of Petroleum Exporting Countries (OPEC), 497
Organs:
  artificial, 482, 483
  copulatory, 230
  (See also Sex organs)
  defined, 153
  formation of, 216—217
  transplants of, 482
    immune responses in, 420
    transplant of salamander eye, 221
  (See also specific organs)
Orgasm, defined, 231—232
Ortega y Gasset, José, 520
Osmosis, 115, 116, 117
Outbreak-crash population cycle, 252, 253, 254
Outputs:
  city, 10—11, 12
  measuring, 13
  fertilizers and pesticides as major farm, 50—52
  to industrialized farms, 43
  soil as major farm, 49
Ova [see Egg(s)]
Ovaries:
  defined, 210, 230
  seed plant, 180
Overdeveloped countries (ODCs), 465—467
Overloading of homeostatic mechanisms, 401—402
Overweight, causes of, 370
Ovulation, human female, 232—234
[See also Egg(s)]
Ovules, seed plant, 182
Oxidation, food, respiration as control of, 356—358
Oxygen:
  molecules of, transported by circulatory system, 364—365
  as not necessary to all forms of respiration, 358—359

Paleontologist, defined, 314
Pancreas, *362*, 363
Paramecium, 154
Parasites:
  defined, 37
  plant, 423
Parasitic protozoa, 194
Parental care:
  after hatching, 247
  leading to group formation, 435—436
Parental generation, 261
Parthenogenesis, defined, 208, 209
Pea plants:
  inheritance of discontinuous traits in, *260, 261*
  results obtained by crossing, *263, 264*
Peck order, 438, *439*
Pedigrees showing occurrence of hemophilia in European royal families, table, 267
Penicillium, 192
Penis:
  described, 230, *232*
  erection of, 231
Penstemon, 183
People (*see* Human beings)
Peppered moths, 297, *298*
  distribution of, *304*
Per capita, term, defined, 22
Per capita energy consumption:
  population size and, 515, *516,* 517
  in U.S. (1850—1970), 24
Perception, 411—412
  cultural differences in, 449—450
  variations in forms of, 413—414
Perennial plants, earth rhythms and, 100—101
Permafrost, 127
Pesticides:
  as major output of farms, 50—52
  plane released, *51*

Pests, management versus extermination of, 496
Petrification, defined, 314
Phenotypes in asexual reproduction, 208
Pheromones, 444
Phloem:
  defined, 184—185
  longitudinal section through, *186*
Photosynthesis, 351, *352,* 353—355
  by algae, 320
  blue-green algae as photosynthetic Monera, 197
  defined, 14
  inorganic environment for, 44—46
Phylum, defined, 158
Phytoplankton, defined, 77
Pig populations, weight distribution in, *273*
Pine tree branches, *184*
Pioneer 10 (space probe), 86, *87*
Placenta, 166
  development of human, *236*
Placental mammals, 166, *167*
Plague, worldwide, 495
Plains Indians, as hunters and gatherers, 60
Planets, 89—94
  (*See also* Solar system)
Plankton:
  defined, 77
  types of, *77*
Plants:
  composition of, 95—96
  defenses of poisonous, *414, 416,* 417
  desert, *124*
  development of, 219
  diseases of: fungal, 193
    parasitic, 423
    viral, 201
  diversity of: in cities, 19
    in forests, 71
  domestication of, *143*

Plants:
  early land, 323—324, *325*
  ferns and seed plants replace, 326—328
  earth rhythms and, 100—101
  effects of day and night lengths on, 104—105
  herbaceous, 76
  interrelationships of, with other living things, 46—48
  major kinds of, 180—186
  manufacturing processes of, 14
  (*See also* Photosynthesis)
  mimicry in, 428—429, *430*
  monoecious and dioecious, defined, 209
  periodic behavior of, 105—108
  providing side benefits for cities, 17, 19
  reproduction in, 244—250
  reproductive strategies, 248—250
  seashore genotype, in inland areas, 273—274
  (*See also* Distribution of life; Flowering plants; Seed plants)
Plasmodium, 194, *195*
Pleistocene overkill, 140
Pluto, 93—94
Poisonous animals, 417, *418*
Poisonous plants, *416,* 417
Poisons in food chains, as increased health hazard, 475—477
  (*See also* Fertilizers; Pesticides; Pollution)
Polarized light, honeybees' use of, 444, *445*
Polio virus, *203*
Political action, 523—525
Pollen grains, *182*
Pollination:
  as coevolutionary system, 429
  in seed plants, 181

Pollinators, insects as, 48, 337
Pollution:
   future industrial, 478
   radioactive, 478–480
   urban, 25–31
      effects on inhabitants, 29–31
   (See also Air pollution; Water pollution)
Pollution control, future, 488
Polymers, defined, 349
Polypeptide, defined, 277
Polypeptide chains, formation of, 384
Polyploid, defined, 214
Polyunsaturated fats, 371
Ponds:
   effects of seasonal changes in temperate zone, 129
   succession in, 129–130
Population crashes:
   defined, 253
   human, 254
Population dynamics, defined, 250
Population growth:
   effects of, 474–475
   effects of agricultural revolution on, 458–459
   forecasts on, 473–474
   future, table, 508
   as outbreak, 253, 254
   U.S., 517
Population-resource-environment problems, careers related to solving, 526–527
Population size, per capita energy consumption and, 515, 516, 517
Populations:
   biological community, 68
   differentiation of, 303–311
   diversity in, 291–293
   gene frequencies in, 300
   replacement of parents in, 254–255
   reproduction and formation of, 250–255

Posture(s):
   as expressions of dominance, 440
   readiness to mate and presentation, 433
   submissive, of coyotes, 439
   upright, of first human fossils, 341–342
Poverty, rural, 53, 54
Power, defined, 22
   (See also Energy)
Precambrian organisms, 319
Predatory animals, defined, 35
Premarital copulation, 240
Premenstrual tension, defined, 235
Presentation posture, readiness to mate and, 433
Primates, 338
   reproduction of, 245
   (See also specific primates)
Producers in food chains, defined, 35
Progesterone, 404
   in menstrual cycle, 235
   table, 405
Prokaryotic cells:
   defined, 151
   internal structure of, 346
Prophase:
   of meiosis, 213
   of mitosis, 208
Prostaglandins, 405
   table, 406
Prosthesis, defined, 482
Protection (defense mechanisms) for steady state, 414–423
Protein malnourishment, 369–370
Protein synthesis:
   control of, 385–386
   first step of, 381, 382
   second step of, 382–384
   as series of enzyme-controlled steps, 384–385
Proteins:
   described, 348, 349
   digestion of, 362

Proteins:
   importance of, 368–370
   in respiration, 360
   structures of, 385
   (See also Enzymes)
Protista, 186–196
   classification of, table, 163, 164
Protoceratops, 332
Protons, defined, 5
Protozoa:
   free living, 196
   get food by ingestion, 194–196
Psychotropic drugs (consciousness-altering drugs), 408
Pterosaurs, 333
Puberty, human male and female, 238
Public Health Service, 487
Public transportation, as urban problem, 21
Pulmonary (lung) circulation, 367
Pupae, 218
Pythons, vestigial legs of, 169

Queen bees, 442, 443
Quercus, leaves and acorns of, 162

Rabbits, coat color in, 270
Races, as arbitrary divisions, 308–311
Racist myths, 309–311
Radiation:
   as cause of mutation, 292–293
   infrared, 97
   reflected, 98
   shortwave, 97
Radioactive dating of fossils, 316–317
Radioactive pollution, 478–480
Rain forests, biomes of tropical, 119, 120, 121, 122
Rain shadow, 101

Ray-finned fishes, 322
Reading aids, 483
Reality:
　culture and perception of, 503—504
　science as study of objectively verifiable, 414
Receptors of reflexes, 409
Recessive traits, defined, 261
Recombination:
　of alleles, 292
　defined, 213
Recycling:
　of essential elements through ecosystems, 38—39
　interference with, 39
Red blood cells, normal, *222, 300*
Reflected radiation, 98
Reflexes, 409
Regeneration, defined, 208
Relative age of fossilized rocks, 316
Replication, mode of DNA, *280*
Repressor molecules, control of protein synthesis and, 385
Reproduction, 209—214, 229—250
　asexual (*see* Asexual reproduction)
　bacterial, 297
　differential, of genotypes, 301—302
　and formation of populations, 250—255
　of fungi, *190*
　human, *210,* 230—244
　　reproduction in other animals compared with, 245—246
　in plants and other animals, 244—250
　sexual (*see* Sexual reproduction)
　strategies of, 229—230, *249*
Reproductive behavior, 431—436

Reptile-hipped dinosaurs, 329, 330
Reptiles, *168, 170*
　classification of, 166, 169
　as dominant land vertebrates, 329—333
　eggs of, 246—247
　evolution of, 329
　give rise to birds and mammals, 333—335
Research, ethical problems posed by, 483—488
　biological research, 483—485
　medical research, 486—488
Resistance:
　to antibiotics, 296—297
　to DDT, 294
Resources (*see* Natural resources)
Respiration, 355—358, *359, 360*
Respiratory cancer, trends in, 478
Respiratory tract, cells on membranes on parts of, 222
Retina of the eye, 410
Revolutions:
　of earth around sun, *99*
　of moon around earth, *103, 104*
Rhinoceros unicornis, 136
Rhizoid, defined, 323
Ribonucleic acid (*see* RNA)
Ribose sugar, chemical structure of, *382*
Ribosomal RNA (rRNA), 382
Ribosomes, described, 382, *384*
Rickets, *373*
Rickettsia, 201
Rickettsial diseases, table, 200
River blindness, described, 419
RNA (ribonucleic acid), transcription of DNA to, 381, *382*
Rockweeds, *189*

Rotations:
　daily cycles and earth, 104, 105
　major atmospheric movements and earth, *101*
Round dance of worker honeybees, 443, *444*
Roundworms, *178*
rRNA, 382
Rural life, 34—55
　farm as ecosystem, 39—43
　impact of the farm, 48—55
　interaction between ecosystems of cities and, 68, 69
　principles of ecosystems in general understanding of, 35—39
　rural ecology, 43—48

Saber-toothed cats, *337*
Sagan, Carl, 86
Salamanders, *171*
　effect of the eye transplant in, *221*
Salinity, defined, 115
Saliva, 361
Salivary glands, *361, 362*
Salmonella, 201
Saturated fats, 371
Saturn, 93—94
Scarcity:
　food: failure of Green Revolution and, 490—492
　　massive future starvation, 489—490
　　table, 508
　of resources in general, 492
Scavengers, defined, 37
Schaller, George, 265
Schistosoma mansoni, life cycle of, *179*
School performance, IQ as measure of, 311
Science, as study of objectively verifiable reality, 414
Scrub forests, 122

Sea animals, early, 319–322
Sea urchins, fertilization of egg of, *216*
Seals, *167*
  group formation of, *436*
  used by Netsilik Eskimos, *62, 63*
Seashore genotype plants in inland areas, 273–274
Seashore habitats:
  growing of inland genotype plants in, 274
  rocky, *115*
Seasons, 100
  effects of changes in, in temperate zone pond, *129*
Second growth forests, 70
Secondary sexual characteristics, hormones responsible for, 404
Seed-eating animals, 336–337
Seed ferns, *328*
Seed plants, 180–185
  basic structure of, *185*
  early land plants replaced by, 326–328
  fertilization of, 181–182
  sprouting seeds of bean plants, *180*
  (*See also* Flowering plants)
Segregation, law of, 264
Selection, selective pressures and geographic variation and, 303
  (*See also* Artificial selection; Natural selection)
Selective agents, defined, 293
Self-sufficient life, 57–67
Semen, defined, 231
Sensory cortex, *447*
Sensory inputs processed by brain, *411*
Serial symbiosis, defined, 151
Serotonin, 408
Sewage treatment plants, *28*
Sex cells, 209–214
Sex chromosomes, defined, 210

Sex-linked inheritance, 266–268
Sex organs:
  human female, 231–232, *233,* 234–236
  human male, 230, 231, *232*
Sexual activities:
  attitudes toward, 239–240
  reproduction distinguished from, 239–240
Sexual characteristics, secondary, hormones responsible for, 404
Sexual maturity, human, 238
Sexual reproduction:
  begins with gamete production, 209–214
  cells vary in chromosome sets, 214–215
  diversity in populations and, 291
  in mammals, 230–231
Sexuality, year-round, single births and, 340
Shapley, Harlow, 86
Sharks, 322
Shklovskii, I. S., 86
"Shooting stars," 89–90
Shortwave radiation, *97*
Siblings, defined, 275
Sickle-cell anemia, 299, *300*
Sickle-cell gene, 298–300
Sickle-cell trait, 299
Signals, as form of communication, 448
Silting, 49
Similarity, classification based on, 154–155
Single birth, year-round sexuality and, 340
Sister chromatids, defined, 213
Skeletal muscle fibers, *222*
Skin cells, *222*
Slash-and-burn agriculture, 41
Slums, urban, 20–21
Small farms, agriculture on, 41
Small intestine, *362, 363*

Smell, sense of, 411
Smoke stacks, *10*
Smooth endoplasmic reticulum, 386
Smooth muscle cells, *222*
Smooth muscles, 391
Snow patterns, *101*
Social behavior, 436–446
  advantages of group life and, 437–441
  in industrial cities, 463–464
Social problems, urban, 30
Social engineering, cultural evolution and, 506–507
Social groups, defined, 437
Social impact of farms, 52–55
Social insects, 442–444
Social responses, defined, 436
Social units, races as, 308
Society:
  consumer, 463
  culture in, 497–509
    (*See also* Cultural evolution; Culture)
  future, 471–510
    ecological prospects in, 488–497
    human prospects in, 472–488
    major trends in variables affecting, table, 508
  human nature and structure of, 499–501
  technological, 460–464
Soil:
  grassland, 76–77
  hardpan, 141
  as major output of farms, 49
  nutrients from, 44–46
  rain forest, 121–122
Soil erosion, 46–48
Soil formation, 46, 47
Solar energy, *480, 481*
  (*See also* Sun)
Solar system, 86–108
  diagram of, *91*
  (*See also* Earth; Moon; Sun *and* specific planets)

Somatotropin (growth hormone), 404
Sound communication, 445
South (cardinal point), finding true, 107
South America, fauna of, 132—133, 135
South Asia, fauna of, 135
Southern solstice, 98
Spaceship, earth as, 467—468
Speciation, defined, 305
Species:
    classification of living, table, 164—165
    defined, 158, 159, 161
    evolution of, 303
        splitting process resulting in new species, 304—305
        (See also Evolution)
    hermaphroditic, 209
    human beings belonging to single, 305—308
        (See also Human beings)
    of plants in cities, 19
    (See also specific species)
Speculations, defined, 472
Sperm:
    defined, 209
    human, 209, 232
    production of, 230
Sphincter muscle, 392
Spiders, courtship behavior of, 433
Spiral galaxy, 85
Spleen, 367
Sponges, 178
Squids, 178
Stability:
    of ecosystems, 80—81
    of temperate zone forests, 74
Stamens of seed plants, 180
Staphylococcus, 199
Star chart, 90
Starches, 350
Starfishes, 178
Starlings, 144
    spread of, 145

Starvation, future massive, 489—490
Steady state, 394—424
    coordination and, 402—414
    effectors and, 390—391
    homeostasis and, 395—402
    protection for, 414—423
Steam engines, 9
Steel girders, assembling, 11
Stegosaurus, 331
Sterilization, as birth control method, 243—244
Stigmas of seed plants, 181
Stimuli, neuron membrane responding to, 406—407
Stomach, human, 362, 363
Stomata, defined, 183
Stone plants, 419
Stone tools, 58
Stratification, dating of fossils and, 316
Stratum, defined, 314
Streams:
    as habitats, 128, 129
    polluted, 30
Streptococcus, 199
Striated muscles, 391, 392
Structural proteins, defined, 277
Sturnus vulgaris (see Starlings)
Subject-object distinction, as example of cultural difference in perception, 449
Submissive posture of coyotes, 439
Subsistence farming, 40—42
Success, IQ and, 311
Sucrose, 349
Sun:
    bee's flight patterns and position of, 106
    earth heated by, 96—98
        effects of, 100
        weather caused by uneven heating, 98—100
    energy from: as energy source for cities, 15—17

Sun:
    energy from:
        produced by conversion of mass to energy, 86—89
    revolution of earth around, 99
Survival:
    behavior and, 425—454
        conceptual behavior, 446—453
        reproductive behavior, 431—436
        social behavior, 436—446
        (See also Coevolution)
    culture and, 455—470
        agricultural revolution and, 457—460
        by hunting and food gathering (see Hunting and food gathering cultures)
        technological culture, 460—469
Sweet peas, 181
Swidden agriculture, 40—41
Symbiosis, defined, 151
Symbols, language, 448—449
Synapse, defined, 213
Synapsis, crossing over of strands of chromosomes during, 214
Synchrony of readiness to mate, 431
Synthesis, energy use and, 380—386
    (See also Photosynthesis)
Synthetic fertilizers, 40
Syphilis, incidence of, table, 244

T cells, 419
Tactile communication, 445
Taiga zone, 126, 127
Tapeworms, 178
Tardigrades, 178
Target cells, 403
Tarsiers, 339
Tasmanian devils, 134

Technological society, cities and, 460–469
Tectonic plates, 136
  six major, *138*
Teeth:
  of browsing animals, *337*
  differentiated, *335*
  of dog and *Homo sapiens*, *165*
Telophase, *208*
Temperate forests:
  biome of, 125–126
  as natural ecosystems, 70–76
Temperate zones, freshwater habitats in, 128–129
Temperature:
  control of body, 395–397
  deterioration and, 389
  mean annual, in Iceland, *489*
Templates, DNA complementary strands as, 279
Tendons, described, 391
Tension, premenstrual, 235
Teratogens, 220
Termites, as social insects, *443*
Territoriality, 441–442
Testes, described, 210, 230
Testosterone, 404
  table, *405*
Theories, defined, 504
Thermal pollution:
  urban, 25–27
  waste heat resulting from, *494*
Thermodynamics, laws of, 8
Three-layered embryos, 216
Threshold of nerve impulses, *412*
Thrombus, defined, 415
Thymus gland, 367
Tides, 103, *104*
Tigris and Euphrates Valleys, *459*
Timberline, 118
Toads, reproductive strategy of, *249*

Tobacco mosaic virus (TMV), *201*
Tonsils, 367
Toxin, defined, 199
Trace element, defined, 44
Tracheae, defined, 368
Tracheal system, 368
  of grasshoppers, *370*
Trades (wind), 99
Trait(s):
  dominant and recessive, 261
  genes controlling inheritance of continuously varying, 271–277
  sickle-cell, 299
  transcription, as first step in protein synthesis, 381, *382*
Transfer RNA (tRNA), 382–384
Translation, as second step of protein synthesis, 382–384
Transmitter substances, *407*, 408
Transpiration, defined, 44
Tree-dwelling ancestors of *Homo sapiens*, 339–340
Tree ferns, *328*
Trees:
  coniferous, 126
  evergreen, *126*, *127*
  pine tree branch, *184*
Triceps muscles, 391
Triceratops, *332*
Trilobites (*see* Arthropods)
Triplets, as combination of three nucleotides, 279
tRNA, 382–384
Tropical rain forests, biome of, 119, *120*, *121*, 122
Tropopause, defined, 95, *96*
Tuatara, *168*
Tubal ligation, 243
Tundra, *127*
  biome of, *127*, 128
Turbines, powerplant, *8*
Turgor, defined, 44
Turtles, *168*

Twins, 274–277
Tyrannosaurus, *330*

Umbilical cord, defined, 237
Underdeveloped countries (UDCs), 465–467
Unicellular, defined, 151
Unicellular organisms, 152, *153*
  reproduction of, 207
Uniqueness of individuals, 512–513
Universe, 84–86
  life elsewhere in, 85, *86*
  (*See also* Solar system)
Unstable ecological systems, 74–76
Upper arm, *391*
Upright posture of first human fossils, 341–342
Uranus, 93–94
Urban air pollution, 27
  effects on inhabitants, 29–30
Urban ecology, 2, 13–19
Urban renewal, failure of, 21
Urbanization:
  defined, 2
  impact of, 19–33
Urea, function of, 398
Uric acid, excretion of, 399
Urine, 398
Uterus, defined, 166
Utility, as basis for classification, 154

Vaccination, 420
Vaccine, defined, 420
Vacuoles, described, 361
Van Lawick-Goodall, Jane, 265
Variability, decay of, 302
Vas deferens:
  described, 231
  vasectomy, 243–244
Vascular plants:
  of marshy areas, 323
  that do not produce seed, *187*

Vascular system (*see* Circulatory system)
Vascular tissues of seed plants, 182–185
Vasectomies, 243–244
VD (veneral disease), table, 244
Vectors:
  control of, 475, *476*
  defined, 418
Vegetation, storied, 119–120 (*See also* Floras)
Veins, 364
Veneral disease (VD), table, 244
Venous blood, 365–366
Venus, 91
Vertebrates:
  courtship behavior of, 432–435
  evolution of brain of, *446*
  examples of classes of, *164*
  fishes as first, 321–322
  land, 328–336
    evolution of, 328–329
    mammals as dominant (*see* Mammals)
    reptiles as dominant (*see* Reptiles)
  lymphatic system of, 367–368
  as most familiar animals, 162–163, 166–170
  movement in, *173*
Vesicles, transmitter substances stored in, *407*
Viruses, 201–204
  diseases caused by, table, 202
  influenza and polio, *203*
Vision:
  binocular, 340
  cortical integration of, 447–448

Vision:
  (*See also* Eyes)
Visual center in brain, 410, *411*
Visual communication, 445–446
Vitamins, 372–374
Voluntary control of muscles, 391–392
Volvox, *154*
Voting, 523–525
Vulvae, described, 231

Waggle dance of worker honeybees, 443, *444*
War, nuclear, 494–495
Wasps, colony of, *176*
Waste disposal:
  in cities, 20
  disposal of waste water polluting waterways, 27–29
  sewage treatment plants, 28
Waste elimination (biological function), as homeostatic problem, 397–401
Waste heat, described, 494
Wastes, urban, 10–11
Water conservation, as homeostatic problem, 397–401
Water, requirement for green plants, 44
Water pollution:
  eutrophication and, 50
  of Lake Erie, *131*
  polluted streams, *30*
  urban waste disposal and, 27–29
Water snakes, differential mortality in, *299*
Water-soluble vitamins, 372
Watt, defined, 22
Wealth, redistribution of, as part of ecological engineering, 497

Weaning:
  in animals other than human beings, 245–246
  of human infants, 237–238
Weight distribution in pig populations, 273
Whales, as example of classification, *157*
Whitenosed monkeys, *167*
Wind erosion, impact of, 49
Wolves, *160*
Woodpeckers:
  acorn, *265*
  nests of, *248*
Work, energy needed for, 8–9
Worker bees, 442, 443
Workforce, urban, 21–22
World ecosystem, future, 488–492
Worldwide plague, 495

X chromosomes, inheritance of, *268*
Xylem:
  defined, 184
  longitudinal section of, *185*

Year-round sexuality, single birth and, 340
Yolk, 215

Zooplankton, defined, 77
Zero Population Growth (ZPG), 517
Ziggurat (Ur), *461*
Zygotes (fertilized eggs):
  defined, 209
  formation of hollow-ball stage after division of, *217*